ENTAILMENT

ENTAILMENT

THE LOGIC OF RELEVANCE AND NECESSITY

by

ALAN ROSS ANDERSON

and

NUEL D. BELNAP, JR.

with contributions by

J. MICHAEL DUNN
ROBERT K. MEYER

and further contributions by

JOHN R. CHIDGEY
J. ALBERTO COFFA
DOROTHY L. GROVER
BAS VAN FRAASSEN
HUGUES LEBLANC
STORRS MCCALL
ZANE PARKS
GARREL POTTINGER
RICHARD ROUTLEY
ALASDAIR URQUHART
ROBERT G. WOLF

VOLUME I

PRINCETON UNIVERSITY PRESS

Published by Princeton University Press
Princeton and London

LCC: 72-14016
ISBN: 0-691-07192-6

Library of Congress Cataloging in Publication Data
will be found on the last printed page of this book

Printed in the United States of America by
Princeton University Press
Princeton, New Jersey

Dedicated to the memory of

WILHELM ACKERMANN (1896–1962)

whose insights in
Begründung einer strengen Implikation
(**Journal of symbolic logic**, 1956)
provided the impetus for
this enterprise.

CONTENTS

ANALYTICAL TABLE OF CONTENTS

VOLUME I

VOLUME II (tentative)

IX. SEMANTIC ANALYSIS OF RELEVANCE LOGICS (with §47 by Alasdair Urquhart and §§48–60 by Robert K. Meyer and Richard Routley)

PREFACE

THIS BOOK is intended as an introduction to what we conceive of, rightly or wrongly, as a new branch of mathematical logic, initiated by a seminal paper of 1956 by Wilhelm Ackermann, to whose memory the book is dedicated. It is also intended as a summary, seventeen years later, of the current state of knowledge concerning systems akin to those of Ackermann's original paper, together with philosophical commentary on their significance.

We argue below that one of the principal merits of his system of *strengen Implikation* is that it, and its neighbors, give us for the first time a mathematically satisfactory way of grasping the elusive notion of *relevance* of antecedent to consequent in "if . . . then —" propositions; such is the topic of this book.

As is well-known, this notion of relevance was central to logic from the time of Aristotle until, beginning in the nineteenth century, logic fell increasingly into the hands of those with mathematical inclinations. The modern classical tradition, however, stemming from Frege and Whitehead-Russell, gave no consideration whatever to the classical notion of relevance, and, in spite of complaints from certain quarters that relevance of antecedent to consequent was important, this tradition rolled on like a juggernaut, recording more and more impressive and profound results in metamathematics, set theory, recursive function theory, modal logic, extensional logic *tout pur*, etc., without seeming to require the traditional notion of relevance at all.

To be sure, even in the modern mathematical tradition, textbooks frequently give some space in earlier pages to the notion of relevance, or logical dependence of one proposition on another, but the mathematical developments in later chapters explicitly give the lie to the earlier demand for relevance by presenting a theory of "if . . . then —" (the classical two valued theory) in which relevance of antecedent to consequent plays no part whatever.

Indeed the difficulty of treating relevance with the same degree of mathematical sophistication and exactness characteristic of treatments of extensional logic led many influential philosopher-logicians to believe that it was *impossible* to find a satisfactory treatment of the topic. And in consequence, many of the most acute logicians in the past thirty years have marched under a philosophical banner reading "*Down* with relevance, meanings, and intensions generally!"

That metaphor is perhaps implausible, but it serves us in pointing out that, in addition to the mathematical bits to follow, there are philosophical battles to be fought. Among them are the principal issues touched on in contemporary philosophical discussions of logic: controversies about extensions and intensions, alethic modalities, and the like. What we have tried to do is to jump into the skirmishes among neo-Platonists, neo-conceptualists, neo-nominalists, and generally exponents of neo-what-have-you in logic, and to hit everyone over the head with a theorem or two.

Such a program seemed to us to require, from an expository view, that the mathematical and philosophical tones of voice be intertwined. Not inextricably, of course: we expect the reader to be able to tell when we are (a) offering serious mathematical arguments, (b) propounding serious philosophical morals, and (c) making *ad hominem* jokes at the expense of the opposition. No doubt some readers will find items under (c) undignified, or in some other way offensive. To such readers we apologize. We suggest that they simply strike from the book those passages in which they find an unseemly lack of solemnity — nothing much in the argument hinges on them anyway, though *ad hominem* arguments are sometimes persuasive, especially if the opposition can be made to look ludicrous enough. For the classical tradition we are attacking, this task is not difficult, but of course we certainly do not demand of our readers that they be entertained by side-comments. Which observation leads us to make another remark about how the book may be read.

We share with many the conviction that the growth of (western) logic from Aristotle to the present day, despite temporal discontinuities in historical development, represents a progressively developing tradition: the more mathematical character of contemporary work in logic does not represent a sharp break with tradition but rather a natural evolution in which more sharp and subtle tools are used in the analysis of the "same subject" — logic. (Mathematical treatment of logic was initiated, so far as we know, by Leibniz in 1679, though his insights did not catch on — which is hardly surprising: they weren't published until 1903; see Łukasiewicz 1951.) We believe in consequence that our hope of supplementing the modern classical mathematical analysis of logic with a sharper, subtler, and more comprehensive analysis of the same topic, whatever its merits, will be thoroughly understood only by those prepared to study both the philosophical and mathematical arguments offered below.

Nevertheless, it is not necessary in many cases to check through the mathematical arguments in detail, provided the sense of the theorems is understood. Proofs frequently, in some mysterious way, illuminate the theorems they prove, and we hope that some readers will read proofs carefully enough to find such errors as are no doubt to be found. But the philosophical thrust of arguments under (b) above can be gathered independently of the compulsive checking of all the mathematical details. Equally, of course, the mathematical arguments under (a) are independent of the philosophical polemics. And we would be *delighted* if someone were to read the book just for the jokes (c).

We have used earlier versions of large parts of this book in advanced undergraduate and graduate courses at Yale and the Universities of Manchester and Pittsburgh. Students with one year of mathematical logic have been able to grasp the material without too much difficulty, though, as one is always supposed to say, there is more here than we have been able to cover in a two-semester course. Enough theorems, lemmas, and the like have been left unproved in the text to provide an ample source of exercises.

As is explained at the outset of §8, the miscellany sections may all be skipped without loss of momentum. A one-semester course designed to touch the most pervasive philosophical and mathematical points in the book might include §§1–5 of Chapter I, §§9–12 of Chapter II, §§15–16 of Chapter III, and §§21–23 of Chapter IV. A second-term continuation should probably include the Gentzen formulations of §§7 and 13, and the algebraic semantics of §§18–19 and 25. The deepest insights into the semantics of these systems will be found in Chapter IX (by Urquhart, Meyer, and Routley), which brings us to the edge of current research in this aspect of the topic.

Sections of the book not mentioned above are designed to bring the reader to the edge of current research in other aspects of the topic. The book is intended to be "encyclopedic," in the modest sense that we have tried to tell the reader everything that is known (at present writing) about the family of systems of logic that grew out of Ackermann's 1956 paper. But there are still many entertaining open questions, chief of which are the decision problems for **R** and **E**, (and perhaps **T**), which have proved to be especially recalcitrant.

Old friends of our project will be surprised to find that we were forced to split the book into two volumes — in order, of course, to avoid weighing the reader down either literally or financially — when we finally realized that the universe of relevance logics had expanded unnoticed overnight. The second volume should appear about a year after this one; we include as part of the Analytical table of contents a tentative listing for Volume II.

Grammatical propaedeutic. A word should be said about the Grammatical propaedeutic, which is placed at the end of the first volume. There has been abroad for a long time the view that one *cannot* discuss the topic of this book without making certain essential mistakes. We complain about this from time to time in an incidental way in the first few chapters, which are devoted mainly to logical discussions near to our hearts. But we can already hear unsympathetic readers whispering as they read that our project has no merit. Rather than try to sprinkle *passim* remarks about our view of the canard that there is no way of talking about "entailment" without making object-language meta-language *bêtises*, we gather our grammatical views in one section, which such an unsympathetic reader should indeed read first, as a propaedeutic. But the reader who has not been moved by listening to *a priori* rejections of our entire topic on grammatical grounds is advised to postpone reading the Grammatical propaedeutic for a long time—maybe indefinitely; and with such a reader in mind, we have relegated the propaedeutic to an appendix, where it is less likely to constitute an obstacle to beginning this book where it should be begun, §1.

Cross-references. The amount of cross-referencing in this book will annoy those readers who feel obliged to look, say, at §27.8 whenever that section (or any other) is mentioned in the text. We have attempted to write and edit in such a way that the reader seldom is *forced* to stick his finger in the book at one page while he refers to another; the aim of the cross-references has simply been to assist those who wish to find other places where the same or similar topics are treated.

Citations. With respect to referencing the literature, we have tried to be liberal in Volume I, except in one respect: our own work has been cited in the text only in the case of joint authorship with another, or occasionally in sections by one of our co-authors. The method of citation, explained at the beginning of the bibliography at the end of this volume, was to the best of our knowledge invented by Kleene 1952, and is the best and most economical we know, simultaneously avoiding footnotes, cross-references among footnotes, and willy-nilly giving (perhaps even forcing on) the reader some sense of history as he reads.

Notation. This has been a headache; as Grover has pointed out to us, the Roman alphabet was not designed to assist logicians, nor were the typesetter's fonts. Those in the market will find a remarkable example on page 206 of Kleene 1952, where Lemma 21 presents us (within two lines) with the first letter of the Roman alphabet in six different typefaces, of different significance; it is an extraordinary tribute to that author's ingenuity, to the typesetters, and to the proofreaders, to have got it to come out right.

We have in fact been sparing in our use of fonts. Our policy is not altogether rational, but it is anyhow straightforward.

Bold face roman only for system names, and variables over systems, as explained below.

Bold face italic for propositional and algebraic (or matrix) constants.

Italic for expression-variables (variables over formulas, individual variables, propositional variables, predicate constants, etc.), and *italic* also for numerical variables.

Γρίκ for sequences of formulas, with occasional other uses.

Plain roman for *all* other variables, temporary constants, etc.

There is of course lots of special notation, explained as it is introduced. We assume, however, familiarity with the standard set-theoretical concepts: $\emptyset$ (the empty set), $\cap$ (intersection of sets), $\cup$ (union of sets), $-$ (complementation of sets — relative to some indicated domain), $\{a_1, \ldots, a_n\}$ (the set containing just $a_1, \ldots, a_n$), $\{x{:}Ax\}$ (the set of x's such that Ax).
For further information, consult the Index under "notation for."

System nomenclature. This book mentions or discusses so many different systems (Meyer claims the count exceeds that of the number of ships in Iliad II) that we have been driven — mostly at Bacon's gentle urging — to try to devise a reasonably rational nomenclature. In sum it goes like this:

1. Bold face: principal system (of propositional logic)
2. Subscripts: fragment (involving subscripted connectives)
3. Superscripts: extension (by adding superscripted connectives)
4. Prefixes: formulations (axiomatic, Fitch, or Gentzen)

Details, examples, additions, and exceptions follow.

1. **Bold face** characters are used to designate *principal systems* of propositional logic, usually involving the connectives $\rightarrow$, $\vee$, &, and $\overline{\ \ }$. Thus: **E, R, RM, T, S4, TV,** etc.

2. *Subscripts* indicate *fragments* of principal systems (of which the principal systems are presumably conservative extensions). We subscript with the connective(s) in the fragment when this does not get out of hand; otherwise, we use some other (hopefully) mnemonic device. The most important subscripts we illustrate with respect to **E**:

$\mathbf{E}_{\rightarrow}$	the pure arrow fragment of **E**
$\mathbf{E}_{\overline{\rightarrow}}$	entailment with negation fragment of **E**
$\mathbf{E}_{+}$	positive (negation-free) fragment of **E**
$\mathbf{E}_{\mathrm{fdf}}$	first degree formula fragment of **E**

3. *Superscripts* indicate (presumably conservative) *extensions* of principal systems by the addition of new pieces of notation, with axioms governing them. We superscript with the actual notation added, to the extent to which this is feasible. We illustrate with respect to **R**:

$\mathbf{R}^{\Box}$	a system which adds a necessity modality to **R**
$\mathbf{R}^{\Box t}$	adds both $\Box$ and the propositional constant t to **R**
$\mathbf{R}^{\forall p}$	adds universal quantification ($\forall$) for propositional variables (p) to **R**
$\mathbf{R}^{\forall \exists x}$	adds universal ($\forall$) and existential ($\exists$) quantification for individual variables (x) to **R**

4. *Prefixes* (not in boldface) distinguish startlingly different *formulations* of the "same" system. Since we use the null prefix for Hilbert (axiomatic) formulations, we obtain as illustrations

$\mathbf{E}_{\rightarrow}$	Hilbert formulation of $\mathbf{E}_{\rightarrow}$
F$\mathbf{E}_{\rightarrow}$	Fitch (natural deduction) formulation of $\mathbf{E}_{\rightarrow}$
L$\mathbf{E}_{\rightarrow}$	Consecution (Gentzen) formulation of $\mathbf{E}_{\rightarrow}$; the "L" derives from Gentzen 1934
L$\mu$$\mathbf{E}_{\rightarrow}$	Merge-style consecution formulation of $\mathbf{E}_{\rightarrow}$

5. *Numerical subscripts* are occasionally used to indicate *mildly different* formulations; e.g.

$\mathbf{E}_{\rightarrow 1}$
$\mathbf{E}_{\rightarrow 2}$

are different axiomatic formulations of $\mathbf{E}_{\rightarrow}$.

6. Various combinations should be self-explanatory; e.g., L$\mathbf{R}_{+}^{\Box t}$ is a consecution formulation of the positive fragment of the $\{\Box, t\}$ extension of **R**.

7. The principal exceptions to our policies occur in Chapter VIII, where we use names used or suggested by Ackermann's original notation for the family of systems with which this book concerns itself. Thus $\mathbf{\Sigma}$, $\mathbf{\Pi}'$, $\mathbf{\Pi}''$ are all due to him (with more thanks than we can, at this late date, muster); principal changes from the policies of 1–5 above are confined to Chapter VIII. But the reader will doubtless recognize cases in which the price of systematic nomenclature was deemed by us to be too high.

8. One of the prices we have already paid — with, let it be said, no cheer — is to change the names of a variety of systems which have been

already used in the literature. We report here the most significant correlations.

EQ, RQ, etc. = $\mathbf{E}^{\forall\exists x}$, $\mathbf{R}^{\forall\exists x}$, etc.
E†, R†, etc. = $\mathbf{E}^{\forall\exists p}$, $\mathbf{R}^{\forall\exists p}$, etc.
E_I, R_I, etc. = $\mathbf{E}_{\rightarrow}$, $\mathbf{R}_{\rightarrow}$, etc.
E^+, R^+, etc. = $\mathbf{E}_+$, $\mathbf{R}_+$, etc.
NR = $\mathbf{R}^{\Box}$
$E_I{}^*$, etc. = $\mathbf{FE}_{\rightarrow}$, etc.

See the Index of systems for a guide to where each system is defined, etc.

Pronouns. We have used "we" because of our essential multiplicity. Then in editing papers by co-authors we have changed "I" to "we" for uniformity; but in these contexts the "we" is editorial.

For example, when a co-author says "we are going to prove X," he or she doesn't mean that *we* are going to prove X, but that the co-author is. In a few cases, however, in which matters of history or opinion have arisen in a particularly delicate way, we have let the first-person singular stand in order to avoid any tinge of ambiguity. (See Meyer's §28.3.2 for highly refined sensitivity to these issues.)

ACKNOWLEDGMENTS

THIS BOOK has been in preparation since 1959. In the ensuing fourteen (see end of §11) years, so many persons have helped us in so many ways that we first considered the possibility of making a list of those who had *not* helped, on the grounds that such a list would be shorter. That policy would however deprive us of the pleasure of expressing our gratitude to those who have contributed so much in producing whatever of value may be found here.

We should mention first our principal teachers, Frederic Brenton Fitch (ARA and NDB), George Henrik von Wright (ARA), and the late Canon Robert Feys (NDB). The help they gave us, both directly, in providing us information about the field, and indirectly, by providing fascination and enthusiasm for techniques in logic, has been immeasurable.

We both profited from Fulbright Fellowships (ARA with von Wright in Cambridge, 1950–52, and NDB with Feys in Louvain, 1957–58), and we wish to express our gratitude to the Fulbright-Hays program for its assistance in helping us learn much from many European friends and colleagues.

Our initial collaboration at Yale in 1958 was sponsored in part by Office of Naval Research (Group Psychology Branch) Contract SAR/Nonr-609 (16) (concerned with Small Groups Sociology), under the direction of Omar Khayyam Moore as Principal Investigator, to whom we are very grateful for initial and continued support and encouragement. With his help, and partially with the assistance of summer programs sponsored by the National Science Foundation through Grants G 11848, G 21871, and GE 2929, we managed to engage the interest of a number of extraordinarily able students at or near Yale, among them John Bacon, Jon Barwise, Neil Gallagher, Saul Kripke (Harvard), David Levin, William Snavely, Joel Spencer (MIT), Richmond Thomason, and John Wallace. We thank each of these for his contributions during those salad days, when almost every week saw the solution of a problem, or the genesis of a fruitful conjecture.

We are also grateful to Yale University for Morse Fellowships (ARA 1960–61, NDB 1962–63), which gave us both time off for research, and to the National Science Foundation, which partially supported both our work and that of many of the students mentioned above and below through Grants GS 190, GS 689, and GS 28478; we further join Robert K. Meyer and J. Michael Dunn in thanking the National Science Foundation for support to them through Grant GS 2648.

The years 1963–65 separated us, NDB being in Pittsburgh (then and since), and ARA at Yale 1963–64, and Manchester (under the auspices of the Fulbright-Hays Program) in 1964–65, a separation which understandably slowed progress. But while separated, both of us were fortunate in finding colleagues and students who combined helpful collaboration with good-natured, abrasive criticism. In addition to those mentioned at Yale, others contributed clarifying insights (both as to agreements and differences) in Manchester, among them E. E. Dawson, Czesław Lejewski, David Makinson, John R. Chidgey (who later spent a helpful year with us in Pittsburgh), and in particular the late Arthur Norman Prior, a close friend with whom we had valuable correspondence on many topics for many years.

Meanwhile NDB had the good fortune to find able allies among students at Pittsburgh, especially J. Michael Dunn, Robert K. Meyer, Bas van Fraassen, and Peter Woodruff.

ARA and NDB rejoined forces in Pittsburgh in 1965, and since then we have enjoyed not only sabbatical leaves, ARA 1971, NDB 1970, but also stimulating discussions with our colleagues Steven Davis, Storrs McCall, Nicholas Rescher, Sally Thomason, and Richmond Thomason, and with a number of delightful and dedicated colleagues-in-students'-clothing. Among the latter we have depended for assistance upon Kenneth Collier, Jose Alberto Coffa, Louis Goble, Richard Goodwin, Dorothy L. Grover, Sandy Kerr, Virginia Klenk, Myra Nassau, Zane Parks, Garrel Pottinger, and Alasdair Urquhart; and more recently Pedro Amaral, Jonathan Broido, Anil Gupta, Martha Hall, Glen Helman, and Carol Hoey.

Word spreads, in this case through Kenneth Collier, who interested his colleague Robert G. Wolf in the enterprise. Wolf has not only lent us his expertise in the later stages of the preparation of the bibliography for this volume, but has generously assumed total responsibility for the Comprehensive bibliography promised for Volume II.

One of our principal debts of course is to those who have consented to stand as co-authors of this book. Two options were open to us: (a) to paraphrase and re-prove their results; (b) to ask their permission simply to include their results verbatim (or very nearly). Both of us were convinced that course (a) would not improve the total result at all, so we resolved on course (b), and our gratitude to those who have allowed their names to appear on the title page with ours is again immeasurably large.

Some of the sections by ourselves and our co-authors were written especially for this work, while others were edited by us — with an eye to cross-referencing, reduction of repetition, reasonable uniformity of notation, and the like — from pieces and parts of pieces which have appeared elsewhere; with respect to the latter we wish to thank not only those who allowed us to include their writings, but also the following journals and publishers. Detailed information appears in the bibliography at the end of this volume under the heading given below; section numbers in parentheses indicate where in this volume (a portion of) the cited material appears.

The journal of philosophy: van Fraassen 1969 (§20.3). **The journal of symbolic logic:** Anderson and Belnap 1962a (§§1–5, 8.1), Belnap 1960b (§§5.1.2, 22.1.3), 1967 (§19); Meyer and Dunn 1969 (§25.2); Dunn 1970 (§29.4). **Australasian journal of philosophy:** Meyer 1974 (§29.12). **Notre Dame journal of formal logic:** Meyer 1972 (§28.4), 1973b (§§24.4.2, 28.3.1). **Philosophical studies:** Anderson and Belnap 1962 (§§15, 16.1). **Zeitschrift für mathematische logik und Grundlagen der Mathematik:** Anderson 1959 (§23); Anderson, Belnap and Wallace 1960 (§8.4.3); Belnap and Wallace 1961 (§13). **Logique et analyse:** Meyer 1971 (§22.3).

Intermittently, many others have made helpful suggestions and offhand illuminating side-comments; to those whose informal conversational ideas, and perhaps words, may have found their way into this book, and gone unwittingly without explicit acknowledgment, we offer apologies — and gratitude.

We have also had the good fortune to secure the services of a large number of superlatively good secretaries, about one of whom we would like to recount the following paradigmatic tale. On her first day, she was given a messy manuscript to turn into typescript, involving many special logical symbols, various underlinings for different typefonts, etc. The typescript came back with *lots* of errors, which we explained to her, finally getting the reply, "Oh . . . I see . . . . I wasn't being careful enough." And for the remainder of the year we had the pleasure of working with her, we could find (with *great* difficulty) only a handful of typographical errors. Our luck in this respect continued to follow us through the following list: Phyllis Buck, Berry Coy, Bonnie Towner, Barbara Care, Catherine Berrett, Margaret Ross, Mary Bender, Rita Levine, Rita Perstac, and (lastly) Rita DeMajistre, who is responsible for the superlatively elegant completion of the whole of the final typescript.

You can well imagine the sinking sensation you would experience were you handed a body of typescript well sprinkled with henscratches and then asked to undertake the horrendous task of faithfully translating it onto the printed page. We are indebted to Trade Composition, and in particular to their remarkable crew of craftsmen, for carrying out this task with a degree of fidelity and sensitivity we would not have believed possible.

Finally, we would like to express our gratitude to two members of the staff of the Princeton University Press: both helped us enormously. Sanford Thatcher opened our neophyte eyes to the possibility of publishing our work with his press, for which we are extremely grateful, and Gail Filion addressed herself beautifully to the task of careful word-by-word editing, which saved us from many infelicities. The problems of editing and proofreading a printed volume of this kind are formidable, as we know well simply from trying to get the typescript right. We will both remember the unfailing helpfulness, kindness, and patience of those we have dealt with at the Princeton University Press with great pleasure.

Two concluding notes: We are of course dissatisfied with the customary, and also apparently inevitable, practice of making the "Acknowledgments" section of a book of this kind appear as a mere catalogue of mentors, colleagues, friends, students, publishers, and the like, who have contributed to the enterprise. The reason is sadly obvious; it leaves out of account entirely the exciting sense of adventure involved in the actual work: sudden unexpected insights and flashes of illumination, as well as disappointments at being unable to prove a true conjecture or disprove a false one. The sense of joy in creation or discovery is lost in a catalogue of names; it is really the close personal interaction with those friends listed above which accounts for the euphoric sense of the enterprise.

Secondly, while it is not common for co-authors to include each other among acknowledgments, we see nothing in principle that makes the practice improper or reprehensible. It is, however, difficult in our case because our respective contributions are intertwined like the "two parts" of a double helix. The closeness of our collaboration is indicated, as an example, by a scribbled manuscript page in which one of us wrote "reduncies," corrected by the other to "redundacies," and finally corrected by the first again to "redundancies." While it is not true, in general, that we have together gone over each word in this work in letter-by-letter fashion, we have certainly looked together at every sentence word-by-word, each paragraph sentence-by-sentence, etc. So we will stand equally convicted for the errors, inaccuracies, and infelicities which are inevitably to be discovered, and it is unlikely, on points of authorship, that either of us will point an accusing finger at the other — unless, perhaps, the book comes under especially brutal or vitriolic attack by the profession at large.

[Note by NDB.] ARA died December 5, 1973. He took pleasure in the fact that he lived to see the completion of the preparation of the manuscript for the printer; each word, save this note, bears his irreplaceable stamp.

ENTAILMENT

CHAPTER I

THE PURE CALCULUS OF ENTAILMENT

§1. The heart of logic. Although there are many candidates for "logical connectives," such as conjunction, disjunction, negation, quantifiers, and for some writers even identity of individuals, we take the heart of logic to lie in the notion "if . . . then —"; we therefore devote the first chapter to this topic, commencing with some remarks about the familiar paradoxes of material and strict "implication."

§1.1. "if . . . then —" and the paradoxes. The "implicational" paradoxes are treated by most contemporary logicians somewhat as follows.

The two-valued propositional calculus sanctions as valid many of the obvious and satisfactory principles which we recognize intuitively as valid, such as

$$(A \rightarrow (B \rightarrow C)) \rightarrow ((A \rightarrow B) \rightarrow (A \rightarrow C))$$

and

$$(A \rightarrow B) \rightarrow ((B \rightarrow C) \rightarrow (A \rightarrow C));$$

it consequently suggests itself as a candidate for a formal analysis of "if . . . then —." To be sure, there are certain odd theorems such as

$$A \rightarrow (B \rightarrow A)$$

and

$$A \rightarrow (B \rightarrow B)$$

which might offend the naive, and indeed these have been referred to in the literature as "paradoxes of implication." But this terminology reflects a misunderstanding. "If A, then if B then A" really means no more than "Either not-A, or else not-B or A," and the latter is clearly a logical truth; hence so is the former. Properly understood there are no "paradoxes" of implication.

Of course this is a rather weak sense of "implication," and one may for certain purposes be interested in a stronger sense of the word. We find a formalization of a stronger sense in semantics, where "A implies B" means that there is no assignment of values to variables which makes A true and B false, or in modal logics, where we consider strict implication, taking "if *A* then *B*" to mean "It is impossible that (*A* and not-*B*)." And, *mutatis mutandis*, some rather odd things happen here too. But again nothing "paradoxical" is going on; the matter just needs to be understood properly — that's all.

And the weak sense of "if . . . then —" can be given formal clothing, after Tarski-Bernays, as in Łukasiewicz 1929 (see bibliography), as follows:

$A \rightarrow (B \rightarrow A)$,
$(A \rightarrow B) \rightarrow ((B \rightarrow C) \rightarrow (A \rightarrow C))$,
$((A \rightarrow B) \rightarrow A) \rightarrow A$,

with a rule of *modus ponens*. (For reference let this system be $\mathbf{TV}_{\rightarrow}$.)

The position just outlined will be found stated in many places and by many people; we shall refer to it as the Official view. We agree with the Official view that there are no paradoxes of implication, but for reasons which are quite different from those ordinarily given. To be sure, there is a misunderstanding involved, but it does not consist in the fact that the strict and material "implication" connectives are "odd kinds" of implication, but rather in the claim that material and strict "implication" are "kinds" of implication at all. In what follows, we will defend in detail the view that material "implication" is *not* an implication connective. Since our reasons for this view are *logical* (and not the usual grammatical pettifoggery examined in the Grammatical propaeduetic appearing as an appendix to this volume), it might help at the outset to give an example which will indicate the sort of criticism we plan to lodge.

Let us imagine a logician who offers the following formalization as an explication or reconstruction of implication in formal terms. In addition to the rule of *modus ponens* he takes as primitive the following three axioms:

$A \rightarrow A$
$(A \rightarrow B) \rightarrow ((B \rightarrow C) \rightarrow (A \rightarrow C))$, and
$(A \rightarrow B) \rightarrow (B \rightarrow A)$.

One might find those who would object that "if . . . then —" doesn't seem to be symmetrical, and that the third axiom is objectionable. But our logician has an answer to that.

> There is nothing paradoxical about the third axiom; it is just a matter of understanding the formulas properly. "If *A* then *B*" means simply "Either *A* and *B* are both true, or else they are both false," and if we understand the arrow in that way, then our rule will never allow us to infer a false proposition from a true one, and moreover all the axioms are evidently logical truths. The implication connective of this system

may not *exactly* coincide with the intuitions of naive, untutored folk, but it is quite adequate for my needs, and for the rest of us who are reasonably sophisticated. And it has the important property, common to all kinds of implication, of *never leading from truth to falsehood.*

There are of course some differences between the situation just sketched and the Official view outlined above, but in point of perversity, muddle-headedness, and downright error, they seem to us entirely on a par. Of course proponents of the view that material and strict "implication" have something to do with implication have frequently apologized by saying that the name "material implication" is "somewhat misleading," since it suggests a closer relation with implication than actually obtains. But we can think of lots of no more "misleading" names for the relation: "material conjunction," for example, or "material disjunction," or "immaterial negation." Material implication is *not* a "kind" of implication, or so we hold; it is no more a kind of implication than a blunderbuss is a kind of buss. (But see §§36.2.3–4.)

§1.2. Program. This brief polemical blast will serve to set the tone for our subsequent formal analysis of the notion of logical implication, variously referred to also as "entailment," or "the converse of deducibility" (Moore 1920), expressed in such logical locutions as "if . . . then —," "implies," "entails," etc., and answering to such conclusion-signaling logical phrases as "therefore," "it follows that," "hence," "consequently," and the like. (The relations between these locutions, obviously connected with the notion of "logical consequence," are considered, in some cases obliquely, in the Grammatical Propaedeutic, and those who are worried about some of the more fashionable views may look there for ours.)

We proceed to the formal analysis as follows:

In the next subsection, we use *natural deduction* (due originally, and independently, to Gentzen 1934 and Jaskowski 1934), in the especially perspicuous variant of Fitch 1952, in order to motivate the choice of formal rules for "→" (taking the arrow as the formal analogue of the connective "that . . . entails that —"). The resulting system, equivalent to the pure implicational part $\mathbf{H}_{\rightarrow}$ of Heyting's intuitionistic logic (§1.4), is seen to have some of the properties associated with the notion of entailment.

In the next two sections we argue that, in spite of this partial agreement, $\mathbf{H}_{\rightarrow}$ is deficient in two distinguishable respects. First, it ignores considerations of *necessity* associated with entailment; in §2, modifications of $\mathbf{H}_{\rightarrow}$ are introduced to take necessity into account, and these are shown to lead to the pure implicational fragment of the system **S4** of strict implication (Lewis and Langford 1932). Second, $\mathbf{H}_{\rightarrow}$ is equally blind to considerations of *relevance;* modifications of $\mathbf{H}_{\rightarrow}$ in §3, designed to accommodate this important feature of the intuitive logical "if . . . then —," yield a calculus $\mathbf{R}_{\rightarrow}$, equivalent to the implicational part of the system of relevant implication first considered by Moh 1950 and Church 1951.

With §4 we are (for the first time) home: combining necessity and relevance leads naturally and plausibly to the pure calculus $\mathbf{E}_{\rightarrow}$ of entailment. §5 proves that $\mathbf{E}_{\rightarrow}$ really does capture the concepts of necessity and relevance in certain mathematically definite senses, and with this we complete the main argument of the chapter. The remaining sections present a number of related results: in §6 we define an even stricter form of entailment, called "ticket-entailment," answering to a conception of entailment as an "inference-ticket"; in §7 we sketch consecution calculuses in the style of Gentzen for various systems; and §8 collects odd bits of information (and a few questions) about the systems thus far considered.

Let us pause briefly to fix some notational matters. We remind the reader that in this chapter we are considering only pure implicational systems, leaving connections between entailment and other logical notions until later. Consequently we can describe the languages we are considering as having the following structure. In the first place, we suppose there is an infinite list of *propositional variables*, which we never display. But we shall often use

$p, q, r, s,$

etc., as variables ranging over them. Then *formulas* are defined by specifying that all propositional variables are formulas, and that whenever A and B are, so is $(A \rightarrow B)$. As variables ranging over formulas we employ

$A, B, C, D,$

etc., often with subscripts. We warn the reader that for the purpose of the present discussion we use the arrow ambiguously in order to compare various proposed formalizations of entailment.

As a further notational convention, we use dots to replace parentheses in accordance with conventions of Church 1956: outermost parentheses are omitted; a dot may replace a left-hand parenthesis, the mate of which is to be restored at the end of the parenthetical part in which the dot occurs (otherwise at the end of the formula); otherwise parentheses are to be restored by association to the left. Example: each of $(A \rightarrow .B \rightarrow C) \rightarrow .A \rightarrow B \rightarrow .A \rightarrow C$ and $A \rightarrow (B \rightarrow C) \rightarrow .A \rightarrow B \rightarrow .A \rightarrow C$ abbreviates $((A \rightarrow (B \rightarrow C)) \rightarrow ((A \rightarrow B) \rightarrow (A \rightarrow C)))$.

§1.3. Natural deduction. The intuitive idea lying behind systems of natural deduction is that there should be, for each logical connective, one rule justifying its introduction into discourse, and one rule for *using* ("eliminating") the connective once it has been introduced. For extended discussions of this motivation see Curry 1963, Popper 1947, or Kneale 1956, and for hazards attendant on careless statements of the leading ideas, see Prior 1960–61 and Belnap 1962.

Since we wish to interpret "$A \rightarrow B$" as "A entails B," or "B is deducible from A," we clearly want to be able to *assert* $A \rightarrow B$ whenever there exists a deduction of B from A, i.e., we will want a rule of *Entailment Introduction*, hereafter "$\rightarrow$I," having the property that if

⊢	A	hypothesis (hereafter "hyp")
│	⋮	
│	B	[conclusion]

is a valid deduction of B from A, then $A \rightarrow B$ shall follow from that deduction. (This sentence contains a lapse from grammar, the first of many. If you did not notice it, or if it did not bother you, please go on; only if our solecism irritates you, consult the Grammatical Propaedeutic for a statement and defense of our policy of loose grammar.)

Moreover, the fact that such a deduction exists, or correspondingly that an entailment $A \rightarrow B$ holds, warrants the inference of B from A. That is, we expect also that a rule of *modus ponens* or *Entailment Elimination*, henceforth "$\rightarrow$E," will obtain for $\rightarrow$, in the sense that whenever $A \rightarrow B$ is asserted, we shall be entitled to infer B from A.

So much is simple and obvious, and presumably not open to question. Problems arise, however, when we ask what constitutes a "valid deduction" of B from A. How may we fill in the dots in the proof scheme above?

At least one rule seems as simple and obvious as the foregoing. Certainly the supposition that A warrants the (trivial) inference that A; and if B has been deduced from A, we are entitled to infer B on the supposition A. That is, we may *repeat* ourselves:

1	⊢	A	hyp
	│	⋮	
i	│	B	?
	│	⋮	
j	│	B	i repetition (henceforth "rep")
	│	⋮	

This rule leads immediately to the following theorem, the *law of identity:*

1	⊢	A	hyp
2	│	A	1 rep
3	$A \rightarrow A$		1–2 $\rightarrow$I

We take the law of identity to be a truth about entailment; $A \rightarrow A$ represents the archetypal form of inference, the trivial foundation of all reasoning, in spite of those who would call it "merely a case of stuttering." Strawson 1952 (p. 15) says that

> a man who repeats himself does not reason. But it is inconsistent to assert and deny the same thing. So a logician will say that a statement has to itself the relationship [entailment] he is interested in.

Strawson has got the cart before the horse: the reason that A and $\bar{A}$ are inconsistent is precisely because A follows from itself, rather than conversely. (We shall in the course of subsequent investigations accumulate a substantial amount of evidence for this view, but the most convincing arguments will have to await treatment of truth functions and propositional quantifiers in connection with entailment. For the moment we observe that the difference between Strawson's view and our own first emerges formally in the system **E** of Chapter IV, where we have $A \rightarrow A \rightarrow \overline{A \& \bar{A}}$ but not $\overline{A \& \bar{A}} \rightarrow . A \rightarrow A$, just as we have $A \rightarrow B \rightarrow \overline{A \& \bar{B}}$ but not $\overline{A \& \bar{B}} \rightarrow . A \rightarrow B$).

But obviously more than the law of identity is required if a calculus of entailment is to be developed, and we therefore consider initially a device contained in the variant of natural deduction of Fitch 1952, which allows us to construct *within* proofs of entailment, further proofs of entailment called "subordinate proofs," or "subproofs." In the course of a deduction, under the supposition that A (say), we may begin a new deduction, with a new hypothesis:

```
1   |_ A        hyp
    |  .
    |  .
i   |  |_ B     hyp
    |  |   .
    |  |   .
```

The new subproof is to be conceived of as an "item" of the proof of which A is the hypothesis, just like A or any other formula occurring in that proof. And the subproof of which B is hypothesis might itself have a consequence (by $\rightarrow$I) occurring in the proof of which A is the hypothesis.

We next ask whether or not the hypothesis A holds also under the assumption B. In the system of Fitch 1952, the rules are so arranged that a step following from A in the outer proof may also be repeated under the assumption B, such a repetition being called a "reiteration" to distinguish it from repetitions within the same proof or subproof:

1	$\vert\underline{\ A\ }$	hyp
	$\vert\ \vdots$	
i	$\vert\ C$	?
	$\vert\ \vdots$	
j	$\vert\ \vert\underline{\ B\ }$	hyp
	$\vert\ \vert\ \vdots$	
k	$\vert\ \vert\ C$	i reiteration ("reit")
	$\vert\ \vert\ \vdots$	
	$\vert\ \vdots$	

We designate as $\mathbf{FH}_{\rightarrow}$ the system defined by the five rules, →I, →E, hyp, rep, and reit. A proof is *categorical* if all hypotheses in the proof have been discharged by use of →I, otherwise *hypothetical;* and A is a *theorem* if A is the last step of a categorical proof. These rules lead naturally and easily to proofs of intuitively satisfactory theorems about entailment, such as the following *law of transitivity*.

1	$\vert\underline{\ A\to B\ }$	hyp
2	$\vert\ \vert\underline{\ B\to C\ }$	hyp
3	$\vert\ \vert\ A\to B$	1 reit
4	$\vert\ \vert\ \vert\underline{\ A\ }$	hyp
5	$\vert\ \vert\ \vert\ A\to B$	3 reit
6	$\vert\ \vert\ \vert\ B$	4 5 →E
7	$\vert\ \vert\ \vert\ B\to C$	2 reit
8	$\vert\ \vert\ \vert\ C$	6 7 →E
9	$\vert\ \vert\ A\to C$	4–8 →I
10	$\vert\ B\to C\to.A\to C$	2–9 →I
11	$A\to B\to.B\to C\to.A\to C$	1–10 →I

Lewis indeed doubts whether

> this proposition should be regarded as a valid principle of deduction: it would never lead to any inference $A\to C$ which would be questionable when $A\to B$ and $B\to C$ are given premisses; but it gives the inference $B\to C\to.A\to C$ whenever $A\to B$ is a premiss. Except as an elliptical statement for "$(A\to B)\&(B\to C)\to.A\to C$ and $A\to B$ is true," this inference is dubious. (Lewis and Langford 1932, p. 496.)

On the contrary, Ackermann 1956 is surely right that "unter der Voraussetzung $A\to B$ ist der Schluss von $B\to C$ auf $A\to C$ logisch zwingend." The mathematician is involved in no ellipsis in arguing that "if the lemma is deducible from the axioms, then this entails that the deducibility of the theorem from the axioms is entailed by the deducibility of the theorem from the lemma."

The proof method sketched above has the advantage, in common with other systems of natural deduction, of motivating proofs: in order to prove $A\rightarrow B$ (perhaps under some hypothesis or hypotheses), we follow the simple and obvious strategy of playing both ends against the middle: breaking up the conclusion to be proved, and setting up subproofs by hyp until we find one with a variable as last step. Only then do we begin applying reit, rep, and $\rightarrow$E.

As a short-cut we allow reiterations directly into subproofs, subsubproofs, etc., with the understanding that a complete proof requires that reiterations be performed always from one proof into another proof immediately subordinate to it. As an example (step 6 below), we prove the *self-distributive* law (**H$_{\rightarrow}$2**, below):

1	$A\rightarrow.B\rightarrow C$	hyp
2	$\quad A\rightarrow B$	hyp
3	$\quad\quad A$	hyp
4	$\quad\quad A\rightarrow B$	2 reit
5	$\quad\quad B$	3 4 $\rightarrow$E
6	$\quad\quad A\rightarrow.B\rightarrow C$	1 reit (twice)
7	$\quad\quad B\rightarrow C$	3 6 $\rightarrow$E
8	$\quad\quad C$	5 7 $\rightarrow$E
9	$\quad A\rightarrow C$	3–8 $\rightarrow$I
10	$A\rightarrow B\rightarrow.A\rightarrow C$	2–9 $\rightarrow$I
11	$(A\rightarrow.B\rightarrow C)\rightarrow.A\rightarrow B\rightarrow.A\rightarrow C$	1–10 $\rightarrow$I

§.1.4. Intuitionistic implication (H$_{\rightarrow}$). Fitch 1952 shows (essentially) that the set of theorems of **FH$_{\rightarrow}$** stemming from these rules is identical with the pure implicational fragment **H$_{\rightarrow}$** of the intuitionist propositional calculus of Heyting 1930 (called "absolute implication" by Curry 1959 and elsewhere), which consists of the following three axioms, with $\rightarrow$E as the sole rule:

H$_{\rightarrow}$1 $A\rightarrow.B\rightarrow A$
H$_{\rightarrow}$2 $(A\rightarrow.B\rightarrow C)\rightarrow.A\rightarrow B\rightarrow.A\rightarrow C$
H$_{\rightarrow}$3 $A\rightarrow A$

(Formulations like that of **H$_{\rightarrow}$** just above, defined by axioms and rules — often just $\rightarrow$E — we refer to sometimes as *Hilbert* systems or formulations, sometimes as *axioms-cum-*$\rightarrow$E, or *axiomatic* formulations. Observe that among H$_{\rightarrow}$1–3, H$_{\rightarrow}$3 is redundant.)

In order to introduce terminology and to exemplify a pattern of argument which we shall have further occasion to use, we shall reproduce Fitch's proof that the two formulations are equivalent.

To see that the subproof formulation $\mathbf{FH}_{\rightarrow}$ contains the Hilbert formulation $\mathbf{H}_{\rightarrow}$, we deduce the axioms of $\mathbf{H}_{\rightarrow}$ in $\mathbf{FH}_{\rightarrow}$ ($\mathrm{H}_{\rightarrow}2$ was just proved and $\mathrm{H}_{\rightarrow}1$ is proved below) and then observe that the only rule of $\mathbf{H}_{\rightarrow}$ is also a rule of $\mathbf{FH}_{\rightarrow}$. It follows that $\mathbf{FH}_{\rightarrow}$ contains $\mathbf{H}_{\rightarrow}$.

To see that the axiomatic system $\mathbf{H}_{\rightarrow}$ contains the subproof formulation $\mathbf{FH}_{\rightarrow}$, we first introduce the notion of a *quasi-proof* in $\mathbf{FH}_{\rightarrow}$; a quasi-proof differs from a proof only in that we may introduce axioms of $\mathbf{H}_{\rightarrow}$ as steps (and of course use these, and steps derived from them, as premisses). We note in passing that this does not increase the stock of theorems of $\mathbf{FH}_{\rightarrow}$, since we may think of a step A, inserted under this rule, as coming by reiteration from a previous proof of A in $\mathbf{FH}_{\rightarrow}$ (which we know exists since $\mathbf{FH}_{\rightarrow}$ contains $\mathbf{H}_{\rightarrow}$); but we do not use this fact in our proof that $\mathbf{H}_{\rightarrow}$ contains $\mathbf{FH}_{\rightarrow}$.

Our object then is to show how subproofs in a quasi-proof in $\mathbf{FH}_{\rightarrow}$ may be systematically eliminated in favor of theorems of $\mathbf{H}_{\rightarrow}$ and uses of $\rightarrow$E, in such a way that we are ultimately left with a sequence of formulas all of which are theorems of $\mathbf{H}_{\rightarrow}$. This reduction procedure always begins with an *innermost subproof*, by which we mean a subproof Q which has no proofs subordinate to it. Let Q be an innermost subproof of a quasi-proof P of $\mathbf{FH}_{\rightarrow}$, where the steps of Q are $A_1, \ldots, A_n$, let Q′ be the sequence $A_1 \rightarrow A_1$, $A_1 \rightarrow A_2, \ldots, A_1 \rightarrow A_n$, and let P′ be the result of replacing the subproof Q of P by the sequence Q′ of formulas. Our task is now to show that P′ is convertible into a quasi-proof, by showing how to insert theorems of $\mathbf{H}_{\rightarrow}$ among the wffs of Q′, in such a way that each step of Q′ may be justified by one of hyp, reit, rep, $\rightarrow$E, or axiomhood in $\mathbf{H}_{\rightarrow}$ (the case $\rightarrow$I will not arise because Q is innermost).

An inductive argument then shows that we may justify steps in Q′ as follows:

$A_1 \rightarrow A_1$ is justified, by $\mathrm{H}_{\rightarrow}3$.

If A_i was by rep in Q, then, by the inductive hypothesis, $A_1 \rightarrow A_i$ is by rep in Q′.

If A_i was by reit in Q, then in Q′ insert $A_i \rightarrow . A_1 \rightarrow A_i$ ($\mathrm{H}_{\rightarrow}1$) and use $\rightarrow$E to get $A_1 \rightarrow A_i$ (the minor premiss being an item of the quasi-proof in P to which Q is subordinate, hence also preceding Q′ in P′).

If A_i was by $\rightarrow$E in Q, with premisses A_j and $A_j \rightarrow A_i$, then in Q′ we have $A_1 \rightarrow A_j$ and $A_1 \rightarrow . A_j \rightarrow A_i$. Then insert $\mathrm{H}_{\rightarrow}2$ and use $\rightarrow$E twice to get $A_1 \rightarrow A_i$ as required.

If A_i was an axiom — recall we are dealing with *quasi*-proofs — then insert $A_i \rightarrow . A_1 \rightarrow A_i$ ($\mathrm{H}_{\rightarrow}1$) and use $\rightarrow$E to obtain $A_1 \rightarrow A_i$.

So every step in Q′ is justified. Now notice that we can conclude that every step in all of P′ is justified, for P′ is exactly the same as P except that Q (in P) has been replaced by Q′. The only possible trouble might be if some step in P were justified through $\rightarrow$I by a reference to the now absent Q; but such a step can be justified in P′ by rep, with a reference to the last line of Q′.

Repeated application of this reduction then converts any proof in $\mathbf{FH}_{\rightarrow}$ into a sequence of formulas all of which are theorems of $\mathbf{H}_{\rightarrow}$; hence the latter system contains the former, and the two are equivalent. Notice incidentally that the choice of axioms for $\mathbf{H}_{\rightarrow}$ may be thought of as motivated by a wish to prove $\mathbf{H}_{\rightarrow}$ and $\mathbf{FH}_{\rightarrow}$ equivalent: they are *exactly* what is required to carry out the inductive argument above.

(We retain the concepts of quasi-proof and innermost subproof, with some sophistications, for use in later arguments which are closely similar to the foregoing.)

The axioms of $\mathbf{H}_{\rightarrow}$ also enable us to prove a slightly different form of the result above. We consider proofs with no subproofs, but with multiple hypotheses, and we define a *proof* of *B from hypotheses* $A_1, \ldots, A_n$ (in the Official way) as a sequence $C_1, \ldots, C_m, B$ of formulas each of which is either an axiom, or one of the hypotheses A_i ,or a consequence of predecessors by $\rightarrow$E. Then we arrive by very similar methods at the Official form of the

Deduction theorem. If there exists a proof of B from the hypotheses $A_1, \ldots, A_n$, then there exists a proof of $A_n \rightarrow B$ on the hypotheses $A_1, \ldots, A_{n-1}$; and conversely.

We return now to consideration of $\mathrm{H}_{\rightarrow}1$, which is proved in $\mathbf{FH}_{\rightarrow}$ as follows:

1	A	hyp
2	B	hyp
3	A	1 reit
4	$B \rightarrow A$	2–3 $\rightarrow$I
5	$A \rightarrow .B \rightarrow A$	1–4 $\rightarrow$I

Thus far the theorems proved by the subordinate proof method have all seemed natural and obvious truths about our intuitive idea of *entailment*. But here we come upon a theorem which shocks our intuitions (at least our untutored intuitions), for the theorem seems to say that anything whatever has A as a logical consequence, provided only that A is true; if the formal machinery is offered as an analysis or reconstruction of the notion of entailment, or formal deducibility, the principle seems outrageous — such at least is almost certain to be the initial reaction to the theorem, as anyone who has taught elementary logic very well knows. Formulas like $A \rightarrow .B \rightarrow A$ and $A \rightarrow .B \rightarrow B$ are of course familiar, and much discussed under the heading of "implicational paradoxes."

Those whose views concerning the philosophy of logic commit them to accept such principles are usually quick to point out that the freshman's objections are founded on confusion. For example, Quine 1950 (p. 37) says that a confusion of use and mention is involved, and that (in effect) although

A implies (*B* implies *A*)

may be objectionable,

if *A* then if *B* then *A*

is not. We have dealt with this sort of grammatical point in the Grammatical Propaedeutic at the end of this volume. But it is worth remarking here that even if Quine and his followers are correct about the grammar of English (or any other natural language), it is still true that the naive freshman objects as much to the second of the two formulations as to the first. So do we.

And Curry 1959 explains that the arrow of $\mathbf{H}_{\rightarrow}$ does not lay any claim to being a definition of logical consequence. "It does not pretend to be anything of the sort" (p. 20). The claim is supported by an argument to the effect that $A \rightarrow .B \rightarrow A$, "far from being paradoxical," is, for any *proper* implication, "a platitude." A "proper" implication is defined by Curry as any implication which has the following properties: there is a proof of B from the hypotheses $A_1, \ldots, A_{n-1}, A_n$ (in the Official sense of "proof from hypotheses") if and only if there is a proof of $A_n \rightarrow B$ from the hypotheses $A_1, \ldots, A_{n-1}$. On these grounds $A \rightarrow .B \rightarrow A$ is indeed a platitude: there is surely a proof of A from the hypotheses A, B; and hence for any "proper" implication, a proof of $B \rightarrow A$ from the hypothesis A; and hence a proof without hypotheses of $A \rightarrow .B \rightarrow A$.

Curry calls this a proof of $A \rightarrow .B \rightarrow A$ "from nothing." We remark that this expression invites the interpretation "there is nothing from which $A \rightarrow .B \rightarrow A$ is deducible," in which case we would seem to have done little toward showing that it is true. But of course Curry is not confused on this point; he means that $A \rightarrow .B \rightarrow A$ is deducible "from" the null set of premisses — in the reason-shattering, Official sense of "from." (These arguments deserve to be taken more seriously than our tone suggests; we will try to do so when the matter comes up again in connection with the notion of relevance, in §3.)

Curry goes on to dub the implicational relation of $\mathbf{H}_{\rightarrow}$ "absolute implication" on the grounds that $\mathbf{H}_{\rightarrow}$ is the minimal system having this property. But we notice at once that $\mathbf{H}_{\rightarrow}$ is "absolute" only relatively, i.e., relatively to the Official definition of "proof from hypotheses." From this point of view, our remarks to follow may be construed as arguing the impropriety of accepting the Official definition of "proof from hypotheses," as a basis for defining a "proper implication"; as we shall claim, the Official view captures neither "proof" (a matter involving logical necessity) nor "from" (a matter requiring relevance). But even those with intuitions so sophisticated that $A \rightarrow .B \rightarrow A$ seems tolerable might still find some interest in an attempt to analyze our initial feelings of repugnance in its presence.

Why does $A \rightarrow .B \rightarrow A$ seem so queer? We believe that its oddness is due to two isolable features of the principle, which we consider forthwith.

§2. Necessity: strict implication ($S4_{\rightarrow}$). For more than two millennia logicians have taught that logic is a *formal* matter, and that the validity of an inference depends not on material considerations, but on formal considerations alone. We here approach a more accurate statement of this condition in several steps, first noting that it amounts to saying that the validity of a valid inference is no accident of nature, but rather a property a valid inference has necessarily. Still more accurately: an entailment, if true at all, is necessarily true.

Because true entailments are necessarily so, we ought to grant, as we do, that truths entailed by necessary truths are themselves one and all necessary; and we then see immediately that $A \rightarrow .B \rightarrow A$ violates this plausible condition. For let A be contingently true, and B necessarily true; then given $A \rightarrow .B \rightarrow A$, A leads to $B \rightarrow A$, and now we have a necessity entailing a contingency, which is no good. That is to say, for such an instance of $A \rightarrow .B \rightarrow A$, the antecedent is true, and the consequent false. Note that this argument is equally an argument against the weaker $A \supset .B \rightarrow A$, where now the horseshoe is material "implication"; i.e., A is true while $B \rightarrow A$ is false.

(We thank Routley and Routley 1969 for pointing out a howler [see §20.2] in the version of this argument in Anderson and Belnap 1962a; Coffa straightens us out on the matter in §22.1.2.)

It might be said in defense of $A \rightarrow .B \rightarrow A$ as an entailment that at least it is "safe," in the sense that if A is true, then it is always safe to infer A from an arbitrary B, since we run no risk of uttering a falsehood in doing so; this thought ("Safety First") seems to be behind attempts, in a number of elementary logic texts, to justify the claim that $A \rightarrow .B \rightarrow A$ has something to do with implication. In reply we of course admit that if A is true then it is "safe" to say so (i.e., $A \rightarrow A$). But saying that A is true on the irrelevant assumption that B, is *not* to deduce A from B, nor to establish that B implies A, in *any* sensible sense of "implies." Of course we *can* say "Assume that snow is puce. Seven is a prime number." But if we say "Assume that snow is puce. *It follows that* (or *consequently*, or *therefore*, or *it may validly be inferred that*) seven is a prime number," then we have simply spoken falsely. A man who assumes the continuum hypothesis, and then remarks that it is a nice day, is not inferring the latter from the former — even if he keeps his supposition fixed firmly in mind while noting the weather. And since a (true) A does not follow from an (arbitrary) B, we reject $A \rightarrow .B \rightarrow A$ as expressing a truth of entailment or implication, a rejection which is in line with the view (shared even by some who hold that $A \rightarrow .B \rightarrow A$ expresses a fact about "if . . . then —") that entailments, if true at all are necessarily true.

How can we modify the formulation of $\mathbf{H}_{\rightarrow}$ in such a way as to guarantee that the implications expressible in it shall reflect necessity, rather than contingency? As a start, picture an (outermost) subproof as exhibiting a mathematical argument of some kind, and reflect that in our usual mathematical or logical proofs, we demand that all the conditions required for the conclusion be stated in the hypothesis of a theorem. After the word "PROOF:" in a mathematical treatise, mathematical writers seem to feel that no more hypotheses may be introduced; and it is regarded as a criticism of a proof if not all the required hypotheses are stated explicitly at the outset. Of course additional machinery *may* be invoked in the proof, but this must be of a logical character, i.e., in addition to the hypotheses, we may use in the argument only propositions tantamount to statements of logical necessity. These considerations suggest that we should be allowed to import into a deduction (i.e., into a subproof by reit) only propositions which, if true at all, are necessarily true: i.e., we should reiterate only entailments. Of course the illustration directly motivates the restriction only for outermost subproofs, but the same reasoning justifies extending the restriction to all subproofs: if at any stage of an argument one is attempting to establish, under a batch of hypotheses, a statement of a logical character — in our case, an entailment — then one should be allowed to bring in from the outside (by reiteration) only those steps which themselves have the appropriate logical character, i.e., entailments. And indeed such a restriction on reiteration would immediately rule out $A \rightarrow .B \rightarrow A$ as a theorem, while countenancing all the other theorems we have proved thus far. We call the system with reiteration allowed only for entailments $\mathbf{FS4}_{\rightarrow}$, and proceed to prove it equivalent to the following axiomatic formulation, which we call $\mathbf{S4}_{\rightarrow}$, since it is the pure strict "implicational" fragment of Lewis's **S4**. (See Hacking 1963).

S4$_{\rightarrow}$1 $A \rightarrow A$
S4$_{\rightarrow}$2 $(A \rightarrow .B \rightarrow C) \rightarrow .A \rightarrow B \rightarrow .A \rightarrow C$
S4$_{\rightarrow}$3 $A \rightarrow B \rightarrow .C \rightarrow .A \rightarrow B$

It is a trivial matter to prove the axioms of $\mathbf{S4}_{\rightarrow}$ in $\mathbf{FS4}_{\rightarrow}$, and the only rule of $\mathbf{S4}_{\rightarrow}$ ($\rightarrow$E) is also a rule of $\mathbf{FS4}_{\rightarrow}$; hence $\mathbf{FS4}_{\rightarrow}$ contains $\mathbf{S4}_{\rightarrow}$. To establish the converse, we show how to convert any quasi-proof of a theorem A in $\mathbf{FS4}_{\rightarrow}$ into a proof of A in $\mathbf{S4}_{\rightarrow}$.

THEOREM. Let $A_1, \ldots, A_n$ be the items of an innermost subproof Q of a quasi-proof P, and let Q′ be the sequence $A_1 \rightarrow A_1, \ldots, A_1 \rightarrow A_n$, and finally let P′ be the result of replacing the subproof Q in P by the sequence of formulas Q′. Then P′ can be converted into a quasi-proof.

PROOF. First we prove that each step of Q′ can be justified, by induction on n. For $n = 1$ we note that $A_1 \rightarrow A_1$ is an instance of $S4_{\rightarrow}1$. Then, assuming the theorem for all $i < n$, consider $A_1 \rightarrow A_n$.

CASE 1. A_n is by repetition in Q of A_i. Then treat $A_1 \rightarrow A_n$ in Q′ as a repetition of $A_1 \rightarrow A_i$.

CASE 2. A_n is a reiteration in Q of B. Then B has the form $C \rightarrow D$, by the restriction on reiteration. Insert $C \rightarrow D \rightarrow .A_1 \rightarrow .C \rightarrow D$ in Q′ by $S4_{\rightarrow}3$, and treat $A_1 \rightarrow .C \rightarrow D$ (i.e., $A_1 \rightarrow A_n$) as a consequence of $C \rightarrow D$ (i.e., B), and $S4_{\rightarrow}3$ by $\rightarrow$E.

CASE 3. A_n follows in Q from A_i and $A_i \rightarrow A_n$ by $\rightarrow$E. Then by the inductive hypothesis we have $A_1 \rightarrow A_i$ and $A_1 \rightarrow .A_i \rightarrow A_n$ in Q′. Then $A_1 \rightarrow A_n$ is a consequence of the latter and $S4_{\rightarrow}2$, with two uses of $\rightarrow$E.

CASE 4. A_n is an axiom. Then A_n has the form $B \rightarrow C$; so it follows from $S4_{\rightarrow}3$ that $A_1 \rightarrow A_n$ is a theorem.

Now we may conclude that P′ is a quasi-proof. For a step $A_1 \rightarrow A_n$, regarded as a consequence of Q in P, may now be regarded as a repetition of the final step $A_1 \rightarrow A_n$ of Q′ in P′.

Hence P′ is convertible into a quasi-proof. And repeated application of this technique to P′ eventually leads to a sequence P″ of formulas each of which is a theorem of $\mathbf{S4}_{\rightarrow}$. Hence $\mathbf{S4}_{\rightarrow}$ includes $\mathbf{FS4}_{\rightarrow}$, and the two are equivalent.

A deduction theorem of the more usual sort is provable also for $\mathbf{S4}_{\rightarrow}$:

THEOREM. If there is a proof of B on hypotheses $A_1, \ldots, A_n$ (in the Official sense), where each A_i, $1 \leq i \leq n$, has the form $C \rightarrow D$, then there is a proof of $A_n \rightarrow B$ on hypotheses $A_1, \ldots, A_{n-1}$. (Barcan Marcus 1946; see also Kripke 1959a.)

Notice again that as in the case of $\mathbf{H}_{\rightarrow}$, the choice of axioms for $\mathbf{S4}_{\rightarrow}$ may be thought of as motivated exactly by the wish to prove an appropriate deduction theorem.

The restriction on reiteration suffices to remove one objectionable feature of $\mathbf{H}_{\rightarrow}$, since it is now no longer possible to establish an entailment $B \rightarrow A$ when A is contingent and B is necessary. But of course it is well known that the "implication" relation of **S4** is also paradoxical, since we can easily establish that an arbitrary irrelevant proposition B "implies" A, provided

A is a necessary truth. $A \rightarrow A$ is necessarily true, and from it and S4$_{\rightarrow}$3 follows $B \rightarrow .A \rightarrow A$, where B may be totally irrelevant to $A \rightarrow A$. Observe that $B \rightarrow .A \rightarrow A$ does *not* violate the intuitive condition laid down at the outset of this section as a basis for dismissing $A \rightarrow .B \rightarrow A$; we cannot by the same device assign values to A and B so that the antecedent of $B \rightarrow .A \rightarrow A$ comes out true, and the consequent false. The presence of $B \rightarrow .A \rightarrow A$ therefore leads us to consider an alternative restriction on $\mathbf{H}_{\rightarrow}$, designed to exclude such fallacies of relevance.

§3. Relevance: relevant implication ($\mathbf{R}_{\rightarrow}$). For more than two millennia logicians have taught that a necessary condition for the validity of an inference from A to B is that A be relevant to B. Virtually every logic book up to the present century has a chapter on fallacies of relevance, and many contemporary elementary texts have followed the same plan. Notice that contemporary writers, in the later and more formal chapters of their books, seem explicitly to contradict the earlier chapters, when they try desperately to bamboozle the students into accepting strict "implication" as a "kind" of implication relation, in spite of the fact that this relation countenances fallacies of relevance. But the denial that relevance is essential to a valid argument, a denial which is implicit in the view that "formal deducibility," in the sense of Montague and Henkin 1956 and others, is an implication relation, seems to us flatly in error.

Imagine, if you can, a situation as follows. A mathematician writes a paper on Banach spaces, and after proving a couple of theorems he concludes with a conjecture. As a footnote to the conjecture, he writes: "In addition to its intrinsic interest, this conjecture has connections with other parts of mathematics which might not immediately occur to the reader. For example, if the conjecture is true, then the first order functional calculus is complete; whereas if it is false, then it implies that Fermat's last conjecture is correct." The editor replies that the paper is obviously acceptable, but he finds the final footnote perplexing; he can see no connection whatever between the conjecture and the "other parts of mathematics," and none is indicated in the footnote. So the mathematician replies, "Well, I was using 'if . . . then —' and 'implies' in the way that logicians have claimed I was: the first order functional calculus *is* complete, and necessarily so, so anything implies that fact — and if the conjecture is false it is presumably impossible, and hence implies anything. And if you object to this usage, it is simply because you have not understood the technical sense of 'if . . . then —' worked out so nicely for us by logicians." And to this the editor counters: "I *understand* the technical bit all right, but it is simply not correct. In spite of what most logicians say about us, the standards maintained by this journal require that the antecedent of an 'if . . . then —' statement must be *relevant* to the conclusion drawn. And you have given no evidence that your conjecture about Banach spaces is relevant either to the completeness theorem or to Fermat's conjecture."

Now it might be thought that our mathematician's footnote should be regarded as true, "if . . . then —" being taken materially or (more likely) strictly — but simply uninteresting because of its triviality. But notice that the editor's reaction was *not* "But heavens, that's trivial" (as the contention that the mathematical "if . . . then —" is the same as material "implication" would require); any such reaction on the part of an editor would properly be judged insane. His thought was rather, "I can't see any reason for thinking that this is *true*."

No, the editor's point is that though the technical meaning is clear, it is simply not the same as the meaning ascribed to "if . . . then —" in the pages of his journal. Furthermore, he has put his finger precisely on the difficulty: to argue from the necessary truth of A to *if* B *then* A is simply to commit a fallacy of relevance. The fancy that relevance is irrelevant to validity strikes us as ludicrous, and we therefore make an attempt to explicate the notion of relevance of A to B.

For this we return to the notion of proof from hypotheses (in standard axiom-cum-$\rightarrow$E formulations), the leading idea being that we want to infer $A\rightarrow B$ from "a proof of B from the hypothesis A." As we pointed out before, in the usual axiomatic formulations of propositional calculuses the matter is handled as follows. We say that $A_1, \ldots, A_n$ is *a proof of* B *from the hypothesis* A, if $A = A_1$, $B = A_n$, and each A_i is either an axiom or else a consequence of predecessors among $A_1, \ldots, A_n$ by one of the rules. But in the presence of a deduction theorem of the form: from a proof of B on the hypothesis A, to infer $A\rightarrow B$, this definition leads immediately to fallacies of relevance; for if B is a theorem independently of A, then we have $A\rightarrow B$ where A may be irrelevant to B. For example, in a system with $A\rightarrow A$ as an axiom, we have

1	B	hyp
2	$A\rightarrow A$	axiom
3	$B\rightarrow .A\rightarrow A$	1–2, deduction theorem

In this example we indeed proved $A\rightarrow A$, but, though our eyes tell us that we proved it *under* the hypothesis B, it is crashingly obvious that we did not prove it *from* B: the defect lies in the definition, which fails to take seriously the word "from" in "proof *from* hypotheses." And this fact suggests a solution to the problem: we should devise a technique for keeping track of the steps used, and then allow application of the introduction rule only when A is relevant to B in the sense that A is *used* in arriving at B.

As a start in this direction, we suggest affixing a star (say) to the hypothesis of a deduction, and also to the conclusion of an application of →E just in case at least one premiss has a star, steps introduced as axioms being unstarred. Restriction of →I to cases where in accordance with these rules both A and B are starred would then exclude theorems of the form $A \rightarrow B$, where B is proved independently of A

In other words, what is wanted is a system, analogous to $\mathbf{H}_{\rightarrow}$ and $\mathbf{S4}_{\rightarrow}$, for which there is provable a deduction theorem to the effect that there exists a proof of B *from* the hypothesis A if and only if $A \rightarrow B$ is provable. And we now consider the question of choosing axioms in such a way as to guarantee this result. In view of the rule →E, the implication in one direction is trivial; we consider the converse.

Suppose we have a proof

$A_1{}^*$	hyp
⋮	
A_i	?
⋮	
$A_n{}^*$	?

of A_n from the hypothesis A_1, in the above sense, and we wish to convert this into an axiomatic proof of $A_1 \rightarrow A_n$. A natural and obvious suggestion would be to consider replacing each starred A_i by $A_1 \rightarrow A_i$ (since the starred steps are the ones to which A_1 is relevant), and try to show that the result is a proof without hypotheses. What axioms would be required to carry the induction through?

For the basis case we obviously require as an axiom $A \rightarrow A$. And in the inductive step, where we consider steps A_i and $A_i \rightarrow A_j$ of the original proof, four cases may arise.

(1) Neither premiss is starred. Then in the axiomatic proof, A_i, $A_i \rightarrow A_j$, and A_j all remain unaltered, so →E may be used as before.

(2) The minor premiss is starred, and the major one is not. Then in the axiomatic proof we have $A_1 \rightarrow A_i$ and $A_i \rightarrow A_j$; so we need to be able to infer $A_1 \rightarrow A_j$ from these (since the star on A_i guarantees a star on A_j in the original proof).

(3) The major premiss is starred, and the minor one is not. Then in the axiomatic proof we have $A_1 \rightarrow .A_i \rightarrow A_j$ and A_i, so we need to be able to infer $A_1 \rightarrow A_j$ from these.

(4) And finally both may be starred, in which case we have $A_1 \rightarrow .A_i \rightarrow A_j$ and $A_1 \rightarrow A_i$ in the axiomatic proof, from which again we need to infer $A_1 \rightarrow A_j$.

Summarizing: the proof of an appropriate deduction theorem where relevance is demanded would require the axiom $A \rightarrow A$ together with the validity of the following inferences:

from $A \rightarrow B$ and $B \rightarrow C$ to infer $A \rightarrow C$;
from $A \rightarrow .B \rightarrow C$ and B to infer $A \rightarrow C$;
from $A \rightarrow .B \rightarrow C$ and $A \rightarrow B$ to infer $A \rightarrow C$.

It then seems plausible to consider the following axiomatic system as capturing the notion of relevance:

$A \rightarrow A$	(*identity*)
$A \rightarrow B \rightarrow .B \rightarrow C \rightarrow .A \rightarrow C$	(*transitivity*)
$(A \rightarrow .B \rightarrow C) \rightarrow .B \rightarrow .A \rightarrow C$	(*permutation*)
$(A \rightarrow .B \rightarrow C) \rightarrow .A \rightarrow B \rightarrow .A \rightarrow C$	(*self-distribution*)

And without further proof we state that for this system $\mathbf{R}_{\rightarrow}'$ ($\mathbf{R}_{\rightarrow}$ gets defined below) we have the following

THEOREM. $A \rightarrow B$ is a theorem of $\mathbf{R}_{\rightarrow}'$ just in case there is a proof of B *from* the hypothesis A (in the starred sense).

Equivalent systems have been investigated by Moh 1950 and Church 1951. (See also Kripke 1959a.) Church calls his system the "weak positive implicational propositional calculus," and uses the following axioms:

$\mathrm{R}_{\rightarrow}1$	$A \rightarrow A$	(*identity*)
$\mathrm{R}_{\rightarrow}2$	$A \rightarrow B \rightarrow .C \rightarrow A \rightarrow .C \rightarrow B$	(*transitivity*)
$\mathrm{R}_{\rightarrow}3$	$(A \rightarrow .B \rightarrow C) \rightarrow .B \rightarrow .A \rightarrow C$	(*permutation*)
$\mathrm{R}_{\rightarrow}4$	$(A \rightarrow .A \rightarrow B) \rightarrow .A \rightarrow B$	(*contraction*)

Following a suggestion which Bacon made to us in 1962, we think of this as a system of "relevant implication," hence the name "$\mathbf{R}_{\rightarrow}$," since relevance of antecedent to consequent, in a sense to be explained later, is secured thereby. The same suggestion was also made by Prawitz, first in a mimeographed version of Prawitz 1964 distributed to those attending the meeting at which the abstracted paper was read, and then in the more extended discussion in Prawitz 1965.

The proof that $\mathbf{R}_{\rightarrow}'$ and $\mathbf{R}_{\rightarrow}$ are equivalent is left to the reader.

A generalization of the deduction theorem above was proved by both Moh 1950 and Church 1951; modified to suit present purposes, it may be stated as follows:

THEOREM. If there exists a proof of B on the hypotheses $A_1, \ldots, A_n$, in which all of $A_1, \ldots, A_n$ are used in arriving at B, then there is a proof of $A_n \to B$ from $A_1, \ldots, A_{n-1}$ satisfying the same condition.

So put, the result acquires a rather peculiar appearance: it seems odd that we should have to use *all* the hypotheses. One would have thought that, for a group of hypotheses to be relevant to a conclusion, it would suffice if *some* of the hypotheses were used — at least if we think of the hypotheses as taken conjointly (see the Entailment theorem of §23.6). The peculiarity arises because of a tendency (thus far not commented on) to confound

$$A_1 \& \ldots \& A_n \to B$$

with

$$A_1 \to .A_2 \to \ldots \to .A_n \to B$$

We would not expect to require that *all* the A_i be relevant to B in order for the first formula to be true, but we shall give reasons presently, deriving from another formulation of $\mathbf{R}_{\to}$, for thinking it sensible that the truth of the nested implication requires each of the A_i to be relevant to B; a feature of the situation which will lead us to make a sharp distinction between the two formulas (see §22.2.2). It is presumably the failure to make this distinction which leads Curry 1959 (p. 13) to say of the relation considered in Moh's and Church's theorem above that it is one "which is not ordinarily considered in deductive methodology at all." (He's right; it's not. But it ought to be, for there is where the heart lies.)

We feel that the star formulation of the deduction theorem makes clearer what is at stake in $\mathbf{R}_{\to}$. On the other hand the deduction theorem of Moh and Church has the merit of allowing for proof of multiply nested entailments in a more direct way than is available in the star formulation. Our next task therefore is to try to combine these approaches so as to obtain the advantages of both.

Returning now to a consideration of subordinate proofs, it seems natural to try to extend the star treatment, using some other symbol for deductions carried out in a subproof, but retaining the same rules for carrying this symbol along. We might consider a proof of contraction in which the inner hypothesis is distinguished by a dagger rather than a star:

1	$A \to .A \to B$	*	hyp
2	A	†	hyp
⋮	⋮		

the different relevance marks reflecting the initial assumption that the two formulas, as hypotheses, are irrelevant to each other (or, equivalently, our initial ignorance as to whether they are irrelevant to each other). Then generalizing the starring rules, we might require that, in application of →E, the conclusion *B* must carry all the relevance marks of both premisses *A* and *A*→*B*, thus:

1	⌊ $A\rightarrow.A\rightarrow B$	*	hyp
2	│ ⌊ A	†	hyp
3	│ │ $A\rightarrow.A\rightarrow B$	*	1 reit
4	│ │ $A\rightarrow B$	*†	2 3 →E
5	│ │ B	*†	2 4 →E

To motivate the restriction on →I, we recall that, in proofs involving only stars, it was required that both *A* and *B* have stars, and that the star was discharged on *A*→*B* in the conclusion of a deduction. This suggests the following generalization: in drawing the conclusion *A*→*B* by →I, we require that the relevance symbol on *A* also be present among those of *B*, and that in the conclusion *A*→*B* the relevance symbol of *A* (like the hypothesis *A* itself) be discharged. Two applications of this rule then lead from the proof above to

6	│ $A\rightarrow B$	*	2–5 →I
7	$(A\rightarrow.A\rightarrow B)\rightarrow.A\rightarrow B$		1–6 →I

But of course the easiest way of handling the matter is to use classes of numerals to mark the relevance conditions, since then we may have as many nested subproofs as we wish, each with a distinct numeral (which we shall write in subscripted set-notation) for its hypothesis. More precisely we allow that: (1) one may introduce a new hypothesis $A_{\{k\}}$, where *k* should be different from all subscripts on hypotheses of proofs to which the new proof is subordinate; (2) from A_{a} and $A\rightarrow B_{b}$ we may infer $B_{a\cup b}$; (3) from a proof of B_{a} from the hypothesis $A_{\{k\}}$, we may infer $A\rightarrow B_{a-\{k\}}$, provided *k* is in a; and (4) reit and rep retain subscripts (where a, b, c, range over sets of numerals).

As an example we prove the *law of assertion:*

1	⌊ $A_{\{1\}}$	hyp
2	│ ⌊ $A\rightarrow B_{\{2\}}$	hyp
3	│ │ $A_{\{1\}}$	1 reit
4	│ │ $B_{\{1,2\}}$	2 3 →E
5	│ $A\rightarrow B\rightarrow B_{\{1\}}$	2–4 →I
6	$A\rightarrow.A\rightarrow B\rightarrow B$	1–5 →I

To see that this generalization of the *† notation, which results in the system we call $\mathbf{FR}_{\rightarrow}$, is also equivalent to $\mathbf{R}_{\rightarrow}$, observe first that the axioms of $\mathbf{R}_{\rightarrow}$ are easily proved in $\mathbf{FR}_{\rightarrow}$; hence $\mathbf{FR}_{\rightarrow}$ contains $\mathbf{R}_{\rightarrow}$. The proof of the converse involves little more than repeated application, beginning with an innermost subproof, of the techniques used in proving the deduction theorem for $\mathbf{R}_{\rightarrow}$; it will be left to the reader. (We call attention in §4 to some of the modifications required by the presence of subscripts.)

If the subscripting device is taken as an explication of relevance, then it is seen that Church's $\mathbf{R}_{\rightarrow}$ does secure relevance since $A \rightarrow B$ is provable in $\mathbf{R}_{\rightarrow}$ only if A is relevant to B. But if $\mathbf{R}_{\rightarrow}$ is taken as an explication of entailment, then the requirement of necessity for a valid inference is lost. Consider the following special case of the law of assertion, just proved:

$$A \rightarrow . A \rightarrow A \rightarrow A.$$

This says that if A is true, then it follows from $A \rightarrow A$. But it seems reasonable to suppose that any logical consequence of $A \rightarrow A$ should be necessarily true. (Note that in the familiar systems of modal logic, it is intended that consequences of necessary truths be necessary.) We certainly do in practice recognize that there are truths which do not follow from any law of logic — but $\mathbf{R}_{\rightarrow}$ obliterates this distinction. It seems evident, therefore, that a satisfactory theory of entailment will require both relevance (like $\mathbf{R}_{\rightarrow}$) *and* necessity (like $\mathbf{S4}_{\rightarrow}$).

§4. Necessity and relevance: entailment ($\mathbf{E}_{\rightarrow}$). We therefore consider the system which arises when we recognize that valid inferences require both necessity and relevance.

§4.1. The pure calculus of entailment: natural deduction formulation. Since the restrictions are most transparent as applied to the subproof format, we begin by considering the system $\mathbf{FE}_{\rightarrow}$ which results from imposing the restriction on reiteration (of $\mathbf{FS4}_{\rightarrow}$) together with the subscript requirements (of $\mathbf{FR}_{\rightarrow}$). We summarize the rules of $\mathbf{FE}_{\rightarrow}$ as follows:

(1) Hyp. A step may be introduced as the hypothesis of a new subproof, and each new hypothesis receives a unit class $\{k\}$ of numerical subscripts, where k is new.

(2) Rep. A_{a} may be repeated, retaining the relevance indices a.

(3) Reit. $(A \rightarrow B)_{\mathrm{a}}$ may be reiterated, retaining a.

(4) $\rightarrow$E. From A_{a} and $(A \rightarrow B)_{\mathrm{b}}$ to infer $B_{\mathrm{a} \cup \mathrm{b}}$.

(5) $\rightarrow$I. From a proof of B_{a} on hypothesis $A_{\{k\}}$ to infer $(A \rightarrow B)_{\mathrm{a}-\{k\}}$, provided k is in a.

It develops that an axiomatic counterpart of $\mathbf{FE}_{\rightarrow}$ has also been considered in the literature, $\mathbf{FE}_{\rightarrow}$ in fact being equivalent to a pure implicational calculus derived from Ackermann 1956. In §8.3.3 we consider various formulations of this system, and in Chapter VIII discuss various aspects of Ackermann's extraordinarily original and seminal paper, which served as the point of departure for all the investigations reported in this book. For the moment we show that $\mathbf{FE}_{\rightarrow}$ is equivalent to the following version of the system $\mathbf{E}_{\rightarrow}$, which gives this chapter its name:

$\mathrm{E}_{\rightarrow}1.$	$A{\rightarrow}A$	(*identity*)
$\mathrm{E}_{\rightarrow}2.$	$A{\rightarrow}B{\rightarrow}.B{\rightarrow}C{\rightarrow}.A{\rightarrow}C$	(*transitivity*)
$\mathrm{E}_{\rightarrow}3.$	$A{\rightarrow}B{\rightarrow}.A{\rightarrow}B{\rightarrow}C{\rightarrow}C$	(*restricted assertion*)
$\mathrm{E}_{\rightarrow}4.$	$(A{\rightarrow}.B{\rightarrow}C){\rightarrow}.A{\rightarrow}B{\rightarrow}.A{\rightarrow}C$	(*self-distribution*)

That $\mathbf{FE}_{\rightarrow}$ contains $\mathbf{E}_{\rightarrow}$ is trivial; the construction of proofs of the axioms by the subproof technique is left to the reader. To reduce proofs in $\mathbf{FE}_{\rightarrow}$ to proofs in $\mathbf{E}_{\rightarrow}$, we introduce, as usual, the mediating notion of a quasi-proof as like an $\mathbf{FE}_{\rightarrow}$ proof except possibly containing axioms of $\mathbf{E}_{\rightarrow}$; it is understood that the null set of subscripts is to be affixed to each such theorem. Then, before getting on with the main job of innermost subproof reduction, we prove a lemma, which, though mysterious in isolation, is required for case (3) of the theorem to follow.

Lemma. If A_{a} is a step of an innermost subproof Q (of a quasi-proof in $\mathbf{FE}_{\rightarrow}$) with hypothesis $H_{\{k\}}$, where k is *not* in a, then for arbitrary B we may with the help of axioms of $\mathbf{E}_{\rightarrow}$ and $\rightarrow$E also obtain $A{\rightarrow}B{\rightarrow}B_{\mathrm{a}}$ in Q.

Proof is by induction, where the inductive hypothesis is that the lemma holds for steps preceding A_{a}. There are three cases for Q innermost, with k not in a. (1) A_{a} is by rep. Then the inductive hypothesis guarantees that $A{\rightarrow}B{\rightarrow}B_{\mathrm{a}}$ may be obtained in Q. (2) A_{a} is either by reit or is an axiom. Then A_{a} has the form $C{\rightarrow}D_{\mathrm{a}}$; insertion of $C{\rightarrow}D{\rightarrow}.C{\rightarrow}D{\rightarrow}B{\rightarrow}B$ ($\mathrm{E}_{\rightarrow}3$) yields $C{\rightarrow}D{\rightarrow}B{\rightarrow}B_{\mathrm{a}}$ (i.e., $A{\rightarrow}B{\rightarrow}B_{\mathrm{a}}$) by $\rightarrow$E. (3) A_{a} is by $\rightarrow$E from C_{c} and $C{\rightarrow}A_{\mathrm{b}}$, where $\mathrm{a} = \mathrm{b} \cup \mathrm{c}$ and k is in neither b nor c. By the inductive hypothesis, we may obtain $C{\rightarrow}A{\rightarrow}A_{\mathrm{c}}$. Then insert $\mathrm{E}_{\rightarrow}2$:

$$C{\rightarrow}A{\rightarrow}A{\rightarrow}.A{\rightarrow}B{\rightarrow}.C{\rightarrow}A{\rightarrow}B,$$

and use $\rightarrow$E to get

$$A{\rightarrow}B{\rightarrow}.C{\rightarrow}A{\rightarrow}B_{\mathrm{c}}.$$

Using $\mathrm{E}_{\rightarrow}2$ again, we have

$$(A{\rightarrow}B{\rightarrow}.C{\rightarrow}A{\rightarrow}B){\rightarrow}.C{\rightarrow}A{\rightarrow}B{\rightarrow}B{\rightarrow}.A{\rightarrow}B{\rightarrow}B,$$

and the preceding two formulas by $\rightarrow$E yield

$$C{\rightarrow}A{\rightarrow}B{\rightarrow}B{\rightarrow}.A{\rightarrow}B{\rightarrow}B_{\mathrm{c}}.$$

But the inductive hypothesis also gives us

$$C \rightarrow A \rightarrow B \rightarrow B_{\mathrm{b}};$$

and the latter two by $\rightarrow$E yield

$$A \rightarrow B \rightarrow B_{\mathrm{b} \cup \mathrm{c}},$$

i.e., $A \rightarrow B \rightarrow B_{\mathrm{a}}$, as required.

Notice that in the proof of the lemma, only $\rightarrow$E and axioms of $\mathbf{E}_{\rightarrow}$ were used; hence the result of applying the lemma to Q is also a quasi-proof.

To prove that $\mathrm{FE}_{\rightarrow}$ is contained in $\mathbf{E}_{\rightarrow}$, we again consider an innermost subproof Q of a quasi-proof P in $\mathrm{FE}_{\rightarrow}$, with steps $A_{1\{k\}}, \ldots, A_{i\mathrm{a}_i}, \ldots, A_{n\mathrm{a}_n}$ and we let $A_{i\mathrm{a}_i}{}'$ be $(A_1 \rightarrow A_i)_{\mathrm{a}_i - \{k\}}$ or $A_{i\mathrm{a}_i}$ according as k is or is not in a_i. Then we replace Q by the sequence Q′ of $A_{i\mathrm{a}_i}{}'$, obtaining P′, and the theorem may be stated as follows:

THEOREM. Under these conditions theorems of $\mathbf{E}_{\rightarrow}$ may be inserted in Q′ in such a way as to make P′ a quasi-proof.

PROOF. First we show that each step of Q′ is justifiable. Basis. $A_{1\{k\}}{}'$ is $A_1 \rightarrow A_1$, and may be treated as an insertion of $\mathrm{E}_{\rightarrow}1$.

If any A_{a} in Q is a consequence of a preceding step by rep or reit, then $A_{\mathrm{a}}{}'$ may be treated in Q′ as a repetition.

If any A_{a} is a consequence of B_{b} and $(B \rightarrow A)_{\mathrm{c}}$ in Q, where $\mathrm{a} = \mathrm{b} \cup \mathrm{c}$, then we distinguish four cases:

(1) k is in neither b nor c. Then $B_{\mathrm{b}}{}'$ is B_{b}, $(B \rightarrow A)_{\mathrm{c}}{}'$ is $(B \rightarrow A)_{\mathrm{c}}$, and $A_{\mathrm{b} \cup \mathrm{c}}$, i.e., $A_{\mathrm{a}}{}'$, follows by $\rightarrow$E in Q′.

(2) k is in b but not c. Then $B_{\mathrm{b}}{}'$ is $(A_1 \rightarrow B)_{\mathrm{b} - \{k\}}$, and $(B \rightarrow A)_{\mathrm{c}}{}'$ is $(B \rightarrow A)_{\mathrm{c}}$. Insert $\mathrm{E}_{\rightarrow}2$, and use $\rightarrow$E twice to get $(A_1 \rightarrow A)_{(\mathrm{b} \cup \mathrm{c}) - \{k\}}$, i.e., $A_{\mathrm{a}}{}'$.

(3) k is in c but not b. Then $B_{\mathrm{b}}{}'$ is B_{b}, and $(B \rightarrow A)_{\mathrm{c}}{}'$ is $(A_1 \rightarrow .B \rightarrow A)_{\mathrm{c} - \{k\}}$. Since k is not in b, we have by the lemma $(B \rightarrow A \rightarrow A)_{\mathrm{b}}$. Then insertion of

$$(A_1 \rightarrow .B \rightarrow A) \rightarrow .B \rightarrow A \rightarrow A \rightarrow .A_1 \rightarrow A \quad (\mathrm{E}_{\rightarrow}2)$$

yields $(A_1 \rightarrow A)_{(\mathrm{b} \cup \mathrm{c}) - \{k\}}$, i.e., $A_{\mathrm{a}}{}'$, by two uses of $\rightarrow$E.

(4) k is in both c and b. Then we have $(A_1 \rightarrow B)_{\mathrm{b} - \{k\}}$ and $(A_1 \rightarrow .B \rightarrow A)_{\mathrm{c} - \{k\}}$. Insert an appropriate instance of $\mathrm{E}_{\rightarrow}4$ and use $\rightarrow$E twice to obtain $(A_1 \rightarrow A)_{(\mathrm{b} \cup \mathrm{c}) - \{k\}}$, i.e., $A_{\mathrm{a}}{}'$, as required.

If any A_{a} is in Q as an axiom, then $A_{\mathrm{a}}{}'$ may be treated in Q′ as a theorem.

So each step of Q′ is justified. To complete the proof, we observe that if any step in P refers to Q as a premiss for $\rightarrow$I, then the restriction on the introduction rule guarantees that k is in a_n, hence the final step $A_{n\mathrm{a}_n}{}'$ of Q′

is $(A_1 \rightarrow A_n)_{a_n - \{k\}}$. But then the conclusion $(A_1 \rightarrow A_n)_{a_n - \{k\}}$ from Q by →I in P can be regarded in P′ as a repetition of the last formula of Q′.

Hence P′ is convertible into a quasi-proof.

Repeated application of these techniques leads ultimately to a sequence of formulas each of which is either an axiom of $\mathbf{E}_{\rightarrow}$ or else a consequence of predecessors by →E, and hence a theorem of $\mathbf{E}_{\rightarrow}$. From this theorem, and the observation preceding the lemma, it follows that the two systems are equivalent.

§4.2. A strong and natural list of valid entailments. The equivalence of $\mathbf{FE}_{\rightarrow}$ with $\mathbf{E}_{\rightarrow}$ gives us an easy proof technique for $\mathbf{E}_{\rightarrow}$, and we now state a summary list of laws of entailment, proofs of which will be left to the reader.

identity: $A \rightarrow A$
transitivity: (*suffixing*) $A \rightarrow B \rightarrow .B \rightarrow C \rightarrow .A \rightarrow C$
(*prefixing*) $A \rightarrow B \rightarrow .C \rightarrow A \rightarrow .C \rightarrow B$
contraction: $(A \rightarrow .A \rightarrow B) \rightarrow .A \rightarrow B$
self-distribution: (*on the major*) $(A \rightarrow .B \rightarrow C) \rightarrow .A \rightarrow B \rightarrow .A \rightarrow C$
(*on the minor*) $A \rightarrow B \rightarrow .(A \rightarrow .B \rightarrow C) \rightarrow .A \rightarrow C$
replacement of the middle: $D \rightarrow B \rightarrow .(A \rightarrow .B \rightarrow C) \rightarrow (A \rightarrow .D \rightarrow C)$
replacement of the third: $C \rightarrow D \rightarrow .(A \rightarrow .B \rightarrow C) \rightarrow (A \rightarrow .B \rightarrow D)$
prefixing in the consequent: $(A \rightarrow .B \rightarrow C) \rightarrow .A \rightarrow .D \rightarrow B \rightarrow .D \rightarrow C$
suffixing in the consequent: $(A \rightarrow .B \rightarrow C) \rightarrow .A \rightarrow .C \rightarrow D \rightarrow .B \rightarrow D$
restricted permutation: $(A \rightarrow .B \rightarrow C \rightarrow D) \rightarrow .B \rightarrow C \rightarrow .A \rightarrow D$
restricted conditioned modus ponens: $B \rightarrow C \rightarrow .(A \rightarrow .B \rightarrow C \rightarrow D) \rightarrow .A \rightarrow D$
restricted assertion: $A \rightarrow B \rightarrow .A \rightarrow B \rightarrow C \rightarrow C$
specialized assertion: $A \rightarrow A \rightarrow B \rightarrow B$

(The terminology is self-explanatory, almost. Though some names were borrowed from the literature, our general intent was to give the formulas names which would give a clue to their structure. In particular, the word "restricted" is used to call attention, e.g., to the fact that $(A \rightarrow .E \rightarrow D) \rightarrow .E \rightarrow .A \rightarrow D$, sometimes called *permutation*, is restricted in $\mathbf{E}_{\rightarrow}$ to cases where E has the form $B \rightarrow C$.)

Note that, by using proof techniques appropriate to the preceding theorems, we can also show that

$$(A \rightarrow .D \rightarrow E) \rightarrow .(B \rightarrow .E \rightarrow F) \rightarrow (A \rightarrow .B \rightarrow .D \rightarrow F)$$

and

$$(A \rightarrow .D \rightarrow E) \rightarrow .(B \rightarrow .E \rightarrow F) \rightarrow (B \rightarrow .A \rightarrow .D \rightarrow F)$$

are both theorems of $\mathbf{E}_{\rightarrow}$; to see what they might mean, consider their relations to their parts $D \rightarrow E$, $E \rightarrow F$, and $D \rightarrow F$. The two formulas above have as a common generalization

Generalized transitivity:

$$(A_m \to .A_{m-1} \to . \ldots \to .A_1 \to .D \to E) \to .$$
$$(B_n \to .B_{n-1} \to . \ldots \to .B_1 \to .E \to F) \to$$
$$(C_{m+n} \to .C_{m+n-1} \to . \ldots \to .C_1 \to .D \to F)$$

Here there are m A_j's, n B_k's, and a total of $m+n$ C_i's, each A_j and B_k counting as a distinct C_i. Moreover the A_j's retain their own initial order as among the C_i's, and so do the B_k's, so that if $i > i'$ and C_i is A_j $\{B_k\}$ and $C_{i'}$ is $A_{j'}$ $\{B_{k'}\}$, then $j > j'$ $\{k > k'\}$. That is, the formulas $C_{m+n}, \ldots, C_1$ may be in any order which could result from a single, simple, somewhat sloppy interlacing shuffle of the sequences of formulas $A_m, \ldots, A_1$ and $B_n, \ldots, B_1$.

Two other useful generalizations:

Generalized contraction:

$$(A_1 \to .A_2 \to . \ldots \to .B \to .B \to C) \to .$$
$$A_1 \to .A_2 \to . \ldots \to .B \to C$$

Generalized restricted permutation:

$$(A_1 \to .A_2 \to . \ldots \to .A_n \to .(B \to C) \to D) \to .$$
$$(A_1 \to .A_2 \to . \ldots \to .(B \to C) \to .A_n \to D$$

All three of these generalizations have close connections with Gentzen formulations and will be used in §7; they admit of easy inductive proofs.

§4.3. That A is necessary $\rightleftarrows .A \to A \to A$. The foregoing seems to us to be a strong and natural list of valid entailments, all of which are *necessarily* true, and in each of which the antecedent is *relevant* to the consequent. We shall consider the matter of relevance more closely in §§5, 12, 16, and 22, but as regards necessity, we observe here that, as might be expected, a theory of *logical necessity* is forthcoming in $\mathbf{E}_{\to}$. We define "it is necessary that A" as follows:

$$\Box A =_{\text{df}} A \to A \to A.$$

Of course we do not mean to suggest that in saying that A is necessarily true we *have in mind* that $A \to A \to A$, any more than Lewis and Langford 1932 thought that in so saying we *had in mind* that $\sim A \to A$ ("$\sim$" for negation); the intent in both cases is rather to find a formula which can play the role of $\Box$, in the sense that it has the right properties. $\sim A \to A$ works tolerably well in the context of strict "implication," but as we point out in §11, it does not work for entailment. We find support for our proposal, however, in the fact that in such systems as **M** (Feys 1937; von Wright 1951), **S4**, and **S5**, $\Box A$ (either taken as primitive, or defined as not-possible-not A) is strictly equivalent to (hence intersubstitutable with) $A \to A \to A$. (This observation for **S4** and **S5** was made by Lemmon *et al.* 1956.)

Further motivation for the definition $A\to A\to A$ of $\Box A$ will be discussed at length in §§10–12 and §31.4. For the moment we simply register the conviction that the notion of logical necessity is indissolubly linked with that of logical consequence (entailment), and in particular that if A is a logical consequence of a *true* entailment, then A ought to be necessary. The following theorem embodies this position: if $B\to C$ is true, then if A is a logical consequence thereof, then A is necessary.

1	$B\to C_{\{1\}}$	hyp
2	$B\to C\to A_{\{2\}}$	hyp
3	$A\to A_{\{3\}}$	hyp
4	$B\to C_{\{1\}}$	1 reit
5	$B\to C\to A_{\{2\}}$	2 reit
6	$A_{\{1,2\}}$	4 5 $\to$E
7	$A_{\{1,2,3\}}$	3 6 $\to$E
8	$A\to A\to A_{\{1,2\}}$	3–7 $\to$I
9	$B\to C\to A\to\Box A_{\{1\}}$	2–8 $\to$I
10	$B\to C\to .B\to C\to A\to\Box A$	1–9 $\to$I

We give as further examples proofs of some theorems involving necessity.

1	$A\to A\to A_{\{1\}}$	hyp
2	$A_{\{2\}}$	hyp
3	$A_{\{2\}}$	2 rep
4	$A\to A$	2–3 $\to$I
5	$A_{\{1\}}$	1 4 $\to$E
6	$\Box A\to A$	1–5 $\to$I

1	$A\to A\to B_{\{1\}}$	hyp
2	$B\to B_{\{2\}}$	hyp
3	$A_{\{3\}}$	hyp
4	$A_{\{3\}}$	3 rep
5	$A\to A$	3–4 $\to$I
6	$A\to A\to B_{\{1\}}$	1 reit
7	$B_{\{1\}}$	5 6 $\to$E
8	$B_{\{1,2\}}$	2 7 $\to$E
9	$B\to B\to B_{\{1\}}$	2–8 $\to$I
10	$A\to A\to B\to\Box B$	1–9 $\to$I

The first says that necessity implies truth (a special case of the *specialized law of assertion*), and the second that, if B follows from the *law of identity*, then B is necessarily true (which accords with previous informal observations). We add some other easily provable theorems.

$A\to B\to.\Box A\to\Box B$
$\Box A\to.A\to B\to.\Box B$
$A\to B\to\Box(A\to B)$
$A\to B\to C\to.A\to B\to\Box C$
$\Box B\to.(A\to.B\to C)\to.A\to C$
$\Box A\to\Box\Box A$

The first expresses distributivity of necessity over entailment, and the second follows by *restricted permutation*. The third (a special case of *restricted assertion*) says that entailments, if true at all, are necessarily true. The fourth that, if C follows from an entailment, then the necessity of C also follows from that entailment (which we may also express by saying that, if C follows from an entailment, then if the entailment is true, C is necessary). The next says that, where B is necessary, we may infer $A\to C$ from $A\to.B\to C$ (see the discussion of Ackermann's rule (δ) in §8.2). And the last says that necessity implies necessary necessity (so to speak): a corollary of $A\to B\to\Box(A\to B)$ together with the fact that necessity is defined in terms of entailment.

As a clinching argument that the definition of necessity above is in some sense "correct," we mention a fact pointed out to us by A. N. Prior (correspondence 1960).

In $\mathbf{E}_\to$ one might, instead of defining $\Box A =_{\text{df}} A\to A\to A$, treat necessity as primitive, taking as axioms

$\Box 1$ $\Box A\to A$
$\Box 2$ $A\to B\to\Box(A\to B)$
$\Box 3$ $\Box A\to.A\to B\to\Box B$

which should all hold, in accordance with the intuitive considerations of §2 and §4. As Prior points out, under these conditions $\Box$A and A→A→A entail each other, since we have:

1	$\Box(A\to A)$	$\Box 2$ *identity* →E
2	$A\to A\to A\to\Box A$	$\Box 3$ 1 →E

and also

3	$A\to A\to\Box A\to.A\to A\to A$	$\Box 1$ *prefixing*
4	$\Box A\to.A\to A\to A$	$\Box 3$ 3 *transitivity*

We are also indebted to Prior for the observation that

$\Box 3'$ $A\to B\to.\Box A\to\Box B$

cannot replace □3 with the same outcome: □1, □2, and □3′ are theorems of $\mathbf{E}_{\rightarrow}$ when □ is interpreted vacuously, but □3 is then not provable in $\mathbf{E}_{\rightarrow}$; essentially what □3 does is to smuggle in a little permutation — just enough to do the job.

We feel that this argument justifies further the choice of $A \rightarrow A$ as the formula which, when it entails A, is tantamount to A's necessity. All this becomes clearer when propositional quantifiers become available in Chapter VI (see §31.4).

§5. Fallacies. The developments of the preceding section (and indeed of two millennia of history) seem to us to provide a set of indisputable truths about entailment. And of course each is provable not only in $\mathbf{E}_{\rightarrow}$ but also in each of the systems $\mathbf{H}_{\rightarrow}$, $\mathbf{S4}_{\rightarrow}$, and $\mathbf{R}_{\rightarrow}$. But the latter all contain in addition paradoxical assertions which we now discuss under two headings: fallacies of relevance and fallacies of modality. We preach first on the former topic, taking our text from Schiller 1930:

> The central doctrine of the most prevalent logic still consists in a flat denial of Relevance and of all the ideas associated with it (p. 75).

§5.1. Fallacies of relevance. The archetype of fallacies of relevance is $A \rightarrow .B \rightarrow A$, which would enable us to infer that Bach wrote the Coffee Cantata from the premiss that the Van Allen belt is doughnut-shaped — or indeed from any premiss you like.

In arguing that $\mathbf{E}_{\rightarrow}$ satisfies principles of relevance, we venture, somewhat gingerly, on new ground, since we know of no previous attempts to lay down formal conditions for the purely logical relevance of A to B, when A entails B. Indeed the Official position is that the only right-minded attitude is to shrug off the concept of relevance by asserting dogmatically that ". . . the notion of connection or dependence being appealed to here is too vague to be a formal concept of logic . . ." (Suppes 1957), or by asking rhetorically, "How is one to characterize such an obscure notion as that of [logical] dependence?" (*Ibid.*)

We, on the contrary, take the question at face value and offer two formal conditions, the first as necessary and sufficient, the second as necessary only.

§5.1.1. Subscripting (in $\mathbf{FE}_{\rightarrow}$). The subscripting technique as applied in $\mathbf{FR}_{\rightarrow}$ and $\mathbf{FE}_{\rightarrow}$ may be construed as a formal analysis of the intuitive idea that for A to be relevant to B it must be possible to *use* A in a deduction of B from A.

Though we know of no other attempt to treat subscripting (or the like) as a systematic and reasonably comprehensive formal analysis of the notion of "using" a proposition in a proof, it is clear that the idea of "use" is of enough importance that people who try to get clear about proofs frequently ask about it, as we all know from our own experiences. A fair examination question on elementary informal set theory is "Where is the axiom of choice used in the proof that the union of denumerably many denumerable sets is denumerable?" and an answer can tell us something about the student's comprehension of the topic. Similarly, many of us are familiar with a question which naturally arises in the course of trying to puzzle through an intricate proof: "Now where did we *use* the assumption that . . . ?" Surely in grooming a proof for public appearance, one of the standard tidying-up maneuvers is to check to see that all the hypotheses are used in the proof.

The straightforward, obvious, simple-minded strategy for keeping track of which premisses have been used in a proof in a formalized calculus is the subscripting technique we have associated with $\mathbf{FR}_{\rightarrow}$ and $\mathbf{FE}_{\rightarrow}$. Curiously enough, the author who asked the rhetorical question quoted in §5.1 goes on to answer himself twenty pages later in words which almost exactly mirror those we have used in describing subscripts: "For each line of the derivation, the list of numbers . . . corresponds to the premisses on which that line depends." As we shall see in the sequel, this obvious and natural approach, in which the notion of dependence is taken from the analysis of the proof, generalizes to truth functions and to quantification theory. The overwhelmingly important difference between our approach and that of the logical tradition we are trying to correct is that, in order to preserve the sanity of the approach, we require that *only* valid entailments be allowed as rules of proof. If, following the Official tradition, we allow the use of fallacious rules, the whole project comes unstuck at the seams, and crashes into chaos. (See §§12, 16, 20.1, 22, 31, and elsewhere *passim.*)

We claim, at any rate, that the analysis of proofs with the help of subscripts gives us a plausible and systematic handle on relevance in the sense of logical dependence.

Of course for A to be relevant to B in this sense it need not be *necessary* to use A in the deduction of B from A — and indeed this is a familiar situation in mathematics and logic. It not infrequently happens that the hypotheses of a theorem, though all relevant to a conclusion, are subsequently found to be unnecessarily strong. An example is provided by the celebrated incompleteness theorem of Gödel 1931, which required the assumption of ω-consistency. Rosser 1936 showed that this condition was not required for the proof of incompleteness — but surely no one would hold that ω-consistency was *irrelevant* to Gödel's result as originally stated. Similarly in the following example (due to Smiley 1959), effort is wasted, since the antecedent is used in the proof of the consequent, though it need not be.

1	$A \rightarrow B_{\{1\}}$	hyp
2	$B \rightarrow A_{\{2\}}$	hyp
3	$B_{\{3\}}$	hyp
4	$B \rightarrow A_{\{2\}}$	2 reit
5	$A_{\{2,3\}}$	3 4 $\rightarrow$E
6	$A \rightarrow B_{\{1\}}$	1 reit
7	$B_{\{1,2,3\}}$	5 6 $\rightarrow$E
8	$A_{\{1,2,3\}}$	4 7 $\rightarrow$E
9	$B \rightarrow A_{\{1,2\}}$	3–8 $\rightarrow$I
10	$B \rightarrow A \rightarrow . B \rightarrow A_{\{1\}}$	2–9 $\rightarrow$I
11	$A \rightarrow B \rightarrow . B \rightarrow A \rightarrow . B \rightarrow A$	1–10 $\rightarrow$I

A similar proof yields

$$B \rightarrow A \rightarrow . A \rightarrow B \rightarrow . B \rightarrow A.$$

The point in both cases is that the antecedent and the antecedent of the consequent can be made to "cycle," producing one or the other as consequent of the consequent.

These two perhaps rather unexpected theorems are of interest in connection with co-entailment (see §8.7), for which we have, as is to be expected, $A \rightarrow B \rightarrow . B \rightarrow A \rightarrow . A \rightleftarrows B$; the two theorems above are consequences of this. And the law $A \rightarrow B \rightarrow . B \rightarrow A \rightarrow . B \rightarrow A$ requires special attention, because it appears to violate a plausible condition due to Smiley 1959, who demands that every true entailment be a "substitution instance of a tautological implication whose components are neither contradictory nor tautological." Smiley's criterion and some related views of von Wright and Geach are plausible and interesting, and, though they appear to us to be incoherent as they stand, they are also very close to the mark when suitably interpreted. A sympathetic understanding of their intent requires more logical apparatus than is available at this stage, and a full-dress discussion will be deferred until §20.1, when truth functions together with entailment will be available.

§5.1.2. Variable-sharing (in $E_{\rightarrow}$). The second formal condition promised above is suggested by the consideration that informal discussions of implication or entailment have frequently demanded "relevance" of A to B as a necessary condition for the truth of $A \rightarrow B$, where relevance is now construed as involving some "meaning content" common to both A and B. This call for common meaning content comes from a variety of quarters. Nelson 1930 (p. 445) says that implication "is a necessary connection between meanings"; Duncan-Jones 1935 (p. 71) that A implies B only when B "arises out of the meaning of" A; Baylis 1931 (p. 397) that if A implies B then "the intensional meaning of B is identical with a part of the intensional meaning of A"; and Blanshard 1939 (vol. 2, p. 390) that "what lies at the root of the common man's objection [to strict implication] is the stubborn feeling that implication has something to do with the *meaning* of propositions, and that any mode of connecting them which disregards this meaning and ties them together in despite of it is too artificial to satisfy the demand of thought."

A formal condition for "common meaning content" becomes almost obvious once we note that commonality of meaning in propositional logic is carried by commonality of propositional variables. So we propose as a *necessary*, but by no means sufficient, condition for the relevance of A to B in the pure calculus of entailment, that A and B must share a variable. (The misunderstanding of this condition exhibited by Woods 1964 has been cleared up in Hockney and Wilson 1965.) If this property fails, then the variables in A and B may be assigned propositional values, in such a way that the resulting propositions have no meaning content in common and are totally irrelevant to each other. $\mathbf{E}_{\rightarrow}$ avoids such fallacies of relevance, as shown by the following

THEOREM. If $A \rightarrow B$ is provable in $\mathbf{E}_{\rightarrow}$, then A and B share a variable.

PROOF. Consider the following matrix (a finite adaptation of a matrix of Sugihara 1955; see §26.9). (*'d values designated; see §8.4.)

$\rightarrow$	-2	-1	$+1$	$+2$
-2	$+2$	$+2$	$+2$	$+2$
-1	-2	$+1$	$+1$	$+2$
$^*+1$	-2	-1	$+1$	$+2$
$^*+2$	-2	-2	-2	$+2$

The axioms of $\mathbf{E}_{\rightarrow}$ take designated values for all assignments of values to the variables, and the rule $\rightarrow$E preserves this property. (For an explanation of the use of such matrices, see §8.4.) But if A and B share no variables, then we may assign the value $+2$ to all the variables of A (yielding $A = +2$), and $+1$ to all the variables of B (yielding $B = +1$), and $+2 \rightarrow +1$ takes the undesignated value -2. Hence if A and B fail to share a variable, $A \rightarrow B$ is unprovable.

We remark that the subscripting condition has to do with entailment in its guise as the converse of deducibility, and in this sense is a proof-theoretical completeness theorem: A is *relevant* to B, when $A \rightarrow B$, just if there exists a proof (satisfying certain relevance conditions) of B *from* the hypothesis A. The variable-sharing condition, however, concerns entailment conceived of as a relation of logical consequence, and is semantical in character, since it has to do with possible assignments of values to the propositional variables.

The matrix given for the second condition above also satisfies $(A \rightarrow .B \rightarrow C) \rightarrow .B \rightarrow .A \rightarrow C$; it follows that $\mathbf{R}_{\rightarrow}$ is also free of fallacies of relevance in the sense of satisfying the necessary condition of sharing variables, as we would expect from the previous discussion. Among fallacies of relevance which are *not* modal fallacies we mention $A \rightarrow A \rightarrow .B \rightarrow B$. ($A \rightarrow \Box A$, mentioned above, has the converse property; it is a fallacy of modality, but not of relevance.)

The same matrix can also be used to rule out certain expressions involving fallacies of relevance even when antecedent and consequent *do* share a variable. We define *antecedent part of A* and *consequent part of A* inductively as follows:

(a) A is a consequent part of A;

(b) if $B \rightarrow C$ is a consequent part of A, then B is an antecedent part of A, and C is a consequent part of A; and

(c) if $B \rightarrow C$ is an antecedent part of A, then C is an antecedent part of A, and B is a consequent part.

EXAMPLE: $A \rightarrow B \rightarrow C \rightarrow .B \rightarrow D$ has as antecedent parts $A \rightarrow B \rightarrow C$, A, C, and the second occurrence of B. It has as consequent parts itself, $B \rightarrow D$, D, $A \rightarrow B$, and the first occurrence of B.

THEOREM. If A is a theorem of $\mathbf{E}_{\rightarrow}$ (or of $\mathbf{R}_{\rightarrow}$), then every variable occurring in A occurs at least once as an antecedent part and at least once as a consequent part of A. (For a related result see Smiley 1959, and §8.8.)

PROOF. If a variable p occurs only as an antecedent part of A, assign p the value $+2$, and if it occurs only as a consequent part, assign it the value -2; assign all other variables in A the value $+1$. It is then possible to prove by an induction on the length of A that every well-formed part B of A has the following property:

(1) If B does not contain p, the value of B is $+1$;

(2) if B contains p and is an antecedent part of A, the value of B is $+2$; and

(3) if B contains p and is a consequent part of A, the value of B is -2.

From (3), together with the fact that A is a consequent part of itself, it will follow that A assumes the value -2, and is therefore unprovable. Details are left to the reader. (For aid see §12.)

The designation "antecedent" and "consequent" parts derives from the Gentzen consecution calculus formulation of $\mathbf{E}_{\rightarrow}$ in §7. The ultimate premisses of a consecution calculus proof of B are axioms $p \vdash p$ (answering to $p \rightarrow p$), where, roughly speaking, the left p appears in B as an antecedent part, and the right p as a consequent part. For this reason, the theorem may be regarded as saying that the "tight" relevance condition satisfied by the axioms is preserved in passing down the Gentzen-like proof tree to *B*.

As an immediate corollary, we see that no variable may occur just once in a theorem of $\mathbf{E}_{\rightarrow}$, which suggests that each variable-occurrence is essential to the theorem; theorems have no loose pieces, so to speak, which can be jiggled about while the rest of the theorem stays put. (Our theorem sharpens a result for $\mathbf{R}_{\rightarrow}$ of Sobociński 1952.) Hence

$$A \rightarrow B \rightarrow . B \rightarrow B$$
$$A \rightarrow B \rightarrow . A \rightarrow A,$$

and Peirce's "law"

$$A \rightarrow B \rightarrow A \rightarrow A,$$

all fail in $\mathbf{E}_{\rightarrow}$. The intuitive sense of Peirce's formula is sometimes expressed in this way: one says "Well, if *A* follows simply from the fact that *A* entails something, then *A* must be true — since obviously *A* must entail something, itself for example." But this reading, designed to make the formula sound plausible, seems also designed to pull the wool over our eyes, since an essential premiss is suppressed. It is of course true that if *A* follows from the fact that it entails *something it does entail* (an assumption hidden in the word "fact" above), *then A* must be true, and a theorem to this effect is provable in $\mathbf{E}_{\rightarrow}$,

$$A \rightarrow B \rightarrow . A \rightarrow B \rightarrow A \rightarrow A$$

(a special case of *restricted assertion*). But if *A* follows from the (quite possibly false) assumption that *A* entails *B*, this would hardly guarantee the truth of *A*.

§5.2. Fallacies of modality. Discussion of this topic requires us to introduce some ugly terminology. We do so forthwith, and apologize just this one time.

It has been customary in the traditional philosophical literature on logic to consider some propositions as positive and some as negative. Without going into the difficulties raised by, e.g., Eaton 1931, concerning "negative properties" such as "baldness," "orphanhood," and the like, it is clear that one sensible thing *might* be meant by describing a proposition as negative: we say that *A* is *negative* just in case there is some proposition *B* such that *A* is equivalent (in the sense of co-entailment) to the denial of *B*. Trivially, of course, all propositions are negative by this criterion (unless one hews to, say, the intuitionistic theory of negation). We might define with equal triviality a "conjunctive" proposition *A* as a proposition equivalent to a conjunction *B*&*C* (whence again every proposition becomes conjunctive, in view of the fact that A is equivalent to $A \& A$). But a more sane way of looking at the matter would require that B and C be distinct from (i.e., not equivalent to) A. (Analogy: in the algebraist's lingo of §18, an element a of a lattice is said to be *meet-reducible* just in case there are b and c, both distinct from a, such that a = b∧c.) Similarly for "disjunctive proposition."

Unfortunately there seems to be nothing parallel to say, in ordinary logical-philosophical language, about a proposition A such that A is equivalent with $\Box B$, for some B. As a result, we are driven to the desperate expedient of coining "necessitive," on the (note: rough) analogy of "negative," "conjunctive," and the like, to refer to such propositions. And we shall also, though we hope not frequently, allow ourselves "possibilitive": A is a possibilitive proposition just in case, for some B, A is equivalent to $\Diamond B$.

Using this terminology, we may distinguish between *necessary* propositions and *necessitive* propositions. Examples:

(1) $A \rightarrow A$ is both necessary (i.e., we have $\Box(A \rightarrow A)$) and necessitive (we also have $A \rightarrow A$ equivalent to $\Box(A \rightarrow A)$);

(2) $\Box A$, for contingent A, is necessitive (since $\Box A$ is equivalent to $\Box\Box A$), but not necessary (indeed it is false);

(3) $A \vee \bar{A}$, though necessary, is not necessitive: as we shall see in §22.1.2, there is no proposition B such that $(A \vee \bar{A})$ is equivalent to $\Box B$; and

(4) plain A is in general neither necessary nor necessitive.

The upshot is that we regard some propositions as both necessary and necessitive, some as one but not the other, and some as neither. It follows from commitments already made that all *true* necessitives are necessary; we shall discuss the failure of the converse later, when we come to consider truth functions in the context of entailment, and give specific arguments for rejecting, e.g., the equivalence of (the necessary) $A \vee \bar{A}$ and (the necessitive) $\Box(A \vee \bar{A})$. Meanwhile we consider fallacies of modality as they arise in pure implicational calculuses.

We have already discussed and dismissed modal fallacies which involve the claim that a non-necessary proposition may be entailed by a necessary one: nothing about the color of pomegranates can be deduced logically from any logical truth, as was explained in §2 (though there will be more about this later). Such fallacies rest on a misconception of the relations between necessary and non-necessary propositions.

But distinguishing between necessary and necessitive propositions enables us to carve out a class of fallacies with different foundations: those which depend on misconceptions involving necessitive and non-necessitive propositions. Since we are at the moment confined to pure implicational calculuses, we are forced to defer thorough consideration of this topic until more machinery is at hand, i.e., until we have notation for such necessary non-necessitives as $A \vee \bar{A}$. At the moment all our necessary propositions are also necessitive, since we have both $A \rightarrow B \rightarrow \Box(A \rightarrow B)$ and $\Box(A \rightarrow B) \rightarrow .A \rightarrow B$. But we do have *one* bit of notation for dealing with non-necessitives in $\mathbf{E}_{\rightarrow}$, namely, propositional variables, which alone among formulas of $\mathbf{E}_{\rightarrow}$ can take non-necessitives as values. It follows that in $\mathbf{E}_{\rightarrow}$, a modal fallacy is a confusion concerning propositional variables and entailments. Such confusions can occur in two distinguishable ways, which we consider forthwith: we may mistakenly suppose that a propositional variable entails an entailment, or we may mistakenly use a propositional variable in establishing an entailment.

§5.2.1. Propositional variables entailing entailments (in $\mathbf{E}_{\rightarrow}$). We argued in §3 that $B \rightarrow .A \rightarrow A$ constitutes a fallacy of relevance. We are now going to argue that in spite of its being a member in good standing of the **S4** Club, whose members are supposed to be Modally Sound, still (as the **S3** Admissions Committee correctly saw) it commits an outrageous modal fallacy. More generally, we shall argue that where p is a propositional variable, $p \rightarrow .A \rightarrow B$ is never valid for any choice of A and B.

What is the modal mistake made in $p \rightarrow .A \rightarrow B$? It is tempting to say that what makes it a fallacy is that non-necessitives can never entail necessitives, and since p can be interpreted as the former and the interpretation of $A \rightarrow B$ is always among the latter, $p \rightarrow .A \rightarrow B$ can always be falsified. But of course some non-necessitives do entail necessitives; for example, we are happy to accept $p\&(A \rightarrow B) \rightarrow .A \rightarrow B$, although the antecedent can be interpreted as a non-necessitive by taking p as a non-necessitive. So it won't do to interpret p just as a non-necessitive. (We are indebted to Routley and Routley 1969 for pointing out errors in some of our earlier remarks on this point.) Rather, we must interpret it as a *pure non-necessitive*, which is to say, a proposition which cannot be expressed as a conjunction one of the propositional conjuncts of which is a necessitive.

Conjunction is essential to the main idea, but we don't have this notion available yet, and consequently we defer a full treatment until we hear from Coffa, who cleans up the general problem in §22.1.2. His discussion, unlike that to follow, is independent of the particular set of assumptions made about entailment; he requires only that *there be* some assumptions about that notion, and some set of necessitives which can be considered within those assumptions. Our remarks to follow are meant to square with his account. Although they do not refer to conjunction, they do depend on the analysis of entailment presently at hand.

In order to suppress reference to an explicit conjunction connective, we take A to be the conjunction of B and C just in case it entails both B and C, and is entailed by anything which entails both B and C. (That is, the conjunction of B and C is the greatest lower bound of B and C, with entailment taken as the less-than-or-equal ordering.) Then the reference to conjunction in the definition above of "pure non-necessitive" can be treated as a detour; and when the detour is eliminated, a pure non-necessitive turns out to be a proposition which does not entail a necessitive.

Everyone can therefore agree that no pure non-necessitive entails a necessitive.

By "everyone" we include even those who like $p \rightarrow .A \rightarrow A$. How would they avoid our attempt to falsify their favorite by interpreting p as a pure non-necessitive and thus as a proposition entailing no necessitive? Of course, by denying that any proposition is a pure non-necessitive. And *this* is the precise fallacy: the denial of the obvious truth that some propositions are in fact pure non-necessitives. In fact we need only one, for just one will do the work of falsifying $p \rightarrow .A \rightarrow B$ for arbitrary A and B.

The Other Side claims that no proposition is a pure non-necessitive, because of the universal validity of $p \rightarrow .A \rightarrow A$; we claim that $p \rightarrow .A \rightarrow A$ is not universally valid, because of the existence of pure non-necessitives. If we are to get this impasse off dead center by making good our existence claim, then it appears to be up to us to submit candidates. We are prepared with one: the proposition that Crater Lake is blue. But having done our part, it now seems to us that the burden lies with the opposition, whose job it is to disqualify our nominee. If he is found to have a necessitive in his pocket, we lose; otherwise he is as advertised, and we win.

We are prepared to hear arguments, and in fact there is one, though it cannot be accommodated within the bounds of $\mathbf{E}_{\rightarrow}$. It is the standard **S5** claim that truth entails necessary possibility. We treat this view several times in what follows, and will give it an elementary airing here, in preparation for a more advanced treatment later (see especially §12).

Crater Lake *is* blue, and it is surely therefore possible that this is so. We agree, and we also agree that in so speaking we are saying that Logic alone does not guarantee the falsity of the proposition in question. **S5**-ers say that the latter fact itself is certainly no accident of nature, and hence is necessarily true, *and add that* the proposition that Crater Lake is blue *does* entail a necessitive after all, namely that it is necessarily possible that Crater Lake is blue.

Of course it is not required of us that we take a stand on the question as to whether or not it is logically necessary that possibly Crater Lake is blue; what we hold is that even if it *is* necessarily possible that Crater Lake is blue, this putative fact does not follow logically from the proposition that Crater Lake is blue. Even those who hold that it *is* necessarily possible rely on logical considerations ("it is no accident of nature") to buttress their claim — they don't go and look at the lake.

These logical considerations cannot be expressed in $\mathbf{E}_{\rightarrow}$, since it lacks negation, but under the restrictions imposed by $\mathbf{E}_{\rightarrow}$, it is hard to conceive of any arguments for the non-existence of pure non-necessitives which would be even remotely persuasive. Consider what the opposition is up against. They must find some necessitive proposition A (which, in the present context, will have to be an entailment $B \rightarrow C$), such that it would be plausible, when asked whether B entails C, to reply, "First go look at Crater Lake; if it turns out to be blue, then it follows from that fact with inexorable logical rigor that it is a logical necessity that B entails C. Otherwise, we'll have to try some other line of investigation." The mind boggles. And if the proposition that Crater Lake is blue is in fact a pure non-necessitive when we are talking about $\mathbf{E}_{\rightarrow}$, it is hard to see how it might become a necessitive if we talk about another topic (truth functions, for example, or limnology).

We recognize that boggling is not conclusive, but we hope that it is persuasive, especially in view of the pervasiveness of the felt distinction between Relations of Ideas and Matters of Fact. Apparently no one claims to be able to deduce that Crater Lake is blue from true necessitives; it would seem to us equally mysterious if deducibility were to go in the other direction. But about the best we can do, in the face of opposition, is to say things such as "Look: suppose snow were blue. Can you *really* believe that any logical claims would be colored by this fact, if it were one?"

The foregoing considerations, unlike those to follow presently, may smack of waffling. If so, we take some comfort in the fact that we waffle in eminently respectable company:

> Confronted with any logical truth or indeed any true statement of mathematics, no matter how complex, we recognize its truth if at all merely by inspecting the statement and reflecting or calculating; observation of craters, test tubes, or human behavior is of no avail (Quine 1940, p. 4).

The extensionalist logical community is so thoroughly inured to the rigors of avoiding modal notions that its members sometimes affect blindness to obvious truths; but if Quine is not here endorsing our claim that there are pure non-necessitives, then either he is expressing himself very badly or else we are. We understand both Quine and ourselves to be glossing Plato: there is an Irreducible Chasm (or something else pretty severe) between Being and Becoming, about which we agree with $\mathbf{E}_{\rightarrow}$.

But we take even more comfort in the fact that our conviction of the existence of pure non-necessitives (recall that our thesis required only that there be at least one) wears, for the full system **E** of Chapter IV, well-tailored formal regalia, as Coffa shows in §22.1.2. For the present limited purpose ($\mathbf{E}_{\rightarrow}$), we summarize the foregoing in the following

THEOREM. $p \rightarrow . B \rightarrow C$ is unprovable in $\mathbf{E}_{\rightarrow}$ whenever p is a propositional variable (Ackermann 1956).

PROOF. Consider the matrix (adapted from Ackermann):

$\rightarrow$	*0*	*1*	*2*
0	*2*	*2*	*2*
1	*0*	*2*	*0*
**2*	*0*	*0*	*2*

For this matrix the theorems of $\mathbf{E}_{\rightarrow}$ always take the value *2*; but for any $p \rightarrow . B \rightarrow C$, where p is a propositional variable, we can assign p the value *1*, giving $p \rightarrow . B \rightarrow C$ the value *0* regardless of the values of B and C. Hence no such $p \rightarrow . B \rightarrow C$ is provable.

§5.2.2. Use of propositional variables in establishing entailments (in $\mathbf{FE}_{\rightarrow}$). Our object here is simply to see intuitively that the restrictions on reit and $\rightarrow$I in $\mathbf{FE}_{\rightarrow}$ combine to have just the same force as is built into the choice of axioms for $\mathbf{E}_{\rightarrow}$ (as of course we know anyway, since $\mathbf{E}_{\rightarrow}$ and $\mathbf{FE}_{\rightarrow}$ have the same theorems). We dispose of the matter briefly as follows.

If in $\mathbf{FE}_{\rightarrow}$ a formula $p \rightarrow . A \rightarrow B$ were provable, we would expect to find a proof in $\mathbf{FE}_{\rightarrow}$ with the skeleton

```
|_ p{1}
|  :
|  |_ B{2}
|  |  :
|  |  C{1,2}
|  B→C{1}
p→.B→C
```

But the restrictions imposed on subscripting show that the only way to get the appropriate subscripts on C would be to *use* p in arriving at C, and the restriction on reiteration precludes this possibility. (Of course this is only heuristic, since it envisages only the "direct" use of p; but a knock-down drag-out argument can be constructed on the basis of §8.20.)

An example showing how considerations of modality and relevance interact to forbid cases of $p \rightarrow . A \rightarrow B$ is given by

$$p \rightarrow . (p \rightarrow . p \rightarrow B) \rightarrow . p \rightarrow B.$$

This expression, discussed in §8.10, is a theorem of $\mathbf{S4}_{\rightarrow}$ (since the consequent is a logical truth — the law of contraction) and also a theorem of $\mathbf{R}_{\rightarrow}$ (since it has the form $A \rightarrow . A \rightarrow C \rightarrow C$ — the law of assertion). But if the *consequent* arrow is taken as expressing a necessitive (unlike the arrow of $\mathbf{R}_{\rightarrow}$), then one ought not to allow oneself to use p in establishing the consequent; and if the *main* arrow is taken as requiring relevance (unlike the arrow of $\mathbf{S4}_{\rightarrow}$), then one ought not to affirm the main formula without using p in establishing the consequent. So be glad to find it no theorem of $\mathbf{E}_{\rightarrow}$.

With this section we complete the main argument of this chapter: we have argued (a) that valid inferences are necessarily valid, and (b) that the antecedent in a valid inference must be relevant to the consequent; we now conclude with two observations.

First, in view of the long history of logic as a topic for investigation, and the near unanimity on these two points among logicians, it is surprising, indeed startling, that these issues should require re-arguing. That they do need arguing is a consequence of the almost equally unanimous contradictorily opposed feeling on the part of contemporary logicians that material and strict "implication" are implication connectives, and that therefore necessity and relevance are not required for true implications. But, if we may be permitted to apply a result of the ingenious Bishop of Halberstadt (Albert of Saxony, *ca.* 1316–1390; see Boehner 1952, pp. 99–100), if *both* of these contradictory views are correct, it follows that Man is a donkey.

Second, the distinctions drawn and the techniques elaborated thus far do *not* "wither in the sun outside the tiny area covered by $\mathbf{E}_{\rightarrow}$," to quote an elegant conceit of Bennett 1965. Quite the contrary: the *logische Weltanschauung* explained here, when applied to familiar extensional logic, flourisheth as doth the green bay tree, basking in the gentle aureola provided by the Natural Light of Reason. The bulk of this book will be devoted to a comprehensive and rigorous mathematical proof of this fact.

§6. Ticket entailment ($\mathbf{T}_{\rightarrow}$). Relevance and necessity are not the only conditions which have been demanded of true "if . . . then —" propositions. These traditional demands have, as we have pointed out, been abandoned in recent years, and it is perhaps partly because of the extraordinarily odd features of the resulting theories of implication that some philosophers have suggested that "if . . . then —" propositions should not be regarded as on the same footing at all with (say) conjunctive propositions (§5.2). A conjunctive proposition in some sense "asserts" both conjuncts, whereas "if A then B" might be taken to mean something more like "the inference of B from A is justified."

While we would not want to charge Ryle 1949 with responsibility for any of the views we explain below, we would still like to borrow some of his terminology. Speaking of "law-statements" that have empirical content, he writes:

> Law-statements are true or false but they do not state truths or falsehoods of the same type as those asserted by the statements of fact to which

> they apply or are supposed to apply. They have different jobs. The crucial difference can be brought out in this way. At least part of the point of trying to establish laws is to find out how to infer from particular matters of fact to other particular matters of fact, how to explain particular matters of fact by reference to other matters of fact, and how to bring about or prevent particular states of affairs. A law is used as, so to speak, an inference-ticket (a season ticket) which licenses its possessors to move from asserting factual statements to asserting other factual statements (p. 121).

We may hazard the guess that these are only "season tickets" because they change according to the prevailing climate of scientific opinion. But Ryle leaves open the possibility that there may also exist tickets with no astronomical or meteorological bounds, i.e., logical or mathematical inference-tickets:

> Law-statements belong to a different and more sophisticated level of discourse from that, or those, to which belong the statements of the facts that satisfy them. Algebraical statements are in a similar way on a different level of discourse from the arithmetical statements which satisfy them (p. 121).

Our intent in this book, at any rate, is to have only "algebraical," or mathematical, or logical, tickets marked with "$A \rightarrow B$," and in this section we formalize a calculus of "ticket entailment" motivated by these considerations.

The terminology "inference-ticket" suggests that an "if . . . then —" proposition should be used, as a premiss in a deduction, only as a major premiss: we don't draw conclusions *from* "if . . . then —" propositions; we only *use* them to further our arguments. But if, as Formal Logicians in the sense of Chapter VIII, Ryle 1954, we are interested in using law-statements of a logical sort to justify other logical law-statements, then some different account must be given, since logicians sometimes want to draw conclusions *from* a proposition to the effect that so-and-so is a valid inference-ticket. For example, we may wish to infer that if not-*B* then not-*A* from the proposition that if *A* then *B*, using the ticket marked "contraposition." Notice that in this context, as you might expect, the rubric "ticket" goes with the major premiss, while the minor premiss, although itself a ticket, and usable as such in some other context, is in *this* inferential context playing the role of a "matter of fact." What we want to do is devise some method of taking seriously the view that entailments are *really* tickets, not embarkation points, in spite of the fact that they are, as in the above example, sometimes used as embarkation points. We would, that is, like to offer some formal theory which would enable us to tell, in a given inferential context, when a ticket is available only as a ticket, and when available as an embarkation point from which another ticket might take us to a destination; or, less fancifully, when an entailment is available as a minor premiss, and when only as a major.

As it appears to us, it would be hopeless to try to make formal sense of this "ticket position" if we had nothing but the ordinary axiom-cum-→E formulations of propositional calculus in terms of which to consider it: the linearity of these formulations provides no foundation on which to rest the hierarchical distinction between "ticket" and "fact." Indeed, most natural deduction systems are of no more help. But if we look at the question from the point of view of Fitch's subproof formulations, then the nesting of subproofs itself provides a natural and obvious formal analogue to the "ticket-fact" hierarchy, as follows:

(1) Recall that the distinction is to be contextual; we identify "inferential context" with "subordinate proof," expecting that what is "ticket" in one context (subproof) may be "fact" in another.

(2) The distinction is not only contextual but also functional: to function as a "ticket" is to be used as a major premiss of *modus ponens*, while the minor premiss is used as a "fact."

(3) Since the point of a subordinate proof is to establish what can be deduced from a hypothesis, we should expect to be able to use the hypothesis as a minor premiss for *modus ponens* ("fact"). The same should be true for anything that follows from the hypothesis, as indicated by the subscript notation.

(4) But the reiteratable hypotheses are to be conceived of as tickets issued by an outside agency, to wit, some proof to which the inferential context (subproof) is subordinate, and therefore usable only *as* tickets, i.e., never as *minor* premisses for →E. Moreover, we should expect this property to be hereditary: if →E allows us to turn in two old tickets for a new one, we still have only a new ticket, and not a new embarkation point.

Let us consider examples of proofs in $\mathbf{E}_{\rightarrow}$ to see how we might best formalize these considerations. $A\rightarrow A\rightarrow A\rightarrow A$ could be proved in $\mathbf{E}_{\rightarrow}$ in the following way:

1	│ $A_{\{1\}}$	hyp
2	│ $A_{\{1\}}$	rep
3	$A\rightarrow A$	1–2 →I
4	│ $A\rightarrow A\rightarrow A_{\{1\}}$	hyp
5	│ $A\rightarrow A$	3 reit
6	│ $A_{\{1\}}$	4 5 →E
7	$A\rightarrow A\rightarrow A\rightarrow A$	4–6 →I

Here at step 5 we import a ticket into a subordinate proof, but then don't *use* it as a ticket, but rather as a minor premiss for →E. If on the other hand the reader will construct a proof of $A\rightarrow B\rightarrow .B\rightarrow C\rightarrow .A\rightarrow C$ in $\mathbf{E}_{\rightarrow}$, he will find that this situation need not arise in that proof: all reiterated entailments are used as tickets, and as tickets only.

Consequently, it would seem that we can maintain the "ticket-fact" distinction by forbidding the use of any item as a minor ("factual") premiss for $\rightarrow$E in a given subproof if that item is a consequence of reiterated items, but not of the hypothesis.

Alas, the matter turns out not to be quite so simple, since one can violate the spirit of this restriction while hewing to its letter, as inspection of the proof of $A\rightarrow A\rightarrow A\rightarrow A$ (i.e., $\Box A\rightarrow A$) in §4.3 shows: we can produce a proof of $A\rightarrow A$ *inside* the subproof having $A\rightarrow A\rightarrow A$ as hypothesis, and then proceed to use $A\rightarrow A$ as a "fact."

We are therefore led to notice another aspect of the notation which might provide some leverage, for clearly the subscripts tell us that $A\rightarrow A$, though proved within the proof with $A\rightarrow A\rightarrow A$ as hypothesis in §4.3, was "really" issued by an outside agency, since $A\rightarrow A$ has an empty class of subscripts. So we want to forbid $\rightarrow$E when $A\rightarrow B$ has a subscript, but A has none. It seems equally clear that we want to forbid the use of $\rightarrow$E for formulas with one subscript in cases where the subscript on A is lower than the subscript on $A\rightarrow B$, at least if we have been careful to let our subscripts on hypotheses increase with the increasing depth of the subordinate proofs, which indeed is the most natural way of handling things. For then a lower subscript will indicate an outside issuing agency.

One way of insuring that subscripts increase with depth is to require that the subscript assigned any hypothesis be *precisely* the number of vertical lines to its left; let us agree to make our assignments in this way here and hereafter (see §8.1).

The application of the ticket-fact principle in cases where the steps A and $A\rightarrow B$ have one or no subscripts (i.e., a unit set or empty set of subscripts) is thereby made straightforward: use of $\rightarrow$E is forbidden if $A\rightarrow B$ has a subscript and A has either no subscript, or a subscript lower than that of $A\rightarrow B$; otherwise use of $\rightarrow$E is allowed. In the general case, however, the sets of subscripts on A and $A\rightarrow B$ may have several members each. For this case, we want to reflect the ticket-fact distinction by forbidding $\rightarrow$E whenever the innermost agency issuing A_{a}, as indicated by the largest subscript in a, is external to the innermost agency issuing $A\rightarrow B_{\mathrm{b}}$, as indicated by the largest subscript in b. Hence, where max(a) is the largest subscript in any set a of subscripts, or zero if a is empty, the form $\rightarrow$E takes for ticket entailment embodies a provisionary clause, which we will refer to as the *ticket-restriction:*

$\rightarrow$E. from A_{a} and $A\rightarrow B_{\mathrm{b}}$ to infer $B_{\mathrm{a}\cup\mathrm{b}}$, provided $\max(\mathrm{b}) \leq \max(\mathrm{a})$.

All of these considerations lead us to formulate a system $\mathbf{FT}_{\rightarrow}$ of ticket entailment, which is exactly like $\mathbf{FE}_{\rightarrow}$ of §4, except that $\rightarrow$E is restricted as above. We remark that the restriction on reiteration in $\mathbf{FE}_{\rightarrow}$ is not really any longer required, since, if one did reiterate a propositional variable in a proof in $\mathbf{FT}_{\rightarrow}$, one would not be able to use it as a minor premiss, hence not at all.

Following the previous patterns of discussion for $\mathbf{FS4}_{\rightarrow}$, $\mathbf{FR}_{\rightarrow}$, and $\mathbf{FE}_{\rightarrow}$, we now turn to the task of characterizing $\mathbf{FT}_{\rightarrow}$ axiomatically, in order to find a linear formulation $\mathbf{T}_{\rightarrow}$ which could be said to embody the intuitions involved in distinguishing inference-tickets from other propositions. The strategy is in general the same: we choose as axioms for $\mathbf{T}_{\rightarrow}$ just those theorems of $\mathbf{FT}_{\rightarrow}$ required for the reduction of proofs in $\mathbf{FT}_{\rightarrow}$ to a linear form.

Reduction of the hypothesis $H_{\{k\}}$ of an innermost subproof uses $A\rightarrow A$. And as in $\mathbf{FR}_{\rightarrow}$ and $\mathbf{FE}_{\rightarrow}$, when $\rightarrow$E is used, we have cases to consider, if we wish to fix things inductively so that the next step can be carried out. For the latter two systems, $B_{\mathrm{a}\cup\mathrm{b}}$ might follow from A_{a} and $A\rightarrow B_{\mathrm{b}}$ where

(i) k is in neither a nor b,
(ii) k is in a but not in b,
(iii) k is in b but not in a, or
(iv) k is in both.

But for $\mathbf{FT}_{\rightarrow}$, (iii) is not to be considered as a case, since the restriction on $\rightarrow$E explicitly rules this "case" out of consideration: because we are in an innermost subproof, (1) the subscript k on the hypothesis of that subproof will be the maximum subscript in any subscript set in which it occurs, and (2) k will be larger than any subscript in any set in which it does not occur. So if k is in b but not in a, then max(b) > max(a), and the ticket restriction prohibits application of $\rightarrow$E. We therefore consider only the cases (i), (ii), and (iv).

Case (i) is by $\rightarrow$E, just as before.

Were we considering $\mathbf{R}_{\rightarrow}$, it is clear that either form of transitivity:

$A\rightarrow B\rightarrow .B\rightarrow C\rightarrow .A\rightarrow C$ (*suffixing*)
$A\rightarrow B\rightarrow .C\rightarrow A\rightarrow .C\rightarrow B$ (*prefixing*)

would do for case (ii), where we want to get $H\rightarrow B_{(\mathrm{a}\cup\mathrm{b})-\{k\}}$ from $H\rightarrow A_{\mathrm{a}-\{k\}}$ and $A\rightarrow B_{\mathrm{b}}$. But for $\mathbf{FT}_{\rightarrow}$ we need both forms of transitivity. For suppose that $\max(\mathrm{a}-\{k\}) \leq \max(\mathrm{b})$. Then insertion of *suffixing*

$H\rightarrow A\rightarrow .A\rightarrow B\rightarrow .H\rightarrow B$

yields

$A\rightarrow B\rightarrow .H\rightarrow B_{\mathrm{a}-\{k\}}$

conforming to the ticket-restriction on $\rightarrow$E. And from the latter, again conforming to the restriction, we get

$$H \rightarrow B_{(\mathrm{a} \cup \mathrm{b})-\{k\}}$$

as required.

But if $\max(\mathrm{b}) < \max(\mathrm{a}-\{k\})$, we need transitivity in the form of *prefixing*,

$$A \rightarrow B \rightarrow .H \rightarrow A \rightarrow .H \rightarrow B,$$

in order to get

$$H \rightarrow A \rightarrow .H \rightarrow B_{\mathrm{b}},$$

and then

$$H \rightarrow B_{(\mathrm{a} \cup \mathrm{b})-\{k\}}.$$

The reader may verify that neither form of transitivity will do the job of the other; so both forms of transitivity are required for case (ii).

Finally, considerations like those of case (ii) lead readily to the conclusion that *both* forms of self-distribution (i.e., on both the major and the minor) are required for case (iv).

So as axioms for $\mathbf{T}_{\rightarrow}$ we choose:

$\mathrm{T}_{\rightarrow}1.$ $A \rightarrow A$
$\mathrm{T}_{\rightarrow}2.$ $A \rightarrow B \rightarrow .B \rightarrow C \rightarrow .A \rightarrow C$
$\mathrm{T}_{\rightarrow}3.$ $A \rightarrow B \rightarrow .C \rightarrow A \rightarrow .C \rightarrow B$
$\mathrm{T}_{\rightarrow}4.$ $(A \rightarrow .B \rightarrow C) \rightarrow .A \rightarrow B \rightarrow .A \rightarrow C$
$\mathrm{T}_{\rightarrow}5.$ $A \rightarrow B \rightarrow .(A \rightarrow .B \rightarrow C) \rightarrow .A \rightarrow C$

and without further proof we state the following

THEOREM. A formula is provable in $\mathbf{FT}_{\rightarrow}$ just in case it is provable in $\mathbf{T}_{\rightarrow}$.

The differences between $\mathbf{T}_{\rightarrow}$ (contained in $\mathbf{E}_{\rightarrow}$), $\mathbf{E}_{\rightarrow}$ (contained in $\mathbf{R}_{\rightarrow}$), and $\mathbf{R}_{\rightarrow}$, suggest some philosophical remarks about modal logic, which in turn suggest a couple of formal problems.

Proponents of the view that the two-valued propositional calculus and its quantificational extensions are the *only* systems of logic worth any sane person's attention, harbor as it seems to us, two confusions.

In the first place, there are questions about intersubstitutability: under what conditions on A and B will we be willing to say that a truth involving A (say f(A)) is also a truth involving B in the same way, f(B)? For the two-valued case, the answer is simple; the material equivalence of A and B is sufficient for the material equivalenee of f(A) and f(B), where "f" is a truth functional context. And as we shall see in §24, this "extensional" condition is retained for the full system **E** of entailment, just as it is for many familiar systems of modal logic. But of course there are in **E**, as in other systems, "intensional contexts," for which material equivalence is not sufficient for intersubstitutability; and for this reason some writers have been led to believe that intensional contexts (involving "belief," or "necessity," or "verified," or the like) are a Bad Thing, and should be banished from the purview of formal logic: "if you can't handle a problem by the most simple-minded techniques available, then throw away the problem."

But we should notice that if intersubstitutability is made an important criterion for identity ("Which items in your theory are distinct from which others? If you can't answer that, you don't know what you're talking about"), then all of the systems $\mathbf{R}_{\rightarrow}$, $\mathbf{T}_{\rightarrow}$, and the modal system $\mathbf{E}_{\rightarrow}$, satisfy the criterion. If A⇄B is provable, then so is f(A)⇄f(B); identity criteria are just as available here as elsewhere, though they are of course different from the extensional criteria, as would be expected of systems intended to deal with intensional notions.

A second complaint, related to but distinct from the first, arises because some standard examples of contexts where material equivalence is *not* a sufficient condition for intersubstitutability are contexts involving necessity. It consequently seems to have followed that, if a system is not purely truth functional, in the extensionalist sense, then it must (apart from psychological or epistemological considerations) be a system involving modalities. The existence of $\mathbf{R}_{\rightarrow}$ and $\mathbf{T}_{\rightarrow}$ shows the falsity of the claim that intensions and modality are ineluctably intertwined. Modalities can be defined in neither (see the very end of this section, and also §8.18), but both have a claim to being intensional, in the sense that relevance is taken into account.

These two systems, both intensional, exhibit two quite different ways of demolishing the theory of necessity enshrined in $\mathbf{E}_{\rightarrow}$: $\mathbf{R}_{\rightarrow}$ by making stronger assumptions about identity or intersubstitutability (and hence having fewer distinct propositional entities), and $\mathbf{T}_{\rightarrow}$ by making weaker assumptions (and hence having more distinct propositional entities). Modal systems, generally being weaker than their cousins, tend to make more distinctions: in $\mathbf{E}_{\rightarrow}$ we can distinguish A from $\Box A$, since, though the latter entails the former, the converse is neither true nor provable. As we saw in §5, adding $A \rightarrow . A \rightarrow A \rightarrow A$, i.e., $A \rightarrow \Box A$, to $\mathbf{E}_{\rightarrow}$ ruins this distinction and produces $\mathbf{R}_{\rightarrow}$: a strong assumption produces fewer distinct propositional entities. And of course further strengthening in the direction of the two-valued calculus produces a system which cannot tell the difference between Bizet's being French and Verdi's being Italian.

This line of thought might make one think that the only way to demodalize a modal system is to make stronger assumptions. But it turns out that there is a different way, e.g., that exhibited by $\mathbf{T}_{\rightarrow}$. That $\mathbf{T}_{\rightarrow}$ makes weaker assumptions about the arrow than $\mathbf{E}_{\rightarrow}$ can be seen by noticing that

$A\to A\to A\to A$, which makes $\Box A\to A$ available in $\mathbf{E}_\to$, is not provable in $\mathbf{T}_\to$, as is shown by the following proof (for which we are indebted to John R. Chidgey, in correspondence, 1968).

Consider the matrix

$\to$	*0*	*1*	*2*	*3*
0	*3*	*3*	*3*	*3*
1	*0*	*2*	*0*	*3*
**2*	*0*	*3*	*2*	*3*
**3*	*0*	*0*	*0*	*3*

This matrix satisfies $\to$E, since if A and $A\to B$ always take designated values, then B must always take a designated value. (That is, the case where $A\to B$ = $\mathit{2}\to\mathit{1}$ = $\mathit{3}$ cannot arise. For if $B = \mathit{1}$, then B must be a propositional variable, since *1* does not occur as a value for the arrow; hence B can be assigned the value *0*, in which case, if A is *always* designated, $A\to B = \mathit{0}$, contrary to hypothesis.) We leave it to the reader to check that the axioms $\mathbf{T}_\to$1–$\mathbf{T}_\to$5 (or, more easily, the axioms $\mathbf{T}_{\to 2}$ of §8.3.2) are also satisfied by the matrix. But the matrix falsifies $A\to A\to A\to A$ (hence also $A\to A\to B\to B$) for $A = \mathit{1}$: $\mathit{1}\to\mathit{1}\to\mathit{1}\to\mathit{1} = \mathit{2}\to\mathit{1}\to\mathit{1} = \mathit{3}\to\mathit{1} = \mathit{0}$.

One gets some further sense of the differences between $\mathbf{T}_\to$ and $\mathbf{E}_\to$ by observing that, in the "strong and natural list of valid entailments" mentioned in §4.2, the following are not provable in $\mathbf{T}_\to$:

restricted permutation
restricted conditioned modus ponens
restricted assertion
specialized assertion
generalized restricted permutation

And indeed it is not hard to show that addition of any one of these to $\mathbf{T}_\to$ yields a system equivalent with $\mathbf{E}_\to$. We consider one of these (in the subproof formulation) in order to help see what is involved.

Restricted permutation. A proof of this theorem of $\mathbf{FE}_\to$ would look as follows:

1	$A\to.B\to C\to D_{\{1\}}$	hyp
2	$\quad B\to C_{\{2\}}$	hyp
3	$\quad\quad A_{\{3\}}$	hyp
4	$\quad\quad B\to C\to D_{\{1,3\}}$	1 reit 3 $\to$E
5	$\quad\quad D_{\{1,2,3\}}$	2 reit 4 $\to$E
6	$\quad A\to D_{\{1,2\}}$	3–5 $\to$I
7	$B\to C\to.A\to D_{\{1\}}$	2–6 $\to$I
8	$(A\to.B\to C\to D)\to.B\to C\to.A\to D$	1–7 $\to$I

But note that step 2 has been reiterated, then used as a minor premiss for step 5, in defiance of the ticket-restriction, since $\max(\{1,3\}) > \max(\{2\})$.

Since *none* of these formulas on which we based our development of the theory of necessity in $\mathbf{E}_{\rightarrow}$ is available in $\mathbf{T}_{\rightarrow}$, one is led to suppose that there is *no* way to define $\Box$ in the latter system.

And indeed, on plausible assumptions, this turns out to be the case. Minimally, we would expect that if necessity were definable in $\mathbf{T}_{\rightarrow}$, there would be some formula f(A) in a single variable A, which satisfied the following conditions. ($\vdash$ {$\dashv$} is for {un}provability.)

(1) $\vdash \mathrm{f}(A) \rightarrow A$,
(2) $\dashv A \rightarrow \mathrm{f}(A)$,
(3) if $\vdash A$ then $\vdash \mathrm{f}(A)$, and
(4) $\vdash A \rightarrow B \rightarrow .\mathrm{f}(A) \rightarrow \mathrm{f}(B)$.

Parks and Chidgey 1972 show that there is no such formula in $\mathbf{T}_{\rightarrow}$; we reproduce their proof verbatim (very nearly).

THEOREM. There is no function f definable in $\mathbf{T}_{\rightarrow}$ satisfying conditions (1)–(4).

PROOF. Assume on the contrary that there is such a function. Consider the matrix above. If f is to be definable in terms of $\rightarrow$, then there must also be a matrix

A	f(A)
0	l
1	m
2	n
3	p

such that each of l, m, n, and p is a member of {***0***, ***1***, ***2***, ***3***}, and such that the two matrices together satisfy (1), (3), and (4). f(A) must be distinct from A to satisfy (2), so f(A) must be an entailment. Since entailments never take the value ***1***, we have it that

(5) l, m, n, p $\in$ {***0***, ***2***, ***3***}.

It is immediate that

(6) n $\neq$ ***3***
(7) m $\neq$ ***3***

if we are to have (1). Now, consider the following row of a truth table for (4)

A	B	$A \rightarrow B \rightarrow .\mathrm{f}(A) \rightarrow \mathrm{f}(B)$
2	***1***	***3*** . n * m

A *3* must be entered in the column under the consequent (starred in the diagram) to insure that (4) takes a designated value for this assignment of values to *A* and *B*. Given (5), (6), and (7), we can have a *3* here only if n = ***0***. But this falsifies (3), since $A \rightarrow A$ is a theorem of $\mathbf{T}_{\rightarrow}$ and f($A \rightarrow A$) is not (for $A = 2$, $A \rightarrow A = 2$, so f($A \rightarrow A$) = ***0***). Thus there is no such f.

The fact that necessity is not definable in the strongly intensional system $\mathbf{T}_{\rightarrow}$ should recommend the system to those truth functionalists who confuse necessity with anathema.

§7. **Gentzen consecution calculuses.** In this section we present formulations in the style of Gentzen 1934 for entailment and some closely related notions. We first discuss the point of the enterprise, secondly some informal motivating considerations; in the third section we describe some calculuses, and in the fourth we prove something.

§7.1. **Perspectives in the philosophy of logic.** It seems to be generally conceded that formal systems are natural or substantial if they can be looked at from several points of view. We tend to think of systems as artificial or *ad hoc* if most of their formal properties arise from some *one* notational system in terms of which they are described.

Consider the contrast between the Lewis systems **S1** and **S4.** The system **S4** can be thought of as a system of modal logic; but, because of connections with the algebra of topology (Kuratowski 1922) established by McKinsey 1941, it can also be looked at as a closure algebra in the topological sense. And even as a modal logic it can rest on several different foundations, as can be seen by inspecting Curry 1950, Fitch 1952, Kripke 1959, Hintikka 1963, Lemmon 1957, and others.

It is just because of these latter-day analyses that **S4** begins to look like an objective reality which was "discovered" rather than "created" by Lewis; and just because **S1** has resisted almost all attempts to provide it with interesting alternative bases, it remains an historical curiosity, an *ad hoc* collection of postulates with nothing substantial behind them.

(Wolf points out in correspondence (1973) that "since this paragraph was written, **S1** has been given a Lemmon-style base, a Genzten system, two algebraic semantics, one Kripke semantics, and a Henkin-type completeness proof," from which he concludes that **S1** is not a good example of something *ad hoc*. "Stupid, perhaps," he writes, "but not *ad hoc*." We retreat a bit, but not much. To a logician, a calculus is as a mountain was to Mallory, but Mallory never said *all* mountains are worth climbing; to anticipate the image we develop below, **S1** may now be considered a physical object, but in our view only a low grade one like, to use an example which Carnap once conscientiously credited to Russell, a dog.)

Substantiality is, of course, something we cannot measure, but it looks as if the possibility of seeing a formal system from different perspectives contributes to our feeling that such a system is more like a tree than a pink elephant. **S1** looks like the latter; intuitionistic propositional calculus looks like a tree (Gentzen 1934). And among the nicest firm formulations of an intuitive concept is the theory of effective computability, a theory which received formal treatment at the hands of Herbrand-Gödel as a new mathematical notion, from Turing as a new analysis of the way in which computers work, and from Church and Post as a new analysis of the way logical systems work. If all these had turned out to be distinct, one might well have felt that the differences hung on *notational* differences; but the fact that all turned out to be equivalent lends support to the attitude that there is something substantial there to view.

It is difficult to be sure why this view is so commonly accepted, but we would like to hazard the conjecture that the indubitable substantiality of the mathematical theory of effective computability arises because the various perspectives make it look more like a physical object, a pool table, say. It's *there*, and we grow more convinced of the fact as we move around it, seeing it and the balls from different angles.

This is a conjecture we do not wish to press, but the fact that there are several equivalent points of view from which to look at entailment makes us feel more comfortable in our conviction that we really have a topic to talk about. We have already discussed the matter from the point of view of Hilbert systems, and that of Fitch's variant of natural deduction, and we now turn to a consideration of Gentzen's point of view.

§7.2. Consecution, elimination, and merge. The fundamental idea of Gentzen 1934 was to consider formal systems which embody the notion of logical consequence as primitive, and to introduce a special symbol to denote this notion. We will use the sign "$\vdash$" for this purpose, and we call it the "turnstile." Gentzen uses the term *sequenz* for a pattern

$$A_1, \ldots, A_n \vdash B$$

affirming that B is a consequence of $A_1, \ldots, A_n$, and he calls the systems based on this idea *sequenzen-kalküle*. Many writers now translate *sequenz* by "sequent," but, as Tryg Ager pointed out to us, there is a substantially more appropriate rendition. The Shorter Oxford English Dictionary 1955 gives two meanings for the unfamiliar word "consecution":

1. Logical sequence; inference; a train of reasoning.
2. Succession, sequence.

This *double entendre* so exactly matches the usage of *sequenz* that we will adopt a departure from the usual translation. Henceforth we say that a statement having the form displayed above is a *consecution*, just as $A\&B$ is a conjunction; and we think of the turnstile as a consecution-sign, just as the ampersand is a conjunction-sign.

Gentzen suggested having axioms $A \vdash A$, and then two groups of rules, some structural rules governing the turnstile itself, and a pair of rules for each connective — one for introducing it on the left, another for introducing it on the right. With trivial changes, his formulation $\mathbf{LJ}_{\rightarrow}$ of intuitionistic implication goes like this, where Greek letters stand for arbitrary (possibly empty) sequences of formulas.

Structural rules:

Identity (*Id*).

$$A \vdash A \text{ (Id)}$$

Weakening (K $\vdash$).

$$\frac{\alpha \vdash \zeta}{\alpha, A \vdash \zeta}(\text{K} \vdash)$$

Permutation (C $\vdash$)

$$\frac{\alpha, A, B, \beta \vdash \zeta}{\alpha, B, A, \beta \vdash \zeta}(\text{C} \vdash)$$

Contraction (W $\vdash$).

$$\frac{\alpha, A, A \vdash \zeta}{\alpha, A \vdash \zeta}(\text{W} \vdash)$$

Elimination Rule (ER).

$$\frac{\alpha \vdash A \qquad \beta, A \vdash \zeta}{\alpha, \beta \vdash \zeta}(\text{ER})$$

Connective rules:

Left rule

$$\frac{\alpha \vdash A \qquad \beta, B \vdash \zeta}{\alpha, \beta, A\rightarrow B \vdash \zeta}(\rightarrow \vdash)$$

Right rule

$$\frac{\alpha, A \vdash B}{\alpha \vdash A\rightarrow B}(\vdash \rightarrow)$$

Then it develops that, for pure "implicational" formulas A, $\vdash A$ (that is, a consecution with empty antecedent, and A as sole consequent) is provable in the Gentzen formulation just in case A is provable in Heyting's formulation of intuitionistic logic.

Among the rules stated above, the elimination rule (ER) stands out as an anomaly. All the other rules have the feature, which one can see by inspection, that every formula that occurs in the premiss(es) of a rule also occurs in the conclusion — not necessarily between commas, but at least as a subformula of the conclusion. It follows that any proof of $\alpha \vdash A$ not involving (ER) has the *subformula property*, according to which every consecution in such a proof is constructed out of subformulas of $\alpha \vdash A$. In this respect such proofs differ radically from proofs involving (ER), or one of its cousins, *modus ponens* or transitivity. A proof in which a formula is eliminated by an application of one of these rules might contain adventitious occurrences of formulas that have no connection with what was to be proved. The subformula property guarantees that, in searching for a proof of a consecution, we need not hunt through all possible consecutions, but only through those that involve subformulas of the consecution in question.

But the elimination rule has the feature that the only thing one can say about the "middle term" A in the two premisses, is that in the conclusion it has vanished; if a consecution comes by application of (ER), inspection of that consecution gives no clue whatever as to what the *spurlos versenkt* term looked like. We can agree with Aristotle, *Posterior Analytics* I, 34, that "quick wit is a faculty of hitting upon the middle term instantaneously," but even quick wit may be of little avail if we are challenged to plow through the infinitely many candidates for A in the hope of finding a winner.

If only subformulas of the consecution we want to prove need be examined, however, one can hope to base a decision procedure on a system in which for every theorem there is a proof having the subformula property. One accordingly hopes that (ER), which would spoil things, is after all redundant, in the sense that it could be dropped without losing any theorems.

There is also a reason to predict that (ER) is redundant: if the axioms and structural rules without (ER) are taken as completely generating all the true "pure turnstile" statements, and if the pair of rules for the horseshoe are taken as completely characterizing the meaning of that connective, then one would expect *a priori* that (ER) would be redundant (see Curry 1963, p. 188); and so it is. Gentzen called this fact his *Hauptsatz;* we follow Curry in calling it the *Elimination theorem*, and agree with him that it "is one of the major theorems of modern mathematical logic" (Curry 1963, p. 166).

Gentzen's own picture of the situation was that his system **LJ** *really had* (ER) as a rule, and his project was to show that one never needed to use it. We tend to prefer an alternative picture, according to which we think of **LJ** as a system which never had (ER) at all, and then show that (ER) holds for the system anyway.

The expression "holds for the system anyway" admits of at least two senses.

We will say that a rule: from $A_1, \ldots, A_n$ to infer B, is *derivable* when it is possible to proceed from the premisses to the conclusion with the help of axioms and primitive rules alone; i.e., there is a list $C_1, \ldots, C_m, B$, each of which is either one of the premisses A_i, or an axiom, or follows from predecessors in the list by one of the primitive rules of the system. Which is to say that there is a proof of B on the hypotheses $A_1, \ldots, A_n$ in the Official sense (§1.4). It can easily be seen that (ER) does not hold in the sense of being derivable, since, by the subformula property, subformulas are never lost in the course of passing down a proof from hypotheses.

But the rule does hold in another sense, namely that it can be shown not to increase the stock of theorems. The result amounts to this: any theorem provable with the help of (ER) is also provable without using (ER). We will say in general (for any system) that if a rule: from $A_1, \ldots, A_n$ to infer B, has the feature that it does not increase the collection of theorems, then that rule is *admissible* (Lorenzen 1955). Evidently every derivable rule is admissible, but (as the proof of the Elimination theorem indicates) not conversely.

Admissibility might also be described as follows: whenever there is a proof of the premisses, there is also a proof of the conclusion. The paradigm arguments for admissibility depend on showing how the proofs of the premisses can be transformed, in some step-by-step fashion, into a proof of the conclusion. Of course there may be other ways of coming to know that if the premisses are provable then the conclusion is also, and it may be that the proof of the conclusion has little or nothing to do with the proofs of the premisses (for an example see §25.2). But Gentzen's procedure was paradigmatic, so that if proofs are given of the premisses of (ER), then a systematic way was given of finding a proof of the conclusion of (ER), where the new proof does not use (ER).

In view of the intrinsic interest of elimination theorems, and the remarks in §7.1, it is of some interest to note that we can provide consecution calculuses for entailment and related notions, for which appropriate elimination theorems can be proved; this can be done in two rather different ways.

The critical element in one of these, the method of Kripke 1959a, consists in placing restrictions on the Gentzen rule for introducing $A \rightarrow B$ on the *right* of the turnstile; we reserve for §13 consideration of this mode of Gentzenization, while in this section we deal with systems having no restrictions on the right-hand rule for $A \rightarrow B$, but having instead restrictions on the rules for introducing $A \rightarrow B$ on the *left* of the turnstile. For both kinds of restrictions, much hangs on the interpretation of the turnstile.

As is well known, Gentzen allowed the possibility that there might be a sequence of formulas on the right of the turnstile as well as on the left, the intended interpretation of

$$A_1, \ldots, A_n \vdash B_1, \ldots, B_m$$

being that the conjunction of the formulas on the left has the disjunction of the formulas on the right as a consequence. Since we are concerned here only with the pure calculus of entailment, this interpretation must be altered: without disjunction a multiple sequence on the right has no interpretation (for which reason we consider here only consecutions with singular right sides), and without conjunction we must reinterpret the sequence on the left of the turnstile.

In view of an application of the rule for the arrow on the right

$$\frac{\alpha,\ A \vdash B}{\alpha \vdash A \rightarrow B}(\vdash \rightarrow)$$

it is clear from the considerations mentioned in §3, and elaborated in §22.2, that the sequence of formulas on the left of the turnstile cannot have the sense of a conjunction of premisses. For on the conjunctive interpretation we would have $A, B \vdash A$, and the rule in question would lead from this to $A \vdash B \rightarrow A$, and thence to $\vdash A \rightarrow .B \rightarrow A$, which is no good. But now the required interpretation must be obvious; we are to think of

$$A_1, \ldots, A_n \vdash B$$

as meaning that the premisses in a *nested* sense imply the conclusion; i.e.

$$A_1 \rightarrow .A_2 \rightarrow . \quad \ldots \quad .A_n \rightarrow B.$$

This situation requires an adjustment in Gentzen's rules, which come in two groups, structural and connective. We recall that the connective rules have to do with the introduction of connectives on the right and left of the turnstile, while the structural rules depend only on the intended interpretation of the turnstile. So that for example the structural rule of contraction

$$\frac{\alpha,\ A,\ A,\ \beta \vdash \zeta}{\alpha,\ A,\ \beta \vdash \zeta}(\mathrm{W} \vdash)$$

holds independently of the logical form of A, whereas $(\vdash \rightarrow)$ gives us a new formula on the right of the arrow, about the logical form of which something can be known.

From the present point of view special interest attaches to the structural rule of permutation

$$\frac{\alpha,\ A,\ B,\ \beta \vdash \zeta}{\alpha,\ B,\ A,\ \beta \vdash \zeta}(\mathrm{C} \vdash)$$

which makes good sense on the conjunctive interpretation of the left of the turnstile, allowing us quite properly to go from $(A \& B) \rightarrow C$ to $(B \& A) \rightarrow C$. Even on the nested interpretation, classical and intuitionistic logicians would allow $(\mathrm{C} \vdash)$, since they think that $A \rightarrow .B \rightarrow C$ "implies" $B \rightarrow .A \rightarrow C$.

This of course involves fallacies of modality. However, retaining modality does enable us to retain some permutations, e.g.,

$$(A \to .B \to C \to D) \to .B \to C \to .A \to D.$$

The problem is to provide some way of keeping the good permutations without the bad ones.

The first thing we do is to discard permutation altogether, in the hope that such permutations as one wants for various systems can be introduced in driblets. As it turns out, this can be done by making the introduction and elimination rules a little more general than Gentzen did, and in effect building some of the features of the structural rules into the logical rules. Eliminating (C $\vdash$) of course forbids the following proof:

1 $A \vdash A \qquad B \vdash B$
(→ $\vdash$)
2 $A, A \to B \vdash B$
(C $\vdash$)
3 $A \to B, A \vdash B$
($\vdash$ →)
4 $A \to B \vdash A \to B$
($\vdash$ →)
5 $\vdash A \to B \to .A \to B$

even though the theorem is provable in $\mathbf{E}_{\to}$. But we notice that the *effect* of moving from the two premisses 1 to the conclusion 3 can be got by allowing this sort of step under the heading of (→ $\vdash$). The general idea is to formulate (→ $\vdash$) so as to take us not to some *particular* consecution (the formulas of which can then be rearranged using (C $\vdash$)), but rather so as to take us directly to *any* member of the "permutation-class" of which the particular conclusion is a member. But one has to be careful. Consider this example.

$$\frac{A_1, A_2 \vdash C \qquad B_1, B_2, E \vdash G}{A_1, A_2, B_1, B_2, C \to E \vdash G}(\to \vdash)$$

It would defeat our purposes to allow passage from the premisses directly to

$$A_2, A_1, B_1, B_2, C \to E \vdash G$$

where A_1 and A_2 have been permuted, since then we would in effect be just allowing unrestricted permutation all over again; for if A_1 and A_2 are not already permutable in the left premiss, we ought not allow a use of (→ $\vdash$) as an excuse to permute them. On the other hand, none of the following conclusions from the same premisses conceals any untoward permutation possibly forbidden in the premisses, since the only shifts allowed are between formulas not occurring together in one of the premisses:

$$A_1, A_2, B_1, C \to E, B_2 \vdash G$$
$$A_1, C \to E, B_1, A_2, B_2 \vdash G$$
$$C \to E, B_1, B_2, A_1, A_2 \vdash G$$

and so forth. The guiding principle is that the following three sequences may be "merged" in any order one likes, so long as the internal order of each is not destroyed: (1) the sequence of antecedents of the left premiss, (2) the sequence of antecedents of the right premiss, and (3) the newly introduced $\rightarrow$-formula.

We shall explain merging in more detail below, and we shall also add certain other subtleties to the exact statement of the rules, but in every case the chief point is to let the merging of sequences of antecedents drawn from different premisses do the work of the rule (C $\vdash$), to the extent that we want that work done. We thus obtain enough control to arrive at formulations with appropriate elimination theorems for several pure implicational calculuses, in which merging plays a role of sufficient importance to enable us to feel justified in calling these *merge formulations.*

§7.3. **Merge formulations.** Our first enterprise is to state rules for merge formulations of the calculuses $\mathbf{T}_{\rightarrow}$, $\mathbf{E}_{\rightarrow}$, $\mathbf{S4}_{\rightarrow}$, $\mathbf{R}_{\rightarrow}$, and $\mathbf{H}_{\rightarrow}$. These systems, unlike $\mathbf{S5}_{\rightarrow}$ and the two-valued "implicational" calculus $\mathbf{TV}_{\rightarrow}$, have natural formulations with singular right side. We shall use the prefix Lμ (e.g., L$\mu\mathbf{E}_{\rightarrow}$) to denote the appropriate merge formulation, and shall otherwise use the terminology of Curry 1950 and Gentzen 1934. Greek letters α, β, γ, δ, ζ, with or without subscripts, stand for (possibly empty) sequences of formulas, such sequences being arbitrary except where conditions are explicitly laid down. $\emptyset$ is the empty sequence, and $[\alpha, \beta]$ is the sequence consisting of the constituents of α followed by those of β. Square brackets will ordinarily be omitted. By a consecution we shall understand an expression having the form

$$\alpha \vdash A.$$

The notion of a merge is made precise as follows: a sequence α is a merge of two sequences β and γ if α can be obtained from a single, simple, somewhat sloppy interlacing shuffle of β and γ (see generalized transitivity under §4.2); i.e., a shuffling of β and γ such that the constituents of each of β and γ retain their internal ordering in the resultant sequence α. In other words, the constituents of γ may be distributed *ad lib.* in the interstices of β so long as the constituents of γ also retain their internal ordering. The class of all sequences which are merges of β and γ will be signified by "$\mu(\beta, \gamma)$." As an example, if β is the sequence

$$B_1, B_2, B_3,$$

and γ is the sequence

$$C_1, C_2, C_3,$$

then the sequence α

$$B_1, C_1, C_2, B_2, B_3, C_3$$

will be one of the merges of β and γ, i.e.,

$$\alpha \in \mu(\beta, \gamma).$$

We may define the notion of a *merge* $\mu(\alpha, \beta)$ as follows: $\mu(\alpha, \beta)$ is the smallest set such that

(i) $[\alpha, \beta] \in \mu(\alpha, \beta)$, and
(ii) if $[\alpha_1, A, B, \alpha_2] \in \mu([\beta_1, A, \beta_2], [\gamma_1, B, \gamma_2])$ then $[\alpha_1, B, A, \alpha_2] \in \mu([\beta_1, A, \beta_2], [\gamma_1, B, \gamma_2])$. We list some of the relevant properties of μ, leaving proofs to the reader.

M1 $\mu(\alpha, \beta) = \mu(\beta, \alpha)$.
M2 $\mu(\alpha, \mu(\beta, \gamma)) = \mu(\mu(\alpha, \beta), \gamma)$. (Here and below we use μ inside μ with the obvious meaning; e.g., $\mu(\alpha, \mu(\beta, \gamma))$ is the set of sequences obtainable by merging α with some sequence in $\mu(\beta, \gamma)$.) Hence we let $\mu(\alpha_1, \ldots, \alpha_n)$ be the class of all merges of the n sequences $\alpha_1, \ldots, \alpha_n$.
M3 $\mu(\alpha, \emptyset) = \{\alpha\}$.
M4 If $\alpha_1 \in \mu(\beta_1, \gamma_1)$ and if $\alpha_2 \in \mu(\beta_2, \gamma_2)$, then $[\alpha_1, A, \alpha_2] \in \mu([\beta_1, A, \beta_2], [\gamma_1, \gamma_2])$.
M5 $\mu(\beta, \gamma)$ is the class of all α such that there are $\beta_1, \beta_2, \gamma_1, \gamma_2$ such that $\beta = [\beta_1, \beta_2]$, $\gamma = [\gamma_1, \gamma_2]$, and $\alpha \in [\mu(\beta_1, \gamma_1), \mu(\beta_2, \gamma_2)]$.
M6 If $[\alpha_1, A, \alpha_2] \in \mu([\beta_1, A, \beta_2], \gamma)$, then there are γ_1, γ_2, such that $\gamma = [\gamma_1, \gamma_2]$ and $\alpha_1 \in \mu(\beta_1, \gamma_1)$ and $\alpha_2 \in \mu(\beta_2, \gamma_2)$.

The various systems will all be described in terms of the certain axioms, rules, and restrictions, all of which we state before defining the systems.

Identity.

$A \vdash A$ (Id)

Arrow on the right.

$$\frac{\alpha, A \vdash B}{\alpha \vdash A \to B}(\vdash \to)$$

Arrow on the left.

$$\frac{\alpha, \gamma \vdash A \qquad \beta, B, \delta \vdash \zeta}{\mu(\alpha, \beta, A \to B), \gamma, \delta \vdash \zeta}(\to \vdash)$$

Contraction.

$$\frac{\alpha, A, A, \beta \vdash \zeta}{\alpha, A, \beta \vdash \zeta}(\mathrm{W} \vdash)$$

Weakening.

$$\frac{\alpha, \gamma \vdash \zeta}{\alpha, A, \gamma \vdash \zeta}(\mathrm{K} \vdash)$$

Ticket restriction: in $(\rightarrow \vdash)$, γ must not be empty.

Modal restriction: in $(\rightarrow \vdash)$ and $(\mathrm{K} \vdash)$, there must be an arrow to the right of α, at least implicitly. Which is to say that *either* γ must not be empty (thus providing an implicit arrow), *or* there must be an explicit arrow on the right of the turnstile of the premiss in which γ occurs; i.e., an arrow in A for $(\rightarrow \vdash)$ and an arrow in ζ for $(\mathrm{K} \vdash)$.

We now define the various systems. All have (Id), $(\vdash \rightarrow)$, $(\mathrm{W} \vdash)$, and the situation as to $(\rightarrow \vdash)$ and $(\mathrm{K} \vdash)$ is as follows:

	Arrow on the left $(\rightarrow \vdash)$	*Weakening* $(\mathrm{K} \vdash)$
$\mathrm{L}\mu\mathbf{T}_{\rightarrow}$:	Ticket restriction	Not allowed
$\mathrm{L}\mu\mathbf{E}_{\rightarrow}$:	Modal restriction	Not allowed
$\mathrm{L}\mu\mathbf{S4}_{\rightarrow}$:	Modal restriction	Modal restriction
$\mathrm{L}\mu\mathbf{R}_{\rightarrow}$:	No restriction	Not allowed
$\mathrm{L}\mu\mathbf{H}_{\rightarrow}$:	No restriction	No restriction

REMARKS. (1) In the statement of the rule $(\rightarrow \vdash)$, the expression below the line does not stand in place of an expression uniquely determined by the premisses. The sense of the rule is rather that from the given premisses, one may infer *any* consecution

$$D_1, D_2, \ldots, D_n, \gamma, \delta \vdash \zeta,$$

provided only that $[D_1, D_2, \ldots, D_n] \in \mu(\alpha, \beta, A \rightarrow B)$.

(2) Note that no rule of permutation

$$\frac{\alpha, A, B, \beta \vdash \zeta}{\alpha, B, A, \beta \vdash \zeta}(\mathrm{C} \vdash)$$

is postulated; however, the statement of the rule $(\rightarrow \vdash)$ in terms of merges gives the effect of permutation-to-the-left of arrow-formulas (the restricted permutation of §4) with the modal restriction, and of arbitrary formulas with no restriction.

(3) With regard to $(\mathrm{W} \vdash)$, the requirement that the like constituents be juxtaposed is important in those systems in which $(\rightarrow \vdash)$ is restricted. We use "*lq*" (left quarter) and "*rq*" (right quarter) for α and β, respectively.

(4) With regard to $(\vdash \rightarrow)$, the requirement that A be at the extreme right of the antecedent is important in those systems in which $(\rightarrow \vdash)$ is restricted. In this connection, it is worthwhile remarking that the designation "antecedent" for the left side of a consecution of the $\mathrm{L}\mu$ systems is something of a

misnomer in view of the fact that the left side cannot be construed as signifying a mere conjunction or class of propositions from which the right side follows. Rather, the interpretation of a consecution $A_1, A_2, \ldots, A_n \vdash B$ is as a nesting of entailments: that A_1 entails that that A_2 entails ... that that A_n entails that B (see §22.2.2).

(5) It is worth remarking that each of these systems has a contraction-free version which is equivalent to the analogous axiomatic version without contraction; indeed, $\mathbf{T}_{\to}$ without contraction is a candidate for a "minimal implication" (Church 1951a). See §8.11 for a few details.

EXAMPLES. Consider first $A\to B\to .A\to .C\to B$, which can be proved in $\mathrm{L}\mu\mathbf{H}_{\to}$ as follows.

$$\dfrac{\dfrac{A \vdash A \qquad \dfrac{B \vdash B}{C, B \vdash B}(\mathrm{K}\vdash)}{A\to B,\ A,\ C \vdash B}(\to\vdash)}{\vdash A\to B\to .A\to .C\to B}(\vdash\to)\text{ thrice}$$

Of course this won't wash in any of the systems in which $(\mathrm{K}\vdash)$ is not allowed. It's no go in $\mathrm{L}\mu\mathbf{S4}_{\to}$ either, since although the use of $(\mathrm{K}\vdash)$ does satisfy the modal restriction, the use of $(\to\vdash)$ does not (γ is empty, but there is no arrow on the right of the $\vdash$); but this proof *would* go through in $\mathrm{L}\mu\mathbf{S4}_{\to}$ were A of the form $A_1\to A_2$, since then there would be an $\to$ on the right. We know furthermore that the same formula is a theorem of $\mathbf{S4}_{\to}$ when B instead of A has the form of an implication, but this assumption about B does not serve to let the displayed proof satisfy the modal restriction on $(\to\vdash)$. However, by simply changing the application of $(\mathrm{K}\vdash)$ to look like

$$\dfrac{B\vdash B}{B,\ C\vdash B}$$

the assumption that B is an implication does suffice to satisfy all restrictions; $(\mathrm{K}\vdash)$ because $B = B_1\to B_2$, and $(\to\vdash)$ because now γ can be taken as A and hence is not empty. (For verification, it is helpful to think of γ as "what replaced B" in introducing $A\to B$.)

Consider next the law of permutation, which may be proved in this way:

$$\dfrac{\dfrac{A\vdash A \qquad \dfrac{B\vdash B \qquad C\vdash C}{B,\ B\to C\vdash C}(\to\vdash)}{A\to .B\to C,\ B,\ A\vdash C}(\to\vdash)}{\vdash (A\to .B\to C)\to .B\to .A\to C}(\vdash\to)\text{ thrice}$$

This proof goes in the systems without the modal restriction on $(\rightarrow\vdash)$, and under the assumption that $B = B_1 \rightarrow B_2$, also in those with the modal restriction. But the ticket restriction rules out anything remotely resembling permutation: there is just no way to obtain anything like permutation without allowing γ to be sometimes empty, which is forbidden by the ticket restriction.

Lastly, consider the following proof of self-distribution, which satisfies every restriction in sight:

$$\dfrac{\dfrac{\dfrac{A \vdash A \qquad \dfrac{A \vdash A \qquad \dfrac{B \vdash B \qquad C \vdash C}{B \rightarrow C,\ B \vdash C}(\rightarrow\vdash)}{A \rightarrow .B \rightarrow C,\ A,\ B \vdash C}(\rightarrow\vdash)}{A \rightarrow .B \rightarrow C,\ A \rightarrow B,\ A,\ A \vdash C}(\rightarrow\vdash)}{A \rightarrow .B \rightarrow C,\ A \rightarrow B,\ A \vdash C}(\mathrm{W}\vdash)}{\vdash (A \rightarrow .B \rightarrow C) \rightarrow .A \rightarrow B.A \rightarrow C}(\vdash\rightarrow)\text{ thrice}$$

We hope that these examples will enable the reader to see what is going on with merge formulations. The demonstrations of equivalence run much alike, but we shall carry out the proof of equivalence of the merge formulations to their axiomatic analogues only in the case of $\mathbf{L}\mu\mathbf{E}_{\rightarrow}$; the chief problem is of course the establishment of an appropriate elimination theorem. In order to make its proof run more smoothly, we use an altered but obviously equivalent formulation of $\mathbf{L}\mu\mathbf{E}_{\rightarrow}$: (1) In (Id) we require A to be a propositional variable (*pv*). (2) The modal restriction on $(\rightarrow\vdash)$ is changed to read "if A is a *pv*, then $\alpha = \emptyset$ and $\gamma = A$, and otherwise $\gamma = \emptyset$." We shall use "*lq*" (left quarter) and "*rq*" (right quarter) to refer, respectively, to β and δ of the second premiss of $(\rightarrow\vdash)$.

As a final preliminary, we list some generalizations of the rules of $\mathbf{L}\mu\mathbf{E}_{\rightarrow}$ useful in the proof of the elimination theorem below. These generalizations are to the effect that if all members of a certain class of consecutions are provable, then so are all members of a certain other class. Expressions of the form "$\mu(\alpha, \beta)$" will occur in various contexts in the statement of these generalizations bearing the following sense: suppose "$\mu(\alpha, \beta)$" occurs in some context above the line and "$\mu(\delta, \gamma)$" occurs in some context below the line; then the entire configuration amounts to the assertion that if all consecutions got by replacing "$\mu(\alpha, \beta)$" by some member of $\mu(\alpha, \beta)$ are provable, then so are all consecutions got by replacing "$\mu(\delta, \gamma)$" by some member of $\mu(\delta, \gamma)$. (Notice that it is not always true that there is a one-to-one correspondence between the class of consecutions represented by the expression above the line and the class of consecutions represented by the expression below the line.) These properties depend on M1–M6. In their application in the proof of the elimination theorem for $\mathbf{L}\mu\mathbf{E}_{\rightarrow}$, use of the principles of commutativity and associativity of merging (M1 and M2) will be tacit.

M*1 $$\frac{\mu([\alpha_1, \beta_1], [\alpha_2, \beta_2]), \delta \vdash \zeta}{\mu(\alpha_1, \alpha_2), \mu(\beta_1, \beta_2), \delta \vdash \zeta}$$

Generalizations of (W $\vdash$).

M*2 $$\frac{\mu(\alpha, \beta), \gamma, A, A, \delta \vdash \zeta}{\mu(\alpha, \beta), \gamma, A, \delta \vdash \zeta}$$

PROOF. By use of (W $\vdash$).

M*3 $$\frac{\mu([\alpha_1, A, A, \alpha_2], \beta), \delta \vdash \zeta}{\mu([\alpha_1, A, \alpha_2], \beta), \delta \vdash \zeta}$$

PROOF. By use of (W $\vdash$).

M*4 $$\frac{\mu(\alpha, \alpha, \beta), \delta \vdash \zeta}{\mu(\alpha, \beta), \delta \vdash \zeta}$$

PROOF. By repeated use of (W $\vdash$).

Generalizations of ($\rightarrow \vdash$).

M*5 $$\frac{\alpha, \gamma \vdash A \qquad \mu(\beta_1, \beta_2), \beta_3, B, \delta \vdash \zeta}{\mu(\alpha, [\mu(\beta_1, \beta_2), \beta_3], A \rightarrow B), \gamma, \delta \vdash \zeta}$$

where if A is a *pv* then $\alpha = \emptyset$ and $\gamma = A$, and otherwise $\gamma = \emptyset$.

PROOF. By ($\rightarrow \vdash$).

M*6 $$\frac{\alpha, \gamma \vdash A \qquad \mu([\beta_1, B, \beta_2], \delta_1), \delta_2 \vdash \zeta}{\mu([\mu(\alpha, \beta_1, A \rightarrow B), \gamma, \beta_2], \delta_1), \delta_2 \vdash \zeta}$$

where . . . (as above). Proof by ($\rightarrow \vdash$).

M*7 $$\frac{\mu(\alpha, \beta), \delta, A \vdash B}{\mu(\alpha, \beta), \delta \vdash A \rightarrow B}$$

PROOF. By ($\vdash \rightarrow$).

§7.4. Elimination theorem. We turn now to a proof of an elimination theorem for $L\mu E_{\rightarrow}$. It turns out to be easiest to divide the question, proving separately an elimination theorem for degree (i.e., number of connectives) = 0 (ER^0), and an elimination theorem for degree ≥ 1 (ER^*). The reader unfamiliar with Curry 1950, or Gentzen 1934, may safely skip this section.

1 ER⁰. We are to prove the following rule redundant, in the sense that, given a derivation of (1) and (2), we can transform this into a derivation of any member of the class of consecutions (3):

$$\frac{(1)\quad \alpha, \gamma \vdash A \qquad (2)\quad \beta, A, \delta \vdash \zeta}{(3)\quad \mu(\alpha, \beta), \gamma, \delta \vdash \zeta}$$

where A is a *pv* and γ contains a single constituent. That A is a *pv* follows from the fact that the degree of $A = 0$. The restriction that γ contain a single constituent is added only to simplify the proof of the elimination theorem for ER⁰; the proof is easily generalized to the case where γ may contain more than a single constituent. A is called the *eliminated constituent* (*ec*) and γ is called the *replacing constituent* (*rc*). The proof is by induction on the left rank (only). We let P(1) be the single premiss of (1), and where applicable, $P_L(1)$ and $P_R(1)$ are, respectively, the left and right premisses of (1); and analogously for P(2), $P_L(2)$, and $P_R(2)$ in the proof of ER* below.

1.1 Left rank = 1; then (1) is by Id, $\alpha = \emptyset$, $\gamma = A$, and hence (3) is the same as (2).

1.2 Left rank > 1. Two main cases.

1.21 (1) is by (W ⊢). Two subcases.

1.211 The *principal constituent* (*pc*) of (1) is in α. Then we have, where $\alpha = [\alpha_1, B, \alpha_2]$,

$$\frac{\dfrac{P(1)\quad \alpha_1, B, B, \alpha_2, \gamma \vdash A}{(1)\quad \alpha_1, B, \alpha_2, \gamma \vdash A}(W \vdash) \qquad (2)\quad \beta, A, \delta \vdash \zeta}{(3)\quad \mu([\alpha_1, B, \alpha_2], \beta), \gamma, \delta \vdash \zeta}ER^0$$

Transformation:

$$\frac{\dfrac{P(1)\quad \alpha_1, B, B, \alpha_2, \gamma \vdash A \qquad (2)\quad \beta, A, \delta \vdash \zeta}{\mu([\alpha_1, B, B, \alpha_2], \beta), \gamma, \delta \vdash \zeta}ER^0}{(3)\quad \mu([\alpha_1, B, \alpha_2], \beta), \gamma, \delta \vdash \zeta}M^*3$$

1.212 The *pc* is the *rc*. Then we have

$$\frac{\dfrac{P(1)\quad \alpha, B, B \vdash A}{(1)\quad \alpha, B \vdash A}(W \vdash) \qquad (2)\quad \beta, A, \delta \vdash \zeta}{(3)\quad \mu(\alpha, \beta), B, \delta \vdash \zeta}ER^0$$

Transformation:

$$\frac{\dfrac{P(1)\quad \alpha, B, B \vdash A \qquad (2)\quad \beta, A, \delta \vdash \zeta}{\mu([\alpha, B], \beta), B, \delta \vdash \zeta}ER^0}{(3)\quad \mu(\alpha, \beta), B, \delta \vdash \zeta}M^*1, M^*2$$

1.22 (1) is by $(\to\vdash)$. Two subcases.

1.221 The *rc* appears in the *rq* of $P_R(1)$. Then we have

$$\dfrac{\dfrac{P_L(1)\quad \alpha_1, \alpha_3 \vdash B \qquad P_R(1)\quad \alpha_2, C, \alpha_4, \gamma \vdash A}{(1)\quad \mu(\alpha_1, \alpha_2, B\to C), \alpha_3, \alpha_4, \gamma \vdash A}(\to\vdash) \qquad (2)\quad \beta, A, \delta \vdash \zeta}{(3)\quad \mu([\mu(\alpha_1, \alpha_2, B\to C), \alpha_3, \alpha_4], \beta), \gamma, \delta \vdash \zeta}\mathrm{ER}^{\circ}$$

where $\alpha \in [\mu(\alpha_1, \alpha_2, B\to C), \alpha_3, \alpha_4]$ and where if B is a *pv*, $\alpha_1 = \emptyset$ and $\alpha_3 = B$, and otherwise $\alpha_3 = \emptyset$.
Transformation:

$$\dfrac{P_L(1)\quad \alpha_1, \alpha_3 \vdash B \qquad \dfrac{P_R(1)\quad \alpha_2, C, \alpha_4, \gamma \vdash A \qquad (2)\quad \beta, A, \delta \vdash \zeta}{\mu([\alpha_2, C, \alpha_4], \beta), \gamma, \delta \vdash \zeta}\mathrm{ER}^0}{(3)\quad \mu([\mu(\alpha_1, \alpha_2, B\to C), \alpha_3, \alpha_4], \beta), \gamma, \delta \vdash \zeta}\mathrm{M}^*6$$

1.222 The *rc* does not appear in the *rq* of $P_R(1)$. Then the *rq* of $P_R(1)$ must be void, and (1) must be some member of the class of consecutions

(1) $\mu(\alpha_1, \alpha_2, B\to C), \alpha_3 \vdash A$

where $P_L(1)$ and $P_R(1)$ are

$P_L(1)$ $\alpha_1, \alpha_3 \vdash B$ and
$P_R(1)$ $\alpha_2, C \vdash A$

(where if B is a *pv* $\alpha_1 = \emptyset$ and $\alpha_3 = B$, and otherwise $\alpha_3 = \emptyset$). However, this information is insufficient to determine just which constituent will be the *rc* — the *rc* could be α_3 or, where $\alpha_3 = \emptyset$, the *rc* could be either $B\to C$, the rightmost constituent of α_1, or the rightmost constituent of α_2. Nevertheless, for all of these cases it will suffice to show the provability of a class of consecutions even larger than the class determined by the result of applying ER^0 to any member of (1) together with (2), namely, the class of consecutions

(3′) $\mu(\alpha_1, \alpha_2, B\to C, \beta), \alpha_3, \delta \vdash \zeta$.

Transformation:

$$\dfrac{P_L(1)\quad \alpha_1, \alpha_3 \vdash B \qquad \dfrac{P_R(1)\quad \alpha_2, C \vdash A \qquad (2)\quad \beta, A, \delta \vdash \zeta}{\mu(\alpha_2, \beta), C, \delta \vdash \zeta}\mathrm{ER}^0}{(3')\quad \mu(\alpha_1, \alpha_2, B\to C, \beta), \alpha_3, \delta \vdash \zeta}\mathrm{M}^*5$$

This completes the proof of ER^0.

2 ER^*. We prove the following rule redundant:

$$\dfrac{(1)\quad \alpha \vdash A \qquad (2)\quad \beta, \delta \vdash \zeta}{(3)\quad \mu(\alpha, \beta), \delta^* \vdash \zeta}$$

where A is of degree ≥ 1, δ contains one or more occurrences of A, and δ^* is like δ except for the elimination of one or more occurrences of A.

REMARKS. (1) That A appear at least once in δ is a non-trivial requirement. (2) We do not require that A fail to appear in δ^*, for there may be occurrences of A which we do not wish to eliminate. Those constituents of δ which are like A and which do not appear in δ^* are called the *eliminated constituents* (*ec*'s). (3) No *ec* appears in β — but there is no requirement that β fail to contain occurrences of A. It is the absence of a rule of weakening which requires us to construe ER* in this way, for $L\mu E_{\to}$ is unlike most consecution calculuses in that, once we have eliminated a constituent which we did not wish to eliminate, we can never recover the loss by weakening. (4) With regard to merge- and star-functions, we shall tacitly employ such evident principles as

$$\mu(\alpha, \beta)^* = \mu(\alpha^*, \beta^*) \text{ and}$$
$$[\alpha, \beta]^* = [\alpha^*, \beta^*].$$

The structure of the proof is essentially like that of Gentzen 1934, the induction being on rank of the derivation and degree of the *ec's*. ER^0, of course, supplies the base clause for the induction on degree.

2.1. Rank of derivation = 2; then, since identities involve only propositional variables and the degree of A is ≥ 1, we must have

$$\dfrac{\dfrac{\text{P}(1)\quad \alpha, B \vdash C}{(1)\quad \alpha \vdash B\to C}(\vdash\to) \qquad \dfrac{\text{P}_{\text{L}}(2)\quad \beta_1, \delta_1, \gamma \vdash B \qquad \text{P}_{\text{R}}(2)\quad \beta_2, \delta_2, C, \delta_3 \vdash \zeta}{(2)\quad \mu(\beta_1, \beta_2), \mu(\delta_1, \delta_2, B\to C), \gamma, \delta_3 \vdash \zeta}(\to\vdash)}{(3)\quad \mu(\alpha, \beta_1, \beta_2), \mu(\delta_1, \delta_2), \gamma, \delta_3 \vdash \zeta}\text{ER}^*$$

where $A = (B\to C)$; if B is a *pv*, $\gamma = B$ and $[\beta_1, \delta_1] = \emptyset$, and otherwise $\gamma = \emptyset$; and where $\beta \in \mu(\beta_1, \beta_2)$, $\delta \in [\mu(\delta_1, \delta_2, B\to C), \gamma, \delta_3]$, and $\delta^* \in [\mu(\delta_1, \delta_2), \gamma, \delta_3]$.

Transformation:

$$\dfrac{\text{P}_{\text{L}}(2)\quad \beta_1, \delta_1, \gamma \vdash B \qquad \dfrac{\text{P}(1)\quad \alpha, B \vdash C \qquad \text{P}_{\text{R}}(2)\quad \beta_2, \delta_2, C, \delta_3 \vdash \zeta}{\mu(\alpha, \beta_2), \delta_2, B, \delta_3 \vdash \zeta}\text{ER}^0 \text{ or ER}^* \text{ and M*1}}{(3)\quad \mu(\alpha, \beta_1, \beta_2), \mu(\delta_1, \delta_2), \gamma, \delta_3 \vdash \zeta}\text{ER}^0 \text{ or ER}^* \text{ and M*1}$$

2.2 The rank of the derivation is >2, right rank arbitrary, left rank >1. Then (1) is either by $(\text{W}\vdash)$ or by $(\to\vdash)$. The required transformations are essentially like *1.211* (using M*3) and *1.221* (using M*6), respectively.

2.3 Rank of derivation is >2, left rank = 1, right rank >1. Three main cases.

2.31 (2) is by $(\text{W}\vdash)$. If the *pc* cf (2) is an *ec*, then (3) follows immediately from (1) and P(2) by ER*. When the *pc* of (2) is not an *ec*, we have two cases.

2.311 The *pc* of (2) is in β. We then have

$$\cfrac{(1)\quad \alpha \vdash A \qquad \cfrac{\mathrm{P}(2)\quad \beta_1, D, D, \beta_2, \delta \vdash \zeta}{(2)\quad \beta_1, D, \beta_2, \delta \vdash \zeta}(\mathrm{W} \vdash)}{(3)\quad \mu(\alpha, [\beta_1, D, \beta_2]), \delta^* \vdash \zeta}\mathrm{ER}^*$$

where $\beta = [\beta_1, D, \beta_2]$.

Transformation: essentially symmetrical to *1.211*, using M*3.

2.312 The *pc* of (2) is in δ. The required transformation is a simple permutation of the application of the rules ER* and (W $\vdash$), using M*2.

2.32 (2) is by ($\vdash \to$). The transformation required is a simple permutation of the application of the rules ER* and ($\vdash \to$), using M*7.

2.33 (2) is by ($\to \vdash$). Two main subcases.

2.331 All of δ appears in the *rq* of $\mathrm{P_R}(2)$; that is, all of δ (in (2)) falls outside of that portion of the conclusion of ($\to \vdash$) which is got by merging the *pc* and the appropriate constituents from the two premisses. Then we have

$$\cfrac{(1)\quad \alpha \vdash A \qquad \cfrac{\mathrm{P_L}(2)\quad \beta_1, \gamma \vdash B \qquad \mathrm{P_R}(2)\quad \beta_2, C, \beta_3, \delta \vdash \zeta}{(2)\quad \mu(\beta_1, \beta_2, B \to C), \gamma, \beta_3, \delta \vdash \zeta}(\to \vdash)}{(3)\quad \mu(\alpha, [\mu(\beta_1, \beta_2, B \to C)\ \gamma, \beta_3]), \delta^* \vdash \zeta}\mathrm{ER}^*$$

where $\beta_1 = \emptyset$ and $\gamma = B$ if B is a *pv*, and otherwise $\gamma = \emptyset$; and $\beta \in [\mu(\beta_1, \beta_2, B \to C), \gamma, \beta_3]$.

Transformation:

$$\cfrac{\mathrm{P_L}(2)\quad \beta_1, \gamma \vdash B \qquad \cfrac{(1)\quad \alpha \vdash A \qquad \mathrm{P_R}(2)\quad \beta_2, C, \beta_3, \delta \vdash \zeta}{\mu(\alpha, [\beta_2, C, \beta_3]), \delta^* \vdash \zeta}\mathrm{ER}^*}{(3)\quad \mu(\alpha, [\mu(\beta_1, \beta_2, B \to C), \gamma, \beta_3]), \delta^* \vdash \zeta}\mathrm{M}^*6$$

2.332 All of δ does not (as opposed to the case *2.331*) appear in the *rq* of $\mathrm{P_R}(2)$; that is, the left part of δ (in (2)) is the right part of that portion of the conclusion of ($\to \vdash$) which is got by merging the *pc* and the appropriate constituents from the two premisses. The following represents the general case:

$$\cfrac{(1)\quad \alpha \vdash A \qquad \cfrac{\mathrm{P_L}(2)\quad \beta_1, \delta_1, \gamma \vdash B \qquad \mathrm{P_R}(2)\quad \beta_2, \delta_2, C, \delta_3 \vdash \zeta}{(2)\quad \mu(\beta_1, \beta_2, \beta_k), \mu(\delta_1, \delta_2, \delta_k), \gamma, \delta_3 \vdash \zeta}(\to \vdash)}{(3)\quad \mu(\alpha, \beta_1, \beta_2, \beta_k), \mu(\delta_1{}^*, \delta_2{}^*, \delta_k{}^*), \gamma^*, \delta_3{}^* \vdash \zeta}\mathrm{ER}^*$$

where if B is a *pv* then $[\beta_1, \delta_1] = \emptyset$ and $\gamma = B$, and otherwise $\gamma = \emptyset$; where either $\beta_k = (B \to C)$ (the *pc*) and $\delta_k = \emptyset$ or else $\delta_k = (B \to C)$ and $\beta_k = \emptyset$; and where $\beta \in \mu(\beta_1, \beta_2, \beta_k)$, $\delta \in [\mu(\delta_1, \delta_2, \delta_k), \gamma, \delta_3]$, and finally $\delta^* \in [\mu(\delta_1{}^*, \delta_2{}^*, \delta_k{}^*), \gamma^*, \delta_3{}^*]$. Now δ_k may or may not be an *ec*; hence, there are two subcases for the transformation.

2.332.1 δ_k is not an *ec*; hence $\delta_k = \delta_k{}^*$. For this case we define a special rule, ER′, in such a way that the conclusion either follows by ER* or else is the same as the right premiss:

$$\frac{\alpha \vdash A \qquad \beta, \delta \vdash \zeta}{\mu(\alpha', \beta), \delta^* \vdash \zeta}\text{ER}'$$

where if δ contains some *ec* then $\alpha' = \alpha$, and otherwise (i.e., if $\delta = \delta^*$) $\alpha' = \emptyset$.

Transformation:

$$\dfrac{\dfrac{(1)\ \ \alpha \vdash A \qquad \text{P}_\text{L}(2)\, \beta_1, \delta_1, \gamma \vdash \text{B}}{\mu(\alpha', \beta_1), \delta_1{}^*, \gamma^* \vdash \text{B}}\text{ER}' \qquad \dfrac{(1)\ \ \alpha \vdash A \qquad \text{P}_\text{R}(2)\, \beta_2, \delta_2, C, \delta_3 \vdash \zeta}{\mu(\alpha', \beta_2), \delta_2{}^*, C, \delta_3{}^* \vdash \zeta}\text{ER}'}{(3')\ \ \mu(\alpha', \alpha', \beta_1, \beta_2, \beta_k), \mu(\delta_1{}^*, \delta_2{}^*, \delta_k{}^*), \gamma^*, \delta_3{}^* \vdash \zeta}\text{M*5, M*1}$$

At least one of the two occurrences of α' will $= \alpha$. If only one occurrence $= \alpha$, then (3′) is the same as (3), while if both $= \alpha$, then (3) follows by M*4.

2.332.2 δ_k is an *ec*; hence $\beta_k = \emptyset$ and $\delta_k{}^* = \emptyset$. The first step of the transformation is as in *2.332.1*. Then we use the fact that since left rank = 1 (see *2.3*) and the *pc* is an *ec*, it follows that A is $B \rightarrow C$ and P(1) is

$$\text{P}(1) \quad \alpha, B \vdash C$$

where degree of B and degree of C are each less than degree of A. The transformation now continues as follows:

$$\dfrac{\dfrac{\mu(\alpha', \beta_1), \delta_1{}^*, \gamma^* \vdash B \qquad \text{P}(1)\ \ \alpha, B \vdash C}{\mu(\alpha, \alpha', \beta_1), \delta_1{}^*, \gamma^* \vdash C}\text{ER}^0\text{ or ER*, and M*1} \qquad \mu(\alpha', \beta_2), \delta_2{}^*, C, \delta_3{}^* \vdash \zeta}{(3')\ \ \mu(\alpha, \alpha', \alpha', \beta_1, \beta_2), \mu(\delta_1{}^*, \delta_2{}^*), \gamma^*, \delta_3{}^* \vdash \zeta}\text{ER}^0\text{ or ER* and M*1}$$

Now one or two applications of M*4 yields (3).

This completes the proof of ER*, and when ER* is taken together with ER^0, the proof of the Elimination theorem for $\text{L}\mu\text{E}_\rightarrow$. We note, finally, that ER^0 can be generalized by a simple inductive argument so that when taken together with ER*, we have that the following is redundant in $\text{L}\mu\text{E}_\rightarrow$:

$$\frac{\alpha, \gamma \vdash A \qquad \beta, \delta \vdash \zeta}{\mu(\alpha, \beta), \delta^* \vdash \zeta}\text{ER}$$

where if A is a *pv* then γ is a single constituent, and otherwise $\gamma = \emptyset$, and where δ^* is the result of replacing some occurrences of A in δ by γ.

§7.5. Equivalence. Having established the Elimination theorem for $\text{L}\mu\text{E}_\rightarrow$, the proof of the equivalence of $\text{L}\mu\text{E}_\rightarrow$ to $\textbf{E}_\rightarrow$ is straightforward:

THEOREM. $L\mu E_{\rightarrow}$ and $E_{\rightarrow}$ are equivalent in the sense that $\vdash A$ is derivable in $L\mu E_{\rightarrow}$ iff A is a theorem of $E_{\rightarrow}$ of §4.

In one direction the proof amounts to the provision of demonstrations — which we leave to the reader — of the axioms $E_{\rightarrow}1$–$E_{\rightarrow}4$ of $E_{\rightarrow}$ in $L\mu E_{\rightarrow}$, since the rule →E is but a special case of the Elimination theorem for $L\mu E_{\rightarrow}$.

For the other direction we show by an inductive proof that if

(S) $A_1, \ldots, A_j \vdash B$

is a theorem of $L\mu E_{\rightarrow}$, then

(S*) $A_1 \rightarrow . \ldots \rightarrow . A_j \rightarrow B$

is a theorem of $E_{\rightarrow}$. The induction is on the length of the proof in $L\mu E_{\rightarrow}$ of (S).

1. The proof of (S) is of length 1; then (S) is by (Id) and (S*) is a theorem of $E_{\rightarrow}$ by $E_{\rightarrow}1$.

2. The proof of (S) is of length >1. Three cases.

2.1. (S) is a consequence of P(S) by (W $\vdash$). Then (S*) is a theorem of $E_{\rightarrow}$ by the hypothesis of the induction, *generalized contraction* (§4.2), and →E.

2.2. (S) is a consequence of P(S) by ($\vdash \rightarrow$). Then (S*) is a theorem of $E_{\rightarrow}$ by the hypothesis of the induction applied to P(S).

2.3. (S) is a consequence of P_L(S) and P_R(S) by ($\rightarrow \vdash$). Two subcases.

2.31. Where the antecedent of the *pc* is a *pv*, we have

$$\frac{P_L(S)\quad A \vdash A \qquad P_R(S)\quad C_1, \ldots, C_m, B, D_1, \ldots, D_n \vdash E}{(S)\ C_1, \ldots, C_{i-1}, A \rightarrow B, C_i, \ldots, C_n, A, D_1, \ldots, D_n \vdash E}(\rightarrow \vdash)$$

By the hypothesis of the induction applied to P_R(S) we have as a theorem of $E_{\rightarrow}$

$P_R(S^*)\ C_1 \rightarrow . \ldots \rightarrow . C_m \rightarrow . B \rightarrow F$

where F is $D_1 \rightarrow . \ldots \rightarrow . D_n \rightarrow E$. Also in $E_{\rightarrow}$ we have

$E_{\rightarrow}1.\quad A \rightarrow B \rightarrow . A \rightarrow B,$

whence (S*) follows from $P_R(S^*)$ and $E_{\rightarrow}1$ by *generalized transitivity* (§4.2).

2.32. Where the antecedent of the *pc* is of the form $G \rightarrow H$ we have

$$\frac{P_L(S)\ A_1, \ldots, A_k \vdash G \rightarrow H \qquad P_R(S)\quad C_1, \ldots, C_m, B, D_1, \ldots, D_n \vdash E}{(S)\ \mu([A_1, \ldots, A_n], [C_1, \ldots, C_n], G \rightarrow H \rightarrow B), D_1, \ldots, D_n \vdash E}(\rightarrow \vdash)$$

By the hypothesis of the induction we have as theorems of $E_{\rightarrow}$

$P_L(S^*)\quad A_1 \rightarrow . \ldots \rightarrow . A_n \rightarrow . G \rightarrow H$, and

$P_R(S^*)\quad C_1 \rightarrow . \ldots \rightarrow . C_n \rightarrow . B \rightarrow F$, where F is as for *2.31*. From $\mathrm{P_R}(S^*)$ and *identity*

$$G\to H\to B\to .G\to H\to B,$$

we have by *generalized transitivity*

$$C_1\to . \; \ldots \to .C_{i-1}\to .G\to H\to B\to .C_i. \; \ldots \to .C_n\to .G\to H\to F,$$

which, after use of *generalized restricted permutation* (§4.2) on $G\to H$, together with $\mathrm{P_L}(S^*)$ yields (S^*) by *generalized transitivity.*

This completes the proof of the equivalence of $\mathrm{L}\mu\mathbf{E}_\to$ and $\mathbf{E}_\to$.

We mention also an equivalence relation between $\mathrm{L}\mu\mathbf{E}_\to$ and the subproof formulation $\mathbf{FE}_\to$. Let a line k of a proof in $\mathbf{FE}_\to$ be translated as follows into a consecution S_k in $\mathrm{L}\mu\mathbf{E}_\to$: $\mathrm{S}_k = H_1, \ldots, H_n \vdash A$, where (1) A is the formula residing at line k of the proof, and (2) $H_1, \ldots, H_n$ is the list, in order of increasing rank (i.e., from outside to in — see §8.1) of hypotheses bearing a subscript contained in the set of relevance subscripts with which A is marked on line k. Then the translation S_k of every line k of every proof in $\mathbf{E}_\to$ is a provable consecution of $\mathrm{L}\mu\mathbf{E}_\to$; and if a consecution S is provable in $\mathrm{L}\mu\mathbf{E}_\to$, then it is the translation of some line of some proof in $\mathbf{FE}_\to$. Similar equivalences hold between other merge formulations and their matching subproof formulations

Oddly enough, though these merge systems prove illuminating and helpful in various ways, it is not yet known how to base a decision procedure upon them — except, of course, in the absence of the rule (W $\vdash$) of contraction. The difficulty is that in the absence of permutation one cannot use what Kleene 1952 calls "cognation classes" in order to convert to finitude an apparent infinity. (See §13.3.) This lacuna is not crucial in the case of most of the merge formulations, since the formulations of Kripke 1959a (§13) *do* lead to decision procedures; however, for the calculus $\mathbf{T}_\to$ which allows no permutation at all ("ticket restriction") there is to date *only* a merge formulation, so that it would seem that finding a way to base a decision procedure upon the merge formulation $\mathrm{L}\mu\mathbf{T}_\to$ offers the readiest route to a decision procedure for that calculus.

PROBLEM. Can decision procedures be based upon the merge formulations?

PROBLEM. Is $\mathbf{T}_\to$ decidable?

§8. Miscellany. In this section, and in others like it at the end of subsequent chapters, we will mention topics of a more specialized nature, including both open problems and minor results, having to do with alternative formulations of systems, independence proofs, special bits of information about particular systems, and the like. Some of the open problems are relatively trivial, not in the sense that they are particularly easy, but rather in the sense that we do not much care whether answers are affirmative or negative; e.g., the questions asked in §8.5. Some are more serious in view of the philosophical stance we have been trying to maintain, such as the question concerning the admissibility of the disjunctive syllogism, which was open at the time we began to write this book; see §25.2, where, to our immense relief, Meyer and Dunn give a satisfactory answer.

At any rate we do not think of the Miscellany sections as part of the connected argument of the book; the last section of each chapter can be skipped without loss of continuity.

§8.1. An analysis of subordinate proofs. Our descriptions of the systems **FH**$_{\rightarrow}$, **FS4**$_{\rightarrow}$, **FR**$_{\rightarrow}$, **FE**$_{\rightarrow}$, and **FT**$_{\rightarrow}$, have been somewhat informal, and were designed to provide natural and easy proof procedures for the various systems under consideration. At this point we pause to mention that an alternative description of Fitch's 1952 subordinate proof format may be given, which relates subproofs to mathematical induction almost as transparently as the usual Hilbert-formulations do.

In order to motivate what follows we notice that with each i-th step A_i in the proofs of §1.3 there is associated first a number of vertical lines to the left of A_i (which we shall call the *rank* of A_i); second, a class of formulas (including A_i) which are candidates for application of the rule of *repetition* to yield a next step for the proof; third, a class of candidates for application of the rule of *reiteration*, to yield the next step; and fourth, a *hypothesis* (if the proof has one) which may together with the final step of the deduction furnish an entailment as next step, as a consequence of the deduction. Accordingly we define a *proof* as consisting of a sequence $A_1, \ldots, A_n$ of formulas, not necessarily distinct, for each A_i of which is defined a *rank* $\mathrm{R}(A_i)$, a class of *repeatable formulas* $\mathrm{Rep}(A_i)$, a class of *reiteratable formulas* $\mathrm{Reit}(A_i)$, and (if $\mathrm{R}(A_i) > 0$) an *immediate hypothesis* $\mathrm{H}(A_i)$. These are all defined by simultaneous induction as follows (the basis case being included only for heuristic reasons):

Basis (hyp):
(a) A_1 is any formula.
(b) $\mathrm{R}(A_1) = 1$.
(c) $\mathrm{Rep}(A_1) = \{A_1\}$.
(d) $\mathrm{Reit}(A_1) = \emptyset$ (the empty set).
(e) $\mathrm{H}(A_1) = A_1$.

Now suppose we have A_j, $\mathrm{R}(A_j)$, $\mathrm{Rep}(A_j)$, $\mathrm{Reit}(A_j)$, and $\mathrm{H}(A_j)$, for every $j < k$. Then $A_1, \ldots, A_k$ is a proof provided $A_1, \ldots, A_{k-1}$ is a proof and A_k satisfies one of the following five conditions:

1 (hyp)
(a) A_k is any formula.
(b) $\mathrm{R}(A_k) = \mathrm{R}(A_{k-1})+1$.
(c) $\mathrm{Rep}(A_k) = \{A_k\}$.
(d) $\mathrm{Reit}(A_k) = \mathrm{Rep}(A_{k-1})$.
(e) $\mathrm{H}(A_k) = A_k$.

2 (rep) If A_j is in $\mathrm{Rep}(A_{k-1})$, then
(a) $A_k = A_j$.
(b) $\mathrm{R}(A_k) = \mathrm{R}(A_{k-1})$.
(c) $\mathrm{Rep}(A_k) = \mathrm{Rep}(A_{k-1}) \cup \{A_k\}$.
(d) $\mathrm{Reit}(A_k) = \mathrm{Reit}(A_{k-1})$.
(e) $\mathrm{H}(A_k) = \mathrm{H}(A_{k-1})$ (if $\mathrm{H}(A_{k-1})$ is defined; otherwise $\mathrm{H}(A_k)$ is undefined).

3 (reit) If A_j is in $\mathrm{Reit}(A_{k-1})$, then
(a) $A_k = A_j$.
(b)–(e) as in 2.

4 ($\rightarrow$E) If A_j and $A_j \rightarrow B$ ($= A_i$, $i < k$) are in $\mathrm{Rep}(A_{k-1})$, then
(a) $A_k = B$.
(b)–(e) as in (2).

5 ($\rightarrow$I)
(a) $A_k = \mathrm{H}(A_{k-1}) \rightarrow A_{k-1}$.
(b) $\mathrm{R}(A_k) = \mathrm{R}(A_{k-1})-1$.
(c) $\mathrm{Rep}(A_k) = \mathrm{Reit}(A_{k-1}) \cup \{A_k\}$.
(d) If $\mathrm{R}(A_k) > 0$, then $\mathrm{Reit}(A_k) = \mathrm{Reit}(A_j)$, where $\mathrm{H}(A_{k-1}) = A_{j+1}$; and if $\mathrm{R}(A_k) = 0$, $\mathrm{Reit}(A_k) = \emptyset$.
(e) If $\mathrm{R}(A_k) > 0$ then $\mathrm{H}(A_k) = \mathrm{H}(A_j)$, where $\mathrm{H}(A_{k-1}) = A_{j+1}$; and if $\mathrm{R}(A_k) = 0$, $\mathrm{H}(A_k)$ is undefined.

Then A is a *theorem* if there is a proof in which A has rank zero. (Notice that this formulation differs slightly from that of the rules as stated at the beginning of §4. There we required that the subscript on a new hypothesis simply be new to the proof; here instead we identify k, on a hypothesis $A_{\{k\}}$, with the number of vertical lines to the left of $A_{\{k\}}$, i.e., with its rank. The change is clearly inessential and simplifies the bookkeeping.)

The basis clause and rule 1 enable us to begin new deductions, and 2–5 correspond to the other four of the five rules for $\mathbf{H}_{\rightarrow}$. (Note that "$\cup\{A_k\}$" is redundant in 2(c), but not in 3(c) or 4(c).) If we use a sequence of n vertical strokes to represent a rank of n, we may arrange proofs as follows:

	$R(A_i)$	Step	Rule	$Rep(A_i)$	$Reit(A_i)$	$H(A_i)$
A_1:	|	$A \to B$	hyp	$\{A_1\}$	$\emptyset$	A_1
A_2:	| |	$B \to C$	hyp	$\{A_2\}$	$\{A_1\}$	A_2
A_3:	| |	$A \to B$	A_1 reit	$\{A_{2-3}\}$	$\{A_1\}$	A_2
A_4:	| | |	A	hyp	$\{A_4\}$	$\{A_{2-3}\}$	A_4
A_5:	| | |	$A \to B$	A_3 reit	$\{A_{4-5}\}$	$\{A_{2-3}\}$	A_4
A_6:	| | |	B	$A_{4-5} \to$E	$\{A_{4-6}\}$	$\{A_{2-3}\}$	A_4
A_7:	| | |	$B \to C$	A_2 reit	$\{A_{4-7}\}$	$\{A_{2-3}\}$	A_4
A_8:	| | |	C	$A_{6-7} \to$E	$\{A_{4-8}\}$	$\{A_{2-3}\}$	A_4
A_9:	| |	$A \to C$	$A_{4-8} \to$I	$\{A_{2-3}, A_9\}$	$\{A_1\}$	A_2
A_{10}:	|	$B \to C \to .A \to C$	$A_{2-9} \to$I	$\{A_1, A_{10}\}$	$\emptyset$	A_1
A_{11}:		$A \to B \to .B \to C \to .A \to C$	$A_{1-10} \to$I	$\{A_{11}\}$	$\emptyset$	—

Then if we connect the lines indicating rank, we have a format that looks much like the proof of the law of transitivity given in §1.3 (provided we disregard all that junk in the three columns on the right).

To obtain a similar description of **FS4**$_\to$, we add to the "if" clause of 3 the requirement that A_j be of the form $B \to C$.

For **FR**$_\to$ we add the following clauses defining a set of "relevance indices" $Rel(A_i)$ for each i-th step.

To Basis	add	(f) $Rel(A_1) = \{1\}$.
To (1)	add	(f) $Rel(A_k) = \{R(A_k)\}$.
To (2)	add	(f) $Rel(A_k) = Rel(A_j)$.
To (3)	add	(f) $Rel(A_k) = Rel(A_j)$.
To (4)	add	(f) $Rel(A_k) = Rel(A_j) \cup Rel(A_i)$.
To (5)	add	If $Rel(H(A_{k-1})) \subseteq Rel\ (A_{k-1})$, then (a)–(e) as before, and (f) $Rel(A_k) = Rel(A_{k-1}) - Rel(H(A_{k-1}))$.

For **FE**$_\to$ we combine the restriction on reiteration of **FS4**$_\to$ with the subscript conditions of **FR**$_\to$, and for **FT**$_\to$ we add to clause 4 of the rules for **FE**$_\to$ another condition, "and if $Rel(H(A_j)) \subseteq Rel(A_j)$."

The foregoing formulations make explicit the techniques involved in constructing subproofs. But making the procedure explicit leads to some loss as regards intuitive naturalness and obviousness, and we shall in the remainder of this book continue to use the rather less formal approach with which we began, relying on the reader to see that the whole matter could be discussed more rigorously, along the lines just indicated.

§8.2. Ackermann's "strengen Implikation" and the rule (δ). So far as is known to us, the first formulation of a calculus equivalent to the system we call **E**$_\to$ was furnished by Ackermann 1956. (See Chapter VIII.) Ackermann did not consider separately the pure implicational part of the systems $\mathbf{\Pi}'$ and $\mathbf{\Pi}''$ of that paper, but it turns out (§24.5) that the pure part of these systems is characterized by his pure implicational axioms and rules:

A1 $A \rightarrow A$
A2 $A \rightarrow B \rightarrow . B \rightarrow C \rightarrow . A \rightarrow C$
A3 $A \rightarrow B \rightarrow . C \rightarrow A \rightarrow . C \rightarrow B$
A4 $(A \rightarrow . A \rightarrow B) \rightarrow . A \rightarrow B$

Rule (α): $\rightarrow$E
Rule (δ): From $A \rightarrow . B \rightarrow C$ and B to infer $A \rightarrow C$

Unlike other formulations of $\mathbf{E}_{\rightarrow}$ which we will consider later, Ackermann's uses *two* primitive rules of inference; we therefore single it out for special consideration, leaving formulations with $\rightarrow$E as sole rule for treatment in §8.3.

As Ackermann points out, the rule (δ) is subject to a rather awkward condition: "Für die Gültigkeit von (δ) im Gegensatz zu (α) . . . ist wesentlich, dass B eine logische Identität ist. Die Regel (δ) könnte also nicht in dieser Form bestehen blieben, falls nichtlogische Zusatzaxiome neben [the axioms of his system] auftreten" (Ackermann 1956, pp. 119–120).

The reason for the requirement that in applications of (δ), the premiss B should be a logical identity (i.e., logical truth) is not far to seek. "Es sei A eine Formel, die aus den Zusatzaxiomen abgeleitet werden kann, aber keine logische Identität ist. Aus A und $A \rightarrow A \rightarrow . A \rightarrow A$ würde man nach (δ) $A \rightarrow A \rightarrow A$ erhalten . . . " (Ackermann 1956, pp. 125–126). This would mean that, adopting the definition of logical necessity in §4.3, if A were any theorem provable in some extension of $\mathbf{E}_{\rightarrow}$ which involved non-logical axioms, we would still, with the help of (δ), be able to prove that A was a logical necessity, "wass," as Ackermann adds, "naturlich nicht sein kann."

We would like to paraphrase Ackermann's distinction by rewriting rules (α) and (δ) as follows (where here, as often below, we use "$\vdash$" for provability; i.e., $\vdash A$ if A is provable in the system under consideration):

Rule (α'): if $A \rightarrow B$ is asserted, then from A to infer B,
Rule (δ'): if $A \rightarrow . B \rightarrow C$ is asserted, then from $\vdash B$ to infer $A \rightarrow C$.

Whether or not this paraphrase does justice to Ackermann's intent may be a moot point, but supposing that it does so, we move on to raise two objections to rule (δ) with Ackermann's qualifications, or, what we take to be the same thing, rule (δ'). The first objection is classical in spirit; the second is idiosyncratic with us.

Everyone demands that for such systems of logic as those thus far considered in this book, which are all fragments of the pure "implicational" part of the two-valued propositional calculus, the notion of proof be *effective* (see e.g., Church 1956, §07, especially pp. 53–54). But we can tell effectively whether an alleged application of rule (δ') actually is one, and hence whether an alleged proof actually is one, only if we can tell effectively whether or not $\vdash B$ in $\mathbf{E}_\rightarrow$. In fact this can be done (§13), but this is simply a piece of luck.

If on the other hand rule (δ') is involved in *defining* the notion of provability for $\mathbf{E}_\rightarrow$, we might appear to be involved in a circularity. No doubt the situation could be fixed, along lines indicated by Smiley 1959, fn. 17, but for reasons which will be apparent from considerations below, we think this an unnecessarily complicated way of approaching the problem.

Our second objection to rule (δ') lies in the conviction that a coherent logical theory requires a certain similarity between the formal logical apparatus and the informal logical talk in the context of which the formal bit is explained. We do not know how to describe, exactly or generally, the "certain similarity" of the preceding sentence, but an example might make clear what we have in mind.

It has been urged classically not only that if $\vdash A$, and $\vdash \bar{A} \vee B$, then $\vdash B$, but also that B follows from A and $\bar{A} \vee B$. It has also been explained classically in modal logic that if $\vdash A$, then $\vdash \Box A$; though *no one* (except perhaps the most unreconstructed of rationalists) has ever held that $\Box A$ *follows from* A. The general point has in some sense been clear all along: the fact that whenever we have $\vdash A_1, \ldots, \vdash A_n$ we also have $\vdash B$, does not justify the inference that B *follows* from (or is entailed by) $A_1, \ldots, A_n$.

The foregoing distinction is paralleled exactly by the difference between rules (α) and (δ). The former, in the guise of (α'), has an exactly parallel theorem in $\mathbf{E}_\rightarrow$.

$$A \rightarrow B \rightarrow . A \rightarrow B$$

"if $A \rightarrow B$, then if A then B." But corresponding to rule (δ) we have in $\mathbf{E}_\rightarrow$ neither

$$(A \rightarrow . B \rightarrow C) \rightarrow . B \rightarrow . A \rightarrow C$$

nor

$$B \rightarrow . (A \rightarrow . B \rightarrow C) \rightarrow . A \rightarrow C.$$

And here it becomes evident that (δ) is in effect a rule of permutation, such as collapses $\mathbf{E}_\rightarrow$ into $\mathbf{R}_\rightarrow$. If we don't really believe in permutation unrestricted, then we ought have neither an axiom nor a primitive rule to the effect that we do. If we accept (δ), we are implicitly saying we believe in permutation, though if we accept the theorems of $\mathbf{E}_\rightarrow$ as an analysis of entailment, some scruples or other are apparently preventing us from writing down our beliefs as theorems. Hence our complaint of incoherence as registered at the beginning of the second objection above; let's have the whole hog ($\mathbf{R}_{\rightarrow}$) or nothing ($\mathbf{T}_{\rightarrow}$). Actually, of course, what we recommend is a policy ($\mathbf{E}_{\rightarrow}$) which allows us to have the hog and eat it too, but this hope demands that we arrive at a formulation of $\mathbf{E}_{\rightarrow}$ which does not require rules we don't believe.

How to improve things? The answer for the pure calculus of entailment should be obvious from inspection of rule (δ'). What is wanted is an axiom to replace that rule, which gives us the same effect when B is a *logische Identität*, and all we need is to rely on the intent to have the turnstile and $\Box$ mean the same thing (apart from Grammatical Considerations), the latter in the formal notation, and the former in the language we are using. And if we replace (δ) by

A5 $(A \rightarrow .B \rightarrow C) \rightarrow . \Box B \rightarrow .A \rightarrow C,$

or by

A6 $\Box B \rightarrow .(A \rightarrow .B \rightarrow C) \rightarrow .A \rightarrow C$

we obtain the desired result. We therefore consider three formulations of Ackermann's system $\mathbf{\Pi}_{\rightarrow}$:

$\Pi_{\rightarrow 1}$: A1–A4, (α), (δ)
$\Pi_{\rightarrow 2}$: A1–A5, (α)
$\Pi_{\rightarrow 3}$: A1–A4, A6, (α)

$\mathbf{E}_{\rightarrow}$ contains all three of these, as is easily seen from §4, and the argument required to show that each of these contains $\mathbf{E}_{\rightarrow}$ (most of the details are implicit in the following section) will be left to the reader.

Upshot: (δ) can be dispensed with, and in what follows we will restrict attention to formulations having (α) as sole rule.

§8.3. Axiom-chopping. For some reason which we do not understand very clearly, it appears that inventing alternative axiomatizations of a system sheds some light on what the system really involves. For this reason we list below a number of alternative ways of splitting axioms while still hewing to the same set of theorems. (We drew the title of this section from terminology for this procedure introduced by Halmos 1962, p. 22.)

§8.3.1. Terminology for derived rules of inference. Proofs here and elsewhere run more smoothly if we allow ourselves to use derivable rules. Just how these will be used will vary from case to case, but we list here a few useful ones:

Suf (for "suffixing"): from $A \rightarrow B$ to infer $B \rightarrow C \rightarrow .A \rightarrow C$. This will be derivable in any system which has transitivity in the form $A \rightarrow B \rightarrow .B \rightarrow C \rightarrow .A \rightarrow C$ as a theorem.

Pre (for "prefixing"): from $A \rightarrow B$ to infer $C \rightarrow A \rightarrow .C \rightarrow B$. This will be derivable in any system which has $A \rightarrow B \rightarrow .C \rightarrow A \rightarrow .C \rightarrow B$.

Trans (for "transitivity"): from $A \rightarrow B$ and $B \rightarrow C$ to infer $A \rightarrow C$. This is derivable in any system which has transitivity in the form of either suffixing or prefixing.

Restr perm (for "restricted permutation"): from $A \rightarrow .B \rightarrow C \rightarrow D$ to infer $B \rightarrow C \rightarrow .A \rightarrow D$. Some of the systems below contain $(A \rightarrow .B \rightarrow C \rightarrow D) \rightarrow .B \rightarrow C \rightarrow .A \rightarrow D$ as an axiom, from which of course the rule follows immediately.

As it turns out, it is sometimes more convenient to prove one of these rules before proving the corresponding theorem; hence the introduction of the terminology. We shall without comment use these rules in any system which has the appropriate formulas mentioned above as axioms.

§8.3.2. Alternative formulations of $\mathbf{T}_{\rightarrow}$. The axioms given in §6, chosen so as to facilitate proof of the equivalence of $\mathbf{T}_{\rightarrow}$ and $\mathbf{FT}_{\rightarrow}$, were as follows:

$\mathbf{T}_{\rightarrow 1}$:	1	$A \rightarrow A$
	2	$A \rightarrow B \rightarrow .B \rightarrow C \rightarrow .A \rightarrow C$
	3	$A \rightarrow B \rightarrow .C \rightarrow A \rightarrow .C \rightarrow B$
	4	$(A \rightarrow .B \rightarrow C) \rightarrow .A \rightarrow B \rightarrow .A \rightarrow C$
	5	$A \rightarrow B \rightarrow .(A \rightarrow .B \rightarrow C) \rightarrow .A \rightarrow C$

1 together with 5 in the form $A \rightarrow A \rightarrow .(A \rightarrow .A \rightarrow B) \rightarrow .A \rightarrow B$ yields *contraction*, which can in fact replace both 4 and 5, showing that 4 is redundant in $\mathbf{T}_{\rightarrow 1}$, and yielding as another formulation:

$\mathbf{T}_{\rightarrow 2}$:	1	$A \rightarrow A$
	2	$A \rightarrow B \rightarrow .B \rightarrow C \rightarrow .A \rightarrow C$
	3	$A \rightarrow B \rightarrow .C \rightarrow A \rightarrow .C \rightarrow B$
	6	$(A \rightarrow .A \rightarrow B) \rightarrow .A \rightarrow B$

The two are equivalent, since from $\mathbf{T}_{\rightarrow 2}$ we get both forms of *self-distribution* (§4); on the major:

7	$((B \rightarrow C \rightarrow .A \rightarrow C) \rightarrow .A \rightarrow .A \rightarrow C) \rightarrow (A \rightarrow B \rightarrow .A \rightarrow .A \rightarrow C)$	2 suf
8	$(A \rightarrow .B \rightarrow C) \rightarrow ((B \rightarrow C \rightarrow .A \rightarrow C) \rightarrow .A \rightarrow .A \rightarrow C)$	2
9	$(A \rightarrow .B \rightarrow C) \rightarrow (A \rightarrow B \rightarrow .A \rightarrow .A \rightarrow C)$	7 8 trans
10	$(A \rightarrow B \rightarrow .A \rightarrow .A \rightarrow C) \rightarrow .A \rightarrow B \rightarrow .A \rightarrow C$	6 pre
4	$(A \rightarrow .B \rightarrow C) \rightarrow .A \rightarrow B \rightarrow .A \rightarrow C$	9 10 trans

And on the minor:

11	$(B \rightarrow C \rightarrow .A \rightarrow C) \rightarrow .(A \rightarrow .B \rightarrow C) \rightarrow .A \rightarrow .A \rightarrow C$	3
12	$A \rightarrow B \rightarrow .(A \rightarrow .B \rightarrow C) \rightarrow .A \rightarrow .A \rightarrow C$	2 11 trans
13	$((A \rightarrow .B \rightarrow C) \rightarrow .A \rightarrow .A \rightarrow C) \rightarrow .(A \rightarrow .B \rightarrow C) \rightarrow .A \rightarrow C$	6 pre
5	$A \rightarrow B \rightarrow .(A \rightarrow .B \rightarrow C) \rightarrow .A \rightarrow C$	12 13 trans

§8.3.3. Alternative formulations of $\mathbf{E}_{\rightarrow}$. The axioms given in §4 for $\mathbf{E}_{\rightarrow}$ were again chosen to make proofs of the equivalence of $\mathbf{E}_{\rightarrow}$ and $\mathbf{FE}_{\rightarrow}$ easy, as follows:

$\mathbf{E}_{\rightarrow 1}$:	1	$A\rightarrow A$
	2	$A\rightarrow B\rightarrow.B\rightarrow C\rightarrow.A\rightarrow C$
	3	$A\rightarrow B\rightarrow.A\rightarrow B\rightarrow C\rightarrow C$
	4	$(A\rightarrow.B\rightarrow C)\rightarrow.A\rightarrow B\rightarrow.A\rightarrow C.$

From these we can easily deduce, by way of $\mathbf{FE}_{\rightarrow}$, the following:

5	$A\rightarrow A\rightarrow B\rightarrow B$
6	$(A\rightarrow.A\rightarrow B)\rightarrow.A\rightarrow B$
7	$A\rightarrow B\rightarrow.B\rightarrow C\rightarrow.A\rightarrow C\rightarrow D\rightarrow D$
8	$(A\rightarrow.B\rightarrow C\rightarrow D)\rightarrow.B\rightarrow C\rightarrow.A\rightarrow D$

The axioms 1–4 of $\mathbf{E}_{\rightarrow 1}$ are necessary and sufficient for $\mathbf{E}_{\rightarrow}$, but among the eight formulas above we can select several other groups which will do the same job. The equivalence of $\mathbf{E}_{\rightarrow}$ and $\mathbf{FE}_{\rightarrow}$ tells us that all of 1–8 are theorems of $\mathbf{E}_{\rightarrow}$; what we shall do now is to show that each of the following sets of axioms yields 1–4, and hence that all these formulations are equivalent.

$\mathbf{E}_{\rightarrow 2}$:	5	$A\rightarrow A\rightarrow B\rightarrow B$
	2	$A\rightarrow B\rightarrow.B\rightarrow C\rightarrow.A\rightarrow C$
	6	$(A\rightarrow.A\rightarrow B)\rightarrow.A\rightarrow B$

1 follows from these by the following maneuver:

9	$((A\rightarrow A\rightarrow.A\rightarrow A)\rightarrow.A\rightarrow A)\rightarrow.A\rightarrow A$	5
10	$(A\rightarrow A\rightarrow.A\rightarrow A)\rightarrow.A\rightarrow A$	5
1	$A\rightarrow A$	9 10 $\rightarrow$E

2 follows from 2. (Our apologies to Strawson for failing to "reason" at this step in the argument; see §1.3.) To continue we need the rule pre:

a	$A\rightarrow B$	premiss
b	$B\rightarrow B\rightarrow.A\rightarrow B$	a suf
c	$(A\rightarrow B\rightarrow.C\rightarrow B)\rightarrow.B\rightarrow B\rightarrow.C\rightarrow B$	b suf
d	$C\rightarrow A\rightarrow.B\rightarrow B\rightarrow.C\rightarrow B$	2 c trans
e	$C\rightarrow A\rightarrow.C\rightarrow B$	5 d trans

Then 3 follows:

11	$A\rightarrow B\rightarrow.B\rightarrow B\rightarrow.A\rightarrow B$	2
12	$(B\rightarrow B\rightarrow.A\rightarrow B)\rightarrow.A\rightarrow B\rightarrow C\rightarrow.B\rightarrow B\rightarrow C$	2
13	$A\rightarrow B\rightarrow.A\rightarrow B\rightarrow C\rightarrow.B\rightarrow B\rightarrow C$	11 12 trans
14	$(A\rightarrow B\rightarrow C\rightarrow.B\rightarrow B\rightarrow C)\rightarrow.A\rightarrow B\rightarrow C\rightarrow C$	5 pre
3	$A\rightarrow B\rightarrow.A\rightarrow B\rightarrow C\rightarrow C$	13 14 trans

4 then follows with the help of 6, as in §8.3.2.

(Storrs McCall pointed out to us that Wajsberg 1937 considers consequences of 5 and 2, and shows that prefixing follows from them; see McCall 1967, p. 293.)

$E_{\to 3}$: 1 $A \to A$
7 $A \to B \to .B \to C \to .A \to C \to D \to D$
6 $(A \to .A \to B) \to .A \to B$

1 follows from 1. 3 follows from 1 and 7 by identifying A and B. 2 seems to require the following lengthy argument:

5 $A \to A \to B \to B$ 1 3 $\to$E

trans

a	$A \to B$	premiss
b	$B \to C$	premiss
c	$B \to C \to \Box(A \to C)$	a 7 $\to$E
d	$\Box(A \to C)$	b c $\to$E
e	$A \to C$	d 5 $\to$E

suf

a	$A \to B$	premiss
b	$B \to C \to \Box(A \to C)$	a 7 $\to$E
c	$B \to C \to .A \to C$	b 5 trans

restr perm

a	$A \to .B \to C \to D$	premiss
b	$B \to C \to D \to D \to .A \to D$	a suf
c	$B \to C \to .A \to D$	b 3 trans

pre

a	$B \to C$	premiss
b	$B \to C \to .A \to B \to \Box(A \to C)$	7 restr perm
c	$A \to B \to \Box(A \to C)$	a b $\to$E
d	$A \to B \to .A \to C$	c 5 trans

Then we prove 2 as follows:

15	$A \to B \to .B \to C \to \Box(A \to C)$	7
16	$\Box(A \to C) \to .A \to C)$	5
17	$B \to C \to \Box(A \to C) \to .B \to C \to .A \to C$	16 pre
2	$A \to B \to .B \to C \to .A \to C$	15 17 trans

In the presence of restr perm, 2 yields prefixing, so we can finally prove 4 with the help of 6, as in §8.3.2.

$\mathbf{E}_{\rightarrow 4}$: 1 $A \rightarrow A$
2 $A \rightarrow B \rightarrow .B \rightarrow C \rightarrow .A \rightarrow C$
8 $(A \rightarrow .B \rightarrow C \rightarrow D) \rightarrow .B \rightarrow C \rightarrow .A \rightarrow D$
6 $(A \rightarrow .A \rightarrow B) \rightarrow .A \rightarrow B$

1 follows from 1 and 2 follows from 2. 3 follows from 1 in the form $A \rightarrow B \rightarrow C \rightarrow .A \rightarrow B \rightarrow C$ by restr perm, which 8 gives us. Also by restr perm from 2 we have prefixing, and all these combined with 6 yield 4, as in §8.3.2.

Though it is hard to speak about these matters without havering, it may be worth trying to get some intuitive grasp of the situation as reflected in these four axiomatizations.

$\mathbf{E}_{\rightarrow 4}$ probably separates most clearly the four chief principles of the pure calculus of entailment: identity, transitivity, restricted permutation, and contraction. In $\mathbf{E}_{\rightarrow 1}$, restricted permutation is reduced to a special case; i.e., 3 will do the work of 8 and is more convenient for comparison with $\mathbf{FE}_{\rightarrow}$. In $\mathbf{E}_{\rightarrow 2}$ identity and restr perm are in effect collapsed into a single formula: $A \rightarrow A \rightarrow B \rightarrow B$ could come by identity and restr perm from $A \rightarrow A \rightarrow B \rightarrow .A \rightarrow A \rightarrow B$; and it turns out, mysteriously, that $A \rightarrow A \rightarrow B \rightarrow B$ yields restr perm in the presence of the other axioms, and yields identity all by itself. In $\mathbf{E}_{\rightarrow 3}$ restr perm is collapsed with transitivity; somehow both of these manage to emerge from the rather odd-looking 7. We remark finally that the proofs seem to show that contraction and self-distribution go hand in glove, and are in some sense an isolable part of $\mathbf{E}_{\rightarrow}$ and $\mathbf{T}_{\rightarrow}$.

§8.3.4. Alternative formulations of $\mathbf{R}_{\rightarrow}$. The axioms of Church 1951 are as follows:

$\mathbf{R}_{\rightarrow 1}$: 1 $(A \rightarrow .A \rightarrow B) \rightarrow .A \rightarrow B$
2 $A \rightarrow B \rightarrow .C \rightarrow A \rightarrow .C \rightarrow B$
3 $(A \rightarrow .B \rightarrow C) \rightarrow .B \rightarrow .A \rightarrow C$
4 $A \rightarrow A$

These give

5 $A \rightarrow .A \rightarrow B \rightarrow B$

from 4 in the form $A \rightarrow B \rightarrow .A \rightarrow B$, and 3; also

6 $A \rightarrow B \rightarrow .B \rightarrow C \rightarrow .A \rightarrow C$

from 2, using 3 again.

$\mathbf{R}_{\rightarrow 2}$: 1 $(A \rightarrow .A \rightarrow B) \rightarrow .A \rightarrow B$
6 $A \rightarrow B \rightarrow .B \rightarrow C \rightarrow .A \rightarrow C$
5 $A \rightarrow .A \rightarrow B \rightarrow B$
4 $A \rightarrow A$

From these we get 2 and 3 as follows:

7	$(B\to C\to C\to .A\to C)\to .B\to .A\to C$	5 suf
8	$(A\to .B\to C)\to .B\to C\to C\to .A\to C$	6
3	$(A\to .B\to C)\to .B\to .A\to C$	7 8 trans

And in the presence of 3, 6 yields 2, so the two formulations are equivalent. And a third formulation $\mathbf{R}_{\to 3}$ could obviously be obtained by replacing 2 by 6 in $\mathbf{R}_{\to 1}$. We note that $\mathbf{R}_{\to 1}$ seems to be related to $\mathbf{R}_{\to 2}$ in much the way that $\mathbf{E}_{\to 4}$ is related to $\mathbf{E}_{\to 1}$, as discussed at the end of §8.3.3: assertion 5 does the work of (unrestricted) permutation 3 for $\mathbf{R}_{\to}$, just as restricted assertion does the work of restricted permutation for $\mathbf{E}_{\to}$. And clearly contraction could be replaced by self-distribution in $\mathbf{R}_{\to 1}$, or $\mathbf{R}_{\to 2}$.

Sobociński 1952 provides, to our astonishment, another formulation $\mathbf{R}_{\to 4}$, in which 2 replaces 6 in $\mathbf{R}_{\to 2}$ (for the source of the surprise, see §8.6): 3 follows from 5 and 2 alone, in the following amusingly self-convoluting way:

9	$B\to C\to C\to .(A\to .B\to C)\to .A\to C$	2
10	$(B\to .B\to C\to C)\to .B\to .(A\to .B\to C)\to .A\to C$	9 pre
11	$B\to .(A\to .B\to C)\to .A\to C$	5 10 →E
12	$(B\to .(A\to .B\to C)\to .A\to C)\to$ $.((A\to .B\to C)\to .(B\to .(A\to .B\to C)\to .A\to C)\to$ $.B\to .A\to C)\to .(A\to .B\to C)\to .B\to .A\to C$	11

And now the antecedent, and the antecedent of the consequent are both forms of 11, from which 3 emerges as a consequence. (Whew!)

The axioms $\mathbf{R}_{\to 1}$ are those of Church 1951, and the axioms $\mathbf{R}_{\to 2}$ are in Moh 1950. We are indebted to Arthur Prior for pointing out that the germ of the equivalence proof can be found in Łukasiewicz 1929, pp. 42–47.

We add that, although we appreciate (we think) the reasons Halmos 1962 had for supplying us with the down-putting title for this subsection, we disagree (perhaps) with him in our belief that one gets important insights by axiom-chopping (or by learning to add, for that matter). Once one has done axiom-chopping for himself, it is easy to be as lofty about the matter as Artur Rubinstein has been about five-finger exercises; but (plea to readers) actually checking such results is instructive (and usually uncovers typographical errors). The enterprise somehow ties in with the remarks in §7.1.

§8.4. Independence. Under this head we make first some elementary remarks about how independence results are secured, leaving until later particular independence results having to do with the systems we consider in this volume.

In dealing with a postulational system S, it is usual to mean by "a proposition A is independent of S" that the postulates of S do not permit a proof of A. A classic example is provided by the proof of the independence of the parallel postulate in Euclidean geometry. We can get a somewhat sharper sense of the notion of independence if we use *the logistic method* (described e.g., in Church 1956, pp. 47ff.) as we have attempted to do, with certain concessions to the habits of ordinary casual philosophical talk, in this book. Inherent in the logistic method is the idea that we specify the axioms of a postulational system, and also the allowable rules of inference, in such a way that the notion of *proof* is effective, i.e., one can always tell by inspection alone whether an arbitrary finite sequence of formulas is a proof. The use of this method involves many details which are not appropriate for discussion here, but it might be worthwhile to state informally a general principle lying behind independence proofs, to discuss its importance, and to give an example of its use, drawn from preceding pages.

PRINCIPLE. Suppose that: (1) all the axioms of a logistic theory S have a certain property P; (2) All the rules of S are such that if their premisses each have P, then so do their conclusions. (3) a certain formula A does not have the property P.

Then we may conclude that A is not a theorem of S, i.e., "the postulates of S do not permit a proof of A," since everything deducible in S has P, by (1) and (2) — but A does not have P, by (3).

The property P may be, of course, anything one likes; so long as (1)–(3) are established, we know that A is independent of S. In the applications we will use here, the property P is generally that of satisfying some matrix or other (see examples below, and also, for a history of this method, the preface to Łukasiewicz 1929). But before turning to these, we comment briefly on the significance of independence results.

Sometimes these are of the highest importance in the philosophy of mathematics. First, many of the important questions in the foundations of mathematics since 1900 have been sparked by the discovery of paradoxes in naive set theory. Second, many systematic reconstructions of this theory, designed to eliminate the paradoxes, have had as part of their underlying propositional logic a principle to the effect that "a contradiction implies anything"; i.e., if $\vdash A$ and $\vdash \sim A$, then $\vdash B$. The consistency of classical mathematics would accordingly be secured if we could show that *just one* formula B of the system were independent, or unprovable (since if we knew that B were not, then we would know that for no A both A and $\sim A$ were). It follows from these facts that *any* independence result in *any* standard system of set-theory would lay to rest an important ghost.

Sometimes these independence results have purely aesthetic interest. One might wish to know, for no other reason than that it would be fun to figure it out, whether $A \rightarrow A$ is independent of *specialized assertion* (§4) and →E (it is not, as we saw in §8.3.3, under $\mathbf{E}_{\rightarrow 2}$). Many of the results under §§8.3.1–8.3.3 have no more interest than this.

Sometimes they can be of help philosophically, and here we would like to draw on the theorem of Ackermann in §5, and give an example of how the property P (in this case, satisfaction of a matrix) might be put to some philosophical use.

If one takes seriously the ideas of §5.2 that (a) entailments are *sui generis*, (b) they should therefore not be entailed by pure non-necessitives, (c) pure non-necessitiveness is carried by propositional variables, and (d) there is some point in trying to say all this precisely enough to allow a proof or disproof of what one has said, then surely one would like to be certain that no propositional variables entail entailments (in the formal theory).

And this is what is shown by taking the property P to be: satisfying the matrix (*'d values, here and hereafter, are designated)

→	*0*	*1*	*2*
0	*2*	*2*	*2*
1	*0*	*2*	*0*
**2*	*0*	*0*	*2*

We say that a formula *satisfies a matrix* if every assignment of values to the variables in it yields, on computation, a designated value. (A linguistic remark: it would be more faithful to Tarski 1933 to speak of a matrix satisfying a formula rather than, as we have it, a formula satisfying a matrix. But to keep to this mode of speech would force us to tiresome passive constructions, etc., and to equally tiresome efforts to remember which should satisfy which. We avoid these annoyances by treating the satisfaction relation as symmetric, agreeing that "formula A satisfies matrix M" is altogether interchangeable with "matrix M satisfies formula A.") Designated values are pre-selected "good guys," or "true-ish values," which we indicate by an asterisk, as in the example above. The computations are carried out on the analogy of (say) a multiplication table, which may be thought of as a generalized truth table. Part of the work of showing, e.g., that transitivity satisfied the matrix above would involve computing its value when $A = \mathit{0}$, $B = \mathit{1}$, and $C = \mathit{2}$, as follows:

$A \rightarrow B \rightarrow .B \rightarrow C \rightarrow .A \rightarrow C$
$\mathit{0} \rightarrow \mathit{1} \rightarrow .\mathit{1} \rightarrow \mathit{2} \rightarrow .\mathit{0} \rightarrow \mathit{2}$
$\mathit{2} \rightarrow .\mathit{0} \rightarrow \mathit{2}$
$\mathit{2} \rightarrow \mathit{2}$
$\mathit{2}$

It is of course also necessary to verify that the rules of inference preserve the property of satisfying the matrix, so that in the systems under consideration (which have →E as sole rule) we must also be sure that the rule preserves "validity" in the matrix. For this it is sufficient to show that if A and $A \to B$ take *only* designated values, then B takes only designated values. (We note that a rule may also have the stronger property of preserving not only validity ("*always*-designated") but designation: whenever A and $A \to B$ take a designated value, so does B. This is most usual, as in the following case.) Suppose then that B takes an undesignated value (to wit, $\mathit{0}$ or $\mathit{1}$). Then inspection of the matrix convinces us quickly that under these circumstances, if $A \to B$ takes the value $\mathit{2}$, then A must be either $\mathit{0}$ or $\mathit{1}$, and if A takes the value $\mathit{2}$, then $A \to B$ must be $\mathit{0}$; i.e., the rule leads from valid formulas (under the matrix) only to valid formulas.

We note then that:

(1) The axioms of $\mathbf{E}_{\to}1 - E_{\to}4$ of §8.3.3 all satisfy the matrix in the sense explained above;

(2) if A and $A \to B$ both satisfy the matrix, so does B; and

(3) if p is a variable, $p \to .B \to C$ does *not* satisfy the matrix: since p is a variable, p can assume the value $\mathit{1}$, whereas $B \to C$ must assume either $\mathit{0}$ or $\mathit{2}$, in either case giving the whole formula the undesignated value $\mathit{0}$.

Hence, in view of the Principle above, $p \to .B \to C$ is not provable in $\mathbf{E}_{\to}$, a fact which we may express by saying that $p \to .B \to C$ is independent of $\mathbf{E}_{\to}$, if p is a variable.

We shall also use the word "independent" in a second sense, common in the literature, which is derivative from that just explained. A set of axioms will be said to be an *independent set*, relative to a set of rules, if each is independent (in the first sense) of all the others. In the sections immediately following, we consider various independent sets of axioms, but before going on we make one observation, and state three theorems, each of which will be useful in what follows.

Observation. As promised in §8.2, we restrict attention here to formulations having →E as sole rule. This rule is itself obviously "independent" in an appropriate sense for all the systems to be considered in this chapter, since otherwise the only theorems would have the form of the axiom schemata. But e.g., $A \to A \to B \to B$, though a theorem of $\mathbf{E}_{\to}$, is not an instance of any of the schemata $E_{\to}1$–$E_{\to}4$ of §4; hence →E is required to prove it. In what follows we always assume this fact tacitly. (In §§26.2 and 29.9, matrix proofs of the independence of →E are given.)

Diamond-McKinsey theorem. It is not possible to give a set of postulates for Boolean algebra where each postulate is a universal sentence containing only two variables (Diamond and McKinsey 1947).

It follows, with the help of a bit of argument connecting propositional calculus and Boolean algebra, that every formulation of the classical two-valued calculus must have at least one axiom containing at least three propositional variables. One such formulation is that of Łukasiewicz 1929:

$$A \to B \to . B \to C \to . A \to C$$
$$A \to . \sim A \to B$$
$$\sim A \to A \to A$$

Since all of the systems we consider here have the first of these as a theorem, and since addition of the second and third turns the arrow into material "implication," it follows that for any system S considered below, if S has only one axiom with three distinct variables, that axiom is independent of the others.

Jaśkowski theorem. In any system of propositional calculus having →E and substitution as only rules of inference, having none but tautologies (in two truth values) as theorems, and having $A \to B \to . B \to C \to . A \to C$ as a theorem, the axioms must include either one axiom at least eleven letters long or two axioms which are each nine letters long (Jaśkowski 1948; the statement is drawn from Church 1948).

Here a "letter" is either a variable, or an arrow, or a truth functional connective. Interpreting the arrow in the Łukasiewicz system described just above as material "implication," we see that Jaśkowski's theorem also guarantees the independence of the first axiom, which has eleven letters, the others having six each.

Lesniewski-Mihailescu theorem. In a system with *modus ponens* as sole rule, if only one axiom contains an odd number of occurrences of one variable, that axiom is independent of the others.

To prove this one needs only to note that if each variable occurs an even number of times in each axiom, →E preserves this property for theorems. This fact was observed by Leśniewski 1929 and Mihailescu 1938, in connection with the study of equivalential calculuses. (See Łukasiewicz 1939.)

§8.4.1. Matrices. We list below a number of matrices which will be used here and elsewhere to prove independence. No attempt will be made to credit these to their originators, for two reasons, the first being that by and large we do not know who the originators were; in some cases it was one of us, or one of our friends. But secondly, most of the matrices have an *ad hoc* flavor, since most of them were designed *simply* to show independence, and have no known independent interest. They come from a variety of writers, live for a brief moment in one or two papers, and then expire; hence there is no standard notation. For example, in a matrix with three elements ***0***, ***1***, ***2***, Church 1951 selects ***0***, or ***0*** and ***1***, as designated elements, whereas other writers sometimes use ***2***, or ***2*** and ***1***, as designated, and still others (e.g. Jaśkowski 1948) use letters rather than numerals to denote elements of a matrix. The historical task of sorting out the equivalent matrices in the literature would obviously be overwhelming, so we attempt to excuse ourselves by saying that though the historical facts may be of some interest, the task of unearthing them would probably be more work than they are worth. On the other hand we can offer one bit of historical information: John R. Chidgey was very helpful in simplifying some of our previous results, in finding errors therein, and in dreaming up some new matrices, especially VI and VII below; the details of §§8.4.3–8.4.4 follow his results.

Matrix I

→	0	1	2
0	2	2	2
1	0	0	2
*2	0	0	2

Matrix II

→	0	1	2
0	2	2	2
1	2	2	2
*2	0	2	2

Matrix III

→	0	1	2
0	2	2	2
1	0	2	2
*2	0	0	2

Matrix IV

→	0	1	2	3
0	3	3	3	3
1	0	2	0	3
*2	0	0	3	3
*3	0	0	0	3

Matrix V

→	0	1	2	3
0	3	3	3	3
1	0	3	3	0
*2	0	0	3	3
*3	0	0	0	3

Matrix VI

→	0	1	2	3	4
0	4	4	4	4	4
*1	0	4	4	4	4
*2	0	0	1	1	4
*3	0	0	0	3	4
*4	0	0	0	0	4

Matrix VII

→	0	1	2	3	4
0	4	4	4	4	4
*1	0	4	4	4	4
*2	0	0	4	1	4
*3	0	0	0	1	4
*4	0	0	0	0	4

In stating independence results below we mention in each case a general theorem from which independence follows, or else a matrix which shows the axiom independent. For notation (see the first example under $\mathbf{E}_{\rightarrow 1}$) we use:

1 $A \rightarrow A$; Matrix I: $A = \mathit{1}$, so 1 = $\mathbf{\mathit{0}}$.

This is to be understood as follows:

The axiom 1 of $\mathbf{E}_{\to 1}$ is $A \to A$; Matrix I satisfies the other axioms of $\mathbf{E}_{\to 1}$, detailed verification of that fact being left to the reader; but if in axiom 1 we give the variable A the value $\boldsymbol{1}$, then the axiom 1 assumes the undesignated value $\boldsymbol{0}$.

Note that we are using plain numerals (e.g. 1) to stand for formulas, and bold-face italic (e.g. $\boldsymbol{1}$) for matrix-elements, so there should be no confusion.

We have made so many matrix-based claims here and there in this book that it seemed to us good to verify them by computer. Accordingly we are grateful to the Provost's Development Fund of the University of Pittsburgh for supporting the creation of the computer program "TESTER." Using TESTER on the matrices as they occurred in the galley proofs, we uncovered a (not very surprising) number of errors by us, and none by the printers.

TESTER was designed by us in conjunction with Dale Isner, and programmed by Isner. For those who might be interested, we add a brief description of its facilities.

The chief feature of interest is that TESTER is made to be used *interactively* and *easily* from a remote terminal attached to a computer. The program is designed for the user who likes matrices but doesn't know anything about computers. TESTER guides the user on how to proceed, what his options are, and how to choose among them. Further, TESTER error-checks everything in sight, and the user is told immediately about his mistakes and how to repair them. Importantly, input-output, which can be such a bore, is minimized in various ways. And lastely, TESTER "saves" all data in permanent storage so that if one wants to return to a problem the next day or week, formulas and matrices do not have to be laboriously typed in again.

What can you do with TESTER? You can type in a formula-name and then define it as naming any formula with up to 50 characters. You can define a formula-set by entering a formula-set name and then either listing names of formulas or adding to or deleting from a previously defined formula-set. You can define a matrix (up to 10×10) by entering a matrix-name and then either (1) entering an appropriate table, (2) using a formula to define a new connective in terms of some already defined ones, or (3) tinkering with an already defined matrix. You can selectively print out formulas, formula-sets, and matrices. And of course you can use the central routine of TESTER, which tests a formula-set against a matrix and prints out any values of the matrix for which any formulas in the set are not designated.

TESTER is written in ANSI Standard FORTRAN so that TESTER can be run on virtually any computer equipped with time-sharing capabilities and sufficient core storage. (The current version of TESTER has been run on a DEC (Digital Equipment Corporation) PDP-10 utilizing 12K core storage.) Furthermore, TESTER has been written in a highly modular fashion so that it can be easily adapted to run on smaller computers or in batch mode. In these ways we have hoped to maximize its usefulness.

To obtain a copy of TESTER, contact Nuel Belnap, Department of Philosophy, University of Pittsburgh, Pittsburgh, PA 15260. Or if he is unavailable, try the University of Pittsburgh Computer Center.

§8.4.2. Independent axioms for $\mathbf{T}_{\to}$. We remarked in §8.3.2 (in effect) that axiom 4 of $\mathbf{T}_{\to 1}$ was not independent. The remaining ones are:

1 $A\to A$	Matrix I: $A = \mathbf{1}$, so $1 = \mathbf{0}$
2 $A\to B\to .B\to C\to .A\to C$	Matrix VI: $A = \mathbf{2}$, $B = C = \mathbf{3}$, so $2 = \mathbf{0}$
3 $A\to B\to .C\to A\to .C\to B$	Matrix VII: $A = C = \mathbf{2}$, $B = \mathbf{3}$, so $3 = \mathbf{0}$
5 $A\to B\to .(A\to .B\to C)\to .A\to C$	Leśniewski-Mihailescu theorem

$\mathbf{T}_{\to 2}$: The same examples show independence; 6 $(A\to .A\to B)\to .A\to B$ being independent by the Leśniewski-Mihailescu theorem.

§8.4.3. Independent axioms for $\mathbf{E}_{\to}$.

$\mathbf{E}_{\to 1}$: 1 $A\to A$;	Matrix I: $A = \mathbf{1}$, so $1 = \mathbf{0}$
2 $A\to B\to .B\to C\to .A\to C$;	Matrix V: $A = \mathbf{1}$, $B = \mathbf{2}$, and $C = \mathbf{3}$, so $2 = \mathbf{0}$.
3 $A\to B\to .A\to B\to C\to C$;	Matrix IV: $A = \mathbf{1}$, $B = \mathbf{1}$, and $C = \mathbf{2}$, so $3 = \mathbf{0}$.
4 $(A\to .B\to C)\to .A\to B\to .A\to C$;	Leśniewski-Mihailescu.
$\mathbf{E}_{\to 2}$: 5 $A\to A\to B\to B$;	Matrix I: $A = \mathbf{1}$, and $B = \mathbf{0}$, so $5 = \mathbf{0}$.
2 $A\to B\to .B\to C\to .A\to C$;	Matrix II: $A = \mathbf{2}$, $B = \mathbf{1}$, and $C = \mathbf{0}$, so $2 = \mathbf{0}$; or Jaśkowski or Diamond-McKinsey.
6 $(A\to .A\to B)\to .A\to B$;	Leśniewski-Mihailescu.
$\mathbf{E}_{\to 3}$: 1 $A\to A$;	Matrix I: $A = \mathbf{1}$, so $1 = \mathbf{0}$.
7 $A\to B\to .B\to C\to .A\to C\to D\to D$;	Matrix II: $A = \mathbf{2}$, $B = \mathbf{1}$, $C = \mathbf{0}$, and $D = \mathbf{0}$, so $7 = \mathbf{0}$; or Jaśkowski or Diamond-McKinsey.
6 $(A\to .A\to B)\to .A\to B$;	Leśniewski-Mihailescu.

$\mathbf{E}_{\to 4}$:	1 $A \to A$;	Matrix I: $A = 1$, so $1 = 0$.
	2 $A \to B \to .B \to C \to .A \to C$;	Matrix II: $A = 2$, $B = 1$, and $C = 0$, so $2 = 0$.
	8 $(A \to .B \to C \to D) \to .B \to C \to .A \to D$;	Matrix IV: $A = 3$ or $A = 1$, $B = 1$, $C = 1$, and $D = 2$, so $8 = 0$.
	6 $(A \to .A \to B) \to .A \to B$;	Leśniewski-Mihailescu.

§8.4.4. Independent axioms for $\mathbf{R}_{\to}$.

$\mathbf{R}_{\to 1}$:	1 $(A \to .A \to B) \to .A \to B$;	Leśniewski-Mihailescu.
	2 $A \to B \to .C \to A \to .C \to B$;	Matrix II: $A = 1$, $B = 0$, and $C = 2$, so $2 = 0$.
	3 $(A \to .B \to C) \to .B \to .A \to C$;	Matrix III: $A = 2$, $B = 1$, and $C = 1$, so $3 = 0$.
	4 $A \to A$;	Matrix I: $A = 1$, so $4 = 0$.
$\mathbf{R}_{\to 2}$:	1 $(A \to .A \to B) \to .A \to B$;	Leśniewski-Mihailescu.
	6 $A \to B \to .B \to C \to .A \to C$;	Matrix II: $A = 2$, $B = 1$, and $C = 0$, so $6 = 0$; or Jaśkowski or Diamond-McKinsey.
	5 $A \to .A \to B \to B$;	Matrix III: $A = 1$, and $B = 1$, so $5 = 0$.
	4 $A \to A$;	Matrix I: $A = 1$, so $4 = 0$.
$\mathbf{R}_{\to 3}$:	1 $(A \to .A \to B) \to .A \to B$;	Leśniewski-Mihailescu.
	6 $A \to B \to .B \to C \to .A \to C$;	Matrix II: $A = 2$, $B = 1$, and $C = 0$, so $6 = 0$.
	3 $(A \to .B \to C) \to .B \to .A \to C$;	Matrix III: $A = 2$, $B = 1$, and $C = 1$, so $3 = 0$.
	4 $A \to A$;	Matrix I: $A = 1$, so $4 = 0$.
$\mathbf{R}_{\to 4}$:	1 $(A \to .A \to B) \to .A \to B$;	Leśniewski-Mihailescu.
	2 $A \to B \to .C \to A \to .C \to B$;	Matrix II: $A = 1$, $B = 0$ and $C = 2$, so $2 = 0$; or Jaśkowski or Diamond-McKinsey.
	5 $A \to .A \to B \to B$;	Matrix III: $A = 1$, and $B = 1$, so $5 = 0$.
	4 $A \to A$;	Matrix I: $A = 1$, so $4 = 0$.

§8.5. Single-axiom formulations. It is well known that for the classical two-valued propositional calculus there are formulations which have $\to$E as sole rule and only one axiom (see e.g. Łukasiewicz 1948).

§8.5.1. Problem. With →E as sole rule, are there single axioms for $\mathbf{H}_{\rightarrow}$, $\mathbf{S4}_{\rightarrow}$, $\mathbf{R}_{\rightarrow}$, $\mathbf{E}_{\rightarrow}$, or $\mathbf{T}_{\rightarrow}$? The most substantial step toward a positive solution for $\mathbf{E}_{\rightarrow}$ is taken in Meredith and Prior 1963, where it is shown that the axioms

$$A\rightarrow A\rightarrow B\rightarrow B$$
$$A\rightarrow B\rightarrow .B\rightarrow C\rightarrow .A\rightarrow C$$

of $\mathbf{E}_{\rightarrow}$ may be collapsed into any one of the following:

$$(A\rightarrow A\rightarrow B\rightarrow B\rightarrow .C\rightarrow .D\rightarrow E)\rightarrow .F\rightarrow D\rightarrow .C\rightarrow .F\rightarrow E$$
$$F\rightarrow D\rightarrow .(A\rightarrow A\rightarrow B\rightarrow B\rightarrow .C\rightarrow .D\rightarrow E)\rightarrow .C\rightarrow .F\rightarrow E$$
$$A\rightarrow B\rightarrow .(D\rightarrow D\rightarrow E\rightarrow E\rightarrow .B\rightarrow C)\rightarrow .A\rightarrow C$$
$$(A\rightarrow .B\rightarrow B\rightarrow .C\rightarrow D\rightarrow E)\rightarrow .F\rightarrow D\rightarrow .C\rightarrow F\rightarrow .A\rightarrow E$$
$$(A\rightarrow .B\rightarrow B\rightarrow .C\rightarrow D\rightarrow E)\rightarrow .C\rightarrow F\rightarrow .F\rightarrow D\rightarrow .A\rightarrow E$$

We seize the occasion to make some remarks about these last five axioms.

(1) Following Leśniewski, we use the term "organic" to describe a theorem A of a system **S** which has the property that no well-formed proper part of A is a theorem. (The definition is due to Wajsberg; see Łukasiewicz and Tarski 1930.) Observe that none of the axioms above is organic.

(2) The first two are restricted permutations of each other; and the fourth and fifth likewise (in the consequent).

(3) The third, while shortest and containing fewest variables, is, like the first two, doubly inorganic; all three contain two theorems of $\mathbf{E}_{\rightarrow}$.

Problem. Are there any organic axioms which do the same job? (For motivation of this enterprise see Sobociński 1956.)

Questions of this kind are delicate, and may well remind one of problems concerning the Knight's tour in Falkener 1892 (lovely puzzles, which shed no light on either chess or number theory). But they can also be stimulating, and produce the following sort of proof, which is the first of its kind we know.

§8.5.2. Solution for $\mathbf{T}_{\rightarrow}$ (by Zane Parks).

Lemma. If A is a theorem of $\mathbf{T}_{\rightarrow}$, then either A is a theorem of the Łukasiewicz three-valued logic, i.e. A satisfies the matrix **I**:

→	0	1	2
0	2	2	2
1	1	2	2
*2	0	1	2

or A satisfies the matrix II:

$\to$	*0*	*1*	*2*	*3*
0	*3*	*3*	*3*	*3*
1	*0*	*2*	*0*	*3*
2	*0*	*0*	*3*	*3*
**3*	*0*	*0*	*0*	*3*

PROOF. We shall call an identity formula $B\to B$ *atomic* if B is a propositional variable. The usual inductive argument shows that if A has a proof in $\mathbf{T}_{\to 2}$ (§8.3.2) in which the only identities that are used are non-atomic, then A satisfies matrix II. Suppose that A is a theorem of $\mathbf{T}_{\to 2}$ but has no proof in $\mathbf{T}_{\to 2}$ in which the only identities that are used are non-atomic. If A is itself an atomic identity, A satisfies matrix I. If A is not an atomic identity, then there is a proof of A in $\mathbf{T}_{\to 2}$, in which each of the atomic identities used is used as a minor premiss of an application of $\to$E. Let $p_1\to p_1, \ldots, p_m\to p_m$ be a complete list of the atomic identities occurring in this proof. We show that A satisfies matrix II by showing that if the k-th line of the proof is not one of the $p_i\to p_i$ then it satisfies matrix II under the assumption that for all $n < k$, if the n-th line is not one of the $p_i\to p_i$, then the n-th line satisfies matrix II. The only interesting case is the one where the k-th line (say) B is an $\to$E consequence of $p_j\to p_j\to B$ and $p_j\to p_j$. By the inductive hypothesis, $p_j\to p_j\to B$ satisfies matrix II. If $B = C\to D$ where C and D are propositional variables, then either $C = D = p_j$ and so the k-th line is one of the $p_i\to p_i$, or $p_j\to p_j\to B$ is not a theorem of $\mathbf{T}_{\to 2}$. If B has the form $C\to D\to E$ or $C\to .D\to E$, then that B satisfies matrix II follows from the fact that $p_j\to p_j\to B$ satisfies matrix II, since each of the rules

$$\frac{A\to A\to .B\to C\to D}{B\to C\to D} \qquad \frac{A\to A\to .B\to .C\to D}{B\to .C\to D}$$

preserves the property of satisfying matrix II.

THEOREM. There is no single axiom formulation of $\mathbf{T}_{\to}$ with $\to$E as sole rule.

PROOF. By the lemma, any theorem of $\mathbf{T}_{\to}$ satisfies matrix I or matrix II. But matrix I falsifies contraction and matrix II falsifies identity. Hence no theorem of $\mathbf{T}_{\to}$ has as $\to$E consequences both contraction and identity.

§8.6. Transitivity. Chidgey 1973 discusses a problem of some interest in relevance logics: what is the difference between prefixing and suffixing (see §4.2 and remarks under $\mathbf{E}_{\to 2}$ in §8.3.3)? Both yield transitivity *tout pur* as a rule, and of course in the presence of permutation they are equivalent.

And if we are allowed a restricted permutation axiom, again nothing is lost. Sobociński 1952 in fact shows that unrestricted assertion $A \rightarrow .A \rightarrow B \rightarrow B$ together with suffixing gives prefixing; details may be found under the talk about $\mathbf{R}_{\rightarrow 2}$ in §8.3.4.

Upshot: Chidgey shows that specialized assertion, prefixing, and contraction (§4.2) are insufficient for suffixing, using the matrix

$\rightarrow$	*0*	*1*	*2*
0	*1*	*1*	*2*
**1*	*0*	*1*	*2*
**2*	*0*	*0*	*2*

which satisfies the axioms mentioned above, but falsifies suffixing for $A = \mathit{0}$, $B = \mathit{2}$, $C = \mathit{0}$ (among other values).

It is also shown that there are contexts in which suffixing is insufficient for prefixing; e.g. if we have

$$A \rightarrow A$$

$$A \rightarrow B \rightarrow .B \rightarrow C \rightarrow .A \rightarrow C$$

and contraction, prefixing does not forthcome (as Chidgey shows by a matrix argument).

§8.7. Co-entailment. $\mathbf{E}_{\rightarrow}$ can be extended in a natural and obvious way to encompass the notion of "if and only if," i.e., co-entailment. The system $\mathbf{E}_{\rightleftarrows}$ arises if to $\mathbf{E}_{\rightarrow}$ we add $\rightleftarrows$ as primitive, with the axioms

$E_{\rightleftarrows}1$ $A \rightleftarrows B \rightarrow .A \rightarrow B$,
$E_{\rightleftarrows}2$ $A \rightleftarrows B \rightarrow .B \rightarrow A$, and
$E_{\rightleftarrows}3$ $A \rightarrow B \rightarrow .B \rightarrow A \rightarrow .A \rightleftarrows B$.

We have no permanent interest in $\mathbf{E}_{\rightleftarrows}$, since the effect of $\rightleftarrows$ can be got equally well in the full system **E** of §21 by defining

$$A \rightleftarrows B =_{\mathrm{df}} (A \rightarrow B) \& (B \rightarrow A).$$

But the natural character of $\mathbf{E}_{\rightleftarrows}$ does make it possible for us to make a philosophical point germane to the von Wright-Geach-Smiley criterion of entailment discussed under the heading of *fallacies of relevance* at the end of §5.1.1, and again at greater length in §20.1.

We notice first that prefixing B→A to $E_{\rightleftarrows}1$ leads to

$$(B \rightarrow A \rightarrow .A \rightleftarrows B) \rightarrow .B \rightarrow A \rightarrow .A \rightarrow B,$$

and that prefixing $A \rightarrow B$ to this formula yields, with the help of $\mathbf{E}_{\rightleftarrows}3$ and →E, the theorem

$$A \rightarrow B \rightarrow .B \rightarrow A \rightarrow .A \rightarrow B.$$

Exactly the same argument leads from $E_{\rightleftarrows}2$ (with the help of prefixing, $E_{\rightleftarrows}3$, and $\to$E) to the theorem

$$A\to B\to.B\to A\to.B\to A.$$

Upshot: In the presence of $E_{\rightleftarrows}1$–$E_{\rightleftarrows}3$, transitivity (in the form of prefixing), and $\to$E, the last two displayed formulas certainly stand or fall together.

What seems to us (a) surprising, then (b) odd, then (c) obviously unsatisfactory (temporal order in reflecting on the fact) is that the von Wright-Geach-Smiley condition appears to sanction $A\to B\to.B\to A\to.A\to B$ but not $A\to B\to.B\to A\to.B\to A$. Again we are led to the conclusion that the condition discussed in §20.1 is simply not coherent.

On the other hand we must admit that these two theorems of $\mathbf{E}_{\rightleftarrows}$ are (a) surprising and perhaps (b) odd. But they do have a rationale, as explained in §5.1.1, which is not shared by such obviously unsatisfactory formulas as

$$A\to B\to.A\to B\to.A\to B,$$

which is not a theorem of $\mathbf{E}_{\rightleftarrows}$, as can be seen from the matrix of §22.1.3, by setting $A = +\boldsymbol{0}$ and $B = -\boldsymbol{0}$.

A related oddity, pointed out by Roger White (and communicated to us by Geoffrey Hunter in 1965), is the following:

$$B\rightleftarrows A\to.A\rightleftarrows A.$$

This arises from $\mathbf{E}_{\rightleftarrows}3$,

$$A\to A\to.A\to A\to.A\rightleftarrows A,$$

by prefixing in the consequent (§4.2), getting

$$A\to A\to.(A\to B\to.B\to A\to.A\to A)\to.A\to B\to.B\to A\to.A\rightleftarrows A,$$

from which two uses of $\to$E give us

$$A\to B\to.B\to A\to.A\rightleftarrows A.$$

Replacement of the middle then gives us

$$A\to B\to.B\rightleftarrows A\to.A\rightleftarrows A,$$

and $E_{\rightleftarrows}2$ and transitivity give us

$$B\rightleftarrows A\to.B\rightleftarrows A\to.A\rightleftarrows A,$$

whence the oddity by contraction.

Symmetry of $\rightleftarrows$ (not hard to prove) also gives us

$$A\rightleftarrows B\to.A\rightleftarrows A,$$

which has the flavor of Peirce's "law": "if A co-entails *anything*, then it co-entails itself." But the flavor is specious, because in the case under discussion there is again a rationale: the (as it were) two facts contained in the antecedent (of the formula just above) spin at each other, and fire out the consequent. It *would* be disappointing if the entailment went in the other direction, but it does not; it is known (see e.g. Mihailescu 1938) that the classical "equivalential" calculus has the feature that every variable must, for provability, occur an even number of times: hence with $\equiv$ for material "equivalence," $A \equiv B \equiv . A \equiv A$ fails, and since $\rightleftarrows$ is a subrelation of $\equiv$, so does $A \rightleftarrows B \rightleftarrows . A \rightleftarrows A$.

§8.8. Antecedent and consequent parts. We owe to John Bacon the following observation.

If we rewrite formulas $(A \rightarrow B)$ as $\rightarrow AB$, in the manner of Łukasiewicz, so that e.g.

$$(A \rightarrow B) \rightarrow ((B \rightarrow C) \rightarrow (A \rightarrow C))$$

becomes

$$\rightarrow \rightarrow AB \rightarrow \rightarrow BC \rightarrow AC,$$

we can find an easy mechanical way of distinguishing antecedent and consequent parts in the sense described in §5.1.2. Writing "a" under a variable, or an arrow, to indicate that the variable, or the subformula of which the arrow is main connective, is an *antecedent part*, and using "c" similarly for *consequent part*, we notice in the case of the example considered in §5 that we obtain a diagram of the situation as follows:

$$\underset{a}{((A} \underset{c}{\rightarrow} \underset{c}{B)} \underset{a}{\rightarrow} \underset{a}{C)} \underset{c}{\rightarrow} \underset{a}{(B} \underset{c}{\rightarrow} \underset{c}{D)}$$

But if we rewrite the example in Łukasiewicz notation, we get

$$\underset{c}{\rightarrow} \underset{a}{\rightarrow} \underset{c}{\rightarrow} \underset{a}{A} \underset{c}{B} \underset{a}{C} \underset{c}{\rightarrow} \underset{a}{B} \underset{c}{D}$$

with a string of c's and a's alternating; it is not hard to show by an induction on the length of formulas that this is generally true.

§8.9. Replacement theorem. $T_{\rightarrow}$, and generally any pure implicational calculus with both $A \rightarrow B \rightarrow . B \rightarrow C \rightarrow . A \rightarrow C$ (suffixing) and $A \rightarrow B \rightarrow . C \rightarrow A \rightarrow . C \rightarrow B$ (prefixing), has the following replacement property: if both $A \rightarrow B$ and $B \rightarrow A$ are theorems, then we may always infer $(\ldots B \ldots)$ from $(\ldots A \ldots)$.

It is possible to sharpen the property by using the language of "antecedent" and "consequent" parts of §5.1.2: if B is a consequent part of

(. . . B . . .) then $A \to B \to .(\ldots A \ldots) \to (\ldots B \ldots)$, and if B is an antecedent part of (. . . B . . .) then $A \to B \to .(\ldots B \ldots) \to (\ldots A \ldots)$. The proof is by a straightforward induction on the depth of the occurrence of B in question, using suffixing and prefixing as required. (If one wants to count the "empty context" of depth zero, then we need $A \to B \to .A \to B$ as well.)

§8.10. $\mathbf{E}_{\to}$ is not the intersection of $\mathbf{R}_{\to}$ and $\mathbf{S4}_{\to}$. The fact that $\mathbf{FE}_{\to}$ arises by combining the subscript restrictions of $\mathbf{R}_{\to}$ with the reiteration restrictions of $\mathbf{FS4}_{\to}$ might lead one to think that $\mathbf{E}_{\to}$ contains as theorems all and only those formulas which are theorems of *both* $\mathbf{R}_{\to}$ and $\mathbf{S4}_{\to}$. We are indebted to Saul Kripke for a counterexample: $A \to .(A \to .A \to B) \to .A \to B$ is provable in both $\mathbf{R}_{\to}$ and $\mathbf{S4}_{\to}$, as is easily seen from the subproof formulations (for $\mathbf{S4}_{\to}$, simply prove contraction "under" the hypothesis A). But as we saw in §5.2, pure non-necessitives never entail necessitives in $\mathbf{E}_{\to}$.

The same conjecture with $\mathbf{S3}_{\to}$ (the pure implicational part of Lewis's **S3**; see §10) replacing $\mathbf{S4}_{\to}$ is also false. $(A \to B \to B \to A) \to .A \to A$ is in $\mathbf{S3}_{\to}$, since $A \to .B \to A$ is provable if A and B are both strict (Hacking 1963). And the same formula can be got in $\mathbf{R}_{\to}$ by suffixing A to $A \to .A \to B \to B$. But the formula is not provable in $\mathbf{E}_{\to}$, as can be seen from the decision procedure of §13; hence $\mathbf{E}_{\to}$ is not the intersection of $\mathbf{R}_{\to}$ and $\mathbf{S3}_{\to}$.

PROBLEM. There is along these lines one unsettled conjecture, raised by Storrs McCall: is $\mathbf{E}_{\to}$ the intersection of $\mathbf{R}_{\to}$ with the system obtained by adding "restricted mingle" $A \to B \to .A \to B \to .A \to B$ (see §8.15) to $\mathbf{E}_{\to}$?

PROBLEM. What interesting matrix shows that $(A \to B \to B \to A) \to .A \to A$ is unprovable in $\mathbf{E}_{\to}$?

§8.11. Minimal logic. Let $\mathbf{T}_{\to} - \mathbf{W}$ be defined by

1 $A \to A$
2 $A \to B \to .B \to C \to .A \to C$
3 $A \to B \to .C \to A \to .C \to B$,

so that $\mathbf{T}_{\to} - \mathbf{W}$ is $\mathbf{T}_{\to}$ minus contraction: $(A \to .A \to B) \to .A \to B$. (Curry associates "**W**" with contraction by analogy with the contractive combinator W, and we associate "$-$" with subtraction.) Though there are even weaker systems for which some kind of deduction theorem is provable (see Curry 1954), it still seems to us that there are some reasons for considering $\mathbf{T}_{\to} - \mathbf{W}$ as a plausible "minimal logic" in something like the sense of Church 1951a. (For other candidates, see §§47.4, 51.1, 51.3.)

In the first place, $\mathbf{T}_{\to} - \mathbf{W}$ has a natural and partly well-motivated subproof formulation, to wit, like that of §6 for $\mathbf{T}_{\to}$, except that it is forbidden to use any step more than once as a premiss. This prohibition means that in the nested implication $C_1 \rightarrow \ldots \rightarrow . C_n \rightarrow B$ associated with a proof of B from hypotheses $A_1, \ldots, A_m$ (in the orthodox sense allowing multiple use), each A_i will appear as a C_j exactly as many times as it is, either mediately or immediately, used.

Second, $\mathbf{T}_{\rightarrow}-\mathbf{W}$ has a neat Gentzen formulation, of a piece with the other "merge formulations" of §7.3. One simply drops the rule (W⊢).

Hence, $\mathbf{T}_{\rightarrow}-\mathbf{W}$ is not *ad hoc* (exactly).

Problem. It is conjectured that $\mathbf{T}_{\rightarrow}-\mathbf{W}$, alone among systems deriving naturally from intuitive considerations, has the following feature indicating minimality:

> If both $A \rightarrow B$ and $B \rightarrow A$ are theorems of $\mathbf{T}_{\rightarrow}-\mathbf{W}$, then A and B are the same formula.

If this conjecture is true, then $\mathbf{T}_{\rightarrow}-\mathbf{W}$ would be free of redundancy in the same sense as in an applied logic with no more than one name for each individual, so that if a = b is a theorem, then a and b are the same expression. For we may think of A and B as expressing the same proposition just in case both $A \rightarrow B$ and $B \rightarrow A$; whence, the truth of the conjecture would entail that in $\mathbf{T}_{\rightarrow}-\mathbf{W}$, distinguishable formulas invariably express distinct propositions.

Although the problem remains open, L. Powers in 1968 sent us a proof of a lemma which might be used in its positive solution: if the conjecture is false, then the offending implications $A \rightarrow B$, $B \rightarrow A$ (with $A \neq B$) can be proved from the two transitivities alone. Powers then observes that in virtue of the instance $A \rightarrow B \rightarrow . B \rightarrow A \rightarrow . A \rightarrow A$ of 2, for the conjecture to be true it suffices that no instance of $A \rightarrow A$ be deducible from the two transitivities alone. Other known facts which do not close the problem but which might help: a matrix of Lemmon *et al.* 1956 shows that no instance of $A \rightarrow A$ in which each variable occurs exactly once in A is provable from the two transitivities alone; and we have a proof that any theorem provable from the two transitivities is a substitution-instance of a theorem provable from them in which each variable occurs exactly twice.

§8.12. Converse Ackermann property. Ackermann proved that his system Π′ (§45) is such that $p \rightarrow . B \rightarrow C$ is unprovable when p is a propositional variable, and we have suggested taking this "Ackermann property" as a partial explication of what it might mean to say of a pure implicational calculus that it avoids fallacies of modality (§5.2). In formal analogy to the Ackermann property is the "converse Ackermann property," i.e., the unprovability of $B \rightarrow C \rightarrow p$ whenever p is a propositional variable. One might wish a system to have this feature if one had the opinion, as we do not, that a non-necessitive could never follow from a necessitive. We therefore report what we know about the property.

$\mathbf{E}_{\rightarrow}$ clearly does not possess the converse Ackermann property, since one of its axioms, in one of its formulations, is $A \rightarrow A \rightarrow B \rightarrow B$, where B may be a propositional variable. We thought for a while that $\mathbf{T}_{\rightarrow}$ (§6) might have the property; but as a counterexample we have $(A \rightarrow .A \rightarrow A) \rightarrow (A \rightarrow A) \rightarrow A \rightarrow A$. (Try proving it by subproofs.)

$\mathbf{T}_{\rightarrow}$ without contraction (§8.11) does indeed have the converse Ackermann property, as can easily be established by appeal to the consecution formulation of this system (§7.3).

Zane Parks has pointed out to us that the following matrix proves the same fact:

→	0	1	2
0	2	1	2
1	1	2	2
*2	1	1	2

§8.13. Converse of contraction. In $\mathbf{R}_{\rightarrow}$, hence also in the weaker systems, one does not have the converse of contraction, $A \rightarrow B \rightarrow .A \rightarrow .A \rightarrow B$, nor even $(A \rightarrow .A \rightarrow B) \rightarrow .(A \rightarrow .A \rightarrow .A \rightarrow B)$. (One proof is via the consecution formulation of §7.) However, even in $\mathbf{T}_{\rightarrow}$, hence also in the stronger systems, one does have as an oddity the latter (but not the former) formula when B is identified with A: $(A \rightarrow .A \rightarrow A) \rightarrow .(A \rightarrow .A \rightarrow .A \rightarrow A)$. The reader may be amused by reconstructing its proof in $\mathbf{FT}_{\rightarrow}$.

Note that $\vdash (A \rightarrow .B \rightarrow B) \rightarrow .A \rightarrow .A \rightarrow .B \rightarrow B$ in $\mathbf{T}_{\rightarrow}$, which fact is used in §14.6.

Last, note that the converse of contraction is precisely the mingle idea of §8.15.

§8.14. Weakest and strongest formulas. A is a weakest formula, with respect to a formal system containing an implicational connective, if A is implied by every formula in the system; and A is a strongest formula if it implies every formula. Sugihara 1955 defines "freedom from paradox" as absence of weakest and strongest formulas, thus giving what is probably the first general analysis of paradoxes of implication. (See §26.9 for more on Sugihara, and §29.3.) We should expect of pure implicational calculuses that they would have no strongest elements since not even the classical two-valued implicational calculus $\mathbf{TV}_{\rightarrow}$ has such; but of course any $A \rightarrow A$ is a weakest element in $\mathbf{TV}_{\rightarrow}$, $\mathbf{H}_{\rightarrow}$, and $\mathbf{S4}_{\rightarrow}$. But none of $\mathbf{R}_{\rightarrow}$, $\mathbf{E}_{\rightarrow}$, and $\mathbf{T}_{\rightarrow}$ has a weakest element, as the variable-sharing property of §5.1.2 tells us at once: any candidate A would fail to be implied by any formula containing no variables in common with A.

The picture changes somewhat when we come to consider formulas in a single variable. Where p is a propositional variable, we call A a p-formula if it contains no other variable than p, and we say that such a formula is p-weakest or p-strongest if it is weakest or strongest among p-formulas. Then $p \to p$ serves as a p-weakest formula for $\mathbf{TV}_{\to}$, $\mathbf{H}_{\to}$, and $\mathbf{S4}_{\to}$, while it can be shown that no p-formula will do as p-weakest for $\mathbf{E}_{\to}$ or $\mathbf{T}_{\to}$.

PROOF: were A a p-weakest formula, we should have to have $p \to A$. But then the matrix of §5.1 showing avoidance of modal fallacies of $\mathbf{E}_{\to}$ (and its subsystem $\mathbf{T}_{\to}$) tells us that A cannot have the form $B \to C$, hence must be p itself. But it is easy to see that p is not p-weakest, since not implied by any two-valued tautology. So no formula is p-weakest for $\mathbf{E}_{\to}$ or $\mathbf{T}_{\to}$.

Does $\mathbf{R}_{\to}$ have a p-weakest formula? We used to conjecture strongly that it does not, but Meyer tells us otherwise; see §29.2.

Turning to p-strongest formulas, it is clear that p itself plays such a role in $\mathbf{TV}_{\to}$ and $\mathbf{H}_{\to}$, and the result of Meyer reported in §29.2 supplies a p-strongest formula for $\mathbf{R}_{\to}$. Permit us to leave the question open for $\mathbf{S4}_{\to}$, $\mathbf{E}_{\to}$, and $\mathbf{T}_{\to}$.

§8.15. Mingle. Ohnishi and Matsumoto 1962 considered Gentzen systems (§7) with a rule

$$\frac{\alpha \vdash C \qquad \beta \vdash C}{\alpha, \beta \vdash C}$$

which suggested strengthening the pure implicational systems above in some way so as to yield a derived rule, called "mingle":

from $A \to C$ and $B \to C$ to infer $A \to .B \to C$.

The following remarks are due to McCall and Dunn, who pursued this suggestion.

Evidently the mingle rule, in the presence of $A \to A$, leads to $A \to .A \to A$. We call this latter formula "the mingle axiom," since it permits the derivation of the mingle rule, given only suffixing:

1	$A \to C$	premiss
2	$B \to C$	premiss
3	$C \to .C \to C$	mingle axiom
4	$A \to .C \to C$	1 3 trans
5	$C \to C \to .B \to C$	2 suf
6	$A \to .B \to C$	4 5 trans

Addition of the mingle axiom to $\mathbf{R}_{\rightarrow}$, yielding $\mathbf{RM0}_{\rightarrow}$, preserves some of its important properties. In particular, variable-sharing is preserved, as is shown by the matrix:

$\rightarrow$	*0*	*1*	*2*
0	*2*	*2*	*2*
**1*	*0*	*1*	*2*
**2*	*0*	*0*	*2*

The matrix satisfies $\rightarrow$E and all axioms of $\mathbf{RM0}_{\rightarrow}$ but if A and B share no variables, we may give all variables in A the value *2*, and all those in B the value *1*, getting $2\rightarrow 1 = 0$.

Just worth mentioning is the "restricted mingle axiom" $A\rightarrow B\rightarrow .A\rightarrow B\rightarrow .A\rightarrow B$. Adding this axiom to $\mathbf{E}_{\rightarrow}$ preserves necessity as well as relevance, as is shown by the following matrix:

$\rightarrow$	*0*	*1*	*2*
0	*1*	*1*	*1*
**1*	*0*	*1*	*1*
**2*	*0*	*0*	*1*

The axioms of $\mathbf{E}_{\rightarrow}$ and restricted mingle all satisfy this matrix, as does $\rightarrow$E, but if p is a propositional variable, then when $p = 2$, $(p\rightarrow .B\rightarrow C) = 0$ regardless of the values assigned to B and C; which shows that fallacies of modality are still avoided in this extension of $\mathbf{E}_{\rightarrow}$.

For more information concerning mingle, consult §14.7, §§27.1.1 and 27.2, §§29.3, 29.4, and 29.5. The last mentioned section reports Meyer's discovery that relevance and mingle are incompatible when truth functions are added.

In §27.1.1 we define the system **RM** by adding the mingle axiom to the full system **R**, involving all of the connectives $\rightarrow$, $\overline{}$, &, and $\vee$. The question then arises as to whether adding the mingle axiom to $\mathbf{R}_{\rightarrow}$ yields the pure implicational fragment $\mathbf{RM}_{\rightarrow}$ of this system. In other words, is **RM** a "conservative extension," in the sense of §14.4, of $\mathbf{RM0}_{\rightarrow}$? Answer: no. Meyer and Parks 1972 show that instead the following is a complete and independent set of axioms for $\mathbf{RM}_{\rightarrow}$:

$A\rightarrow B\rightarrow .B\rightarrow C\rightarrow .A\rightarrow C$ (*suffixing*)
$A\rightarrow .A\rightarrow B\rightarrow B$ (*assertion*)
$(A\rightarrow .A\rightarrow B)\rightarrow .A\rightarrow B$ (*contraction*)
$(A\rightarrow B\rightarrow B\rightarrow A\rightarrow C)\rightarrow .B\rightarrow A\rightarrow A\rightarrow B\rightarrow C\rightarrow C$ (*?*)

The simplest way of seeing that the mingle axiom does not suffice is due to Meyer: the axioms of $\mathbf{RM0}_{\rightarrow}$ are all intuitionistically acceptable; but the last axiom displayed above is not. Perhaps $\mathbf{RM0}_{\rightarrow}$ axiomatizes "constructive mingle," which needs inventing, possibly along the lines of Woodruff 1969.

§8.16. $\mathbf{FR}_{\rightarrow}$ without subscripts. Fitch (in a letter of June 20, 1969) has suggested that $\mathbf{FR}_{\rightarrow}$ can be formulated without using subscripts, as follows (we quote):

> There are two fundamental requirements beyond the usual ones for proofs and subproofs.
>
> First requirement: Every item of a subproof S must satisfy at least one of the following three conditions: (1) It is an hypothesis item of S, or it is the last item of S. (2) It serves as a premiss of a rule of direct consequence for justifying some subsequent item of S. (3) It is reiterated into some subsequent subproof S′ which is an item of S. (Hence any item reiterated into a subproof is directly or indirectly used by the last item of that subproof, or is the last item.)
>
> Second requirement: If a subproof S serves as the premiss of the rule of entailment introduction, then every item of S must satisfy at least one of the following four conditions: (1) It is the hypothesis of S. (2) It has been reiterated into S but is not the last item of S. (3) It is a direct consequence of one or more preceding items of S, at least one of which was not reiterated into S. (4) It is a subproof into which some preceding item of S has been reiterated which was not itself reiterated into S. (Hence the last item of the subproof must directly or indirectly use the hypothesis of the subproof, or must be the hypothesis.)

Of course Fitch's requirements would rule out uses of →I and →E in which the conclusion does not depend on the premiss of the subproof in which the conclusion lies, although $\mathbf{R}_{\rightarrow}$ permits such uses; but that would not change anything significant.

§8.17. No finite characteristic matrix. The argument of §29.2, second paragraph, suffices to show that $\mathbf{T}_{\rightarrow}$, $\mathbf{E}_{\rightarrow}$, and $\mathbf{R}_{\rightarrow}$ have no finite characteristic matrix. For an earlier related argument, see Pahi 1966.

§8.18. Indefinability of necessity in $\mathbf{R}_{\rightarrow}$ (by Zane Parks). Meyer 1970b can be used to solve a problem concerning $\mathbf{R}_{\rightarrow}$ analogous to the problem concerning $\mathbf{T}_{\rightarrow}$ raised at the end of §6. That is, is there a function f of a single variable A definable in $\mathbf{R}_{\rightarrow}$ making f(A) look like: necessarily A? Such a function would presumably allow us to prove

1 $f(A) \rightarrow A$
2 $f(A \rightarrow B) \rightarrow .f(A) \rightarrow f(B)$

but not

3 $A \rightarrow f(A)$.

Meyer shows (see §29.2) that there are just six non-equivalent functions of a single variable in $\mathbf{R}_{\rightarrow}$, i.e.,

4 $A\rightarrow(A\rightarrow(A\rightarrow A)\rightarrow A\rightarrow A\rightarrow A)\rightarrow A$
5 A
6 $A\rightarrow(A\rightarrow A)\rightarrow A\rightarrow A$
7 $A\rightarrow(A\rightarrow A)\rightarrow A$
8 $A\rightarrow A$
9 $A\rightarrow.A\rightarrow(A\rightarrow A)\rightarrow A$

It is clear from Meyer's work that 5, 7, and 9 won't work since they fail to reject 3. Further, neither 6 nor 8 works since they fail to yield 1. Finally, 4 won't work since it fails to yield 2. This may be seen as follows. Using the matrix for → in §22.1.3, which satisfies the axioms and rules of $\mathbf{R}_{\rightarrow}$, the matrix for f, when f(A) is defined as in 4, is

A	f(A)
-3	-3
-2	-2
-1	-1
-0	-3
$+0$	$+0$
$+1$	$+1$
$+2$	$+2$
$+3$	$+3$

so that for $A = +1$, $B = -0$, the formula $2 = \mathrm{f}(+1\rightarrow -0)\rightarrow.\mathrm{f}(+1)\rightarrow \mathrm{f}(-0) = \mathrm{f}(-1)\rightarrow.+1\rightarrow -3 = -1\rightarrow -3 = -3$ (which is not designated).

§8.19. Necessity in $\mathbf{T}_{\rightarrow}$. At the end of §4.3, we remarked that (as was pointed out by A. N. Prior in correspondence), addition of □ as primitive to $\mathbf{E}_{\rightarrow}$, with plausible axioms, leads to the theorem $\Box A \rightleftarrows.A\rightarrow A\rightarrow A$. Good.

One is then led to wonder what happens when one adds □ as primitive to $\mathbf{T}_{\rightarrow}$ (the question for $\mathbf{R}_{\rightarrow}$ is discussed at the end of §8.20, and indirectly in §27.1.3).

We are indebted to Robert K. Meyer (correspondence, 1971) for the following remarks. He notes in the first place that the §4.3 move of adding

1 $\Box A \rightarrow A$
2 $A\rightarrow B\rightarrow\Box(A\rightarrow B)$
3 $\Box A \rightarrow.A\rightarrow B \rightarrow \Box B$

to $\mathbf{T}_{\rightarrow}$ won't do, since then $\mathbf{T}_{\rightarrow}$ would collapse to $\mathbf{E}_{\rightarrow}$: start by choosing A in 3 as $C\rightarrow C$, and continue to obtain $C\rightarrow C\rightarrow B\rightarrow B$, thus yielding $\mathbf{E}_{\rightarrow}$. So yet one more time we are indebted to Meyer for closing a problem left open in earlier drafts of this book. Second, Meyer suggests that we might start as in §31.5 or §32 with the addition of propositional quantifiers, and then adopt the definition of §31.4:

$$\Box A =_{\text{df}} (p)(p\rightarrow p)\rightarrow A.$$

The collapse to $\mathbf{E}_\rightarrow$ at which we hinted above is blocked by the absence, under this definition, of $\Box A \rightarrow A$. Meyer expresses the hope that we like this system, but we remain noncommittal. Third, he allows us to consider the system defined by adding to $\mathbf{T}_\rightarrow$ all of 1, 2, and 3′

$$A\rightarrow B\rightarrow.\Box A\rightarrow\Box B,$$

which certainly does not collapse to $\mathbf{E}_\rightarrow$ — the vacuous interpretation leaves proofs proofs, even if $\Box A \rightarrow \Box\Box A$ is added as well. We rather suspect a deeper understanding of necessity in the context of ticket entailment will incline in the direction of this third alternative; but we leave the matter for future investigation.

§8.20. The $\mathbf{C_I}$ systems: an irenic theory of implications (by Garrel Pottinger). Entailment and its sidekick, relevant implication, are the heroes of the bulk of this treatise, and various other notions of implication are the villains. Surprisingly, it turns out to be possible to make the good guys get along perfectly well with two of the bad guys, namely, intuitionist "implication" and **S4** strict "implication." Indeed, it is possible to do this in *two* ways, which reflect different underlying intuitions and lead to logically different results. One way is motivated in §36.1 and explained by Meyer in §36.2. The aim of this section is to explain the other as briefly as possible. To this end we shall (1) define two systems in which all four implicational connectives are present as primitives, and (2) state some theorems which justify the claim that the connectives get along perfectly well in the systems so defined. Proofs of the results to be quoted under (2) and various other results about the systems to be defined under (1) may be found in Pottinger 1972.

§8.20.1. The systems $\mathbf{FC_I}$ and $\mathbf{C_I}$. Both systems have $\Rightarrow$, $\rightarrow$, $⥽$, and $\supset$ as primitive connectives, where $\Rightarrow$ stands for entailment, $\rightarrow$ stands for relevant implication, $⥽$ stands for **S4** strict "implication," and $\supset$ stands for intuitionist "implication." In each system, *formulas* are built up from these connectives, variables, and grouping indicators in the usual way. The connectives $\Rightarrow$ and $⥽$ are *modal*, and the connectives $\Rightarrow$ and $\rightarrow$ are *relevant*. "ρ", "σ", "τ", . . . will be used as metalinguistic variables which may take $\Rightarrow$, $\rightarrow$, $⥽$, and $\supset$ as values. The formula A is *modal* iff A has the form $A_1\rho A_2$, where ρ is modal.

FC_I is a Fitch style system that involves both (a modified version of) the subscripting technique used in $FE_{\rightarrow}$ and $FR_{\rightarrow}$ and a distinction between *modal* and *non-modal* subproofs as in Fitch 1952. A *modal* subproof is distinguished by writing □ to the left of the hypothesis. In the statement of the rules which follows, M, M_1, . . . , are to be either occurrences of □ or the empty expression.

Postulates for FC_I

Structural Rules

hyp
$$\left|\begin{array}{l} \quad\vdots \\ \mathrm{M}\,\Big|\underline{A_{\{k+1\}}} \end{array}\right.$$

where k is the number of subproofs to which the new subproof is subordinate.

reit
$$\left|\begin{array}{l} \quad A_{\mathrm{a}} \\ \quad\vdots \\ \mathrm{M}\,\left|\begin{array}{l} \underline{B_{\{k\}}} \\ \vdots \\ A_{\mathrm{a}} \end{array}\right. \end{array}\right.$$

provided that if M is not empty, then A is modal.

Operational Rules

ρE
$$\left|\begin{array}{l} \vdots \\ A_{\mathrm{a}} \\ \vdots \\ A\rho B_{\mathrm{b}} \\ \vdots \\ B_{\mathrm{c}} \end{array}\right.$$

where (1) if ρ is relevant, then $\mathrm{c} = \mathrm{a} \cup \mathrm{b}$, and (2) if ρ is not relevant, then $\mathrm{c} = \mathrm{b}$. (Motivation for ρE where ρ is non-relevant is a long story, for which see Pottinger 1972. We here note only that it is rooted in a *constructivist* account of relevance.)

ρI
$$\left|\begin{array}{l} \mathrm{M}\,\left|\begin{array}{l} \underline{A_{\{k\}}} \\ \vdots \\ B_{\mathrm{a}} \end{array}\right. \\ A\rho B_{\mathrm{a}-\{k\}} \end{array}\right.$$

provided (1) if ρ is modal, then M is not empty, and (2) if ρ is relevant, then $k \in \mathrm{a}$.

REMARK. Rep is suppressed by counting inferences of the forms

$$\begin{array}{|l} \mathrm{M}\,\underline{A}_{\{k\}} \\ A\rho A \end{array}$$

as uses of ρI, and by allowing that the premisses for ρE as stated above may appear either in the order displayed *or* with the major premiss first — either form being equally valid (as in all other similar cases for **F** systems in this book).

Next, we define the Hilbert-style system $\mathbf{C}_{\mathrm{I}}$.

POSTULATES FOR $\mathbf{C}_{\mathrm{I}}$

Axioms.

AS1 $A \Rightarrow .B\rho A$

where (1) if ρ is modal, so is A, and (2) ρ is not relevant.

AS2 $(A\rho.B\sigma C) \Rightarrow .(A\tau B)\rho_1(A\sigma_1 C)$

where (1) if ρ is not modal, neither is ρ_1 or σ_1; (2) if τ is not modal, neither is σ_1; (3) if σ_1 is relevant, then either ρ is relevant, or both σ and τ are relevant; and (4) if σ is not relevant, neither is ρ_1.

Rules.

$$\rho\mathrm{E} \quad \frac{A \qquad A\rho B}{B}$$

§8.20.2. Some theorems.

THEOREM 1. A is a theorem of $\mathbf{FC}_{\mathrm{I}}$ iff A is a theorem of $\mathbf{C}_{\mathrm{I}}$.

PROOF as in §§1–4.

The theorems which show that the various connectives of these systems get along nicely with each other are derivable from a Prawitz 1965 style normal form theorem that holds for $\mathbf{FC}_{\mathrm{I}}$. A few definitions and some notation are required for the statement of the theorems.

Let D, D_1, . . . be proofs of $\mathbf{FC}_{\mathrm{I}}$. Henceforth, it is assumed that an *analysis* of D constitutes part of D, i.e. it is assumed that at each line of D it is specified which rule of $\mathbf{FC}_{\mathrm{I}}$ is applied in order to obtain that line, and, in the case of rules other than hyp, it is specified which items are the premisses of the inference in question.

Consider a line l of D, and let Dl be the portion of D which ends with line l. Define *ancestor* as follows:

(i) l and all lines j such that j is occupied by an undischarged hypothesis of Dl are ancestors of l.

(ii) If l is a conclusion by ρE or reit, then the lines which are the premisses of the inference in question are ancestors of l.

(iii) If l is a conclusion by ρI, then the last and first lines of the subproof terminated by the inference in question are ancestors of l.

(iv) If l_1 is an ancestor of l, and l_2 is an ancestor of l_1, then l_2 is an ancestor of l.

D is *single-minded* iff every line of D is an ancestor of the last line of D.

The line l of D is *maximal* in D iff line l is produced by zero or more reiterations from the conclusion of a ρI, and line l is the major premiss of a ρE. D is *normal* iff D is single-minded, and no line of D is maximal.

Let $B_{1\{1\}}, \ldots, B_{k\{k\}}$ be the undischarged hypotheses of D, and let M_i be either $\Box$ or empty, according as the subproof headed by $B_{i\{i\}}$ is modal or not. Define:

$$\mathrm{H}_{\mathrm{D}} = B_1{}^{\mathrm{M}_1}, \ldots, B_k{}^{\mathrm{M}_k}$$

$\mathrm{D}_2 \lesssim \mathrm{D}_1$ iff A_{a_1} appears at the last line of D_1, A_{a_2} appears at the last line of D_2, $\mathrm{a}_1 \subseteq \mathrm{a}_2$, and $\mathrm{H}_{\mathrm{D}_1} = \mathrm{H}_{\mathrm{D}_2}$.

THEOREM 2 (NORMAL FORM THEOREM). Given a proof D of $\mathbf{FC}_{\mathrm{I}}$, one can find a proof D′ of $\mathbf{FC}_{\mathrm{I}}$ such that D′ is normal and $\mathrm{D}' \lesssim \mathrm{D}$.

We can show

THEOREM 3 (SUBFORMULA THEOREM). If D is a normal proof of $\mathbf{FC}_{\mathrm{I}}$, then every formula which occurs in D either is a subformula of the end-formula of D or is a subformula of some undischarged hypothesis of D.

Combining Theorems 2 and 3 one obtains

THEOREM 4 (SEPARATION THEOREM). If D is a derivation of $\mathbf{FC}_{\mathrm{I}}$ with A_{a} as its last item and the connective ρ does not occur in A or in any formula of H_{D}, then one can find a derivation D′ of $\mathbf{FC}_{\mathrm{I}}$ such that $\mathrm{D}' \lesssim \mathrm{D}$ and such that ρE and ρI are not applied in D′.

From Theorem 4 one gets

THEOREM 5. $\mathbf{FC}_{\mathrm{I}}$ and (in view of Theorem 1) $\mathbf{C}_{\mathrm{I}}$ are conservative extensions (in the sense of §14.4) of $\mathbf{FE}_{\rightarrow}$, $\mathbf{FR}_{\rightarrow}$, $\mathbf{FS4}_{\rightarrow}$, and $\mathbf{FH}_{\rightarrow}$.

This theorem obviously supports the contention that the connectives of $\mathbf{FC}_\mathrm{I}$ and $\mathbf{C}_\mathrm{I}$ get along well together. Additional information about how the various connectives of these systems interact (or don't, as the case may be) is contained in

THEOREM 6. (a) All formulas of the following forms are theorems of $\mathbf{FC}_\mathrm{I}$ and $\mathbf{C}_\mathrm{I}$.

(i) $(A \Rightarrow B)\rho(A\sigma B)$
(ii) $(A \rightarrow B)\rho(A \supset B)$
(iii) $(A \dashv B)\rho(A \supset B)$

(b) If ρ is not relevant and τ is relevant, then $(p\rho q)\sigma(p\tau q)$ is not a theorem of $\mathbf{FC}_\mathrm{I}$ or $\mathbf{C}_\mathrm{I}$.

(c) If ρ is not modal and τ is modal, then $(p\rho q)\sigma(p\tau q)$ is not a theorem of $\mathbf{FC}_\mathrm{I}$ or $\mathbf{C}_\mathrm{I}$.

Furthermore, $\Rightarrow$ and $\rightarrow$ have the properties one would expect them to have in view of the results of §§3–5, as one can see from

THEOREM 7. If $A\rho B$ is a theorem of $\mathbf{FC}_\mathrm{I}$ or $\mathbf{C}_\mathrm{I}$ and ρ is relevant, then A and B share a variable.

THEOREM 8. If one defines

$$\Box A =_{\text{df}} A \Rightarrow A \Rightarrow A$$

then the theory of necessity which arises in both $\mathbf{FC}_\mathrm{I}$ and $\mathbf{C}_\mathrm{I}$ is the same as what one gets by taking $\Box$ as a primitive, counting formulas of the form $\Box A$ among the modal formulas, and adding as axioms all formulas of the following forms:

$\Box 1$ $\Box A\rho A$
$\Box 2$ $(A\rho B)\sigma\Box(A\rho B)$,

where ρ is modal, and

$\Box 3$ $\Box A\rho.(A\sigma B)\tau\Box B$,

where σ is modal, and if ρ is relevant, so is σ.

THEOREM 9. No theorem of $\mathbf{FC}_\mathrm{I}$ or $\mathbf{C}_\mathrm{I}$ has the form $p\rho.A\sigma B$, where ρ is relevant and σ is modal.

Finally, define:

$A \rightleftarrows B$ iff both $A \Rightarrow B$ and $B \Rightarrow A$ are

theorems of FC_I (or $\mathbf{C}_\mathrm{I}$).

THEOREM 10. (a) An appropriate replacement theorem holds for $\rightleftarrows$.

(b) $(A \Rightarrow B) \rightleftarrows \Box(A \rightarrow B)$
(c) $(A \prec B) \rightleftarrows \Box(A \supset B)$

So it turns out that, according to the $\mathbf{C_I}$ systems, entailment is the necessitation of relevant implication, and **S4** strict implication is the necessitation of intuitionist implication. (It is observed at the end of §28.1, however, that considerable caution is required in extending the former result to the case in which truth functions are present.)

§8.21. Fogelin's restriction. In a letter which we have characteristically lost, Fogelin proposes the following restriction governing Fitch proofs: "No step in a proof shall involve any funny business." The reader is at liberty either to take Fogelin's *Rule of no funny business* as a summary of our work, or to construe the first eight sections of this volume as glosses on his.

CHAPTER II

ENTAILMENT AND NEGATION

§9. **Preliminaries ($\mathbf{E}_{\overline{\rightarrow}}$).** In this chapter we undertake a study of entailment with negation. In the present section we show that a calculus of entailment and negation can be developed independently of other truth functional connectives. In §10 we consider the structure of modalities (in the sense of Parry 1939) of $\mathbf{E}_{\overline{\rightarrow}}$, and in §11 we discuss relations between $\mathbf{E}_{\overline{\rightarrow}}$ and some other standard systems embodying modal distinctions, making in this context some philosophical remarks about our definition of necessity. §12 shows that $\mathbf{E}_{\overline{\rightarrow}}$ is free of fallacies of modality and relevance; and we attempt there to clinch our argument that calculuses of material and strict "implication" are not even close to providing a reasonable account of the converse of deducibility, by exhibiting some plausible entailments which flatly contradict some theorems about material and strict "implication." §13 provides a Gentzen-style decision procedure for $\mathbf{E}_{\overline{\rightarrow}}$, a matter of some interest in view of the fact that no way is yet known of extending these techniques beyond this fragment of the calculus **E** of Chapter IV; and §14, like §8, adds some miscellaneous results and problems.

We use $\bar{A}$ and $\sim A$ interchangeably for the negation of A, and we begin by considering the axioms of Ackermann 1956 for negation:

$$A \rightarrow B \rightarrow . \bar{B} \rightarrow \bar{A}$$
$$A \& \bar{B} \rightarrow \overline{A \rightarrow B}$$
$$A \rightarrow \bar{\bar{A}}$$
$$\bar{\bar{A}} \rightarrow A$$

The second of these is not a theorem of the pure calculus of entailment with negation, since it involves a conjunction in the antecedent; however, substituting $\bar{A}$ for B yields $A \& \bar{\bar{A}} \rightarrow \overline{A \rightarrow \bar{A}}$, where the antecedent collapses into A, thus giving by contraposition $A \rightarrow \bar{A} \rightarrow \bar{A}$, which turns out to bear the full import of the axiom for the pure calculus of entailment with negation. Some slight economy may also be obtained without loss of perspicuity by replacing $A \rightarrow B \rightarrow . \bar{B} \rightarrow \bar{A}$ and $A \rightarrow \bar{\bar{A}}$ by $A \rightarrow \bar{B} \rightarrow . B \rightarrow \bar{A}$, and we therefore choose as axioms for the calculus $\mathbf{E}_{\overline{\rightarrow}}$ of entailment with negation the following:

$E_{\overset{\sim}{\rightarrow}}1$ $A \rightarrow A \rightarrow B \rightarrow B$

$E_{\overset{\sim}{\rightarrow}}2$ $A \rightarrow B \rightarrow .B \rightarrow C \rightarrow .A \rightarrow C$

$E_{\overset{\sim}{\rightarrow}}3$ $(A \rightarrow .A \rightarrow B) \rightarrow .A \rightarrow B$

$E_{\overset{\sim}{\rightarrow}}4$ $A \rightarrow \overline{B} \rightarrow .B \rightarrow \overline{A}$

$E_{\overset{\sim}{\rightarrow}}5$ $A \rightarrow \overline{A} \rightarrow \overline{A}$

$E_{\overset{\sim}{\rightarrow}}6$ $\overline{\overline{A}} \rightarrow A$

The axioms $E_{\overset{\sim}{\rightarrow}}1$–$E_{\overset{\sim}{\rightarrow}}3$ were chosen from the groups of axioms mentioned in §8.4.3 for no better reason than that we are fond of them (see Anderson, Belnap and Wallace 1960), but of course any other group from §8.4.3 would do as well. And we should add that $E_{\overset{\sim}{\rightarrow}}4$ was chosen rather than the more straightforward $A \rightarrow B \rightarrow .\overline{B} \rightarrow \overline{A}$ because it then turns out that $E_{\overset{\sim}{\rightarrow}}1$–$E_{\overset{\sim}{\rightarrow}}5$ are intuitionistically valid, under the obvious translation, while $E_{\overset{\sim}{\rightarrow}}6$ is not. This fact might be of some aid to anyone who might want to know what happens to $E_{\rightarrow}$ if one adds a style of negation which is intuitionistic rather than classical.

The system $\mathbf{FE}_{\overset{\sim}{\rightarrow}}$ arises if we add to $\mathbf{FE}_{\rightarrow}$ the following three rules:

Negation introduction (∼I). From a proof of $\overline{A}_a$ on the hypothesis $A_{\{k\}}$, to infer $\overline{A}_{a-\{k\}}$, provided k is in a.

Contraposition (contrap). From B_a, and a proof of $\overline{B}_b$ on the hypothesis $A_{\{k\}}$, where k is in b, to infer $\overline{A}_{(a \cup b)-\{k\}}$.

Double negation elimination (∼∼E). From $\overline{\overline{A}}_a$ to infer A_a.

The reader may easily verify that adding these three rules to $\mathbf{FE}_{\overset{\sim}{\rightarrow}}$ is equivalent to adding $E_{\overset{\sim}{\rightarrow}}4$–$E_{\overset{\sim}{\rightarrow}}6$ as axioms to $\mathbf{FE}_{\rightarrow}$. For example, in the context of $\mathbf{FE}_{\overset{\sim}{\rightarrow}}$, $E_{\overset{\sim}{\rightarrow}}5$ can be made to do the work of ∼I as follows:

1	⎿ $A_{\{k\}}$	hyp	premiss for ∼I
⋮	⋮		
i	$\overline{A}_a$	?	
$i+1$	$A \rightarrow \overline{A}_{a-\{k\}}$	→I	
$i+2$	$A \rightarrow \overline{A} \rightarrow \overline{A}$	$E_{\overset{\sim}{\rightarrow}}5$	
$i+3$	$\overline{A}_{a-\{k\}}$	$i+1$ $i+2$ →E	

Conversely, $E_{\overset{\sim}{\rightarrow}}5$ can be proved via ∼I:

1	⎿ $A \rightarrow \overline{A}_{\{1\}}$	
2	│ ⎿ $A_{\{2\}}$	
3	│ │ $\overline{A}_{\{1,2\}}$	1 reit 2 →E
4	│ $\overline{A}_{\{1\}}$	2–3 ∼I
5	$A \rightarrow \overline{A} \rightarrow \overline{A}$	1–4 →I

The other two rules are equally easy.

Now observe that the proof of the equivalence of $E_{\overset{\sim}{\rightarrow}}$ and $\mathbf{FE}_{\rightarrow}$ of §4 is not upset by the addition of new axioms to each, provided, for applicability of the lemma of §4, such new axioms each have the form $A\rightarrow B$. Hence, the proof may be taken over bodily in order to establish the equivalence of $\mathbf{E}_{\overline{\rightarrow}}$ and $\mathbf{FE}_{\overline{\rightarrow}}$.

For ease in carrying out proofs in $\mathbf{E}_{\overline{\rightarrow}}$ we also use the following obviously equivalent forms of the rules, all with the same designations.

Negation introduction. (1) From $A\rightarrow\overline{A}_{\mathrm{a}}$ to infer $\overline{A}_{\mathrm{a}}$. (2) From a proof with hypothesis $A_{\{k\}}$, containing steps B_{a} and $\overline{B}_{\mathrm{b}}$, to infer $\overline{A}_{(\mathrm{a}\cup\mathrm{b})-\{k\}}$, provided k is in either or both of a and b. (The "both" case arises from the fact that we could then construct proofs of both $A\rightarrow B_{\mathrm{a}-\{k\}}$ and $A\rightarrow\overline{B}_{\mathrm{b}-\{k\}}$, whence $A\rightarrow\overline{A}_{(\mathrm{a}\cup\mathrm{b})-\{k\}}$ by contraposition and transitivity.)

Contraposition. (1) From $A\rightarrow B_{\mathrm{a}}$ $\{A\rightarrow\overline{B}_{\mathrm{a}},\ \overline{A}\rightarrow B_{\mathrm{a}},\ \overline{A}\rightarrow\overline{B}_{\mathrm{a}}\}$ to infer $\overline{B}\rightarrow\overline{A}_{\mathrm{a}}$ $\{B\rightarrow\overline{A}_{\mathrm{a}},\ \overline{B}\rightarrow A_{\mathrm{a}},\ B\rightarrow A_{\mathrm{a}}\}$. (2) From $A\rightarrow B_{\mathrm{a}}$ $\{A\rightarrow\overline{B}_{\mathrm{a}},\ \overline{A}\rightarrow B_{\mathrm{a}},\ \overline{A}\rightarrow\overline{B}_{\mathrm{a}}\}$ and $\overline{B}_{\mathrm{b}}$ $\{B_{\mathrm{b}},\ \overline{B}_{\mathrm{b}},\ B_{\mathrm{b}}\}$ to infer $\overline{A}_{\mathrm{a}\cup\mathrm{b}}$, $\{\overline{A}_{\mathrm{a}\cup\mathrm{b}},\ A_{\mathrm{a}\cup\mathrm{b}},\ A_{\mathrm{a}\cup\mathrm{b}}\}$.

We also have as a derivable rule:

Double negation introduction ($\sim\sim$I). From A_{a} to infer $\overline{\overline{A}}_{\mathrm{a}}$. In $\mathbf{FE}_{\overline{\rightarrow}}$ we have:

1	A_{a}		
2		$\overline{A}_{\{1\}}$	hyp
3		$\overline{A}_{\{1\}}$	rep
4	$\overline{\overline{A}}_{\mathrm{a}}$		1 2–3 contrap

And in $\mathbf{E}_{\overline{\rightarrow}}$ we may replace A by $\overline{B}$ in $\mathbf{E}_{\overline{\rightarrow}}4$, getting $\overline{B}\rightarrow\overline{B}\rightarrow.B\rightarrow\overline{\overline{B}}$, whence the corresponding theorem by identity and $\rightarrow$E.

The following theorems give some idea of the strength of the calculus of entailment and negation:

$$\overline{A}\rightarrow A\rightarrow A$$
$$A\rightarrow B\rightarrow.A\rightarrow\overline{B}\rightarrow\overline{A}$$
$$A\rightarrow\overline{B}\rightarrow.A\rightarrow B\rightarrow\overline{A}$$
$$A\rightarrow B\rightarrow.A\rightarrow\overline{C}\rightarrow.A\rightarrow\overline{B\rightarrow C}$$

We also have both forms of double negation

$$A\rightarrow\overline{\overline{A}}$$
$$\overline{\overline{A}}\rightarrow A$$

and all forms of contraposition

$$A\rightarrow B\rightarrow.\overline{B}\rightarrow\overline{A}$$
$$A\rightarrow\overline{B}\rightarrow.B\rightarrow\overline{A}$$
$$\overline{A}\rightarrow B\rightarrow.\overline{B}\rightarrow A$$
$$\overline{A}\rightarrow\overline{B}\rightarrow.B\rightarrow A$$

Lastly, we may extend to $\mathbf{E}_{\overset{\sim}{\rightarrow}}$ the replacement theorem of §8.9 by adding to the definition of "antecedent" and "consequent" parts of §5.1.2 the following clauses:

(d) if $\bar{B}$ is a consequent part of A, then B is an antecedent part of A;
(e) if $\bar{B}$ is an antecedent part of A, then B is a consequent part of A.

The statement of the theorem of §8.9 in terms of these kinds of parts remains exactly the same for $\mathbf{E}_{\overset{\sim}{\rightarrow}}$; and the only difference in the proof lies in the use for $\mathbf{E}_{\overset{\sim}{\rightarrow}}$ of the first form of contraposition listed above. One then has as a consequence a standard replacement theorem to the effect that if $A \rightarrow B$ and $B \rightarrow A$ are provable and if C' results from C by replacing an occurrence of A by B, then so are both $C \rightarrow C'$ and $C' \rightarrow C$.

A final word. The postulates for $\mathbf{FE}_{\overset{\sim}{\rightarrow}}$ appear (even to us) *ad hoc;* we think the reason is that our subproofs are all supposed to be logical in character — as witnessed by the restriction on reiteration — whereas for negation introduction (second form) it would seem that, while one wants a relevant connection between hypothesis A and contradictory steps B and $\bar{B}$ in order to be able to infer $\bar{A}$, it is not further necessary that the connection be logical in character (*that* would make A *impossible*). So suppose we relax the restriction on reiteration for subproofs used in connection with negation introduction.

And suppose further we use the inverse form of negation introduction appropriate to a two-valued concept of negation — call it "reductio":

$$
\begin{array}{r|l}
n & \underline{\ \bar{A}_{\{k\}}\ } \qquad \text{hyp (for reductio)} \\
\vdots & \ \vdots \\
n+j & \ B_{\mathrm{a}} \\
n+j+1 & \ \bar{B}_{\mathrm{b}} \\
\multicolumn{2}{r}{A_{(\mathrm{a} \cup \mathrm{b})-\{k\}} \qquad n-(n+j+1) \text{ reductio}}
\end{array}
$$

provided, of course, $k \in (\mathrm{a} \cup \mathrm{b})$. We observe that this rule *alone* is enough to give us all the properties of negation, a fact which may be confirmed by the reader. (But we have no proof that the calculus so defined is not too strong; it isn't, though.)

What makes this rule appear less *ad hoc*, is that it (without the subscripts) is *precisely* a way — and a standard way — of obtaining the full two-valued calculus from the intuitionistic calculus $\mathbf{FH}_{\overset{\sim}{\rightarrow}}$.

§10. Modalities. The concept of a "modality" was introduced by Parry 1939 in connection with the study of the modal calculuses of Lewis 1918 and Lewis and Langford 1932. The intuitive motivation amounted to the following. In some of the Lewis systems, for example, **S2**, we know that we do not have $\Box A \rightleftarrows \Box\Box A$; in others, for example **S4**, we do. So it looks immediately as if **S4** makes fewer modal distinctions than **S2**: since a replacement theorem is provable for both, **S4** identifies $\Box A$ and $\Box\Box A$, but **S2** does not.

The question then arises naturally, concerning the structure of the relations among these modal notions, as to which entail which in $\mathbf{E}_{\rightleftarrows}$. In order to facilitate the work, we define $\Diamond A$ ("it is possible that A") as $\sim\Box\sim A$, and proceed to list some properties and derived rules concerning $\Diamond$, $\Box$, and $\sim$. It will also be convenient at this point (and also later) to introduce the sign "$\rightleftarrows$," *not* as a connective (as in §8.7), but rather as a metalinguistic predicate: we will write "$A\rightleftarrows B$" to mean that both $A\rightarrow B$ and $B\rightarrow A$ are theorems of $\mathbf{E}_{\rightleftarrows}$.

It is easy to verify the following:

$$\Diamond A \rightleftarrows \sim\Box\sim A$$
$$\Box A \rightleftarrows \sim\Diamond\sim A$$
$$\Diamond\sim A \rightleftarrows \sim\Box A$$
$$\Box\sim A \rightleftarrows \sim\Diamond A$$
$$\Diamond\Diamond A \rightleftarrows \Diamond A$$
$$\Box\Box A \rightleftarrows \Box A$$
$$\vdash A\rightarrow B\rightarrow.\Diamond A\rightarrow\Diamond B$$
$$\vdash A\rightarrow B\rightarrow.\Box A\rightarrow\Box B$$

all of which are as we would expect. And of course the obvious introduction and elimination rules for $\Diamond$ and $\Box$ are also easily forthcoming:

Necessity introduction ($\Box$I). From a proof of A_{a} on the hypothesis $A\rightarrow A_{\{k\}}$, to infer $\Box A_{a-\{k\}}$, provided $k\in a$.

Necessity elimination ($\Box$E). From $\Box A_{a}$ to infer A_{a}.

Possibility introduction ($\Diamond$I). From A_{a} to infer $\Diamond A_{a}$.

Possibility elimination ($\Diamond$E). From $\Diamond A_{a}$ and $A\rightarrow B_{b}$ to infer $\Diamond B_{a\cup b}$.

We shall also find it convenient to have rules for collapsing double possibilities ($\Diamond\Diamond$) and for doubling necessities ($\Box\Box$):

Double possibility elimination ($\Diamond\Diamond$E). From $\Diamond\Diamond A_{a}$ to infer $\Diamond A_{a}$.

Double necessity introduction ($\Box\Box$I). From $\Box A_{a}$ to infer $\Box\Box A_{a}$.

Following the lead of Parry 1939, we are now in a position to introduce the notion of a *modality*, as follows: if A is a propositional variable, A is a modality; and if A is a modality, so are $\Box A$, $\Diamond A$, and $\sim A$. If A and B are modalities, then we say A and B are *reducible to* each other if $A\rightleftarrows B$; otherwise they are *distinct*.

We also say that a modality is in *normal form* if it has either the form of a variable preceded by an alternation (perhaps empty) of $\Box$'s and $\Diamond$'s, e.g., $\Box\Diamond\Box\Diamond\Box A$, or the form of the negation of such, e.g., $\sim\Box\Diamond A$. Clearly every modality is reducible to a modality in normal form since (1) $\Box\sim A \rightleftarrows \sim\Diamond A$ and $\Diamond\sim A \rightleftarrows \sim\Box A$ permits us to drive all negations to the left; (2) $A \rightleftarrows \sim\sim A$ permits the elimination of all but at most one of the thus collected negation signs; and (3) $\Box\Box A \rightleftarrows \Box A$ and $\Diamond\Diamond A \rightleftarrows \Diamond A$ allow the removal of any doubling of $\Box$'s or $\Diamond$'s. This argument, as well as several below, depends on the replacement theorem for $\mathbf{E}_{\rightarrow}$ described at the end of §9.

Did every modality in normal form represent a distinct modality, we should have exhausted our reductive resources; but in view of the following theorems, a further reduction is possible.

1	$\Box\Diamond A_{\{1\}}$	hyp
2	$\quad \Diamond\Box\Diamond A \rightarrow \Diamond\Box\Diamond A_{\{2\}}$	hyp
3	$\quad \Box\Diamond A_{\{1\}}$	1 reit
4	$\quad \Diamond\Box\Diamond A_{\{1\}}$	3 $\Diamond$I
5	$\quad \Diamond\Box\Diamond A_{\{1,2\}}$	2 4 $\rightarrow$E
6	$\Box\Diamond\Box\Diamond A_{\{1\}}$	2–5 $\Box$I
7	$\Box\Diamond A \rightarrow \Box\Diamond\Box\Diamond A$	1–6 $\rightarrow$I

(Recall that as defined, $\Box A$ is always an entailment, and hence reiterable in $\mathbf{E}_{\rightarrow}$.)

1	$\Box\Diamond\Box\Diamond A_{\{1\}}$	hyp
2	$\quad \Diamond A \rightarrow \Diamond A_{\{2\}}$	hyp
3	$\quad \Box\Diamond\Box\Diamond A_{\{1\}}$	1 reit
4	$\quad \Diamond\Box\Diamond A_{\{1\}}$	3 $\Box$E
5	$\quad\quad \Box\Diamond A_{\{3\}}$	hyp
6	$\quad\quad \Diamond A_{\{3\}}$	5 $\Box$E
7	$\quad \Box\Diamond A \rightarrow \Diamond A$	5–6 $\rightarrow$I
8	$\quad \Diamond\Diamond A_{\{1\}}$	4 7 $\Diamond$E
9	$\quad \Diamond A_{\{1\}}$	8 $\Diamond\Diamond$E
10	$\quad \Diamond A_{\{1,2\}}$	2 9 $\rightarrow$E
11	$\Box\Diamond A_{\{1\}}$	2–10 $\Box$I
12	$\Box\Diamond\Box\Diamond A \rightarrow \Box\Diamond A$	1–11 $\rightarrow$I

These two theorems lead to

$$\Box\Diamond\Box\Diamond A \rightleftarrows \Box\Diamond A.$$

And, taking A as $\sim A$, we have

$$\Box\Diamond\Box\Diamond{\sim}A \rightleftarrows \Box\Diamond{\sim}A,$$

from which by the replacement theorem one obtains

$$\sim\Box\Diamond\Box\Diamond{\sim}A \rightleftarrows {\sim}\Box\Diamond{\sim}A,$$

which reduces to its normal form equivalent,

$$\Diamond\Box\Diamond\Box A \rightleftarrows \Diamond\Box A.$$

Having both $\Box\Diamond\Box\Diamond A \rightleftarrows \Box\Diamond A$ and $\Diamond\Box\Diamond\Box A \rightleftarrows \Diamond\Box A$ at our disposal, we can evidently reduce every modality in normal form to a modality in normal form with no more than three consecutive modal operators; hence we can so reduce every modality. $\mathbf{E}_{\rightarrow}$ therefore distinguishes no more than the following fourteen modalities, which we arrange in a way that exhibits which are known to entail which:

Positive modalities:

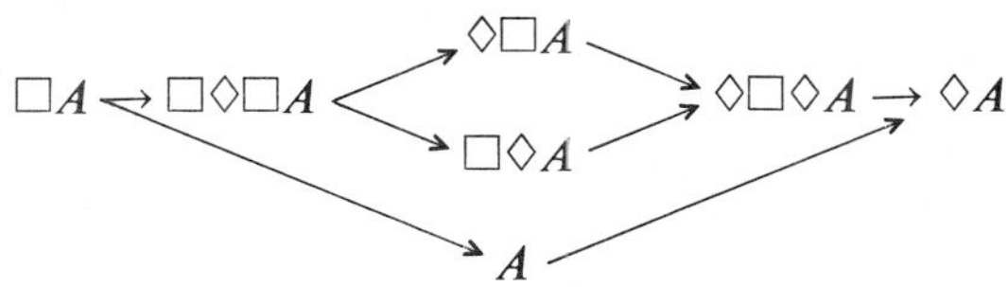

Negative modalities:

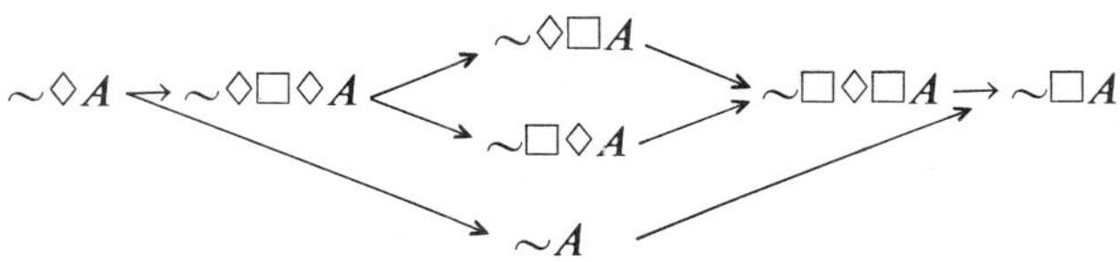

That this diagram is complete, in the sense that no entailments hold among the fourteen modalities except those exhibited, can be made to follow from a result of Parry 1939 to the effect that the same fourteen modalities are distinguished in the system **S4** of Lewis and Langford 1932. But for the system here considered, we give a just slightly different proof, using a property available in $\mathbf{E}_{\rightarrow}$ but not in **S4**. We first simplify the problem by making two observations.

In the first place, that none of the positive modalities is reducible to any of the negative modalities is shown by the ordinary two-valued interpretation of $\rightarrow$ and $\sim$; since with this interpretation both $\Box$ and $\Diamond$ are identity-operators: i.e., $\Diamond A = \Box A = A$.

Second, since the table for the negative modalities is just the contrapositive of the table for the positive modalities, the validity of contraposition in $\mathbf{E}_{\rightarrow}$ implies that any non-entailments in the latter table will be reflected in the former — so that we need to show for the positive modalities alone that no entailments hold beside those exhibited.

All the facts required for the positive modalities can be summed up in the following four-part observation: where φ and ψ are sequences, perhaps empty, of $\Box$'s and $\Diamond$'s, for some A (namely, any variable p), each of the following is unprovable.

a $\varphi\Diamond A \rightarrow \psi\Box A$
b $\Diamond\varphi A \rightarrow \Box\psi A$
c $\varphi_1\Diamond\varphi_2 A \rightarrow \psi A$ if ψ contains no $\Diamond$
d $\varphi A \rightarrow \psi_1\Box\psi_2 A$ if φ contains no $\Box$

The last two observations are of course equivalent by contraposition, and by relations between $\sim\Diamond$ and $\Box\sim$, and between $\sim\Box$ and $\Diamond\sim$.

Proof of a is by the following matrix of Wajsberg and (independently) Parry, as reported in Lewis and Langford 1932, p. 492. In this example, as in those following, the tables for $\Box$ and $\Diamond$ are computed according to the definitions $\Box A = A \rightarrow A \rightarrow A$ and $\Diamond A = \sim\Box\sim A$ (and of course the matrices below satisfy the axioms and rules of $\mathbf{E}_{\overset{\sim}{\rightarrow}}$).

$\rightarrow$	*1*	*2*	*3*	*4*	$\sim$	$\Diamond$	$\Box$
**1*	*1*	*4*	*4*	*4*	*4*	*1*	*1*
**2*	*1*	*1*	*4*	*4*	*3*	*1*	*4*
3	*1*	*4*	*1*	*4*	*2*	*1*	*4*
4	*1*	*1*	*1*	*1*	*1*	*4*	*4*

When A has the value *2*, $\varphi\Diamond A = 1$ while $\psi\Box A = 4$; but $1 \rightarrow 4 = 4$.

For b we may use the matrix of Ackermann 1956, which satisfies $\mathbf{E}_{\overset{\sim}{\rightarrow}}$, though not **S4**:

$\rightarrow$	*0*	*1*	*2*	*3*	*4*	*5*	$\sim$	$\Diamond$	$\Box$
0	*3*	*3*	*3*	*3*	*3*	*3*	*5*	*2*	*0*
1	*0*	*3*	*3*	*0*	*3*	*3*	*4*	*2*	*0*
2	*0*	*0*	*3*	*0*	*0*	*3*	*3*	*2*	*0*
**3*	*0*	*0*	*0*	*3*	*3*	*3*	*2*	*5*	*3*
**4*	*0*	*0*	*0*	*0*	*3*	*3*	*1*	*5*	*3*
**5*	*0*	*0*	*0*	*0*	*0*	*3*	*0*	*5*	*3*

When A has the value *0*, $\Diamond\varphi A = 2$ while $\Box\psi A = 0$; but $2 \rightarrow 0 = 0$.

For c and d we require the matrix of Parry 1939;

$\rightarrow$	*0*	*1*	*2*	*3*	*4*	*5*	*6*	*7*	$\sim$	$\Diamond$	$\Box$
0	*7*	*7*	*7*	*7*	*7*	*7*	*7*	*7*	*7*	*0*	*0*
1	*2*	*7*	*2*	*7*	*2*	*7*	*2*	*7*	*6*	*5*	*1*
2	*1*	*1*	*7*	*7*	*1*	*1*	*7*	*7*	*5*	*6*	*2*
3	*0*	*1*	*2*	*7*	*0*	*1*	*2*	*7*	*4*	*7*	*3*
**4*	*3*	*3*	*3*	*3*	*7*	*7*	*7*	*7*	*3*	*4*	*0*
**5*	*2*	*3*	*2*	*3*	*2*	*7*	*2*	*7*	*2*	*5*	*1*
**6*	*1*	*1*	*3*	*3*	*1*	*1*	*7*	*7*	*1*	*6*	*2*
**7*	*0*	*1*	*2*	*3*	*0*	*1*	*2*	*7*	*0*	*7*	*7*

For c, set $A = 3$, whence $\psi A = 3$ and $\varphi_1\Diamond\varphi_2 A = 7$. And for d, let $A = 4$, whence $\varphi A = 4$ and $\psi_1\Box\psi_2 A = 0$. But $7 \rightarrow 3 = 4 \rightarrow 0 = 3$. (This matrix, incidentally, also suffices to prove b. Let $A = 5$; then φA and ψA are both either *5* or *1*; in either case $\Diamond\varphi A = 5$ and $\Box\psi A = 1$, and $5 \rightarrow 1 = 3$.)

That these four facts do all the work required can be read off from the following table, where an arrow indicates a provable entailment and the letters indicate the observations in virtue of which the entailment fails. (When more than one letter appears in a column, what is meant is that any one of these observations will suffice to prove the point.)

$\rightarrow$	A	$\Box A$	$\Box\Diamond\Box A$	$\Diamond\Box A$	$\Box\Diamond A$	$\Diamond\Box\Diamond A$	$\Diamond A$
A	$\rightarrow$	d	d	d	d	d	$\rightarrow$
$\Box A$	$\rightarrow$	$\rightarrow$	$\rightarrow$	$\rightarrow$	$\rightarrow$	$\rightarrow$	$\rightarrow$
$\Box\Diamond\Box A$	c	c	$\rightarrow$	$\rightarrow$	$\rightarrow$	$\rightarrow$	$\rightarrow$
$\Diamond\Box A$	c	b, c	b	$\rightarrow$	b	$\rightarrow$	$\rightarrow$
$\Box\Diamond A$	c	a, c	a	a	$\rightarrow$	$\rightarrow$	$\rightarrow$
$\Diamond\Box\Diamond A$	c	a, b, c	a, b	a	b	$\rightarrow$	$\rightarrow$
$\Diamond A$	c	a, b, c, d	a, b, d	a, d	b, d	d	$\rightarrow$

§11. Necessity: historical remarks. In the modal systems of Lewis and Langford 1932, necessity has the property (which may be taken as a defining property) that $\Box A$ is interchangeable with $\bar{A}\rightarrow A$. This definition (let us use temporarily $\Box' A$ as short for $\bar{A}\rightarrow A$) is not without plausibility. In $\mathbf{E}_{\rightarrow}$ for example we have

$$\Box' A \rightarrow A,$$
$$A\rightarrow B\rightarrow.\Box' A\rightarrow\Box' B,$$

and even

$$\Box' A \rightarrow \Box A$$

Proofs of the latter two are as follows:

1	$A\rightarrow B_{\{1\}}$	hyp
2	$\quad\bar{A}\rightarrow A_{\{2\}}$	hyp
3	$\quad\quad\bar{B}_{\{3\}}$	hyp
4	$\quad\quad\bar{A}_{\{1,3\}}$	1 reit 3 contrap
5	$\quad\quad A_{\{1,2,3\}}$	2 reit 4 $\rightarrow$E
6	$\quad\quad B_{\{1,2,3\}}$	1 reit 5 $\rightarrow$E
7	$\quad\bar{B}\rightarrow B_{\{1,2\}}$	3–6 $\rightarrow$I
8	$\bar{A}\rightarrow A\rightarrow.\bar{B}\rightarrow B_{\{1\}}$	2–7 $\rightarrow$I
9	$A\rightarrow B\rightarrow.\Box' A\rightarrow\Box' B$	1–8 $\rightarrow$I

1	$\bar{A}\rightarrow A_{\{1\}}$	hyp
2	$\quad A\rightarrow A_{\{2\}}$	hyp
3	$\quad\quad\bar{A}_{\{3\}}$	hyp
4	$\quad\quad A_{\{1,3\}}$	1 reit 3 $\rightarrow$E
5	$\quad\bar{\bar{A}}_{\{1\}}$	3–4 $^{-}$I
6	$\quad A_{\{1\}}$	5 $^{=}$E
7	$\quad A_{\{1,2\}}$	2 6 $\rightarrow$E
8	$\Box A_{\{1\}}$	2–7 $\Box$I
9	$\Box' A \rightarrow \Box A$	1–8 $\rightarrow$I

So $\Box'$ has *some* of the properties of necessity; furthermore there are many logically necessary propositions A that do satisfy the condition $\bar{A}\rightarrow A$. If, for example, in any axiom of the full system **E** of Chapter IV, we replace every subformula $A\rightarrow B$ by $\bar{A}\vee B$, then the resulting formula C is such that $\bar{C}\rightarrow C$ is a theorem of **E** — a fact which lends plausibility to the Lewis definition. Moreover, there is a sense in which the definition is entirely suitable in systems of strict "implication," where the "implicative" connective is understood as the necessity of material "implication." And this sense can be explained in **E**. For let us define

$$A \dashv B =_{\text{df}} \Box(\bar{A}\vee B)$$

and

$$\Box''A =_{\text{df}} (\bar{A}\dashv A)$$

Then trivially $\Box''$ is the same as $\Box$, since in **E**,

$$\Box''A \rightleftarrows (\bar{A}\dashv A) \rightleftarrows \Box(\bar{\bar{A}}\vee A) \rightleftarrows \Box A.$$

The only error lies in interpreting the fishhook as implication.

When the arrow is entailment, however, the condition $\bar{A}\rightarrow A$ is much too strong for necessity. On the grounds that $\bar{A}\rightarrow A\rightarrow A$ is known as the "Law of Clavius" in the Polish Quarter of Dublin (see Prior 1956), let us agree to follow Storrs McCall in calling a proposition *Clavian* if it is entailed by its own negation. Then the conjecture we wish to refute is that all necessary propositions are Clavian, and two rather different classes of counterexamples suggest themselves.

First, there are truth functional counterexamples: not every tautology is Clavian, i.e., entailed by its own denial. The shortest example of such a tautology is doubtless the non-Clavian $(A\vee\bar{A})\&(B\vee\bar{B})$; but this topic belongs to the next chapter. (See Fact 9 of §16.2.2 for necessary and sufficient conditions under which a truth functional formula is Clavian.)

A class of counterexamples falling within the scope of the present limitation to formulas in entailment and negation is not, however, hard to come by, for, as we shall see below, though some entailments are necessarily true, *no* entailment is implied by its own denial; more generally, the denial of an entailment never entails an entailment. And this is as it should be. We herewith offer the reader a blanket invitation to find a case in the literature where someone seriously presents an argument having the form "A doesn't follow from B; hence C follows from D." No one argues in this way, and for good reason; hence $\overline{A\rightarrow B}\rightarrow.A\rightarrow B$ is not a necessary condition for the necessity of $A\rightarrow B$ (though it is a sufficient condition: $\Box'A\rightarrow\Box A$).

It is therefore reasonable that $\mathbf{E}_{\overline{\rightarrow}}$ should have $\Box(A\rightarrow A)$ as a theorem without also having $\overline{A\rightarrow A}\rightarrow.A\rightarrow A$, since $A\rightarrow A$ is necessary, but not Clavian.

Defining $\Box A$ as $A \rightarrow A \rightarrow A$ also shows that modal notions have more stability than they have been given credit for. Of the familiar systems of truth functions with modality, **S1** and **S2** of Lewis and Langford 1932, and **M** of Feys 1937 and von Wright 1951, all distinguish infinitely many modalities. **S3** distinguishes forty-two, **S4** distinguishes fourteen, **S5** distinguishes six, and Parry 1939 (among others) discusses other systems with other finite numbers of modalities. The difficulty of making a satisfactory choice among all these options has led some writers to the conclusion that our modal notions are altogether unstable, and that modal logic is a Bad Thing.

From the present point of view the decision is not difficult to make. We may rule out immediately the weak modal structures of **S1** and **S2** — and any weaker structures; any system which does not admit the necessity of its own theorems simply has no theory of logical necessity. (We exempt from this consideration systems like those of Łukasiewicz 1952 and Lemmon 1957, which are designed explicitly to exhibit relations among modalities, and have no truths provably necessary.) The system **M** of Feys-von Wright does not have this defect, but the primitive rule of necessitation is non-normal, in the sense that the corresponding implication $A \rightarrow \Box A$ is not a theorem, and cannot be replaced by a normal rule in the presence of finitely many axiom schemata; hence the system as usually formulated lacks elegance. (On this point see Kripke 1965.)

But **S3**, **S4**, and systems in between, *all* have a structure of modalities identical with that of **E**, when necessity is properly defined. To be sure, Parry 1939 has shown that **S3** has forty-two distinct modalities, and that **S4** has fourteen modalities — but this result is based on the unsatisfactory definition $\Box'$ of necessity considered above. (The collapse of **S3**'s forty-two $\Box'$ modalities into **S4**'s fourteen $\Box'$ modalities is connected with $\Box A \rightarrow \Box' A$, which is provable in **S4**, but fails in **S3**.) Indeed **E** itself, like **S3**, has forty-two $\Box'$-modalities.

However, in Lewis's system **S3**, which has as axioms for the calculus of strict implication with negation

$A \rightarrow A$

$(A \rightarrow .B \rightarrow C) \rightarrow .A \rightarrow B \rightarrow .A \rightarrow C$

$A \rightarrow B \rightarrow .C \rightarrow A \rightarrow .C \rightarrow B$

$A \rightarrow B \rightarrow .C \rightarrow D \rightarrow .A \rightarrow B$

$A \rightarrow B \rightarrow .A \rightarrow \bar{B} \rightarrow .A \rightarrow C$

$A \rightarrow \bar{\bar{A}}$

$\bar{B} \rightarrow \bar{A} \rightarrow .A \rightarrow B$

(see Hacking 1963), we have as theorems the group of formulas $\mathbf{E}_{\rightarrowtail}1$–$\mathbf{E}_{\rightarrowtail}6$ listed at the beginning of §9 as the axioms of $\mathbf{E}_{\rightarrowtail}$. Hence, under the obvious mapping, $\mathbf{E}_{\rightarrowtail}$ is contained in **S3** — and **S3** (like **S4**) has fourteen modalities (properly understood). And in general, all the systems between **E** and **S4** (inclusive) have fourteen modalities — a fact which seems to make the situation more stable than it at first appeared.

The structure of modalities in **S5**, however, admits modal fallacies. In **S4** we have:

$$\Box A \to \Box\Box A,$$

i.e., "if A follows from a law of logic, then that fact itself follows from a law of logic." We take this to be an obvious truth of entailment. But in **S5** we have

$$\Diamond A \to \Box\Diamond A,$$

i.e., "If $\overline{A}$ *fails* to follow from the law of logic $\overline{A}{\to}A$, then that fact itself follows from a law of logic," which (as it seems to us) is not generally true. In special cases it is true that the denial of an entailment follows from a law of logic; for example we have

$$(\overline{A\to A\to\overline{A\to A}}\to\overline{A\to A\to\overline{A\to A}})\to\overline{A\to A\to\overline{A\to A}}.$$

But denials of entailments, unlike entailments themselves, are sometimes entailed by purely non-necessitive facts. We know for example that if A is *true* and B is *false*, then $A{\to}B$ is *false*. In contrast with this situation, we have that no purely non-necessitive proposition ever entails a necessitive; our reasons for holding a proposition logically necessary are always logical reasons. One would therefore expect $\Box A \to \Box\Box A$, but not $\Diamond A \to \Box\Diamond A$, since the latter leads to $A \to \Box\Diamond A$ in violation of the condition that pure non-necessitives do not entail entailments; and we observe that the first property does, and the second does not, flow in a natural way from the axioms of $\mathbf{E}_{\overset{\sim}{\to}}$ (without *ad hoc* assumptions about modalities) *provided* we mean by "A is necessary" that A follows from the true entailment $A{\to}A$.

We should like to add that there is nothing (much) circular in this rejection of the **S5** modal structure. Our intent from the word *go* was to provide a theory of logical consequence, or entailment, rather than a theory of necessarily true propositions, or of necessitives (§5.2). To be sure, a theory about necessary propositions and necessitives emerges from the study of entailment, which is to be expected, since entailment has to do with inferences which are logically necessary. But the initial motivation of Lewis, though in some sense similar to ours ("Find a decent theory of 'if . . . then — ' ") is also different in that he confused "if . . . then — " with the necessity of material "implication."

We have offered some reason to believe that our serious intuitions about modalities are reasonably stable after all, and that they settle down on the number fourteen. If further evidence is required, we point out that the system **E** of Chapter IV itself has fourteen axioms, and that we have it on no less an authority than Matthew 80(?) that

> . . . all the generations from Abraham to David are fourteen generations; and from David until the carrying away into Babylon are fourteen generations; and from the carrying away into Babylon unto Christ are fourteen generations.

(We remark that $3 \times 14 = 42$.)

We also record that there are fourteen radial lines in the millstone embedded in the courtyard of Yale's Saybrook College, just outside the office in which these truths first became self-evident. And Schiller 1931, commenting on two of his critics (A. C. Ewing and Rex Knight), says of the latter, "The first of his nine points (adding up with Dr. Ewing's five to President Wilson's famous fourteen!) is fundamental." No one can fly in the face of evidence like this. (We should note that as this is being written, more evidence of a similar sort keeps flowing in. It develops that even the astrological tradition is wrong; Schmidt 1970, points out that there *should* be fourteen constellations in the zodiac, because of a slight change in the tilt of the earth's axis over the past 2,000 years.)

Finally we acknowledge that there is something a trifle odd about the definition of necessity as $A \rightarrow A \rightarrow A$. This gets fixed in §31.4.

§12. Fallacies. Theorems of the previous chapter concerning fallacies of modality and relevance may all be extended to $\mathbf{E}_{\rightarrow}$. We have:

THEOREM. If A contains no occurrence of $\rightarrow$, then for no B and C is $A \rightarrow .B \rightarrow C$ provable in $\mathbf{E}_{\rightarrow}$.

THEOREM. If $A \rightarrow B$ is provable in $\mathbf{E}_{\rightarrow}$, then A and B share a variable.

These are corollaries of similar theorems for the full system **E** (proved in §22.1). We also have another relevance theorem which extends a result of the previous chapter.

THEOREM. If A is a theorem of $\mathbf{E}_{\rightarrow}$, then every variable in A occurs at least once as an antecedent part and at least once as a consequent part, where as in §§5.1.2 and 9, these terms are defined as follows: (1) A is a consequent part of A. (2) If $\bar{B}$ is a consequent {antecedent} part of A, then B is an antecedent {consequent} part of A. (3) If $B \rightarrow C$ is a consequent {antecedent} part of A, then B is an antecedent {consequent} part of A, and C is a consequent {antecedent} part of A.

The proof uses the matrix of §5.1.2, which satisfies all theorems of $\mathbf{E}_{\rightarrow}$, with $+$ values designated.

$\rightarrow$	-2	-1	$+1$	$+2$	$\sim$
-2	$+2$	$+2$	$+2$	$+2$	$+2$
-1	-2	$+1$	$+1$	$+2$	$+1$
$*+1$	-2	-1	$+1$	$+2$	-1
$*+2$	-2	-2	-2	$+2$	-2

PROOF. Let p occur in A only as a consequent {antecedent} part. Then, when p is assigned -2 $\{+2\}$ and all other variables are assigned the value $+1$, A assumes the value -2. We will show this by proving that for every subformula B of A (including A), for this assignment of values to the variables of A,

(a) if B does not contain p, then $B = \pm 1$;

(b) if B contains p and is a consequent {antecedent} part of A, then $B = -2$ $\{+2\}$.

The proof is by induction on the length of parts.

The base clause is obvious. For formulas $\overline{B}$ there are two cases: (i) B does not contain p. Then neither will $\overline{B}$, and $\overline{\pm 1} = \mp 1$. (ii) B contains p. Then $\overline{B}$ is a consequent {antecedent} part if B is an antecedent {consequent} part, and $\overline{+2} = -2$ $\{\overline{-2} = +2\}$.

For formulas $B \rightarrow C$, we require four cases: (i) Neither B nor C contains p. Then neither will $B \rightarrow C$, and since $B = \pm 1$ and $C = \pm 1$, we have $B \rightarrow C = \pm 1$. (ii) B contains p, C does not. Then $B \rightarrow C$ contains p, and if B is a consequent {antecedent} part, $B \rightarrow C$ is an antecedent {consequent} part, and $-2 \rightarrow \pm 1 = +2$ $\{+2 \rightarrow \pm 1 = -2\}$. (iii) like (ii). (iv) both B and C contain p. Then $B \rightarrow C$ contains p, and if B is a consequent {antecedent} part and C is an antecedent {consequent} part, then $B \rightarrow C$ is an antecedent {consequent} part, and $-2 \rightarrow +2 = +2$ $\{+2 \rightarrow -2 = -2\}$.

Hence, since A is a consequent part of A, and contains p, A takes -2.

This property obviously does not extend to the full system $\mathbf{E}$ of §21, in which such theorems as $A \& B \rightarrow A$ and $A \rightarrow A \vee B$ may have one variable occurring just once. But see the last theorem of §22.1.3.

We pointed out at the beginning of this section that, if $A \rightarrow .B \rightarrow C$ is provable in $\mathbf{E}_{\rightarrow}$, then A contains an arrow; hence fallacies of modality of this type are excluded from $\mathbf{E}_{\rightarrow}$. But we also argued in §11 that entailments do not follow from *denials* of entailments. Such fallacies are ruled out in $\mathbf{E}_{\rightarrow}$ by the following

THEOREM. For no A, B, C, and D, is $\overline{A \rightarrow B} \rightarrow .C \rightarrow D$ provable in $\mathbf{E}_{\rightarrow}$.

Notice that the theorem goes well beyond claiming that $\overline{A\to B}\to .C\to D$ is unprovable for *some* A, B, C, and D; rather, the fact is that it is unprovable for *every* A, B, C, and D.

PROOF. Consider the following matrix (adapted by us from Ackermann 1956; it turns out to be the same as Group III of Lewis and Langford 1932) with *2* and *3* designated.

$\to$	*0*	*1*	*2*	*3*	$\sim$
0	*2*	*2*	*2*	*2*	*3*
1	*0*	*2*	*0*	*2*	*2*
**2*	*0*	*0*	*2*	*2*	*1*
**3*	*0*	*0*	*0*	*2*	*0*

These matrices satisfy the axioms and rules of $\mathbf{E}_{\overline{\to}}$. $C\to D$ assumes as values only *0* and *2*, and $\overline{A\to B}$ assumes only *3* and *1*, but $1\to 0 = 1\to 2 = 3\to 0 = 3\to 2 = 0$; hence $\overline{A\to B}\to .C\to D$ is unprovable.

This result also shows that $\mathbf{E}_{\overline{\to}}$ is, in a plausible sense, incomplete with respect to negation. Namely, we say that $\mathbf{E}_{\overline{\to}}$ is *Lindenbaum complete* (see McKinsey 1941, especially p. 122) if it has the following property: if no substitution instance of A is provable, then $\overline{A}$ is provable. (We offer this as a plausible syntactical condition which ought to be satisfied by a semantically complete system.) And A is *satisfiable* if some substitution instance A' of A is a theorem of $\mathbf{E}_{\overline{\to}}$. We then note that *both* $\overline{A\to A}\to .A\to A$ and $\overline{\overline{A\to A}\to .A\to A}$ are unsatisfiable in this sense, the latter by the obvious two-valued matrix, the former by the Ackermann matrix above. Hence $\mathbf{E}_{\overline{\to}}$ is not Lindenbaum complete.

The formula

$$\overline{\overline{A\to A}\to .A\to A},$$

a descendant of $\overline{\overline{A\to B}\to .C\to D}$, is of particular interest in showing how the theory of entailment diverges from the classical theories of implication. It is *true* of entailment, but if we read the arrow as material, strict, or intuitionistic "implication," the formula comes out *false*. Which is (again) as it should be: the classical theories are not simply (and vaguely) "inadequate to meet the demands of our intuitive ideas about if . . . then —"; they're wrong.

The above considerations reinforce our hunch that $\overline{\overline{A\to B}\to .C\to D}$ is true, and we have considered adding it to $\mathbf{E}_{\overline{\to}}$ as a further axiom. (Note that as a special case we would then have $\Diamond A\to\Box\Diamond A$.) Such a step would be a move in the direction of completing $\mathbf{E}_{\overline{\to}}$. But we have no reason to believe that the resulting system is semantically complete — indeed we do not even know what it is best to mean by the latter term. So we defer adding this axiom until the semantics of the situation is better understood. (This paragraph is dated, having been written a decade or so before the semantics and completeness results of Chapter IX became available. At this writing, however, the newer techniques have not been turned back on the problem of $\overline{A \to B} \to . C \to D$, and we remain perplexed.)

But there is still a little philosophical hay to be made out of these facts, namely, that the structure of modalities in **S5** is false to our ordinary, casual sense of necessity. It might be observed that our argument is in a sense circular: we reject

$$\Diamond A \to \Box\Diamond A$$

on the grounds that $\Diamond A$ (the denial of an entailment) does not entail $\Box\Diamond A$ (an entailment), and also on the grounds that if the principle in question held, we would have

$$A \to \Box\Diamond A,$$

which would allow us to reason from a propositional variable to a necessitive, a policy already rejected in §5. And there is a certain sense in which the charge of circularity is just, since the two conditions laid down above for rejecting $\Diamond A \to \Box\Diamond A$ were already built into $\mathbf{E}_{\rightarrow}$.

It is therefore gratifying to have an argument against the **S5** structure of modalities that is completely independent of **E** or $\mathbf{E}_{\rightarrow}$. For the following we are indebted to the late J. C. C. McKinsey, both for his paper, McKinsey 1945, and for a letter of Feb. 8, 1952, explaining the motivation for the paper.

He mentions in the paper that one of our reasons for believing that

$\Diamond$(Lions are indigenous to Alaska)

is that

Lions are indigenous to Africa

is true. That is, we can replace the descriptive term "are indigenous to Alaska" in the statement of possibility, by another descriptive term which makes the sentence come out true. Which leads one naturally to the following *principle:*

> If a sentence A is necessary, and if B results from A by replacing some or all of the descriptive terms in A by other terms, then B is true.

As is pointed out in McKinsey 1945, we need not be clear about which terms are "descriptive" and which are not; we need only assume that the distinction has been made, and that the standard logical relations ($\sim$, &, $\vee$, entailment, and the like) are not among the descriptive terms. As McKinsey shows, this leads to a system at least as strong as **S4**, but not necessarily as strong as **S5**. (Drake 1962 improves the result: the issue is exactly **S4**.) McKinsey's argument against **S5** (we quote from the letter referred to above) proceeds as follows (supposing that if we can distinguish between "descriptive" and "non-descriptive" terms at all, then "sugar" and "vinegar" are as good candidates for the former as any):

> Now consider the sentence:
>
> (1) $\Diamond$(Sugar is sweet, and vinegar is not sweet).
>
> This is clearly true, since the sentence
>
> (2) Sugar is sweet, and vinegar is not sweet
>
> is true. On the other hand I do not think that the sentence
>
> (3) $\sim\Diamond\sim\Diamond$(Sugar is sweet, and vinegar is not sweet)
>
> is true. For if (3) were true, then (1) would be necessary, and hence, by the principle stated above, the sentence
>
> (4) $\Diamond$(Sugar is sweet, and sugar is not sweet)
>
> would be true, since (4) results from (1) by replacing the descriptive term "vinegar" by the term "sugar."

The principle seems to us plausible, to the extent that we understand it, and renders even more plausible the rejection of $\Diamond A \to \Box\Diamond A$ in $\mathbf{E}_{\to}$ and **E** (or indeed anywhere).

If you're keeping track of principles, notice that McKinsey's declares not only against $\Diamond A \to \Box\Diamond A$ but even against $\overline{\Diamond A} \vee \Box\Diamond A$. But the arguments of the earlier part of this section and of §§5 and 11 militate only against the former, relying as they do not only on modal considerations but also on relevance.

We wish to make it clear that in discussing the fallacies above we have been concerned primarily with relations between entailment and modality, where the situation appears to be pellucid. But we are still perplexed as to the proper attitude to take toward modalities and truth functions. It would be possible for example to buy our earlier arguments, and still be **S5**-ish in having $\Diamond A \supset \Box\Diamond A$ (i.e., $\overline{\Diamond A} \vee \Box\Diamond A$) in an extension of $\mathbf{E}_{\to}$ (say **E** or **R**). Meyer has shown that $\Box(A \supset B) \supset . \Box A \supset \Box B$ is not a theorem of **E** (thus reinforcing the contention that $\Box(\bar{A} \vee B)$ is a bewildering notion), and it is unclear what the effect of adding this is. We will return to these topics briefly in Chapter IV, but our understanding of the relation between truth functions and modality is, at present, unsatisfactory. (This remains to a certain extent true even though the profound discussion of Chapter IX has done much to illuminate the matter. And though we shan't talk about the topic in serious detail in what follows, the discussion of this section illustrates how damnably, elusively, *slippery* these ideas are. We have in this book been occasionally (e.g. in Chapter I) just as puffy as the Officers have been for many dreary years. But we would like at this point — after all, we're in parentheses — to let down our hair and admit that we would love it enormously if someone could *please* set us all straight on these matters.)

§13. Gentzen consecution calculuses: decision procedure. In §7 we presented some merge-style consecution calculuses, which are distinguished by their placing restrictions on the rule for introducing $A \to B$ on the *left* of the turnstile; here we formulate a version having no restrictions on this rule, but instead having a restricted rule for introducing $A \to B$ on the *right*. We mentioned at the end of §7 that merge-formulations are not known to lead to decision procedures. On the other hand there are consecution calculuses which do lead to decision procedures, and in particular we present in this section one which does the job not only for $\mathbf{E}_{\to}$ but for $\mathbf{E}_{\overset{\sim}{\to}}$.

The rules for entailment given below and the fundamental combinatorial lemma needed to establish the decidability of a system based on these rules are both due to Kripke 1959a, who was able to establish the equivalence of his Gentzen formulation of pure entailment with the usual axiomatic formulation; hence, applying his combinatorial lemma, he established the decidability of the pure calculus $\mathbf{E}_{\to}$ of entailment. When standard rules for negation are added to the consecution calculus of entailment, the proofs of the elimination theorem and decidability both remain valid; but Kripke's proof of the equivalence of his consecution formulation to an axiomatic formulation does not extend to the systems with negation. The bulk of this section, drawn from Belnap and Wallace 1961, will be devoted to overcoming this difficulty.

The proof of decidability makes use of two equivalent consecution calculuses (see §7.2), the first of which, $\mathrm{L_r}\mathbf{E}_{\overset{\sim}{\to}}$, is more easily proved to be equivalent to $\mathbf{E}_{\overset{\sim}{\to}}$. The second system, $\mathrm{L}\mathbf{E}_{\overset{\sim}{\to}}$, is more readily seen to be decidable. After having established the equivalence of the axiomatic and consecution calculuses, we outline the essential elements in the decision procedure.

§13.1. Calculuses. We use Greek letters to range over sequences (possibly empty) of formulas, and add the conventions (for each Greek letter) that

(1) $\tilde{\alpha}$ is a sequence of formulas each of which has the form $\overline{A}$,

(2) $\vec{\alpha}$ is a sequence of formulas each of which has the form $A \to B$, and

(3) $\overset{\sim}{\vec{\alpha}}$ is a sequence of formulas each of which has the form $\overline{A \to B}$.

The first calculus $\mathrm{L_r}\mathbf{E}_{\overset{\sim}{\to}}$ is so named because sequences of formulas always appear only on the *right* of the turnstile, unlike $\mathrm{L}\mathbf{E}_{\overset{\sim}{\to}}$, which more closely resembles the usual Gentzen calculuses.

THE CALCULUS $\mathbf{L_rE}_{\overset{\sim}{\rightarrow}}$.

Axioms.

$$\vdash \overline{A}, A \qquad \vdash \overline{A}, \overline{A\rightarrow B}, B$$

Structural rules of inference.

$$\frac{\vdash \alpha, A, B, \beta}{\vdash \alpha, B, A, \beta}(\vdash \mathrm{C}) \qquad \frac{\vdash \alpha, A, A, \beta}{\vdash \alpha, A, \beta}(\vdash \mathrm{W})$$

$$\frac{\vdash \alpha, A \qquad \vdash \beta, \overline{A}}{\vdash \alpha, \beta}(\mathrm{ER})$$

Logical rules of inference.

$$\frac{\vdash \overset{\sim}{\vec{\alpha}}, \overline{A}, B}{\vdash \overset{\sim}{\vec{\alpha}}, A\rightarrow B}(\vdash \rightarrow) \qquad \frac{\vdash \alpha, A}{\vdash \alpha, \overline{\overline{A}}}(\vdash \sim\sim)$$

THE CALCULUS $\mathbf{LE}_{\overset{\sim}{\rightarrow}}$ (Kripke 1959a).

Axioms.

$$A \vdash A$$

Logical rules of inference (generalized to give the effect of structural rules).

$$\frac{\vec{\alpha}, A \vdash B}{\vec{\alpha} \vdash A\rightarrow B}(\vdash \rightarrow) \qquad \frac{\alpha \vdash A, \gamma \qquad \beta, B \vdash \delta}{[\alpha, \beta, A\rightarrow B] \vdash [\gamma, \delta]}(\rightarrow \vdash)$$

where $[\alpha, \beta, A\rightarrow B]$ is any permutation and contraction of $\alpha, \beta, A\rightarrow B$, within the following limits: if a formula occurs n times in $[\alpha, \beta, A\rightarrow B]$, then it occurs no more than n times in α and no more than n times in β.

$$\frac{\alpha, A \vdash \beta}{\alpha \vdash [\overline{A}, \beta]}(\vdash \sim)$$

where $[\overline{A}, \beta]$ is any permutation of $\overline{A}, \beta$ with the following contraction permitted on $\overline{A}$: if $\overline{A}$ occurs n times in $[\overline{A}, \beta]$, then it occurs no more than n times in β.

$$\frac{\alpha \vdash A, \beta}{[\alpha, \overline{A}] \vdash \beta}(\sim \vdash)$$

where the notational conventions are exactly like those for the preceding rule.

The requirement that parameters for the rule $(\vdash \rightarrow)$ have certain shapes has a logical point which the reader will easily discern as being a "modal" restriction like that of §7.3. Contrariwise, the multiplicity restrictions on the rules amount only to the requirement that a logical rule not be used to perform any contraction in its conclusion which could instead have been carried out in its premisses. They have no point other than to render finite the number of possible premisses from which a conclusion might follow, a feature essential, as we shall see, to our decision method.

We show the equivalence of $\mathbf{E}_{\rightarrow}$, $\mathrm{L}\mathbf{E}_{\rightarrow}$, and $\mathrm{L_r}\mathbf{E}_{\rightarrow}$ by the method Tinker-to-Evers-to-Chance (see Turkin and Thompson 1970, p. 19).

First we need a standard elimination theorem for $\mathrm{L}\mathbf{E}_{\rightarrow}$:

THEOREM. If $\alpha \vdash \gamma$ and $\beta \vdash \delta$ are both provable in $\mathrm{L}\mathbf{E}_{\rightarrow}$, then so is $[\alpha, \beta^*] \vdash [\gamma^*, \delta]$, where β^* and γ^* are the results of deleting one or more occurrences of some formula A from β and γ, respectively, and with the number of occurrences of formulas limited as in the rules of $\mathrm{L}\mathbf{E}_{\rightarrow}$.

This can be proved after the manner of Gentzen 1934.

THEOREM. $\mathbf{E}_{\rightarrow}$ is contained in $\mathrm{L}\mathbf{E}_{\rightarrow}$, in the sense that if A is a theorem of $\mathbf{E}_{\rightarrow}$, then $\vdash A$ is a theorem of $\mathrm{L}\mathbf{E}_{\rightarrow}$.

PROOF. Analogues of the axioms of $\mathbf{E}_{\rightarrow}$ may be proved easily in $\mathrm{L}\mathbf{E}_{\rightarrow}$, and the availability of an analogue of $\rightarrow$E follows from the elimination theorem for $\mathrm{L}\mathbf{E}_{\rightarrow}$.

THEOREM. $\mathrm{L}\mathbf{E}_{\rightarrow}$ is contained in $\mathrm{L_r}\mathbf{E}_{\rightarrow}$, in the sense that if $\alpha \vdash \beta$ is provable in $\mathrm{L}\mathbf{E}_{\rightarrow}$, then $\vdash \bar{\alpha}, \beta$ is provable in $\mathrm{L_r}\mathbf{E}_{\rightarrow}$.

PROOF. This is immediate for the axiom schema of $\mathrm{L}\mathbf{E}_{\rightarrow}$ and for all the rules except $(\rightarrow\vdash)$; for the latter, use the axiom $\vdash \overline{A}, \overline{A\rightarrow B}, B$ and the rule (ER).

The last two theorems take us two-thirds the way round the circle: $\mathbf{E}_{\rightarrow} \subseteq \mathrm{L}\mathbf{E}_{\rightarrow} \subseteq \mathrm{L_r}\mathbf{E}_{\rightarrow}$.

§13.2. Completing the circle. To complete the circle we need to show that $\mathrm{L_r}\mathbf{E}_{\rightarrow}$ is contained in $\mathbf{E}_{\rightarrow}$, in the sense that if $\vdash A$ is a theorem of the former, then A is a theorem of the latter. The analogous problem for $\mathbf{R}_{\rightarrow}$ and $\mathrm{L_r}\mathbf{R}_{\rightarrow}$ can be solved readily by interpreting $\vdash A_1, \dots, A_n$ as $\overline{A_1}\rightarrow.\ \dots \rightarrow .A_{n-1}\rightarrow A_n$, and then showing by an easy inductive argument that provability of the former in $\mathrm{L_r}\mathbf{R}_{\rightarrow}$ guarantees provability of the latter in $\mathbf{R}_{\rightarrow}$. But in the case of $\mathrm{L_r}\mathbf{E}_{\rightarrow}$ we are unable to interpret $\vdash A_1, \dots, A_n$ straightforwardly as signifying $\overline{A_1}\rightarrow.\ \dots \rightarrow .A_{n-1}\rightarrow A_n$, since the former can be provable in $\mathrm{L_r}\mathbf{E}_{\rightarrow}$ while the latter is unprovable in $\mathbf{E}_{\rightarrow}$. Example:

$$\vdash \bar{p}, \overline{p\rightarrow q}, q$$

is provable in $\mathrm{L_r}\mathbf{E}_{\rightarrow}$, but

$$\bar{\bar{p}}\rightarrow.\overline{\overline{p\rightarrow q}}\rightarrow q$$

is of course unprovable in $\mathbf{E}_{\to}$. What happens is that permutation can be used in "illegitimate" ways in the consecution calculus in the course of proving some theorem, although its pernicious effects are offset toward the end of the proof by means of the modal restriction on the rule ($\vdash\to$). The problem is further complicated by the possibility of using contraction to conflate constituents which could not be brought together for purposes of contraction except by such "unlawful" permutations. Example: suppose

$$\overline{A}\to.\overline{p}\to.\overline{B}\to.\overline{p}\to C$$

is a theorem of $\mathbf{E}_{\to}$, so that

$$\vdash A, p, B, p, C$$

is a theorem of $\mathbf{L_rE}_{\to}$; then by permutation in $\mathbf{L_rE}_{\to}$ we can obtain

$$\vdash A, p, p, B, C,$$

hence by contraction

$$\vdash A, p, B, C.$$

But neither

$$\overline{A}\to.\overline{p}\to.\overline{B}\to C$$

nor any permutation thereof is likely to be a theorem of $\mathbf{E}_{\to}$. Fortunately these difficulties can be overcome by introducing a more general notion of interpretation. After stating some preliminary definitions and some facts which follow therefrom, we prove a theorem which has as a corollary the containment required to close the circle.

The definition of "Γ-formula" to follow helps us to get clear about the relations between consecutions in $\mathbf{L_rE}_{\to}$ and formulas of $\mathbf{E}_{\to}$. Let Γ be $\vdash A_1, \ldots, A_n$; then we say that (a) each A_i $(1\leq i\leq n)$ is a Γ-*formula*, and (b) if B and C are Γ-formulas, then so is $\overline{B}\to C$. Example: if Γ is $\vdash A_1, A_2, A_3$, then each node of the following tree is a Γ-formula.

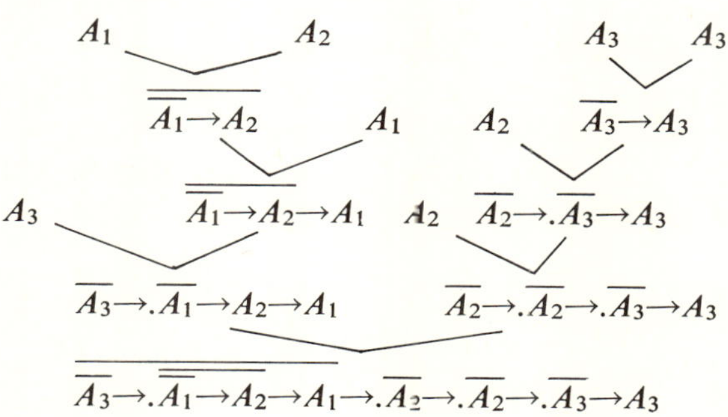

We will need to count *occurrences* of constituents of Γ in a Γ-formula; for this purpose, each Γ-formula is to have associated with it a tree diagram such as that above, with each tip labeled as to the particular constituent of Γ of which it is an occurrence. Thus, should some of the constituents in Γ look alike, as in $\vdash A, B, A$, some of the tips may be labeled as occurrences of the first A and others as occurrences of the second A. Further, *only* tips will count as occurrences, so that, if Γ is $\vdash A, B, \bar{A}\to B$, the expression $\bar{A}\to B$ is ambiguous as to whether it has the tree diagram

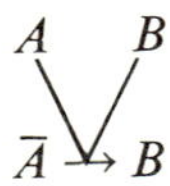

and hence does not contain any occurrence of the third constituent, $\bar{A}\to B$, of Γ, or has instead the single-node tree diagram with $\bar{A}\to B$ as both tip and bottom and hence contains occurrences of neither A nor B. In short, we are using "occurrence" in a special sense derived from labeled tree diagrams; if B is a Γ-formula, then a constituent of Γ *occurs* in B at just those places associated with a tip of the tree diagram of B which is labeled as an occurrence of that constituent. We shall not, except at one place in the argument to follow, recur to the language of tree diagrams, leaving it to the reader to interpret us in accordance with these strictures.

We may now give the following definitions of "interpretation" and "valid interpretation": if B is a Γ-formula containing at least one occurrence of every constituent of Γ, then B is an *interpretation* of Γ. An interpretation of Γ which is a theorem of $\mathbf{E}_{\to}$ is said to be *valid*. Our work is made more difficult by the fact that not every interpretation of every consecution provable in $\mathbf{L_rE}_{\to}$ is valid, but we do have the following

THEOREM. If Γ is a provable consecution of $\mathbf{L_rE}_{\to}$, then among all the interpretations of Γ, at least one is valid.

Our object now is to prepare for the proof of this theorem, from which the completion of the circle follows easily. The preparation requires a definition of "reduction" for the purposes at hand, and proof of eight facts about reductions. We may recursively define this notion (which corresponds to the ancestral of certain "legitimate" (by the standards of $\mathbf{E}_{\to}$) permutations and contractions) by the following conditions on *reduces to:*

R0. B reduces to B.
R1. $\bar{B}\to C$ reduces to $\bar{C}\to B$.
R2. $\bar{B}\to\overline{\overline{E\to F}}\to C$ reduces to $\overline{\overline{E\to F}}\to.\bar{B}\to C$.
R3. $\bar{B}\to B$ reduces to B.

R4. $\overline{B}\rightarrow.\overline{B}\rightarrow C$ reduces to $\overline{B}\rightarrow C$.

R5. If B reduces to B', B is a node in the tree diagram for A, and A' results from putting B' for B in A, then A reduces to A'.

R6. If B reduces to C and C to D, then B reduces to D.

Uses of R1–R4 will be sometimes explicit, but uses of R0, R5, and R6 are always tacit. To say that B' is a *reductum* of B is to say that B reduces to B', and it is easy to see that in these circumstances the following must hold:

(1) If B is provable in $\mathbf{E}_{\rightarrow}$, so is B' (indeed, $B\rightarrow B'$ is provable).

(2) If B is a Γ-formula, then B' is a Γ-formula containing occurrences of exactly the same constituents A_i of Γ as B.

(3) If B is a Γ-formula, then in passing from B to B' the number of occurrences of a constituent A_i of Γ is always reduced or kept the same, but never increased.

Note that (2) means reduction preserves interpretationhood, and (1) and (2) together imply the preservation of valid interpretationhood. (3) is needed to keep control of some inductions. In general we will not cite (1)–(3) explicitly, leaving it to the reader to observe that in establishing that B reduces to B' we are also showing that (1)–(3) hold for B and B'.

We will need some facts designed to show that Γ-formulas of various kinds have reducta of certain prescribed forms. We suppose $\Gamma = \vdash A_1, \ldots, A_n$, and we use A_i and A_k accordingly.

(4) If B is a Γ-formula and A_i is the one and only constituent of Γ occurring in B, then B reduces to A_i. Hence if B is an interpretation of $\vdash A$, then B reduces to A.

PROOF. Obvious from the match between R3 and clause (b) of the definition of Γ-formula.

(5) If B is a Γ-formula and A_i and A_k are the only two constituents of Γ occurring in B, then B reduces to $\overline{A_i}\rightarrow A_k$.

PROOF. B must have the form $\overline{B_1}\rightarrow B_2$, and we assume as inductive hypothesis that (5) holds for B_1 and B_2. There are three cases.

CASE 1. B_1 contains occurrences of A_i alone. Then by (4) B_1 reduces to A_i, so $B = (\overline{B_1}\rightarrow B_2)$ reduces to $\overline{A_i}\rightarrow B_2$. If B_2 contains occurrences of only A_k, another application of (4) yields the required $\overline{A_i}\rightarrow A_k$, while if B_2 contains occurrences of A_i as well, the inductive hypothesis guarantees that B_2 reduces to $\overline{A_i}\rightarrow A_k$ and accordingly that B reduces to $\overline{A_i}\rightarrow.\overline{A_i}\rightarrow A_k$, hence by R4 to $\overline{A_i}\rightarrow A_k$.

CASE 2. B_1 contains occurrences of A_k only, or B_2 of A_i only, or B_2 of A_k only. Similar to Case 1.

CASE 3. B_1 and B_2 both contain occurrences of each of A_i and A_k. Then by the inductive hypothesis each reduces to $\overline{A_i}\to A_k$, so B reduces to $\overline{\overline{A_i}\to A_k}\to.\overline{A_i}\to A_k$ and accordingly by R3 to $\overline{A_i}\to A_k$.

(6) We shall use "$\varphi\to A$" to abbreviate formulas having the form $C_1\to.C_2\to.\ \dots\to.C_n\to A$. (We include the case $n = 0$, when $\varphi\to A$ is simply A. Note that unless $n = 1$, φ is not a formula and $\varphi\to A$ does not have the form $B\to A$.) This notation is useful for expressing the following fact: every constituent A_i of Γ occurring in a Γ-formula B can be squeezed out to the right, in the sense that we can define a function *Squeez* (B, A_i), which has the effect of transforming B into a reductum with the form $\varphi\to A_i$. Indeed, *Squeez*(B, A_i) can be defined for a Γ-formula B with an occurrence of A_i quite simply as follows: (a) *Squeez*$(A_i, A_i) = A_i$; (b) if D contains an occurrence of A_i, then *Squeez*$(\overline{C}\to D, A_i) = \overline{C}\to$*Squeez*$(D, A_i)$, and (c) otherwise *Squeez*$(\overline{C}\to D, A_i) = \overline{D}\to$*Squeez*$(C, A_i)$. Obviously *Squeez*$(B, A_i)$ has the form $\varphi\to A_i$, and is by R1 a reductum of B, as required.

EXAMPLE. The following sequence finds

$$Squeez(\overline{\overline{\overline{C}\to B}\to.\overline{\overline{A}\to B}\to.\overline{C}\to C}\to.\overline{C}\to.\overline{\overline{B}\to B}\to C,\ A)$$

according to the above recipe:

$$\overline{\overline{\overline{C}\to B}\to.\overline{\overline{A}\to B}\to.\overline{C}\to C}\to.\overline{C}\to.\overline{\overline{B}\to B}\to C$$
$$\overline{\overline{C}\to.\overline{\overline{B}\to B}\to C}\to.\overline{\overline{C}\to B}\to.\overline{\overline{A}\to B}\to.\overline{C}\to C$$
$$\overline{\overline{C}\to.\overline{\overline{B}\to B}\to C}\to.\overline{\overline{C}\to B}\to.\overline{\overline{C}\to C}\to.\overline{A}\to B$$
$$\overline{\overline{C}\to.\overline{\overline{B}\to B}\to C}\to.\overline{\overline{C}\to B}\to.\overline{\overline{C}\to C}\to.\overline{B}\to A$$

(7) For one part of our argument below it is necessary to reduce B to a more specialized version of $\varphi\to A_i$, to wit, the form $\varphi\to.\overline{H}\to A_i$, where H is known to contain no occurrences of A_i. (This makes sense only when constituents of Γ other than A_i also occur in B.) Since the definition involves some uses of contraction, we shall use the notation

Sqz(B, A_i)

for this more specialized version of *Squeez*(B, A_i). Suppose, then, that B is a Γ-formula containing occurrences of some constituent A_i of Γ and also of some constituent of Γ other than A_i. First write *Squeez*(B, A_i) in the form

$$\varphi\to.\overline{F}\to G$$

where G contains *only* A_i's and where F, which might or might not contain an occurrence of A_i, is known to contain some other constituent of Γ; this representation of $Squeez(B, A_i)$ is unique. The definition of Sqz now depends on whether or not F contains any occurrence of A_i: if A_i does not occur in F, then $Sqz(B, A_i)$ is just $\varphi\to.\overline{F}\to A_i$, which by (4) is a reductum of $\varphi\to.\overline{F}\to G$; on the other hand, if A_i does occur in F, so that F contains both A_i and also a constitutent of Γ other than A_i, we let $Sqz(B, A_i) = \varphi\to.\overline{G}\to Sqz(F, A_i)$. Evidently, $Sqz(B, A_i)$ so defined has the proper form $\varphi\to.\overline{H}\to A_i$ with A_i not in H, and is by (4) and R1 a reductum of B.

EXAMPLE. The following sequence finds

$$Sqz(\overline{\overline{\overline{\overline{\overline{A}\to A}\to A}\to.\overline{B}\to B}\to.\overline{A}\to A}\to.\overline{B}\to C,\ A)$$

according to the above recipe:

$$\overline{\overline{\overline{\overline{\overline{A}\to A}\to A}\to.\overline{B}\to B}\to.\overline{A}\to A}\to.\overline{B}\to C$$
$$\overline{B}\to C\to.\overline{\overline{\overline{\overline{A}\to A}\to A}\to.\overline{B}\to B}\to.\overline{A}\to A$$
$$\overline{B}\to C\to.\overline{\overline{A}\to A}\to.\overline{\overline{\overline{A}\to A}\to A}\to.\overline{B}\to B$$
$$\overline{B}\to C\to.\overline{\overline{A}\to A}\to.\overline{\overline{B}\to B}\to.\overline{\overline{A}\to A}\to A$$
$$\overline{B}\to C\to.\overline{\overline{A}\to A}\to.\overline{\overline{B}\to B}\to A$$

(8) Let $\Gamma = \vdash\tilde{\vec{\alpha}}, \beta$, with $\tilde{\vec{\alpha}}$ all negated entailments and with β non-empty. Let B be a Γ-formula containing at least one occurrence of a formula in β. Then B has a reductum

$$\overline{\overline{E}}_1\to.\ \dots\ \to.\overline{\overline{E}}_n\to G$$

with each negated entailment $\overline{E_i}$ a constituent of $\tilde{\vec{\alpha}}$, and with no constituent of $\tilde{\vec{\alpha}}$ occurring in G.

PROOF: Suppose inductively that (8) holds for Γ-formulas with fewer occurrences of constituents of $\tilde{\vec{\alpha}}$ than B. The thing is trivial when B contains no occurrences of constituents of $\tilde{\vec{\alpha}}$; otherwise, select one of the constituents of $\tilde{\vec{\alpha}}$, say $\overline{E}$, and consider the reductum $Sqz(B, \overline{E})$ of B, with the form $\varphi\to.\overline{G}\to\overline{E}$. This reduces by R1 to $\varphi\to.\overline{\overline{E}}\to G$, hence by R2 to $\overline{\overline{E}}\to.\varphi\to G$. Since $\varphi\to G$ will contain fewer occurrences of constituents of $\tilde{\vec{\alpha}}$ than B, we may apply the inductive hypothesis to obtain a reductum of $\varphi\to G$, say $\overline{\overline{E}}_1\to.\ \dots\ \to\overline{\overline{E}}_n\to G'$, with each $\overline{E_i}$ a constituent of $\tilde{\vec{\alpha}}$ and with no constituent of $\tilde{\vec{\alpha}}$ occurring in G'. Hence, $\overline{\overline{E}}\to\overline{\overline{E}}_1\to.\ \dots\ \to.\overline{\overline{E}}_n\to G'$ is a reductum of B having the required form.

We now use these eight facts to prove the theorem we stated a long time ago: every provable consecution of $L_rE_{\overset{\sim}{\to}}$ has at least one valid interpretation.

Proof. We proceed by induction on the proof of Γ in $\mathbf{L}_r\mathbf{E}_{\overline{\rightarrow}}$. For the basis, we observe that when Γ is an axiom $\vdash \overline{A}, A$, the formula $\overline{\overline{A}}\rightarrow A$ will serve as a valid interpretation, while the formula $\overline{\overline{A\rightarrow B}}\rightarrow.\overline{\overline{A}}\rightarrow B$ validly interprets the axiom $\vdash \overline{A}, \overline{A\rightarrow B}, B$. For the induction, we need to show that, whenever the theorem holds for the premisses of a rule of $\mathbf{L}_r\mathbf{E}_{\overline{\rightarrow}}$, it holds for the conclusion as well. There are five cases, corresponding to the five rules of the system.

Cases 1 and 2. The rules ($\vdash$ C) and ($\vdash$ W) are particularly simple, since the valid interpretation known to exist for the premisses will serve equally well for the conclusion.

Case 3. The rule (ER) presents the only case of much complexity. There are various subcases depending on the emptiness of α and β, but since they cannot both be empty (for then, by the argument of the corollary, both A and $\overline{A}$ would be provable in $\mathbf{E}_{\overline{\rightarrow}}$, which they are not), it suffices to consider the two cases α non-empty or β non-empty; and since these two cases are exactly parallel, we will simply suppose without further ado that α is non-empty. Suppose then as hypothesis of the case that the left premiss $\vdash \alpha, A$ of the rule (ER) has B as its valid interpretation, and that the right premiss $\vdash \beta, \overline{A}$ has C. We need to find a valid interpretation of the conclusion $\vdash \alpha, \beta$. To this end we define a sequence of formulas, beginning with B, (i) each of which is provable, and (ii) each of which, except for the first, contains occurrences of every constituent of α and β, and no others except possibly A, $\overline{A}$, or both. As the sequence progresses, the number of occurrences of A and $\overline{A}$ will gradually be reduced according to a double inductive pattern, each step either reducing the number of occurrences of A (possibly increasing the $\overline{A}$'s) or reducing the number of occurrences of $\overline{A}$ while not increasing the number of A's. (There will be no control over the number of occurrences of constituents of α and β.) The last member of the sequence will contain occurrences of neither A nor $\overline{A}$ and will hence be the required valid interpretation of $\vdash \alpha, \beta$.

In preparation for defining the sequence, we let

$$Squeez(C, \overline{A}) = \psi\rightarrow\overline{A}.$$

We remark that B will be used only to get the sequence started, while we shall recur again and again to C via $Squeez(C, \overline{A})$. In defining the sequence we shall find it convenient to refer to a formula as *$\overline{A}$-free* if, like B, it contains no occurrences of $\overline{A}$. The sequence $D_1, \ldots, D_i, \ldots$ can now be defined inductively as follows.

(i) $D_1 = B$.

(ii) D_i is $\bar{A}$-free. Two cases.

(iia) D_i contains no occurrences of A. Then the sequence terminates in D_i.

(iib) D_i contains occurrences of A. We let $Sqz(D_i, A) = \varphi_i \rightarrow .\overline{F_i} \rightarrow A$ and recall that $Squeez(C, \bar{A}) = \psi \rightarrow \bar{A}$; then we define

$$D_{i+1} = \varphi_i \rightarrow .\psi \rightarrow F_i.$$

(iii) D_i is not $\bar{A}$-free. First let D_k ($k < i$) be the most recent $\bar{A}$-free predecessor of D_i, and let $Sqz(D_k, A) = \varphi_k \rightarrow .\overline{F_k} \rightarrow A$. Also let D_i be represented in the form $\varphi_k \rightarrow G$ (with the same front end as $Sqz(D_k, A)$) so that $Squeez(D_i, \bar{A}) = \varphi_k \rightarrow Squeez(G, \bar{A}) = \varphi_k \rightarrow .\varphi_i \rightarrow \bar{A}$. Then define

$$D_{i+1} = \varphi_k \rightarrow .\varphi_i \rightarrow F_k.$$

REMARKS. 1. Each formula in the sequence, except the first, is constructed from two other formulas, which can serve as its "premisses." This is illustrated in the example below.

2. It should be borne in mind that an occurrence of a constituent of α cannot be an occurrence of A, or of $\bar{A}$, or of a constituent in β; and so forth.

3. $D_1 = B$ must fall under the hypothesis of (iib), so that D_2 is bound to start a long line of formulas each of which contains at least one occurrence of each constituent of α and β, and of no other except possibly A, $\bar{A}$, or both. In particular, at no stage do we altogether omit occurrences of any constituent except possibly A or $\bar{A}$. Property (2) of reductions is important here.

4. To show that the provability of D_{i+1} in $\mathbf{E}_{\rightarrow}$ follows from the provability of the two formulas used in constructing it in each of (iib) and (iii) is but an exercise in the use of contraposition together with generalized transitivity and, in the case of (iii), repeated use of generalized contraction (§4.2) to conflate the identical front ends.

5. We next observe that the various functions employed in constructing the sequence are well defined. Thus, in (iib) $Sqz(D_i, A)$ is bound to be well defined since by assumption α is nonempty, so that $D_1 (= B)$ and all its successors will contain occurrences of constituents other than A. With respect to (iii), we can be sure that D_k is well defined, since there is always an $\bar{A}$-free predecessor of D_i if D_i is not $\bar{A}$-free ($D_1 = B$ will be such), hence always a most recent. And lastly, again for (iii), we can be sure that D_i can always be represented as $\varphi_k \rightarrow G$, with its front end the same as $Sqz(D_k, A)$, since (iib) guarantees this for the immediate successor of any $\bar{A}$-free formula, and (iii) keeps things that way.

6. The last and crucial observation is that the sequence is bound to terminate in accordance with (iia). The argument amounts to a double induction on the number of occurrences of A and the number of occurrences

of $\bar{A}$, since what we show is that at every stage *either* the number of occurrences of A is reduced (with no control over the $\bar{A}$'s) *or* that the number of $\bar{A}$'s is reduced while the number of A's is not increased (i.e., either held constant or reduced). Both numbers must then go to zero, at which time the sequence terminates by (iia).

There are two cases to the argument.

Consider first (iib). That the number of occurrences of A in D_{i+1} is in this case less than the number in D_i (but with no control on the number of $\bar{A}$'s) is due largely to the fact that the reduction from D_i to $Sqz(D_i, A)$ does not increase the number of occurrences of any formula, hence not the number of A's. That F_i in $Sqz(D_i, A)$ contains no A's is not used here.

Next consider (iii). Here we observe that in passing from $D_i = \varphi_k \to G$ to $Squeez(D_i, \bar{A}) = \varphi_k \to Squeez(G, \bar{A}) = \varphi_k \to .\varphi_i \to \bar{A}$ and with the help of $Sqz(D_k, A) = \varphi_k \to .\overline{F_k} \to A$ to $D_{i+1} = \varphi_k \to .\varphi_i \to F_k$, we reduce the number of $\bar{A}$'s (since D_k, hence $Sqz(D_k, A)$, hence its part F_k, is $\bar{A}$-free), while not increasing the number of A's (since the machinery deployed in the definition of $Sqz(D_k, A)$ was specifically designed to guarantee that F_k is free of A's). Herein lay the necessity for all the complications.

Example. Suppose we have the following in $\mathbf{L_rE_{\to}}$:

$$\frac{\vdash L, H, A \qquad \vdash K, E, \bar{A}}{\vdash L, H, K, E}\,(\mathrm{ER})$$

Suppose further that B is a formula

$$\bar{A} \to .\bar{L} \to .\overline{\bar{\bar{A}} \to A} \to H,$$

which validly interprets $\vdash L, H, A$ and that C is a formula

$$\overline{\overline{\bar{\bar{A}} \to K} \to .\bar{E} \to \bar{A}} \to .\bar{E} \to \bar{A}$$

which validly interprets $\vdash K, E, \bar{A}$. Our task is to construct the sequence $D_1, \ldots, D_i \ldots$. We will set off above and to the right of each D_i the formulas on which it immediately depends.

($\bar{A}$-free) $D_1 = B = \bar{A} \to .\bar{L} \to .\overline{\bar{\bar{A}} \to A} \to H$ (results by (i))

$$Sqz(D_1, A) = \bar{A} \to .\bar{L} \to .\bar{H} \to A$$

$$Squeez(C, \bar{A}) = C = \overline{\overline{\bar{\bar{A}} \to K} \to .\bar{E} \to \bar{A}} \to .\bar{E} \to \bar{A}$$

which yield by (iib)

$$D_2 = \bar{A} \to .\bar{L} \to .\overline{\overline{\bar{\bar{A}} \to K} \to .\bar{E} \to \bar{A}} \to .\bar{E} \to H$$

$$Sqz(D_1, A) = \bar{A} \to .\bar{L} \to .\bar{H} \to A$$

$$Squeez(D_2, \bar{A}) = \bar{A} \to .\bar{L} \to .\overline{\bar{E} \to H} \to .\overline{\bar{\bar{A}} \to K} \to .\bar{E} \to \bar{A}$$

which yield by (iii)

$$D_3 = \bar{A}\to.\bar{L}\to.\overline{\bar{E}\to H}\to.\overline{\bar{\bar{A}}\to K}\to.\bar{\bar{E}}\to H$$

$$Sqz(D_1, A) = \bar{A}\to.\bar{L}\to.\bar{H}\to A$$

$$Squeez(D_3, \bar{A}) = \bar{A}\to.\bar{L}\to.\overline{\bar{E}\to H}\to.\overline{\bar{E}\to H}\to.\bar{K}\to\bar{A}$$

which yield by (iii)

$$(\bar{A}\text{-free})\ D_4 = \bar{A}\to.\bar{L}\to.\overline{\bar{E}\to H}\to.\overline{\bar{E}\to H}\to.\bar{K}\to H$$

$$Sqz(D_4, A) = \overline{\bar{L}\to.\overline{\bar{\bar{E}}\to H}\to.\overline{\bar{\bar{E}}\to H}.\to\bar{K}\to H}\to A$$

$$Squeez(C, \bar{A}) = C = \overline{\overline{\bar{\bar{A}}\to K}\to.\bar{\bar{E}}\to\bar{A}}\to.\bar{\bar{E}}\to\bar{A}$$

which yield by (iib)

$$D_5 = \overline{\overline{\bar{\bar{A}}\to K}\to.\bar{\bar{E}}\to\bar{A}}\to.\bar{\bar{E}}\to.\bar{L}\to.\overline{\bar{\bar{E}}\to H}\to.\overline{\bar{\bar{E}}\to H}\to.\bar{K}\to H$$

$$Sqz(D_4, A) = \overline{\bar{L}\to.\overline{\bar{\bar{E}}\to H}\to.\overline{\bar{\bar{E}}\to H}\to.\bar{K}\to H}\to A$$

$$Squeez(D_5, \bar{A}) = \overline{\bar{E}\to.\bar{L}\to.\overline{\bar{\bar{E}}\to H}\to.\overline{\bar{\bar{E}}\to H}\to.\bar{K}\to H}\to .\overline{\bar{\bar{A}}\to K}\to.\bar{\bar{E}}\to\bar{A}$$

which yield by (iii)

$$D_6 = \overline{\bar{E}\to.\bar{L}\to.\overline{\bar{\bar{E}}\to H}\to.\overline{\bar{\bar{E}}\to H}\to.\bar{K}\to H}\to.\overline{\bar{\bar{A}}\to K}\to.\bar{\bar{E}}\to.\bar{L}\to.\overline{\bar{\bar{E}}\to H}\to .\overline{\bar{\bar{E}}\to H}\to.\bar{K}\to H.$$

Note that the front end "φ_k" of $Sqz(D_4, A)$ is empty in this example. The reader is invited to find D_7, which will terminate the sequence in the desired valid interpretation of $\vdash L, H, K, E$; he should use $Sqz(D_4, A)$ as most recent $\bar{A}$-free predecessor together with $Squeez(D_6, \bar{A})$. He is also invited to find a simpler routine for finding a valid interpretation of the conclusion of the rule (ER).

We assume the reader is sufficiently convinced of the correctness of our construction for Case 3 by the example, if not by the argument, and so proceed at once to

Case 4. The rule ($\vdash\to$). We suppose B a valid interpretation of the premiss $\vdash\tilde{\bar{\alpha}}, \bar{C}, D$, and need to find a valid interpretation of the conclusion $\vdash\tilde{\bar{\alpha}}, C\to D$. By (8) B has a reductum $\overline{\overline{E_1}}\to.\ \dots\to.\overline{\overline{E_n}}\to G$ with each E_i a constituent of $\tilde{\bar{\alpha}}$, and with no constituent of $\tilde{\bar{\alpha}}$ occurring in G. Hence the constituents occurring in G are just $\bar{C}$ and D, so that by (5) G reduces to $\overline{\overline{C}}\to D$ and accordingly B to $\overline{\overline{E_1}}\to.\ \dots\to.\overline{\overline{E_n}}\to.\overline{\overline{C}}\to D$. By double negation (in $\mathbf{E}_{\to}$)

we obtain $\overline{\overline{E_1}}\rightarrow. \dots \rightarrow.\overline{\overline{E_n}}\rightarrow.C\rightarrow D$, which is the desired valid interpretation of the conclusion $\vdash \tilde{\alpha}, C\rightarrow D$.

CASE 5. The rule ($\vdash\sim\sim$). Suppose B validly interprets the premiss $\vdash \alpha, A$; then the result of replacing every occurrence of A in B by $\overline{\overline{A}}$ validly interprets the conclusion $\vdash \alpha, \overline{\overline{A}}$.

This completes the proof of the theorem. We can now show quickly that $\vdash A$ in $\mathbf{L_rE}_{\rightarrow}$ implies A in $\mathbf{E}_{\rightarrow}$.

Suppose that $\vdash A$ is provable in $\mathbf{L_rE}_{\rightarrow}$. Then by the theorem, some interpretation of $\vdash A$ is provable in $\mathbf{E}_{\rightarrow}$. But all interpretations of $\vdash A$ must contain occurrences of only a single constituent — i.e., A — so that by fact (4) above, any interpretation of $\vdash A$ reduces to A itself. Hence, since reduction preserves provability in $\mathbf{E}_{\rightarrow}$, A is provable.

The circle is now complete: we have shown that all the systems $\mathbf{E}_{\rightarrow}$, $\mathbf{LE}_{\rightarrow}$, and $\mathbf{L_rE}_{\rightarrow}$, are equivalent, so that providing a decision procedure for any one of them automatically does so for the others.

§13.3. Decision procedure. We outline a decision procedure for $\mathbf{LE}_{\rightarrow}$, thus indirectly providing a procedure for $\mathbf{E}_{\rightarrow}$.

We say that a consecution Γ_1 *CW-reduces* to a consecution Γ_2 if Γ_2 can be obtained from Γ_1 by zero or more applications of the rules ($\vdash$ C) and ($\vdash W$) (for $\mathbf{L_rE}_{\rightarrow}$) or their analogues for the left side. We have concerning $\mathbf{LE}_{\rightarrow}$ the following.

CURRY'S LEMMA. If a consecution Γ is provable in $\mathbf{LE}_{\rightarrow}$, then there is a proof of Γ containing no Γ' such that Γ' CW-reduces to Γ. This follows from the fact, easily provable by induction as in Curry 1950, that if Γ' CW-reduces to Γ and Γ' is provable in m steps, then Γ is provable in $\leq m$ steps.

By a *tree* we mean a configuration of elements arranged in levels, which has a first (or bottom) level containing only one element and such that lines are drawn between elements of consecutive levels in such a way that every element in every level (except the first) is connected to exactly one element in the preceding (lower) level. A *branch* of the tree is a sequence of elements $a_1, a_2, a_3, \dots$ (which may or may not terminate), such that a_1 is the element at the first level and a_i is some member of the i-th level which is connected to a_{i-1} by a line. A branch may terminate in an element a_n such that no line connects a_n to an element of the next higher level; or a branch may be infinite. We say that a tree is *finite* if the number of elements in it is finite.

A tree has the *finite fork property* if each level of the tree has only a finite number of elements; a tree has the *finite branch property* if every branch of the tree is finite.

KÖNIG'S LEMMA. A tree is finite if and only if it has both the finite fork property and the finite branch property (König 1927).

Now we turn our attention from trees to finitary logical systems having formulas, axioms, and rules. Let **S** be such a system. Evidently any proof in **S** can be written in the form of a finite tree. We put the formula proved at the bottom; above each formula D of the k-th level we put the formulas from which D follows by means of one of the rules, with lines drawn from these formulas to D. Such a tree will be called a *proof tree*. All branches of a proof tree must terminate in axioms.

By a *proof search tree for* (*a formula*) C we shall mean a tree whose elements are formulas of **S**, which has C at the bottom level. The idea is that given a proof search tree for C we may hunt for a proof of C among the subtrees (where "subtree" is defined in the obvious way) of the proof search tree. By a *complete proof search tree for* (*a formula*) C we shall mean a proof search tree for C which has a proof of C as a subtree, if a proof of C exists.

For an arbitrary finitary logical system **S** there is clearly an effective procedure which when applied to an arbitrary formula C yields a complete proof search tree for C; such a complete proof search tree will not, in general, be finite. On the other hand, if for a system **S** there is an effective procedure which when applied to an arbitrary formula C yields a *finite* complete proof search tree for C, then **S** is decidable. This follows from the fact that a finite complete proof search tree has only a finite number of subtrees — one of which is a proof of C, if such a proof exists.

It follows from the above together with König's lemma that, if for a finitary logical system **S** there is an effective procedure which when applied to an arbitrary formula C yields a complete proof search tree having both the finite fork property and the finite branch property, then **S** is decidable.

We shall show that $\mathbf{LE}_{\to}$ is decidable by describing an effective procedure which when applied to an arbitrary consecution Γ of $\mathbf{LE}_{\to}$ yields a proof search tree for Γ which

(i) is complete,
(ii) has the finite fork property,
(iii) has the finite branch property.

Given a consecution Γ we form the tree (called *the distinguished proof search tree for* Γ) as follows:

We put Γ at the bottom.

Above every consecution Γ' occurring in the k-th level we put (a) nothing, if Γ' is an axiom, (b) otherwise, all Γ'' such that

(i) Γ'' can serve as a premiss for Γ',

(ii) Γ'' does not CW-reduce to Γ' nor to any consecution below Γ' in the same branch as Γ'.

That the distinguished proof search tree for Γ is complete is a consequence of Curry's lemma, which shows that if there is a proof of Γ, then there is a proof of Γ which satisfies condition (ii) just above.

That the distinguished proof search tree for Γ has the finite fork property is a result of the following property of $\mathrm{LE}_{\rightarrow}$: there are only a finite number of formulas in Γ which can serve as principal constituent, and the choice of principal constituent determines a finite set of premisses from which Γ can follow. (The restrictions on the number of formulas occurring in the premisses and conclusions of applications of rules of $\mathrm{LE}_{\rightarrow}$ were designed to secure this property.)

It remains to show that the distinguished proof search tree for Γ has the finite branch property. Following Kleene 1952 we will say that two consecutions $\alpha \vdash \beta$ and $\alpha' \vdash \beta'$ are *cognate* if exactly the same formulas (not counting multiplicity) occur in α $\{\beta\}$ as in α' $\{\beta'\}$. We call the class of consecutions which are cognate with a given consecution the *cognation class* of Γ.

We define the class of *subformulas* of a given formula as follows:

(1) A is a subformula of A.

(2) If C is a subformula of A or B, then C is a subformula of $A \rightarrow B$; if C is a subformula of A, then C is a subformula of $\overline{A}$.

Subformula theorem. If $\alpha \vdash \beta$ is a provable consecution in $\mathrm{LE}_{\rightarrow}$ then any formula occurring in any consecution in the proof of $\alpha \vdash \beta$ is subformula of some formula occurring in $\alpha \vdash \beta$.

Proof. Inspection of the rules of $\mathrm{LE}_{\rightarrow}$ shows that no subformulas are lost in passing from premiss(es) to conclusion. From the subformula theorem, it follows that the number of cognation classes occurring in any branch is finite.

If we can show that only a finite number of members of each cognation class occurs in any branch, then it will follow that each branch is finite. To show this we introduce the following terminology. A sequence of cognate consecutions $\Gamma_0, \Gamma_1, \ldots$ is *irredundant* if for no $i, j, j > i$, Γ_j CW-reduces to Γ_i. From condition (ii) in the definition of the complete proof search tree for Γ it follows that every sequence of cognate consecutions occurring in any given branch is irredundant. Hence the finite branch property will follow from

Kripke's lemma. If a sequence of cognate consecutions is irredundant, it is finite.

This lemma, which is referred to obliquely in Kripke 1959a, was communicated to one of us by Kripke in a letter dated September 1959.

PROOF. Let $\Sigma = \Gamma_0, \Gamma_1, \ldots$ be a sequence (finite or infinite) of cognate consecutions. We shall show that if there are no $i, j, j>i$, such that Γ_j CW-reduces to Γ_i (i.e., if the sequence Σ is irredundant), then the sequence Σ is finite. The proof is by induction on the number n of distinct formulas having one or more occurrences in Σ. If $n = 1$ the theorem is obvious. Suppose inductively that the theorem is true for Σ' with n formulas and let Γ have $n+1$ formulas. Pick out one of the formulas in Σ, say A. Γ_k is *critical* in a sequence Σ^* if and only if, for all $m > k$, the number of occurrences of A in Γ_k is less than or equal to the number of occurrences of A in Γ_m. Let Γ_i' be the result of deleting all A's in Γ_i. We define a new sequence of consecutions $\Sigma' = \Delta_0, \Delta_1, \ldots$ in the following manner. Let Γ_k be the first consecution critical in Σ. Δ_0 is Γ_k'. Suppose Δ_k has been defined and is Γ_i'. If Γ_{i+1} exists, find the first critical consecution in the sequence $\Gamma_{i+1}, \Gamma_{i+2}, \ldots$; call it Γ_j and let Δ_{k+1} be Γ_j'. Note that Σ' terminates only if Σ terminates.

Σ' is a sequence of cognate consecutions which contains occurrences of only n distinct formulas. Hence by the hypothesis of the induction, if Σ' is irredundant, Σ' is finite. But by choice of critical consecutions in the construction of Σ', if Σ is irredundant, Σ' is irredundant; and if Σ' is finite then Σ is finite. Hence, if Σ is irredundant, Σ is finite. This completes the proof of Kripke's lemma.

It follows that distinguished proof search trees have the finite branch property, and, in turn, that constructing such a tree for Γ constitutes a decision procedure for $\mathbf{LE}_{\overline{\rightarrow}}$, and hence for $\mathbf{E}_{\overline{\rightarrow}}$.

§14. Miscellany. This section is analogous to §8; see the Analytical table of contents for a list of topics treated.

§14.1. Axiom-chopping. Here we consider the results of adding negation to the relevance logics $\mathbf{T}_{\rightarrow}$, $\mathbf{E}_{\rightarrow}$, and $\mathbf{R}_{\rightarrow}$, where negation is added to the first and third of these on analogy with its addition to $\mathbf{E}_{\rightarrow}$.

§14.1.1. Alternative formulations of $\mathbf{T}_{\overline{\rightarrow}}$.

$\mathbf{T}_{\overline{\rightarrow}1}$:

1 $A \rightarrow A$

2 $A \rightarrow B \rightarrow .B \rightarrow C \rightarrow .A \rightarrow C$

3 $A \rightarrow B \rightarrow .C \rightarrow A \rightarrow .C \rightarrow B$

4 $(A \rightarrow .A \rightarrow B) \rightarrow .A \rightarrow B$

5 $A \rightarrow \overline{B} \rightarrow .B \rightarrow \overline{A}$

6 $A \rightarrow \overline{A} \rightarrow \overline{A}$

7 $\overline{\overline{A}} \rightarrow A$

1–4 are the axioms of $\mathbf{T}_{\to 2}$ of §8.3.2. The proof that $\mathbf{T}_{\overset{\sim}{\to}1}$ is equivalent to the following is left to the reader.

$\mathbf{T}_{\overset{\sim}{\to}2}$: Replace 5 in $\mathbf{T}_{\overset{\sim}{\to}1}$ by

8 $A\to B\to.\bar{B}\to\bar{A}$

9 $A\to\bar{\bar{A}}$

It is obvious also that in either formulation (or indeed any of those considered in this section), 6 may be replaced by

10 $\bar{A}\to A\to A$

which contains fewer connectives than 6. What is not so obvious is the observation, which we owe to Bacon, that in $\mathbf{T}_{\overset{\sim}{\to}1}$, we may replace both 5 and 7 by the following oddity:

11 $\sim A\to\sim\sim\sim B\to.B\to A.$

For from this, by substituting $\sim\sim A$ for A, A for B, and using 1, we get

9 $A\to\sim\sim A$

from which we get both

$\sim A\to\sim\sim\sim A$

and

$\sim\sim\sim A\to\sim\sim\sim\sim\sim A,$

whence by transitivity

12 $\sim A\to\sim\sim\sim\sim\sim A.$

But

13 $\sim A\to\sim\sim\sim\sim\sim A\to.\sim\sim A\to A$

is a case of 11, so 12 and 13 yield 7. And in the presence of these facts it is then easy to verify that the following is equivalent to the other formulations:

$\mathbf{T}_{\overset{\sim}{\to}3}$: 1–4, 10, 11.

We note that since what is required to get $\mathbf{T}_{\overset{\sim}{\to}1}$ from $\mathbf{T}_{\overset{\sim}{\to}3}$ is also available in the stronger systems $\mathbf{E}_{\overset{\sim}{\to}}$ and $\mathbf{R}_{\overset{\sim}{\to}}$ considered below, 11 could replace 5 and 7 in any of these.

It has further been pointed out to us — rather to our surprise — by Chidgey and (independently) by Broido that 1 $A\to A$ is redundant in $\mathbf{T}_{\overset{\sim}{\to}1}$; it is indeed provable from 3 (or 2 — all we need is the rule trans), 5, and 7 as follows.

Use the instance $\sim\sim\sim A\rightarrow\sim A$ of 7 and the instance $\sim\sim\sim A\rightarrow\sim A\rightarrow .A\rightarrow\sim\sim\sim\sim A$ of 5 to obtain

$$A\rightarrow\sim\sim\sim\sim A.$$

Then by two uses of 7 together with trans obtain

$$\sim\sim\sim\sim A\rightarrow A.$$

The archetypal form of all inference is now a consequence of $A\rightarrow\sim\sim\sim\sim A$ and $\sim\sim\sim\sim A\rightarrow A$ by trans. Note that in $\mathbf{T}_{\rightarrow}2$, the redundancy of 1 $A\rightarrow A$ follows immediately from 9 and 7 by trans.

$\mathbf{T}_{\rightarrow}1$ has further surprises up its sleeve. Chidgey also informed us of the redundancy of 2. We summarize his proof as follows: $A\rightarrow\bar{\bar{A}}$ follows from 5 and the instance $\bar{A}\rightarrow\bar{A}$ of derived 1. The rules pre and trans are available from 3, and these together with instances of 5 provide

14 $A\rightarrow B\rightarrow .\bar{B}\rightarrow\bar{A}$

and 15 $\bar{A}\rightarrow\bar{B}\rightarrow .B\rightarrow A.$

The rule suf is then derived as follows:

a	$A\rightarrow B$	premiss
b	$\bar{B}\rightarrow\bar{A}$	a 14 $\rightarrow$E
c	$\bar{C}\rightarrow\bar{B}\rightarrow .\bar{C}\rightarrow\bar{A}$	b pre
d	$B\rightarrow C\rightarrow .\bar{C}\rightarrow\bar{A}$	14 c trans
e	$B\rightarrow C\rightarrow .A\rightarrow C$	15 d trans

2 is then derived by the following proof:

a	$\bar{B}\rightarrow\bar{A}\rightarrow .\bar{C}\rightarrow\bar{B}\rightarrow .\bar{C}\rightarrow\bar{A}$	3
b	$A\rightarrow B\rightarrow .\bar{C}\rightarrow\bar{B}\rightarrow .\bar{C}\rightarrow\bar{A}$	14 a trans
c	$(\bar{C}\rightarrow\bar{B}\rightarrow .\bar{C}\rightarrow\bar{A})\rightarrow .B\rightarrow C\rightarrow .\bar{C}\rightarrow\bar{A}$	14 suf
d	$A\rightarrow B\rightarrow .B\rightarrow C\rightarrow .\bar{C}\rightarrow\bar{A}$	b c trans
e	$(B\rightarrow C\rightarrow .\bar{C}\rightarrow\bar{A})\rightarrow .B\rightarrow C\rightarrow .A\rightarrow C$	15 pre
2	$A\rightarrow B\rightarrow .B\rightarrow C\rightarrow .A\rightarrow C$	d e trans

Naturally the redundancy of 1 holds equally for any of the formulations below of $\mathbf{E}_{\rightarrow}$ or $\mathbf{R}_{\rightarrow}$ containing 2 or 3, 5, and 7; or (more obviously) containing 2 or 3, 7 and 9. Also 2 is redundant in any formulation containing 3, 5, and 7. But we do not dignify with a title any such formulation of any of these systems, for all of them fail to be well-axiomatized in the sense that the pure arrow axioms by themselves completely determine the pure arrow theorems, without any detour through the negation axioms. That is, if we left $A\rightarrow A$ or $A\rightarrow B\rightarrow .B\rightarrow C\rightarrow .A\rightarrow C$ out of $\mathbf{T}_{\rightarrow}1$, then it would not be a conservative extension — in the sense of §14.4 — of the calculus determined by its pure implicational axioms.

§14.1.2. Alternative formulations of $\mathbf{E}_{\overline{\rightarrow}}$. One could tinker around with the various formulations of pure entailment in §8.3.3, using these in combination with the results of §14.1.1, to obtain a large variety of formulations of entailment with negation, but nothing of interest emerges, since there seems to be no interaction between the pure entailment tinkering and the negation tinkering.

We list for reference the axioms given at the beginning of §9.

$\mathbf{E}_{\overline{\rightarrow}1}$
1 $A \rightarrow A \rightarrow B \rightarrow B$
2 $A \rightarrow B \rightarrow . B \rightarrow C \rightarrow . A \rightarrow C$
3 $(A \rightarrow . A \rightarrow B) \rightarrow . A \rightarrow B$
4 $A \rightarrow \overline{B} \rightarrow . B \rightarrow \overline{A}$
5 $A \rightarrow \overline{A} \rightarrow \overline{A}$
6 $\overline{\overline{A}} \rightarrow A$

And the perhaps more perspicuous formulation

$\mathbf{E}_{\overline{\rightarrow}2}$ replace 1 by the following two axioms:
7 $A \rightarrow A$
8 $(A \rightarrow . B \rightarrow C \rightarrow D) \rightarrow . B \rightarrow C \rightarrow . A \rightarrow D$

See §14.1.1 for the redundancy of 7 in this formulation.

§14.1.3. Alternative formulations of $\mathbf{R}_{\overline{\rightarrow}}$. Church 1951 investigates various possible kinds of negation in connection with $\mathbf{R}_{\rightarrow}$, but he does not happen to consider the system $\mathbf{R}_{\overline{\rightarrow}}$, obtained by adding negation axioms to $\mathbf{R}_{\rightarrow 1}$ of §8.4.4 in analogy with those added to $\mathbf{T}_{\rightarrow}$ and $\mathbf{E}_{\rightarrow}$, as follows:

$\mathbf{R}_{\overline{\rightarrow}}$
1 $(A \rightarrow . A \rightarrow B) \rightarrow . A \rightarrow B$
2 $A \rightarrow B \rightarrow . C \rightarrow A \rightarrow . C \rightarrow B$
3 $(A \rightarrow . B \rightarrow C) \rightarrow . B \rightarrow . A \rightarrow C$
4 $A \rightarrow A$
5 $A \rightarrow \overline{B} \rightarrow . B \rightarrow \overline{A}$
6 $A \rightarrow \overline{A} \rightarrow \overline{A}$
7 $\overline{\overline{A}} \rightarrow A$

Using only 2, 4, 5, and 7, it is easy to prove a replacement theorem of the usual sort; and in particular that if B' is obtained from B either by adding or subtracting an outermost negation sign, and A' similarly from A, then $A \rightarrow B$ and $B' \rightarrow A'$ are everywhere interchangeable.

But here, unlike the previous case, negation and implication do seem to interact. As is pointed out in Meyer 1966, we can show that 1–7 contain a redundancy (besides the redundancy of $A \rightarrow A$ noted in §14.1.1): either of 1 or 6 can be proved from the other.

Suppose we have

6 $A\to\bar{A}\to\bar{A}$

Then prefixing yields $(\bar{B}\to.A\to\bar{A})\to.\bar{B}\to\bar{A}$, which by permutation in the antecedent, followed by a couple of contrapositions as replacements, leads to 1 $(A\to.A\to B)\to.A\to B$. Suppose instead we start with 1. First permute and contrapose $A\to\bar{A}\to.A\to\bar{A}$ to obtain $A\to.A\to\overline{A\to\bar{A}}$, and use 1 to obtain $A\to\overline{A\to\bar{A}}$. Then 6 follows by contraposition.

What this means is that in the context of $\mathbf{R}_{\to}$, a single axiom for negation suffices to obtain all the relevant properties: double negation introduction and elimination, contraposition, and *reductio;* to wit, Bacon's

$\sim A\to\sim\sim\sim B\to.B\to A$

of §14.1.1.

§14.2. Independence (by John R. Chidgey). We follow the notational conventions of §8.4.1, and with the help of some additional matrices develop an independent axiom set for each of the systems $\mathbf{T}_{\stackrel{\sim}{\to}}$, $\mathbf{E}_{\stackrel{\sim}{\to}}$ and $\mathbf{R}_{\stackrel{\sim}{\to}}$. As these formulations all have →E as sole rule, this rule is "independent" in an appropriate sense (see the Observation in §8.4).

§14.2.1. Matrices.

Matrix I

→	*0*	*1*	*2*	*3*	*4*	*5*	~
0	*5*	*5*	*5*	*5*	*5*	*5*	*5*
1	*0*	*5*	*0*	*0*	*0*	*5*	*4*
2	*0*	*0*	*4*	*0*	*0*	*5*	*3*
**3*	*0*	*0*	*0*	*4*	*0*	*5*	*2*
**4*	*0*	*0*	*0*	*0*	*5*	*5*	*1*
**5*	*0*	*0*	*0*	*0*	*0*	*5*	*0*

Matrix II

→	*0*	*1*	*2*	*3*	~
0	*2*	*2*	*2*	*2*	*3*
1	*1*	*2*	*2*	*2*	*2*
**2*	*0*	*1*	*2*	*2*	*1*
**3*	*0*	*0*	*1*	*2*	*0*

Matrix III

→	*0*	*1*	*2*	~
0	*2*	*2*	*2*	*2*
**1*	*0*	*2*	*2*	*2*
**2*	*0*	*2*	*2*	*0*

Matrix IV

→	*0*	*1*	*2*	~
0	*2*	*2*	*2*	*2*
1	*0*	*2*	*2*	*1*
**2*	*0*	*0*	*2*	*0*

Matrix V

→	*0*	*1*	*2*	~
0	*2*	*2*	*2*	*2*
1	*0*	*2*	*2*	*2*
**2*	*0*	*2*	*2*	*2*

Matrix VI

→	*0*	*1*	*2*	~
0	*2*	*2*	*2*	*2*
1	*2*	*2*	*2*	*2*
**2*	*0*	*2*	*2*	*0*

Matrix VII

→	*0*	*1*	*2*	*3*	*4*	*5*	~
0	*5*	*5*	*5*	*5*	*5*	*5*	*5*
**1*	*0*	*1*	*3*	*3*	*3*	*5*	*4*
**2*	*0*	*0*	*3*	*3*	*3*	*5*	*3*
**3*	*0*	*0*	*0*	*3*	*3*	*5*	*2*
**4*	*0*	*0*	*0*	*0*	*1*	*5*	*1*
**5*	*0*	*0*	*0*	*0*	*0*	*5*	*0*

§14.2.2. Independent axioms for $\mathbf{T}_{\overrightarrow{}}$.

$\mathbf{T}_{\rightarrow 1}$: 1	$A\rightarrow A$	Redundant (§14.1.1)
2	$A\rightarrow B\rightarrow.B\rightarrow C\rightarrow.A\rightarrow C$	Redundant (§14.1.1)
3	$A\rightarrow B\rightarrow.C\rightarrow A\rightarrow.C\rightarrow B$	Matrix VI: $A=\mathit{1}$, $B=\mathit{0}$, $C=\mathit{2}$, so $3=\mathit{0}$
4	$(A\rightarrow.A\rightarrow B)\rightarrow.A\rightarrow B$	Matrix II: $A=\mathit{1}$, $B=\mathit{0}$, so $4=\mathit{0}$
5	$A\rightarrow\overline{B}\rightarrow.B\rightarrow\overline{A}$	Matrix III: $A=\mathit{2}$, $B=\mathit{1}$, so $5=\mathit{0}$
6	$A\rightarrow\overline{A}\rightarrow\overline{A}$	Matrix IV: $A=\mathit{1}$, so $6=\mathit{0}$
7	$\overline{\overline{A}}\rightarrow A$	Matrix V: $A=\mathit{0}$, so $7=\mathit{0}$.

§14.2.3. Independent axioms for $\mathbf{E}_{\rightarrow}$.

$\mathbf{E}_{\rightarrow 1}$: 1	$A\rightarrow A\rightarrow B\rightarrow B$	Matrix I: $A=\mathit{2}$ or $\mathit{3}$, $B=\mathit{4}$, so $1=\mathit{0}$
2	$A\rightarrow B\rightarrow.B\rightarrow C\rightarrow.A\rightarrow C$	Diamond-McKinsey theorem
3	$(A\rightarrow.A\rightarrow B)\rightarrow.A\rightarrow B$	Matrix II: $A=\mathit{1}$, $B=\mathit{0}$, so $3=\mathit{0}$
4	$A\rightarrow\overline{B}\rightarrow.B\rightarrow\overline{A}$	Matrix III: $A=\mathit{2}$, $B=\mathit{1}$, so $4=\mathit{0}$
5	$A\rightarrow\overline{A}\rightarrow\overline{A}$	Matrix IV: $A=\mathit{1}$, so $5=\mathit{0}$
6	$\overline{\overline{A}}\rightarrow A$	Matrix V: $A=\mathit{0}$, so $6=\mathit{0}$.

Note also that $\mathbf{E}_{\rightarrow 1}1$–$\mathbf{E}_{\rightarrow 1}5$ are valid intuitionistically when the arrow is interpreted as intuitionistic implication and the bar as intuitionistic negation; but $\mathbf{E}_{\rightarrow 1}6$ is not, and hence is independent.

§14.2.4. Independent axioms for $\mathbf{R}_{\rightarrow}$.

$\mathbf{R}_{\rightarrow 1}$: 2	$A\rightarrow B\rightarrow.C\rightarrow A\rightarrow.C\rightarrow B$	Matrix VI: $A=\mathit{1}$, $B=\mathit{0}$, $C=\mathit{2}$, so $2=\mathit{0}$
3	$(A\rightarrow.B\rightarrow C)\rightarrow.B\rightarrow.A\rightarrow C$	Matrix VII: $A=\mathit{1}$, $B=\mathit{4}$, $C=\mathit{4}$ or $A=\mathit{3}$, $B=\mathit{1}$ or $\mathit{2}$, $C=\mathit{2}$, so $3=\mathit{0}$
4	$A\rightarrow A$	Redundant (§14.1.1)
5	$A\rightarrow\overline{B}\rightarrow.B\rightarrow\overline{A}$	Matrix III: $A=\mathit{2}$, $B=\mathit{1}$, so $5=\mathit{0}$
7	$\overline{\overline{A}}\rightarrow A$	Matrix V: $A=\mathit{0}$ so $7=\mathit{0}$ or Intuitionism (§14.2.3)
and either 1	$(A\rightarrow.A\rightarrow B)\rightarrow.A\rightarrow B$	Leśniewski-Mihailescu theorem
or 6	$A\rightarrow\overline{A}\rightarrow\overline{A}$	Leśniewski-Mihailescu theorem.

§14.3. Negation with das Falsche. Meyer 1966 observes that it is possible to formulate $\mathbf{R}_{\overline{\rightarrow}}$ with a propositional constant f, defining $\overline{A}$ as $A{\rightarrow}f$. Then the axiom $A{\rightarrow}f{\rightarrow}f{\rightarrow}A$ (called $f2$ in §27.1.2) does all the work of $\mathrm{R}_{\overline{\rightarrow}1}5$–$\mathrm{R}_{\overline{\rightarrow}1}7$: the unpacking of $\mathrm{R}_{\overline{\rightarrow}1}5$ under the definition is $(A{\rightarrow}.B{\rightarrow}f){\rightarrow}.B{\rightarrow}.A{\rightarrow}f$, which is an instance of permutation, while the unpacking of $\mathrm{R}_{\overline{\rightarrow}1}7$ is the new axiom $A{\rightarrow}f{\rightarrow}f{\rightarrow}A$ itself; finally, $\mathrm{R}_{\overline{\rightarrow}}6$ unzips into $(A{\rightarrow}.A{\rightarrow}f){\rightarrow}.A{\rightarrow}f$, an instance of contraction.

A similar procedure in the context of $\mathbf{E}_{\rightarrow}$, as opposed to $\mathbf{R}_{\rightarrow}$, is not available. For suppose we defined $\overline{A}$ as $A{\rightarrow}f$ in $\mathbf{E}_{\rightarrow}$, giving f enough properties to secure that in the combined entailment-f calculus we have all the negation properties present in $\mathbf{E}_{\overline{\rightarrow}}$. Then by double negation and the proposed definition of $\overline{A}$ as $A{\rightarrow}f$, we should have A equivalent to $A{\rightarrow}f{\rightarrow}f$. Also in $\mathbf{E}_{\rightarrow}$ we have $A{\rightarrow}f{\rightarrow}f{\rightarrow}.A{\rightarrow}f{\rightarrow}f{\rightarrow}B{\rightarrow}B$ by restricted assertion. So by the replacement theorem, we should have, willy-nilly, $A{\rightarrow}.A{\rightarrow}B{\rightarrow}B$, thus reducing the pure entailment part of the calculus to $\mathbf{R}_{\rightarrow}$.

§14.4. Conservative extensions. We follow Post 1943 in using this term to refer to the following sort of situation.

Let **S** be a system whose grammar involves only the $\rightarrow$, and which has $\rightarrow$E together with the single axiom

1 $A{\rightarrow}B{\rightarrow}.B{\rightarrow}C{\rightarrow}.A{\rightarrow}C.$

Then let **S′** be obtained from **S** by adding a sign of negation to the grammar, and also adding two new axioms governing this new sign:

2 $A{\rightarrow}\overline{\overline{A}}$

and

3 $\overline{\overline{A}}{\rightarrow}A.$

Then we can prove

4 $A{\rightarrow}A.$

This means that in the new system **S′** we can prove a new theorem 4, involving only the notation of the old system **S**, which was not provable in **S** before the introduction of new notation and axioms. In order to prove 4, which involves *only* the arrow, we must take a detour through negation, so to speak. Clearly **S′** is an *extension* of **S**, in the sense that we have added new connectives (in this case negation), and new axioms (2 and 3); but this extension **S′** is *not* "conservative" in Post's sense for precisely the reason just mentioned.

We call **S′** a *conservative extension* of **S** if it is an extension of the grammar, axioms or rules, having the following feature. Let A be a formula in the notation of the smaller system **S**; then if A is provable in the larger system

$\mathbf{S}'$, then A is also provable in $\mathbf{S}$. Equivalently, we may say that every formula provable in the new system $\mathbf{S}'$ but not in the old system $\mathbf{S}$ involves some of the new notation; or that the addition of the new machinery doesn't infect the set of theorems already present; or that it never happens that a proof of a formula in the old notation needs to take a detour through the machinery of the new system. One need not of course always want extensions to be conservative. The example system $\mathbf{S}$ at the beginning of this section, with $\to$E as sole rule and transitivity as sole axiom, needs (on *everyone's* account) more furniture in the way of axioms for the arrow. So if we add $A \to \overline{\overline{A}}$ and $\overline{\overline{A}} \to A$, say, as further axioms, we get an extension which is not conservative, but does have an additional feature we want (to wit, $A \to A$). But when we have postulated all we want to about "if . . . then —," i.e. when we have a *complete* theory, then we don't want to have that theory mucked up by assumptions having to do with other topics (e.g. truth functions). So we want any extension of the pure calculus of entailment designed to incorporate negation to leave the pure bit alone, which is to say that we want any such extensions to conserve all and only those properties entailment already had.

We have already proved, in effect, several theorems about conservative extensions of the implicational systems thus far considered. We list them below:

1. $\mathbf{E}_{\stackrel{\to}{\neg}}$ is a conservative extension of $\mathbf{E}_{\to}$. The subformula theorem in §13.3 guarantees that any formula involving only arrows, can be proved from axioms involving only arrows, with the help of rules involving only arrows: it follows that negation will not affect the pure entailment part of the calculus $\mathbf{E}_{\stackrel{\to}{\neg}}$.

2. Exactly the same considerations show that $\mathbf{R}_{\stackrel{\to}{\neg}}$ is a conservative extension of $\mathbf{R}_{\to}$.

3. If we add to $\mathbf{R}_{\to}$ the axiom

$$f2 \quad (A \to f \to f) \to A,$$

we get all the required properties of negation as a conservative extension (see §14.3).

4. But if we add $f2$ together with

$$A \to .A \to f \to f$$

as axioms to $\mathbf{E}_{\to}$, $\mathbf{E}_{\to}$ collapses into $\mathbf{R}_{\to}$, as was shown in §14.3; so that addition of these axioms to $\mathbf{E}_{\to}$ does *not* produce a conservative extension, since $A \to .A \to B \to B$ is not provable in $\mathbf{E}_{\to}$ from implication axioms and rules alone, but is provable in the system consisting of $\mathbf{E}_{\to}$ together with $f2$ and $A \to .A \to f \to f$ by taking a detour through negation.

We (in the Larger Sense; see §28.3.2) return to the topic of conservative extensions in §24.4.2 and §28.3.

§14.5. $\mathbf{E}_{\rightleftarrows}$ and $\mathbf{R}_{\rightleftarrows}$ with co-entailment. To the consecution calculus $LE_{\rightleftarrows}$ of §13.1, we may add the following rules for the double-arrow of co-entailment, where now the rules for $(\vdash\to)$ and $(\to\vdash)$ must be adjusted for co-entailment; i.e. $\vec{\alpha}$ means that the sequence α is a sequence of formulas each of which has either an arrow or a double arrow as main connective.

$$(\rightleftarrows\vdash)\quad \frac{\alpha\vdash A,\gamma \qquad \beta, B\vdash\delta}{[\alpha,\beta,A\rightleftarrows B]\vdash[\gamma,\delta]}$$

$$(\rightleftarrows\vdash)\quad \frac{\alpha\vdash B,\gamma \qquad \beta, A\vdash\delta}{[\alpha,\beta,A\rightleftarrows B]\vdash[\gamma,\delta]}$$

$$(\vdash\rightleftarrows)\quad \frac{\vec{\alpha}, A\vdash B \qquad \vec{\alpha}, B\vdash A}{\vec{\alpha}\vdash A\rightleftarrows B}$$

It then turns out that this system, decidable by techniques of §13.3, is equivalent to the result of adding as axioms to $\mathbf{E}_{\rightleftarrows}$ of §9, the axioms $E_{\rightleftarrows}1$–$E_{\rightleftarrows}3$ of §8.7.

The following observation is due to Meyer. In $\mathbf{R}_{\rightleftarrows}$ — but not $\mathbf{E}_{\rightleftarrows}$ — one may *define* $A\rightleftarrows B$ in terms of negation and implication:

$$(\mathrm{A}\rightleftarrows\mathrm{B}) =_{\mathrm{df}} \overline{A\to B\to.\overline{\overline{B\to A}}}.$$

The reader is left with showing that, on this definition, the characteristic properties $E_{\rightleftarrows}1$–$E_{\rightleftarrows}3$ of §8.7 are forthcoming in $\mathbf{R}_{\rightleftarrows}$.

§14.6. Paradox regained. Are there *any* examples of theorems of $\mathbf{E}_{\rightleftarrows}$ with either of the paradoxical forms

1 $B\to.A\to B$

or

2 $\bar{A}\to.A\to B$?

We already considered a formula having the form 1 in §5.2: $B\to A\to.A\to B\to.B\to A$. But it is more exciting to note that one can find an A and B such that both of these hold simultaneously. First let A be $\overline{\overline{\overline{C\to C}}\to.C\to C}$, and then with A so chosen, let B be $A\to.A\to A$. Then $B\to.A\to B$ is the converse of contraction mentioned in §8.13. The proof of $\bar{A}\to.A\to B$ is left to the reader, who may also want to verify that for this choice of A and B, we also have

$B\to.A\to A$,
$B\to.B\to B$,
$\bar{A}\to.B\to B$, and
$\bar{A}\to.A\to A$.

The weird character of A and B is evident also from the fact that in addition to these paradoxical results, it is also provable that $\bar{A}\to B$.

§14.7. Mingle again. We remarked in §8.15 that adding the mingle axiom to $\mathbf{R}_{\to}$ preserves the variable-sharing property required for relevance, and it may be of some interest to see how far this property can be extended.

In the first place, adding the mingle axiom $A\to.A\to A$ to $\mathbf{R}_{\overset{\sim}{\to}}$ does preserve the variable-sharing property; use the following matrix from Parks 1972.

$\to$	*0*	*1*	*2*	*3*	$\sim$
0	*3*	*3*	*3*	*3*	*3*
**1*	*0*	*1*	*0*	*3*	*1*
**2*	*0*	*0*	*2*	*3*	*2*
**3*	*0*	*0*	*0*	*3*	*0*

Given $A\to B$ without sharing, just assign the A-variables *1* and the B-variables *2*.

But the system so defined does not constitute the implication-negation fragment of **RM**, the full mingle system mentioned in §8.15 and defined in §27.1.1. Instead, Parks 1972 shows that the following axioms, with rule $\to$E, constitute exactly the implication-negation fragment $\mathbf{RM}_{\overset{\sim}{\to}}$ of **RM**:

1 $A\to B\to.B\to C\to.A\to C$
2 $A\to.A\to B\to B$
3 $(A\to.A\to B)\to.A\to B$
4 $A\to.B\to.\bar{B}\to A$
5 $\bar{A}\to\bar{B}\to.B\to A$

As Parks shows by the matrix exhibited above, 4 does not follow from $\mathbf{R}_{\overset{\sim}{\to}}$ together with the mingle axiom.

Now Parks did not invent these axioms; he discovered them in Sobociński 1952. There, Sobociński proved that the axioms are independent and that they exactly axiomatize the matrix of §8.15 together with the usual three-valued negation table. We repeat:

$\to$	*0*	*1*	*2*	$\sim$
0	*2*	*2*	*2*	*2*
**1*	*0*	*1*	*2*	*1*
**2*	*0*	*0*	*2*	*0*

That is, this matrix is characteristic for the displayed axiom set; so that $\mathbf{RM}_{\overset{\sim}{\to}}$ is in this sense three-valued. Since **RM** as a whole is far from three-valued (see §§29.3–4), this comes to us as a distinct surprise; and we wonder if it did to Meyer, who independently (and earlier) proved that the displayed three-valued matrix is characteristic for the implication-negation fragment of **RM**.

In virtue of axiom 4, it is immediate that variable-sharing breaks down for $\mathbf{RM}_{\overset{\rightarrow}{\sim}}$.

There is however something in the vicinity which holds. In §29.6 we discuss Parry's concept of "analytic implication" according to which for $A \to B$ to be provable, B must contain no variables foreign to A. This property too certainly fails for the systems we have so far discussed, but an exceptionally special case of the converse holds, as observed in effect by Meyer in correspondence: for *theorems* A and B, for $A \to B$ to be provable in any system contained in $\mathbf{RM}_{\overset{\rightarrow}{\sim}}$ — including $\mathbf{T}_{\overset{\rightarrow}{\sim}}$, $\mathbf{E}_{\overset{\rightarrow}{\sim}}$, and $\mathbf{R}_{\overset{\rightarrow}{\sim}}$ — it is required that A contain no variables foreign to B.

PROOF. Use the matrix above. Assign some variable in A but not in B the value $\mathit{2}$, and all other variables the value $\mathbf{1}$. Then A (being a theorem) takes the value $\mathit{2}$, and B takes the value $\mathit{1}$; but $\mathit{2} \to \mathit{1} = \mathit{0}$.

What does this mean?

CHAPTER III

ENTAILMENT BETWEEN TRUTH FUNCTIONS

§15. Tautological entailments. In Chapter I we offered a formal calculus of pure entailment, and in Chapter II we extended our results to include negation. In the foregoing discussions we took seriously the nestability of entailments, explicating their behavior one within another by means of the method of subproofs. In the present chapter we take a different tack, which ignores the possibility and problems of nested entailments in favor of considering *first degree entailments* $A \rightarrow B$, where A and B can be truth functions of any degree but cannot contain any arrows. That is, we are trading complexity of nested arrows for complexity of nested truth functions; but until Chapter IV, we are not going to consider the simultaneous occurrence of both kinds of complexity. Because of the absence of nested arrows, we shall be treating only the relational properties of entailment, ignoring its life as a connective. Accordingly, the reader is entitled to interpret the arrow of this chapter as signifying a metalinguistic relation of logical consequence standing between truth functional expressions, just as those with extensionalist tendencies would prefer. In this way the reader may be able to make more direct comparisons between our ideas and those of orthodoxy, since apparent violations of grammatical niceties will no longer disturb his sense of propriety. As for ourselves, we shall continue our maddening policy of grammatical perversity, taking the arrow ambiguously as a sentence-predicate standing for a metalinguistic relation, as a proposition-predicate standing for a relation between propositions, and as a connective (see Grammatical Propaedeutic).

Throughout this chapter, eschewal of nested arrows will carry with it eschewal of commitments concerning modality, so that our concern is simply with the *relevance* of antecedent to consequent when both are purely truth functional. It follows that nested "if . . . then —" statements can be added to such relevant implications in several ways, each of which will preserve the properties of first degree arrow-sentences (i.e., those with but one arrow, namely, the major connective). In Chapter IV we consider in detail the system **E** (§21.1), which arises when the nested superstructure on first degree formulas is that of $\mathbf{E}_{\overset{\sim}{\rightarrow}}$, but we shall also from time to time advert to the system **R** which results when the nesting is non-modal, as in $\mathbf{R}_{\overset{\sim}{\rightarrow}}$. As we shall see, **R** (§27.1.1) bears to its modal counterpart **E** approximately the same relation that the classical two-valued calculus bears to **S4**.

For the present we simply point out that the results of this section will be common to both **E** and **R**, and indeed to the system **T** (§27.1.1) of ticket entailment as well.

NOTATION. The language we consider is based on entailment, together with symbols for the truth functional notions of negation, conjunction, and disjunction. We already have notation for entailment and negation, and we now introduce "$A\&B$" (sometimes "AB") for the conjunction "A and B," and "$A\vee B$" for the disjunction "A or B." That is, we suppose as before that we have an (undisplayed) collection of propositional variables as *formulas*, and that whenever A and B are formulas, so are $\bar{A}$, $(A\&B)$, $(A\vee B)$, and most importantly, $(A\rightarrow B)$.

In this chapter, we are not concerned with *all* formulas, a matter we make clear as follows.

We think of "degree" as meaning the degree of nesting of arrows. Accordingly, we define A to be a *zero degree formula* (*zdf*) if A contains no arrows at all, and we define a *first degree entailment* (*fde*) as a formula $A\rightarrow B$, with both A and B zero degree ("purely truth functional") formulas. It is these on which we concentrate for a while. In this chapter we use A, B, C, . . . to range over zero degree formulas.

We add some more conventions.

"$A\supset B$" is short for "$\bar{A}\vee B$," and "$A\equiv B$" for "$(A\supset B)\&(B\supset A)$." Although retaining use of a dot as left parenthesis together with a principle of association to the left (§1.2), we modify the rule of left association in order to allow ourselves to omit the pair of parentheses on a purely truth functional formula following a single arrow. Hence, while left association already secures that $A\vee B\rightarrow C$ is $(A\vee B)\rightarrow C$, the new rule says that $A\rightarrow B\&C$ is $A\rightarrow(B\&C)$. The point is that the arrow should always be taken as the major connective.

Finally, $A_1\&\ldots\&A_m$ is any m-termed conjunction, all associated to the left, and similarly for $B_1\vee\ldots\vee B_m$.

§15.1. Tautological entailments. We take the problem of this section to be the discovery of plausible criteria for picking out from among first degree entailments (i.e., entailments of the form $A\rightarrow B$, where A and B are purely truth functional) those that are valid. We refer to such valid entailments as *tautological entailments*.

Clearly none of the orthodox "implication" relations will do as a criterion, since if any of these relations were sufficient for valid entailment, we would have $A\&\bar{A}\rightarrow B$. But as the discussion of the preceding chapters would indicate, we regard a contradiction $A\&\bar{A}$ as in general irrelevant to an arbitrary proposition B, and we accordingly think of the principle "(A and not-A) implies B" as embodying a fallacy of relevance. On the other hand, "(A and not-A) implies A" seems true to our preanalytic idea of conjunction, since it is a special case of the plausible principle "(A and B) implies A." What is wanted, *inter alia*, is a way of distinguishing these cases.

Von Wright 1957 (p. 181) has proposed a criterion of the sort we seek: "A entails B, if and only if, by means of logic, it is possible to come to know the truth of $A \supset B$ without coming to know the falsehood of A or the truth of B," where of course the horseshoe is to be read as material "implication." Geach 1958, following von Wright, proposes a similar criterion: "I maintain that A entails B if and only if there is an *a priori* way of getting to know that $A \supset B$ which is not a way of getting to know whether A or whether B." These proposals seem to us to be on the right track, but they need improvement, for two reasons.

In the first place, the expression "come to know" is loose. One might imagine a person's "coming to know" the truth of $A \supset . B \vee \bar{B}$ without coming to know the truth of $B \vee \bar{B}$, owing to the fact (say) that the formula was fed into a computer programmed to test tautologies. Strawson 1958 finds a similar difficulty:

> It appears that von Wright has overlooked the implications of one familiar way of arriving at the paradoxes. Consider $p \supset (q \vee \bar{q})$. Von Wright would of course wish to deny that p entails $q \vee \bar{q}$. Now the following is demonstrable independently of demonstrating the truth of $q \vee \bar{q}$ or, of course, the falsity of p:
>
> (1) $p \supset ((p \& q) \vee (p \& \bar{q}))$;
>
> for in the truth-table proof of (1) there is only one "unmixed" column (i.e. column consisting purely of Ts or Fs) which is the last column showing the whole expression to be a tautology. Still more obviously the following are demonstrable independently of demonstrating the falsity of p or the truth of $q \vee \bar{q}$:
>
> (2) $((p \& q) \vee (p \& \bar{q})) \supset (p \& (q \vee \bar{q}))$
> (3) $(p \& (q \vee \bar{q})) \supset (q \vee \bar{q})$
> (4) $(p \supset q) \supset [(q \supset r) \supset ((r \supset s) \supset (p \supset s))]$.
>
> For although (2) and (3) both contain $q \vee \bar{q}$, they are respectively substitution instances of $((p \& q) \vee (p \& r)) \supset (p \& (q \vee r))$ and of $(p \& q) \supset q$. Hence, by substituting the second halves of (1), (2), and (3) for q, r, and s respectively in (4) and by repeated applications of Modus Ponens, we obtain a demonstration of $p \supset (q \vee \bar{q})$ which is independent of demonstrating the falsity of p or the truth of $q \vee \bar{q}$. Consequently, on von Wright's definition, p entails $q \vee \bar{q}$. But this is one of the paradoxical cases which his theory is intended to avoid.

We remark that, of the statements (1)–(4), all but (1) are valid when $\supset$ is replaced by $\rightarrow$. (1) itself contains the seeds of paradox, as will be evident from considerations below.

In reply, von Wright 1959 distinguishes several senses of "coming to know," in one of which, he claims, Strawson's alleged counterexample is not a counterexample. But we do not pursue these distinctions, because, as von Wright indicates, the situation remains in an unsatisfactory state.

Smiley 1959, while retaining the spirit of von Wright's proposal, modifies it in such a way as to eliminate the looseness. He holds that $A_1\& \ldots \& A_n$ should entail B just in case $(A_1\& \ldots \& A_n)\supset B$ is a substitution instance of a tautology $(A_1'\& \ldots \& A_n')\supset B'$, such that neither B' nor the denial of $A_1'\& \ldots \& A_n'$ is provable. He cites as an example: ". . . for any A, $A\&\bar{A}$ entails A, because $A\&\bar{A}\rightarrow A$ is a substitution instance of $A\&B\rightarrow A$; but $A\&\bar{A}$ does not entail just any B, because there is in general no way of deriving $A\&\bar{A}\rightarrow B$ from an implication which is itself tautologous but whose antecedent is not self-contradictory."

Smiley's criterion gives rise to a definition of entailment which is effectively decidable, and seems also to capture the intent of von Wright and Geach. But there is an application of it which leads to a second objection (as do the proposals of von Wright and Geach, under at least one interpretation). Since $A\rightarrow A\&(B\vee\bar{B})$ satisfies the criterion, and $A\&(B\vee\bar{B})\rightarrow B\vee\bar{B}$ does also, we find that the paradox $A\rightarrow B\vee\bar{B}$ can be avoided only at the price of giving up transitivity of entailment. This unwelcome course has in fact been recommended by Lewy 1958, Geach 1958, and Smiley 1959. Smiley considers the matter as follows:

> It is true that "connexion of meaning" is not as simple as might be thought: "it has been plausibly argued that any proposition asserts (at least implicitly) something about all objects whatsoever. 'Grass is green,' for instance, says among other things that it is not the case that grass is not-green and roses are red, and so on. This follows simply from the fact that any proposition constitutes a denial of some other propositions and therefore of all conjunctions of which these propositions are members" [Bennett 1954]. But to conclude from this that "thus there is a connexion of meanings between any two propositions; and a necessary or impossible proposition has with any other proposition a connexion of meanings such as will validate one or other of the paradoxical inferences" is to assume that "connexion of meanings" is a transitive relation, and it is only necessary to examine the derivation of one of the paradoxical principles to see that it is not.

It is of course correct that "connexion of meanings" is not transitive, at least under one interpretation: there is a meaning connection between A and

$A\&B$, also between $A\&B$ and B—but there need be no connection of meaning between A and B. And what this shows is that connection of meaning, though necessary, is not a sufficient condition for entailment, since the latter relation *is* transitive. Any criterion according to which entailment is not transitive, is *ipso facto* wrong. It seems in fact incredible that anyone should admit that B follows from A, and that C follows from B, but feel that some further argument was required to establish that A entails C. What better evidence for $A \rightarrow C$ could one want?

The failure of these proposals (see §20.1) arises in part from an attempt to apply them indiscriminately to all formulas. For there is a class of entailments for which Smiley's criterion is absolutely unarguably both a necessary and a sufficient condition; namely, the class of primitive entailments, which, after introducing some auxiliary notions, we proceed to define.

We shall use p, q, and r as variables ranging over propositional variables. Then an *atom* is a propositional variable or the negate of such, i.e., an atom has either the form p or the form $\bar{p}$. A *primitive conjunction* is a conjunction $A_1\&A_2\& \ldots \&A_m$, where each A_i is an atom. A *primitive disjunction* is a disjunction $B_1 \vee B_2 \vee \ldots \vee B_n$, each B_j being an atom. We allow $m = 1$ and $n = 1$, so that atoms are both primitive conjunctions and disjunctions. $A \rightarrow B$ is a *primitive entailment* if A is a primitive conjunction and B is a primitive disjunction. We take it as obvious that if A and B are both atoms, then $A \rightarrow B$ should be a valid entailment if and only if A and B are the same atom; e.g., we would want $p \rightarrow p$ and $\bar{p} \rightarrow \bar{p}$ but not $p \rightarrow q$ or $p \rightarrow \bar{p}$. We think it equally obvious that if $A_1\&A_2 \ldots \&A_m$ is a primitive conjunction and $B_1 \vee \ldots \vee B_n$ is a primitive disjunction, then $A_1\& \ldots \&A_m \rightarrow B_1 \vee \ldots \vee B_n$ should be a valid entailment if and only if some atom A_i is the same as some atom B_j; e.g., we want $p\&q \rightarrow q \vee r$, and $\bar{p}\&q\&r \rightarrow s \vee \bar{p} \vee \bar{r}$, but neither $p\&\bar{q} \rightarrow r$, nor $\bar{p} \rightarrow q \vee p \vee r$, nor (it need hardly be added) $p\&\bar{p} \rightarrow q$.

These considerations lead us to collect a few instructive examples of good guys and bad guys among primitive entailments, which we sort out in pairs as follows:

Valid	*Invalid*
$p \rightarrow p$	$p \rightarrow q$
$\bar{p} \rightarrow \bar{p}$	$\bar{p} \rightarrow p$
$p\&q \rightarrow q$ (hence, $p\&\bar{p} \rightarrow \bar{p}$)	$p\&\bar{p} \rightarrow q$
$p \rightarrow p \vee q$ (hence, $p \rightarrow p \vee \bar{p}$)	$p \rightarrow q \vee \bar{q}$
$p\&q \rightarrow r \vee q$ (hence, $p\&\bar{p} \rightarrow p \vee \bar{p}$)	$p\&\bar{p} \rightarrow q \vee \bar{q}$
$p\&\bar{p}\&t\&\bar{q}\&r \rightarrow s \vee t \vee \bar{s} \vee q \vee \bar{r}$	$p\&\bar{p}\&\bar{q}\&r \rightarrow s \vee \bar{s} \vee q \vee \bar{r}$

Notice that the last example in the right column exhibits all the ways in which a primitive entailment can remain invalid while treading on the brink of validity, in the sense that no atoms made up of p, q, r, and s, can be added on either the right or left of the arrow (unless they are already there) without rendering the formula valid.

We shall say that a primitive entailment $A \rightarrow B$ is *explicitly tautological* if some (conjoined) atom of A is identical with some (disjoined) atom of B. Such entailments may be thought of as satisfying the classical dogma that for A to entail B, B must be "contained" in A.

Some (e.g. Nelson 1930) have objected to taking $p\&q \rightarrow q$ as valid on the grounds that a portion of the antecedent (i.e., p) is not relevant to the conclusion; but surely there *is* a sense of relevance in which, if any of the conjoined premisses are used in arriving at the conclusion, then the conjoined premisses are relevant to the conclusion. (See §3 and, for a full discussion, §22.2.2.)

The principle of "containment" is of course familiar from Kant 1781. Some, e.g. Parry 1933 (see §29.6), have understood the dogma in such a way that all variables in the consequent of a valid entailment must also occur in the antecedent. But surely Kant would have regarded "all brothers are siblings" as an analytic truth, and if "sibling" is defined in the natural way, we have (without the quantifiers) a case of $p \rightarrow p \vee q$ (i.e., brothers are either brothers or sisters). So although $p\&q \rightarrow p$ *clearly* meets Kant's criterion, there may be some doubt (an historical consideration into which we will not enter) as to whether $p \rightarrow p \vee q$ is correct in the Kantian sense. We think it is.

At any rate it is clear that explicitly tautological entailments satisfy the requirements of von Wright, Geach, and Smiley (see §20.1): every explicitly tautological entailment answers to a material "implication" which is a substitution instance of a tautologous material "implication" with non-contradictory antecedent and non-tautologous consequent; and evidently we may ascertain the truth of the entailment without coming to know the truth of the consequent or the falsity of the antecedent. Certainly all explicitly tautological entailments are valid, and there is obviously not the slightest way in which the stock of valid primitive entailments could plausibly be enlarged; we take it therefore that explicitly tautological entailmenthood is both necessary and sufficient for the validity of a primitive entailment.

We now return to a consideration of the non-primitive entailment $A \rightarrow A\&(B \vee \overline{B})$, which, as was pointed out before, satisfies the criteria of von Wright, Geach, and Smiley. Lewy 1958 remarks (in effect) that $A \rightarrow A\&(B \vee \overline{B})$ seems "very nearly, if not quite," as counterintuitive as $A \rightarrow B \vee \overline{B}$. We agree in substance with Lewy, but we think his estimate is too high: $A \rightarrow A\&(B \vee \overline{B})$ is exactly 50% as counterintuitive as $A \rightarrow B \vee \overline{B}$. That is, the non-primitive entailment $A \rightarrow A\&(B \vee \overline{B})$ is valid just in case both the primitive entailments $A \rightarrow A$ and $A \rightarrow B \vee \overline{B}$ are valid; the former is valid, but the latter is not — hence $A \to A\&(B \vee \bar{B})$ does not represent a valid inference. Dually, $(A\&\bar{A}) \vee B \to B$ is valid if and only if both $A\&\bar{A} \to B$ and $B \to B$ are valid; and again one is valid and the other not.

These considerations suggest criteria for evaluating certain first degree entailments other than primitive ones: $A \to B\&C$ is valid if and only if both $A \to B$ and $A \to C$ are valid; and $A \vee B \to C$ is valid if and only if $A \to C$ and $B \to C$ are both valid. This gives us a technique for evaluating *entailments in normal form*, i.e., entailments $A \to B$ having the form $A_1 \vee \ldots \vee A_m \to B_1 \& \ldots \& B_n$, where each A_i is a primitive conjunction and each B_j is a primitive disjunction. Such an entailment is valid just in case each $A_i \to B_j$ is explicitly tautological. For example, $(p\&q) \vee \bar{p} \to (\bar{p} \vee p)\&(\bar{p} \vee q)$, $(p\&q) \vee (p\&r) \to p\&(p \vee r)$, $(p \vee q) \vee r \to p \vee (q \vee r)$, and $p\&q \to q\&(r \vee p)$, are all valid entailments in normal form; but the following are invalid: $(p\&\bar{p}) \vee q \to q$, and $p \to p\&(q \vee \bar{q})$.

We therefore call an entailment $A_1 \vee \ldots \vee A_m \to B_1 \& \ldots \& B_n$ in normal form *explicitly tautological* (extending the previous definition) iff for every A_i and B_j, $A_i \to B_j$ is explicitly tautological (sharing); and we take such entailments to be valid iff explicitly tautological.

The proposal as stated is still not complete, however, since there are combinations of negation, disjunction and conjunction which the rule fails to cover; there is no way to apply it directly to entailments such as $A\&(\bar{A} \vee B) \to B$ or $\bar{A} \to \overline{\overline{\overline{A\&B}}}$, which are not in normal form. But all that is needed to make the criterion everywhere applicable is the ability to convert any first degree entailment into normal form. This in turn will require converting truth functional formulas into *disjunctive* and *conjunctive normal forms*, i.e., into disjunctions of one or more primitive conjunctions, and conjunctions of one or more primitive disjunctions.

We therefore propose adopting the following replacement rules (all of which we take to preserve validity), which enable us to find, for any first degree entailment, at least one equivalent entailment in normal form:

Commutation: replace a part $A\&B$ by $B\&A$; replace a part $A \vee B$ by $B \vee A$;

Association: replace a part $(A\&B)\&C$ by $A\&(B\&C)$, and conversely; replace a part $(A \vee B) \vee C$ by $A \vee (B \vee C)$, and conversely;

Distribution: replace a part $A\&(B \vee C)$ by $(A\&B) \vee (A\&C)$, and conversely; replace a part $A \vee (B\&C)$ by $(A \vee B)\&(A \vee C)$, and conversely;

Double negation: replace a part $\bar{\bar{A}}$ by A, and conversely;

De Morgan's laws: replace a part $\overline{A\&B}$ by $\bar{A} \vee \bar{B}$, and conversely; replace a part $\overline{A \vee B}$ by $\bar{A}\&\bar{B}$, and conversely.

That these rules suffice to reduce any formula to an equivalent one in (say) conjunctive normal form is readily seen: first, De Morgan's laws and Double negation can be used to drive signs of negation inward until each rests over a propositional variable; and then Distribution (second form) can be used — prefaced, if necessary, by an application of Commutation — to move all disjunction signs from outside to inside signs of conjunction; last, Association can be used to group things toward the left. (Recall that $B_1\& \ldots \&B_n$ stands for $(\ldots ((B_1\&B_2)\&B_3) \ldots B_n)$, since conventions introduced earlier say that for two-place connectives, parentheses are to be replaced by association to the left.) More detailed accounts of the proof may be found in almost any elementary text, e.g. Copi 1954.

We call an entailment $A\rightarrow B$ where A and B are purely truth functional, a *tautological entailment*, if $A\rightarrow B$ has a normal form $A_1 \vee \ldots \vee A_m \rightarrow B_1\& \ldots \&B_n$ which is explicitly tautological.

We note that although $A\rightarrow B$ has in general more than one normal form, these will differ only in the order and association of conjuncts and disjuncts; and that consequently one normal form of $A\rightarrow B$ will be explicitly tautological iff all of them are. We appeal to this fact silently whenever we cite as sufficient evidence that $A\rightarrow B$ is Bad, that some one of its normal forms is not explicitly tautological.

We propose tautological entailmenthood as a necessary and sufficient condition for the validity of first degree entailments. (The property is obviously decidable.)

As an example, we show that $(p\supset q)\&(q\supset r) \rightarrow p\supset r$ is invalid. By the definition of "$\supset$," we have $(\bar{p}\vee q)\&(\bar{q}\vee r) \rightarrow \bar{p}\vee r$ which has a normal form, $(\bar{p}\&\bar{q})\vee(\bar{p}\&r)\vee(q\&\bar{q})\vee(q\&r) \rightarrow \bar{p}\vee r$. But $q\&\bar{q} \rightarrow \bar{p}\vee r$ is not an explicitly tautological entailment; hence the candidate fails.

The foregoing example shows that material "implication" is not transitive, if by saying that ρ is transitive we mean that $A\rho B$ and $B\rho C$ jointly entail $A\rho C$.

The replacement rules listed above are obviously classically valid, so that the orthodox logician agrees with us that the validity question concerning a candidate A→B reduces to a question concerning a formula $A_1 \vee \ldots \vee A_m \rightarrow B_1\& \ldots \&B_n$, in normal form. He also agrees that the latter is valid just in case each $A_i\rightarrow B_j$ is valid, so that the *sole difference* between the classical proposal and the correct one concerns standards of acceptability for primitive entailments $A_1\& \ldots \&A_m \rightarrow B_1 \vee \ldots \vee B_n$, (each A_i and B_j an atom). Several criteria have been proposed.

(1) We have a contradiction on the left, in the sense that some variable p and its negate $\bar{p}$ both appear as conjuncts.

(2) There is an excluded middle on the right; i.e., some variable p and its negate $\bar{p}$ both appear as disjuncts.

(3) Sharing: some atom occurs as conjunct on the left and as disjunct on the right.

According to our proposal, (3) alone is sufficient for the validity of a primitive entailment. Other logicians, e.g. Fitch 1952, are more relaxed about the matter and allow both (1) and (3) as good, though not going so far as to countenance (2); they like $p \rightarrow p$ and $p \& \bar{p} \rightarrow q$, but not $p \rightarrow q \vee \bar{q}$. And S. K. Thomason 197+ allows (2) and (3), but not (1): $p \rightarrow q \vee \bar{q}$ and $p \rightarrow p$ but not $p \& \bar{p} \rightarrow q$. (We trust that no one has yet investigated the system obtained by taking as "valid" primitive entailments those satisfying (1) and (2) *only*.)

But a classical logician can be readily identified by his claim not to be able to tell the difference between $p \& \bar{p} \rightarrow q$, $p \rightarrow q \vee \bar{q}$, and $p \rightarrow p$, and by indiscriminately tolerating all of (1)–(3). As against the classical logician, then, our plea is for decidedly less tolerance: it may be well, as a recent head of the Johns Hopkins University is said to have remarked, to keep our minds open — but not so far open that our brains fall out.

§15.2. A formalization of tautological entailments ($\mathbf{E}_{fde}$). In this section we illuminate the claims just made by exhibiting a Hilbert-style formalism $\mathbf{E}_{fde}$ which exactly matches the intuitive considerations above. ("$\mathbf{E}_{fde}$" stands for the first degree entailment fragment of the calculus **E** of Chapter IV.) This formulation suffers, from a proof-theoretical point of view, in having lots of rules; Hilbert would have preferred just one. But the spirit is still the same: we introduce seven axioms and four rules, the postulates being arranged for easy comparison with other formulations. As before, the variables A, B, C, . . . , range over truth functions of variables.

POSTULATES FOR THE SYSTEM $\mathbf{E}_{fde}$

Entailment:

Rule: from $A \rightarrow B$ and $B \rightarrow C$ to infer $A \rightarrow C$

Conjunction:

Axiom: $A \& B \rightarrow A$

Axiom: $A \& B \rightarrow B$

Rule: from $A \rightarrow B$ and $A \rightarrow C$ to infer $A \rightarrow B \& C$

Disjunction:

Axiom: $A \rightarrow A \vee B$

Axiom: $B \rightarrow A \vee B$

Rule: from $A \rightarrow C$ and $B \rightarrow C$ to infer $A \vee B \rightarrow C$

Distribution:

Axiom: $A \& (B \vee C) \rightarrow (A \& B) \vee C$

Negation:

Axiom: $A \rightarrow \bar{\bar{A}}$

Axiom: $\bar{\bar{A}} \rightarrow A$

Rule: from $A \rightarrow B$ to infer $\bar{B} \rightarrow \bar{A}$

We shall now demonstrate that $\mathbf{E}_{\text{fde}}$ is, as advertised, a formalization of tautological entailments. In order to accomplish this, we need two facts: (a) every tautological entailment is provable in $\mathbf{E}_{\text{fde}}$, and (b) nothing else is.

(a) Every tautological entailment is provable in $\mathbf{E}_{\text{fde}}$.

We point out first that from the rule for entailment, and the axioms for negation, we get (by cheating) as a theorem:

$$A \to A.$$

Hence, in virtue of the axioms for conjunction and disjunction, and the rule for entailment, all primitive entailments are provable. Notice, incidentally, that negation is totally irrelevant to validity as among explicitly tautological entailments, which fact reinforces our view that truth and falsity are in general irrelevant to validity of the consequence relation, and that it is therefore silly to say that a contradiction implies any old thing, or that any old thing implies the excluded middle: valid primitive entailments have the feature that all negation signs can simply be erased without affecting validity. For example, $(p\&\bar{q}) \to (\bar{q}\vee r)$ is valid and provable, just as is $(p\&q) \to (q\vee r)$.

Second, in view of the rules for conjunction and disjunction, all tautological entailments in normal form are provable.

The reader may next verify that the following equivalences are provable for $\mathbf{E}_{\text{fde}}$ (for the notation "$\rightleftarrows$" see the beginning of §10):

$$\begin{aligned}
A\&B &\rightleftarrows B\&A \\
A\vee B &\rightleftarrows B\vee A \\
(A\&B)\&C &\rightleftarrows A\&(B\&C) \\
(A\vee B)\vee C &\rightleftarrows A\vee(B\vee C) \\
A\&(B\vee C) &\rightleftarrows (A\&B)\vee(A\&C) \\
A\vee(B\&C) &\rightleftarrows (A\vee B)\&(A\vee C) \\
A &\rightleftarrows \bar{\bar{A}} \\
\overline{A\&B} &\rightleftarrows \bar{A}\vee\bar{B} \\
\overline{A\vee B} &\rightleftarrows \bar{A}\&\bar{B}
\end{aligned}$$

The last point to be made is that a standard replacement theorem of the usual sort: if $A\rightleftarrows B$ then $(\ldots A \ldots) \rightleftarrows (\ldots B \ldots)$, can be made to follow by the usual induction on the length of formulas. And in consequence we have available in $\mathbf{E}_{\text{fde}}$ the replacement rules listed above, thus completing the proof that all tautological entailments are provable in $\mathbf{E}_{\text{fde}}$, the proposition we called (a) above.

Each $A\to B$ has a normal form $A_1\vee \ldots \vee A_m \to B_1\& \ldots \&B_n$, of which the antecedent is in disjunctive normal form, and the consequent in conjunctive normal form. But notice that it does not follow that we can rewrite formulas in *Boolean* (or *distinguished*, or *expanded*, or *ausgezeichnete*) normal form (for which see §16.2). If we could do so, we would have $A \rightleftarrows (A\&B) \vee (A\&\bar{B})$, from which would follow $A \rightleftarrows A\&(B\vee\bar{B})$, whence $A \rightarrow (B\vee\bar{B})$, and we would be back in the soup.

(b) Only tautological entailments are provable in $\mathbf{E}_{\mathrm{fde}}$.

For this it is required to see that all the axioms are tautological entailments and that the rules preserve this property.

That all the axioms are so can easily be proved by induction on the length of formulas. If A, B, C in the axioms are atoms, the result is trivial; and if for example (first axiom for conjunction) we have $(p\vee\bar{p})\&(q\vee\bar{q}) \rightarrow (p\vee\bar{p})$, then the normal form $(p\&q)\vee(\bar{p}\&q)\vee(p\&\bar{q})\vee(\bar{p}\&\bar{q}) \rightarrow p\vee\bar{p}$ obviously has the required property. Proof of the general case is left to the reader.

It is equally easy to show that the rules for conjunction and disjunction preserve the property. The only items to worry about are the rule of transitivity for entailment and the rule of contraposition.

The argument for contraposition relies on the observation that, given a conjunctive normal form of C, we obtain a disjunctive normal form of $\bar{C}$ by: trading all $\vee$'s for &'s and conversely, and trading all variables p for their negates $\bar{p}$ and conversely. Hence, if a disjunctive normal form of A has the right sharing property with a conjunctive normal form of B, so also will a disjunctive normal form of $\bar{B}$ with a conjunctive normal form of $\bar{A}$.

For transitivity, we need to show that if $A\rightarrow B$ and $B\rightarrow C$ are tautological entailments, so also is $A\rightarrow C$. First put $A\rightarrow B$, $B\rightarrow C$, and $A\rightarrow C$ each in normal form:

$$A\rightarrow B\colon \quad A_1\vee \ldots \vee A_m \rightarrow B_1\& \ldots \&B_n,$$
$$B\rightarrow C\colon \quad B_1'\vee \ldots \vee B_k' \rightarrow C_1\& \ldots \&C_p, \text{ and}$$
$$A\rightarrow C\colon \quad A_1\vee \ldots \vee A_m \rightarrow C_1\& \ldots \&C_p.$$

If we can show that each of the A_i's shares an atom with each of the C_j's, e.g. A_1 with C_1, we will be home. It is known that A_1 shares an atom with each B_i; let $B_1{}^*\& \ldots \&B_n{}^*$ be a conjunction of the atoms $B_i{}^*$ which A_1 shares with B_i. Evidently, $B_1{}^*\& \ldots \&B_n{}^*$ is a (proper or improper) subconjunction of A_1.

Now observe that, by a generalized principle of distributivity, $B_1{}^*\& \ldots \&B_n{}^*$ is *precisely* one of the terms in the disjunctive normal form $B_1'\vee \ldots \vee B_k'$ of B — for these terms are generated exactly by taking conjunctions of elements selected one each from $B_1\& \ldots \&B_n$ in all possible ways. But each B_i' shares an atom with each C_j, since $B\rightarrow C$ is a tautological entailment. Therefore, $B_1{}^*\& \ldots \&B_n{}^*$ and accordingly A_1, of which it is a subconjunction, must share an atom with each C_j.

Similarly for the other A_i's.

It follows then that no formulas other than tautological entailments are provable in $\mathbf{E}_{fde}$, and hence $\mathbf{E}_{fde}$ is both complete and consistent with respect to the guiding intuitions.

Thinking along these lines leads us to state a nice property of $\mathbf{E}_{fde}$: the

PERFECT INTERPOLATION THEOREM. If $\vdash A \rightarrow C$ in $\mathbf{E}_{fde}$, then there is an "interpolation formula" B such that (1) $\vdash A \rightarrow B$, (2) $\vdash B \rightarrow C$, and (3) B has no variables not in both A and C.

This is a "perfecting" of Craig 1957's justly celebrated theorem in that we do not have to exclude, as does he, the special cases when A is contradictory or B is valid. The proof is easy and obvious: represent $A \rightarrow C$ in the form above; choose from each Ai a subconjunction Ai' containing just those atoms needed for Ai to share with each C_j, and let B $= A_i' \vee \ldots \vee A_m'$.

PROBLEM. Does the theorem hold for the full system **E** of §21? Or for **R** or **T** of §27.1.1? (See §29.3.3 for a negative result for **RM**.)

§15.3. Characteristic matrix. Until this point we have used matrices only for proving independence (see §8.4 and examples which follow). But of course matrices have other uses as well, a principal use being to solve decision problems. We have already given a solution to the decision problem for tautological entailments and for the system $\mathbf{E}_{fde}$, since clearly the problem for $A \rightarrow B$ can be reduced to that for its normal form $A_1 \vee \ldots \vee A_n \rightarrow B_1 \& \ldots \& B_m$; and the decidability of each $A_i \rightarrow B_j$ can be determined by inspection.

But it may be of interest to give a decision method based on matrices in a more classical way. The classical two-valued case is of course the most familiar: A is provable in a proof-theoretical formulation of the two-valued propositional calculus just in case A is a two-valued tautology in the semantical sense (on this point see §24.1). We call a matrix (like the two-valued matrix for two-valued logic) *characteristic* for a calculus when a formula A is provable just in case it assumes designated values for every assignment of values to its variables.

The following matrices, due to T. J. Smiley (in correspondence), prove to be characteristic for the system $\mathbf{E}_{fde}$.

&	*1*	*2*	*3*	*4*
**1*	*1*	*2*	*3*	*4*
2	*2*	*2*	*4*	*4*
3	*3*	*4*	*3*	*4*
4	*4*	*4*	*4*	*4*

∨	*1*	*2*	*3*	*4*
**1*	*1*	*1*	*1*	*1*
2	*1*	*2*	*1*	*2*
3	*1*	*1*	*3*	*3*
4	*1*	*2*	*3*	*4*

$\rightarrow$	*1*	*2*	*3*	*4*
**1*	*1*	*4*	*4*	*4*
2	*1*	*1*	*4*	*4*
3	*1*	*4*	*1*	*4*
4	*1*	*1*	*1*	*1*

$\sim$	
**1*	*4*
2	*2*
3	*3*
4	*1*

It is easy to see that the rules required for reducing entailments $A \rightarrow B$ (A, B both truth functional) to primitive entailments are all satisfied by the matrices, so it suffices to show that the matrices are characteristic for primitive entailments.

To the reader we leave the task of verifying that sharing an atom suffices for primitive entailments $A \rightarrow B$ assuming always the designated value *1*; and for a candidate primitive entailment $A \rightarrow B$ without sharing of atoms, there is an assignment of values to the variables which will cause $A \rightarrow B$ to assume the undesignated value, *4*. We tabulate the assignment as follows.

p is, $\bar{p}$ is a conjunct of A: give p the value *2*.
p is, $\bar{p}$ is not a conjunct of A: give p the value *1*.
p is not, $\bar{p}$ is a conjunct of A: give p the value *4*.
p is not, $\bar{p}$ is not a conjunct of A: give p the value *3*.

Again we leave it to the reader to verify that with these assignments, and the tables given above, any primitive entailment $A_i \rightarrow B_j$ with no sharing can take the value *4*, thus falsifying $A \rightarrow B$. (Hint: show that every conjunct of A has the value *1* or *2*, and every disjunct of B has the value *3* or *4*; then use the matrix for the arrow. Notice that this arrow matrix is used only once, and then only at the end of the procedure; it sheds no light at all when we come to consider nested entailments.)

It is occasionally interesting to try to figure out what the intuitive sense of such a matrix is, and when we come to the problem of adding truth functional axioms to the pure calculus of entailment, we will treat this matter again. For the moment we will simply observe that *2* and *3* have the odd property that each is equivalent to its own negation, though neither implies the other. This situation is familiar, but if examples fail to spring to mind, we will help the reader a little by mentioning the paradoxical statements of Russell and Epimenides. In these two cases we have clearly independent statements (neither implies the other), each of which is equivalent to its own denial.

§16. Fallacies. As remarked at the beginning of §15, we confine ourselves in this chapter to contexts where modality is irrelevant; such fallacies as arise must therefore necessarily be fallacies of relevance, of which the paradigm cases are $A\&\bar{A} \rightarrow B$ and its dual, $A \rightarrow B\vee\bar{B}$.

As logicians have always taught, logic is a formal matter, and the validity of an inference has nothing to do with the truth or falsity of premisses or conclusion. But the view that the orthodox concept of logical consequence is an implication relation flies squarely in the face of this teaching and leads directly and immediately to fallacies of relevance. $A\&\bar{A} \rightarrow B$ has been defended on the ground that, although it is useless, it is harmless, since the antecedent can never be realized. We grant that it is harmless in this sense, but still contend that it is harmful in another sense, namely, in being false. To be sure, there is a somewhat odd sense in which we "lose control" in the presence of enough contradictions. Namely: we define a *manifest repugnancy* as a primitive conjunction having the property that for every propositional variable p occurring therein, both p and $\bar{p}$ occur as conjuncts. An example is $p\&\bar{p}\&q\&\bar{q}\&r\&\bar{r}$. And for such expressions we have the following

THEOREM. Manifest repugnancies entail every truth function to which they are analytically relevant.

PROOF. We follow Parry 1933 (see §29.6) in saying that A is *analytically relevant* to B if all variables of B occur in A. The theorem then states that a manifest repugnancy entails every truth functional compound of its own variables. And this may be readily seen as follows. Let A be a manifest repugnancy, and let B be any truth function of the variables in A. Rewrite B equivalently in conjunctive normal form $B_1\& \ldots \&B_n$. Then each B_i contains at least one of the atoms in A; hence each $A \rightarrow B_i$ is an explicitly tautological entailment.

Dually, we have that every truth functional expression entails a (very weak) tautology, consisting of a disjunction of various special cases of $A\vee\bar{A}$. But admitting these obvious logical truths is a far cry from admitting that a contradiction entails any old thing, or that any old thing entails an excluded middle.

It is of course sometimes said that the "if . . . then —" we *use* admits that false or contradictory propositions imply anything you like, and we are given the example "If Hitler was a military genius, then I'm a monkey's uncle." But it seems to us unsatisfactory to dignify as a principle of logic what is obviously no more than rhetorical figure of speech, and a facetious one at that; one might as well cite Cicero's use of *praeteritio* as evidence that one can do and not do the same thing at the same time (and in the same respect).

§16.1. The Lewis argument. Lewis and Langford 1932 (pp. 248–251), however, have explicitly argued that the paradoxes of strict "implication"

state "a fact about deducibility," and have presented "independent proofs" of their validity. Since there is a clear opposition between our position and that of Lewis and Langford (and practically everyone else), we will examine one of these proofs in detail.

The argument concerns $A\&\overline{A} \rightarrow B$, and has two steps, (i) "*A* entails *B*," or "$A \rightarrow B$," means that *B* is deducible from *A* "by some mode of inference which is valid," and (ii) there is a "valid mode of inference" from $A\&\overline{A}$ to *B*.

We may accept (i) without cavil. Arguments for (ii), that is, for the proposition that there is a valid mode of inference from a contradiction to any arbitrary proposition, were known to several logicians flourishing circa the year 1350, and are found in extant writings of the astute Bishop of Halberstadt, Albert of Saxony (see Boehner 1952, pp. 99–100). Lewis and Langford's presentation of the argument does not differ significantly from Albert's, although it is almost certain that the modern appearance of the argument represents a rediscovery rather than a continuity of tradition. The argument has also been accepted by a variety of other modern logicians — e.g. Popper 1940 and 1943 — and, indeed, as Bennett 1954 points out, "this acceptance has not been an entirely academic matter. Kneale 1945–46 and Popper 1947 have both used the paradoxes as integral parts of their respective accounts of the nature of logic." (Their idea is to give an "explanation" of why contradictions are so Bad: they yield everything. We think it goes the other way around, as we make clear in §33.5.)

The following is a convenient presentation of the baffling argument. Grant that the following are "valid modes of inference":

1 from $A\&B$ to infer *A*,
2 from $A\&B$ to infer *B*,
3 from *A* to infer $A \vee B$, and
4 from $A \vee B$ and $\overline{A}$ to infer *B*.

The argument then proceeds in this way:

a	$A\&\overline{A}$	premiss
b	A	from a by 1
c	$\overline{A}$	from a by 2
d	$A \vee B$	from b by 3
e	B	conclusion: from c and d by 4.

Than which nothing could be simpler: if the four rules above are "valid modes of inference" and if "$A \rightarrow B$" means that there is a valid mode of inference from *A* to *B*, then a contradiction such as $A\&\overline{A}$ surely does entail any arbitrary proposition, *B*, whence $A\&\overline{A} \rightarrow B$ represents a fact about deducibility.

We agree with those who find the argument from a to e self-evidently preposterous, and from the point of view we advocate it is immediately obvious where the fallacious step occurs: namely, in passing from c and d to e. The principle 4 (from $A \vee B$ and $\overline{A}$ to infer B), which commits a fallacy of relevance, is not a tautological entailment. We therefore reject 4 as an entailment and as a valid principle of inference.

We seem to have been pushed into one of the "peculiar positions" of which Prior 1955 (p. 195) speaks, for we are explicitly denying that the principle of the disjunctive syllogism or detachment for material "implication" is a valid mode of inference. The validity of this form of inference is something Lewis never doubts (see, for example, Lewis and Langford 1932, pp. 242–43) and is something which has perhaps never been seriously questioned before, though the possibility of dispensing with the disjunctive syllogism is raised by Smiley 1959. Nevertheless, we do hold that the inference from $\overline{A}$ and $A \vee B$ to B is in error: it is a simple inferential mistake, such as only a dog would make (see §25.1, The Dog). Such an inference commits nothing less than a fallacy of relevance. We shall first anticipate possible misinterpretations of this thesis and then proceed (in §16.2.2) to an "independent proof" of the invalidity of $\overline{A}\&(A \vee B) \rightarrow B$.

In the first place, we do not deny that the inference from $\vdash \overline{A}$ and $\vdash A \vee B$ to $\vdash B$ is valid, where "$\vdash A$" means "A is a theorem of the two-valued propositional calculus." However, from this it does not follow that B follows from $\overline{A}$ and $A \vee B$. We even admit that if $\vdash B$ then B is necessarily true, and still hold that the argument from $\overline{A}$ and $A \vee B$ is invalid even when $\vdash \overline{A}$ and $\vdash A \vee B$ (and hence $\vdash B$). Such a claim would be senseless on Lewis' doctrine, for to admit that B is necessarily true is to admit that any argument for B is valid.

Second, we do not say that the inference from $\overline{A}$ and $A \vee B$ is invalid for all choices of A and B; it will be valid at least in an enthymematic sense (see §35) when A entails B (in which case $\overline{A}$ is not required) or when $\overline{A}$ entails B (in which case $A \vee B$ is not required); more generally, it will be valid when $A\&\overline{A}$ entails B.

Furthermore, in rejecting the principle of the disjunctive syllogism, we intend to restrict our rejection to the case in which the "or" is taken truth functionally. In general and with respect to our ordinary reasonings this would not be the case; perhaps always when the principle is used in reasoning one has in mind an intensional meaning of "or," where there is relevance between the disjuncts. But for the intensional meaning of "or," it seems clear that the analogues of $A \rightarrow A \vee B$ are invalid, since this would hold only if the simple truth of A were sufficient for the relevance of A to B; hence, there is a sense in which the real flaw in Lewis's argument is not a fallacy of relevance but rather a fallacy of ambiguity. The passage from b to d is valid only if the "$\vee$" is read truth functionally, while the passage from c and d to e is valid only if the "$\vee$" is taken intensionally. We shall further consider the intensional "or" below, in §§16.3 and 27.1.4.

Our final remark concerns what Lewis might have meant by "some valid form of inference." It is hardly likely that he meant that a form of inference is valid if and only if either the premisses are false or the conclusion true ("material validity"); more plausibly, he might have meant that a form of inference is valid if and only if it is *necessary* that either the premisses are false or the conclusion true ("strict validity"). If this is what Lewis meant, then we agree at once that the inference from A and $\bar{A}\vee B$ to B is valid *in this sense*. However, if this is *all* that Lewis meant by "some valid form of inference," then his long argument for $A\&\bar{A}\rightarrow B$ is a quite unnecessary detour, for *in this sense* we should have agreed at once that there is a valid form of inference from $A\&\bar{A}$ to B: it is surely true that necessarily either the premiss is false or the conclusion is true inasmuch as the premiss is necessarily false. In short, Lewis's "independent proof" of $A\&\bar{A}\rightarrow B$ is convincing if "valid inference" is defined in terms of strict implication; but in that case it is superfluous and circular. And his argument serves a useful purpose only if "valid inference" is thought of in some other sense, in which case he has failed to prove — or even to argue for — his premisses. Finally, should he wish to escape the horns of this dilemma by remarking that the various forms of inference used in the argument are valid in the sense of having always been accepted and used without question, then we should rest our case on the fallacy of ambiguity noted above.

Such a thesis so strongly stated will seem hopelessly naive to those logicians whose logical intuitions have been numbed through hearing and repeating the logicians' fairy tales of the past half-century, and hence it stands in need of further support. It will be insisted that to deny detachment for material and strict implication, as well as to deny the principle of the disjunctive syllogism, surely goes too far: "from $\bar{A}$ and $A\vee B$ to infer B," for example, is surely valid. For one of the premisses states that at least one of A and B is true, and since the other premiss, $\bar{A}$, says that A isn't the true one, the true one must be B (see Popper 1943). Our reply is to remark again that this argument commits a fallacy of ambiguity. There are indeed important senses of "or," "at least one," etc., for which the argument from $\bar{A}$ and A-or-B is perfectly valid, namely, senses in which there is a true relevance between A and B, for example, the sense in which "A-or-B" means precisely that $\bar{A}$ entails B. However, in *this* sense of "or," the inference from A to A-or-B is fallacious, and therefore this sense of "or" is not preserved in the truth functional constant translated by the same word. As Lewis himself argued in some early articles, there are *intensional* meanings of "or," "not both," "at least one is true," etc., as well as of "if . . . then —." Those who claim that only an intensional sense of these words will support inferences are right — Lewis's error was in supposing he captured this sense by tacking a modal operator onto a fundamentally truth functional formula.

§16.2. Distinguished and undistinguished normal forms. Nevertheless the inference from $\bar{A}$ and A-or-B to B is sometimes valid even when the "or" is truth functional, for it will be valid in every case in which $A\&\bar{A} \rightarrow B$. For example, although the decision procedure of §15 shows that $\overline{A\&B} \rightarrow (\bar{A}\&B)\vee(A\&\bar{B})\vee(\bar{A}\&\bar{B})$ is not valid, nevertheless the inference from $\overline{A\&B}$ and the tautology $(A\&B)\vee(\bar{A}\&B)\vee(A\&\bar{B})\vee(\bar{A}\&\bar{B})$ to $(\bar{A}\&B)\vee(A\&\bar{B})\vee(\bar{A}\&\bar{B})$ is valid since, as is easily verified, $(A\&B)\&\overline{A\&B} \rightarrow (\bar{A}\&B)\vee(A\&\bar{B})\vee(\bar{A}\&\bar{B})$.

On the other hand, it is equally easily verified that the inference from $\overline{(A\&B)\vee(\bar{A}\&B)}$ and $(A\&B)\vee(\bar{A}\&B)\vee(A\&\bar{B})\vee(\bar{A}\&\bar{B})$ to $(A\&\bar{B})\vee(\bar{A}\&\bar{B})$ is *not* valid. This pair of facts leads us to ask when in fact one can validly perform a disjunctive syllogism on a "distinguished" (*ausgezeichnete*), or "expanded," or "full," or "Boolean," disjunctive normal form of a "perfect" tautology in $\mathbf{E}_{\text{fde}}$.

We pause for some definitions of these terms. By a *state-description in* p_1, . . . , p_n we mean a conjunction of atoms $A_1\& \ldots \&A_n$ such that the jth conjunct A_j $(1 \leq j \leq n)$ is either the jth variable p_j or its negate $\overline{p_j}$. By a *distinguished disjunctive normal form* we mean a formula having the form $A_1\vee \ldots \vee A_k$, such that for some list of variables $p_1, \ldots, p_n$, each A_i $(1 \leq i \leq k)$ is a distinct state description. By a *perfect tautology* we mean a formula A whose conjunctive normal form looks like

$$1 \quad (p_1\vee\overline{p_1})\& \ldots \&(p_n\vee\overline{p_n}),$$

where we assume for convenience that the p's, all distinct, come in alphabetic order. Evidently not all formulas, not even all tautologies, co-entail in $\mathbf{E}_{\text{fde}}$ a distinguished disjunctive normal form; but every perfect tautology does; namely, 1 co-entails

$$2 \quad A_1\vee \ldots \vee A_{2^n}$$

where each A_i $(1 \leq i \leq 2^n)$ is a distinct state-description in $p_1, \ldots, p_n$. (We expressly leave the order of disjuncts arbitrary.)

We wish to inquire into the conditions under which the denial of some of the disjuncts of 2, when conjoined with 2, entails the remainder of 2. That is, with reference to 2, and with $1 \leq m \leq 2^n$, we ask when the disjunctive syllogism

$$3 \quad \overline{A_1\vee \ldots \vee A_m}\&(A_1\vee \ldots \vee A_m\vee A_{m+1}\vee \ldots \vee A_{2^n}) \rightarrow. A_{m+1}\vee \ldots \vee A_{2^n}$$

holds. The answer is satisfying:

THEOREM. 3 holds just in case no pair of state-descriptions in $A_1 \vee \ldots \vee A_m$ differ in exactly one place.

We postpone proof of this theorem until §16.2.3, first making some more general observations about distinguished normal forms and stating some slightly more general facts involving them.

The results of §15 tell us how the land lies in $\mathbf{E}_{\text{fde}}$ for undistinguished, or garden variety, disjunctive and conjunctive normal forms: given a purely truth functional formula A, we can always find a disjunctive normal form ("dnf") D of A, and a conjunctive normal form ("cnf") C of A, such that $A \rightleftarrows D$ and $A \rightleftarrows C$. We can also, by uses of co-entailments $A \rightleftarrows A \& A$ and $A \rightleftarrows A \vee A$, eliminate redundancies in C and D (in the future we will assume that redundancies have been eliminated), and if we like we can make the forms unique.

In view of the associativity and commutativity of disjunction and conjunction, making the forms unique is not of much interest; it can be done in any of several ways, which we leave to the reader to figure out. But however it is done, the disjunctive form D and the conjunctive form C remain undistinguished.

By this we mean that, since the co-entailments $A \rightleftarrows A \& (p \vee \bar{p})$ and $A \rightleftarrows A \vee (p \& \bar{p})$ both fail (for good reason: they lead to $A \rightarrow p \vee \bar{p}$ and $p \& \bar{p} \rightarrow A$), the classical dodge of calling, e.g., $(p \& q) \vee (p \& \bar{q})$ an "equivalent" distinguished disjunctive normal form of p won't work. Classically, in order to acquire distinction, we must have, in addition to the normalizing principles available in $\mathbf{E}_{\text{fde}}$, $p \equiv . \, p \& (q \vee \bar{q})$, which leads via distribution to $p \equiv . \, (p \& q) \vee (p \& \bar{q})$. This gives us classically a distinguished disjunctive normal form ("Ddnf"), by *expanding* p. Dually, we can classically *contract* a cnf $(q \vee p) \& (q \vee \bar{p})$ to $q \vee (p \& \bar{p})$, and then to q, which is the distinguished conjunctive normal form ("Dcnf") thereof. As is familiar, tautological Ddnfs (in n variables) expand classically to 2^n-termed disjunctions, and tautological Dcnfs classically vanish (to use archaic mathematical terminology).

None of these classical facts are of much interest from the point of view of $\mathbf{E}_{\text{fde}}$, however, for two reasons. A trivial reason is that no purely truth functional formulas are provable in $\mathbf{E}_{\text{fde}}$, theorems of which always have one intensional arrow as main connective, so that where A and B are purely truth functional, $A \equiv B$, being arrow-free, is not a theorem. More important is the fact that in the full system $\mathbf{E}$ of Chapter IV, material "equivalence" is not sufficient for intersubstitutability, as will become evident later, so that, from a logical point of view, material "equivalences" have minimal interest.

But there is something akin to Dcnfs and Ddnfs in $\mathbf{E}_{\text{fde}}$; and, in view of the remarks made at the outset, it appears that classical Ddnfs have some sort of privileged position. Our program is to look into the situation generally in §16.2.1, to recite an uneasy mixture of odd bits of fact and philosophy in §16.2.2, and finally, in §16.2.3, to prove the happy theorem stated above.

§16.2.1. Set-ups. When a classical logician asks us for the Ddnf of a formula A in n variables (throughout this discussion let's have $n = 2$), he wants in effect a list of the assignments to variables in A which make A come out true. So that if A is contradictory (i.e., no assignments make it true), then the list is empty, and so is the Ddnf; if some assignments make it come out true, e.g., just when p is true and q is false, or when p is false and q is true, then the list looks like "$p = \boldsymbol{T}$ and $q = \boldsymbol{F}$, or $p = \boldsymbol{F}$ and $q = \boldsymbol{T}$," and the Ddnf is $(p\&\bar{q})\vee(\bar{p}\&q)$; and if it is a tautology, . . . etc. We can of course read all this off from a truth table.

Moreover, the only cases he cares to contemplate form a mutually exclusive and jointly exhaustive set, namely, those rows given by the truth table.

From the point of view of $\mathbf{E}_{\text{fde}}$, however, we are either more fortunate (e.g. in being able to distinguish $p\&\bar{p}$ from $p\&\bar{p}\&q$), or else less fortunate (e.g. in being unable to confuse $p\&\bar{p}$ with $p\&\bar{p}\&q$); anyway, we are different.

And the difference becomes most striking when we observe that for a truth functional compound A of n variables, a classical logician distinguishes just 2^n possible cases, whereas we distinguish $2^{2n}-1$ possible cases. For the case $n = 2$ at hand, the classical table with $2^2 = 4$ entries, looks like

$p\&q$
$p\&\bar{q}$
$\bar{p}\&q$
$\bar{p}\&\bar{q}$

whereas ours, with $2^{2\times 2}-1 = 15$ entries, looks like

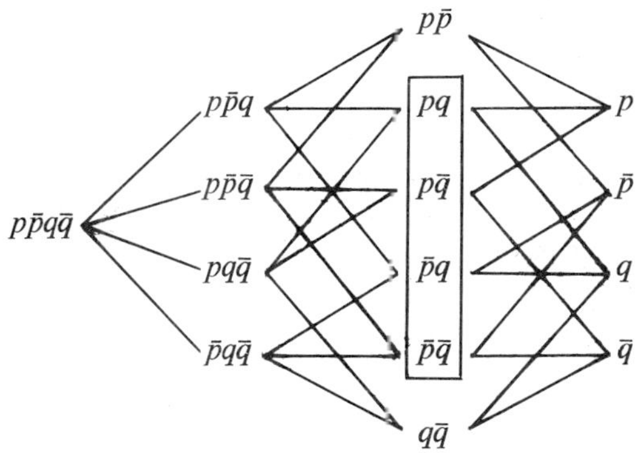

(where now and hereafter we suppress ampersands in favor of adjunction where convenient).

Entailments obviously follow lines from left to right, and equally obviously these are the only valid entailments. The four classical cases are contained in the little box.

Now it will be objected immediately that some of the possible cases we consider, e.g. $\bar{p}q\bar{q}$, are not possible, hence presumably are not possible cases; we avoid this punning argument by borrowing the terminology of Routley and Routley 1972, and referring to the fifteen distinct entries in the second table as "set-ups": a *set-up* in $p_1, \ldots, p_n$ is a conjunction of atoms drawn from among the $p_1, \ldots, p_n$, without repetition and in alphabetic order (the order being given by $p_1, \overline{p_1}, \ldots, p_n, \overline{p_n}$). Some of these set-ups are underdetermined (those on the extreme right in the diagram above), some are completely and consistently determined (those in the box), and some are overdetermined (the rest). The first give us too little information (about p and q), the second just the right amount, and the third too much. The 15 are, nonetheless, all distinct in $\mathbf{E}_{\text{fde}}$. (For more about under- and overdetermination see §§47, 49.)

The differences between classical *state-descriptions* (in the sense of Carnap) and *set-ups* become more apparent if we look at the matter in the following way.

Classically, the four complete and consistent state-descriptions answer to the four classical subsets of a two-element set $\{p, q\}$ of variables: $\{p, q\}$ itself answers to $p\&q$, $\{p\}$ (since each state-description is complete) answers to $p\&\bar{q}$, $\{q\}$ (similarly) to $\bar{p}\&q$, and the null set to $\bar{p}\&\bar{q}$.

Set-ups, on the other hand, need be neither consistent nor complete, and answer rather to theories than to state-descriptions. (Many theories, from the time of Thales on, have, notoriously, been inconsistent and/or incomplete; here again we borrow from Meyer, and Quine.) For set-ups, we consider rather the set of non-empty subsets of a four-element set $\{p, \bar{p}, q, \bar{q}\}$ of atoms in two variables. The maximal subset of this set characterizes the beliefs of the most uninhibited (see beginning of §16), and the march from left to right in the fifteen-entry table above characterizes the temporal flow from youth to age, as we become more conservative.

The upshot of all this is that set-ups answer intensionally to extensional state-descriptions, and disjunctions of set-ups do the work intensionally that disjunctions of state-descriptions do extensionally. Set-ups and state-descriptions disagree in that no state-description classically "implies" another, whereas some set-ups entail others. They agree in that no disjunction of one set of state-descriptions classically "co-implies" a disjunction of a distinct set of state-descriptions, and no disjunction of one set of set-ups co-entails a disjunction of a distinct set of set-ups. We set forth these facts, and some others, in the next section.

§16.2.2. Facts, and some philosophical animadversions. Between our account of set-ups in §16.2.1 and our proof of the main theorem in §16.2.3, we insert here a group of facts, interspersed with some observations for which we can find no better location. Among the staccato of facts, we star (*) those used in §16.2.3 to prove the theorem mentioned toward the beginning of §16.2.

FACT 1. Every truth functional compound of n variables co-entails an *undistinguished disjunctive normal form* ("udnf"), i.e., a disjunction of at most $2^{2n}-1$ set-ups.

Let us collect here our various normal form notations: *dnf* (disjunction of primitive conjunctions), *Ddnf* (disjunction of state-descriptions), *udnf* (disjunction of set-ups). For obvious reasons, we call the duals of set-ups, i.e., disjunctions of atoms, *put-downs* (with no redundancies, and alphabetically unique).

FACT 2. Every truth functional compound of n variables co-entails an *undistinguished conjunctive normal form* ("ucnf"), i.e. a (non-empty) conjunction of at most $2^{2n}-1$ put-downs. These "improve" over §15.2 only in suppressing repetition and fixing alphabetic order by the definition of "set-up" (all set-ups are primitive conjunctions, but not conversely).

We can now re-state a slightly improved version of the results of §15.2 in the following way: consider truth functional compounds A and B; rewrite A as its co-entailed udnf A', and B as its co-entailed ucnf B'; then $\vdash A \rightarrow B$ in $\mathbf{E}_{\mathrm{fde}}$ iff each disjoined set-up in A' shares a variable with each conjoined put-down in B'.

FACT 3. Rewrite A and B (as above) in udnf A' and B'; then $\vdash A \rightarrow B$ iff each disjoined set-up in A' is a (proper or improper) superconjunction of at least one disjoined set-up in B'.

PROOF. It suffices to prove this fact for the case where A is a set-up. Let B' be $B_1 \vee \ldots \vee B_m$. Trivially, if for some B_i, A is a superconjunction of B_i, then $\vdash A \rightarrow B_1 \vee \ldots \vee B_m$ in $\mathbf{E}_{\mathrm{fde}}$.

For the converse, suppose that A is not a superconjunction of any of the B_i. Then every B_i contains some atom not contained in A. Hence the conjunctive normal form $C_1 \& \ldots \& C_n$ of $B_1 \vee \ldots \vee B_m$ contains a disjunction C_j, none of the atoms of which occurs in A; so $A \rightarrow C_j$ is not provable. But $A \rightarrow (B_1 \vee \ldots \vee B_m)$ is provable iff $A \rightarrow (C_1 \& \ldots \& C_n)$ is provable, i.e., iff $A \to C_j$ is provable for each j, as we saw in §15. Hence $A \to (B_1 \vee \ldots \vee B_m)$ is unprovable, as required.

Straightforward duality considerations then enable us to prove

FACT 4. Rewrite A and B (as above) in ucnf A' and B'; then $\vdash A \to B$ iff each conjoined put-down in B' has at least one (proper or improper) subdisjunction among the conjoined put-downs in A'.

Facts 3 and 4 add to our arsenal of methods for checking the validity of first degree entailments. For example, the falsehood of

$$\overline{pq \vee \bar{p}q}(pq \vee \bar{p}q \vee p\bar{q} \vee \bar{p}\bar{q}) \to p\bar{q} \vee \bar{p}\bar{q}$$

could be checked in the *old* way as in §15.2, by observing that when the left is in udnf and the right in ucnf we have (after some computation):

$$p\bar{p}q \vee p\bar{p}\bar{q} \vee p\bar{p}q\bar{q} \vee \bar{p}q\bar{q} \vee \bar{p}\bar{q} \vee p\bar{q} \vee pq\bar{q} \to \bar{q}(p \vee \bar{p}).$$

Evidently each disjunct on the left entails $p \vee \bar{p}$, but the first fails to entail $\bar{q}$.

But Fact 3 tells us that when both right and left are in udnf:

$$p\bar{p}q \vee p\bar{p}\bar{q} \vee p\bar{p}q\bar{q} \vee \bar{p}q\bar{q} \vee \bar{p}\bar{q} \vee p\bar{q} \vee pq\bar{q} \to p\bar{q} \vee \bar{p}\bar{q},$$

then the entailment fails because $p\bar{p}q$ is not a superconjunction of either $p\bar{q}$ or $\bar{p}\bar{q}$, and Fact 4 tells us that when both right and left are in ucnf:

$$(\bar{p} \vee \bar{q})(p \vee \bar{q})(p \vee \bar{p})(q \vee \bar{q}) \to \bar{q}(p \vee \bar{p})$$

then the entailment fails because $\bar{q}$ has no subdisjunction among the antecedent put-downs.

For reasons no doubt arising from a certain compulsiveness about making catalogues of Facts reasonably complete, we extract the following from the proof of Fact 3:

*FACT 5. If $A, B_1, \ldots, B_m$ are set-ups, or indeed primitive conjunctions, then $\vdash A \to B_1 \vee \ldots \vee B_m$ iff $\vdash A \to B_i$ for some i, i.e., iff A is a superconjunction of some B_i, i.e., iff A *entails* some B_i.

The dual goes as follows:

FACT 6. If $A_1, \ldots, A_m, B$ are put-downs, or indeed primitive disjunctions, then $\vdash A_1 \& \ldots \& A_m \to B$ iff for some A_i, A_i is a subdisjunction of B.

As advertised earlier, none of the six facts above is deep, though they have some deeper applications. But we seize this opportunity to make some philosophical remarks about Facts 1–6, anticipating in part later discussions of material "implication."

We have all been told from infancy that the recognition of zero as a "perfectly good" number was an advance in the history of mathematics. We agree. We have also been told that the recognition of the empty set as a "perfectly good" set was an equally important advance in the history of set-theory. We agree again; the recognition of such a set involves certain admirable simplifications, e.g. that every set with n elements has exactly 2^n subsets (as classically defined).

We are not blind to these considerations, but neither should such considerations blind us to others. It is familiar from elementary set-theory that one of the 2^n members of the power-set of A is funny: the empty set. The reason it is funny is that it is held to be a subset of *every* set, because set-inclusion was defined with the help of material "implication." And as we all know (or all knew, in palmier days), a false proposition "implies" any, hence $\forall x(x \in \emptyset \supset . x \in \mathrm{A})$ is "vacuously" (or more accurately "jocosely") true. Probably the originators of the idea that logic involved relevance had no clear conception of empty sets (or non-referring terms), just as the originators of "number sense" (birds and wasps, according to Dantzig 1930) probably had no clear concept of zero. But it does not follow that the Aristotelian logical intuitions of those demanding relevance had no merit. There are no doubt those who will object to the fact that, though there are classically empty state-descriptions, there are no relevantly empty set-ups or put-downs; retreating from the tidy 2^n to the less easily manipulated $2^{2n}-1$ may seem to be a major defeat, but this is one of the costs of paying attention to relevance. We rest our case on the received metaphysical principle that *non ex aliquo fit nihil.*

The remaining facts apply particularly to state-descriptions and to Ddnfs of perfect tautologies. We are considering (left over from the beginning of §16.2)

2 $A_1 \vee \ldots \vee A_{2^n}$

where for some fixed $p_1, \ldots, p_n$, each A_i is a distinct state-description in $p_1, \ldots, p_n$.

*Fact 7. If $i \neq j$, then $\vdash A_i \to \overline{A_j}$, hence, $\vdash A_i \& \overline{A_j} \rightleftarrows A_i$.

Proof. Since $i \neq j$, A_i and A_j must differ in at least one place; hence A_i and $\overline{A_j}$, when the latter is De Morganized, agree in that place. For example, $p_1\overline{p_2}p_3\overline{p_4} \to . \overline{p_1 p_2 \overline{p_3} p_4}$, since by De Morgan this comes to $p_1\overline{p_2}p_3\overline{p_4} \to . \overline{p_1} \vee \overline{p_2} \vee p_3 \vee \overline{p_4}$.

Fact 8. $\vdash A_1 \vee \ldots \vee A_m \to \sim(A_{m+1} \vee \ldots \vee A_{2^n})$.

Proof. Immediate from Fact 7.

But notice that whereas the right hand side of 8 classically "implies" the left, the entailment in that direction in general fails. For example, $\vdash \overline{pq} \to .\bar{p}q \vee p\bar{q} \vee \bar{p}\bar{q}$ is No Good, as is easily checked by available methods.

And here another important difference between material "implication" and entailment emerges. Classically, where $A \vee B$ is a Ddnf, we have not only $\bar{A} \supset B$ but, equally trivially, $\bar{A}(A \vee B) \supset B$. As one might expect (generalizing from our other observations about the classical view), this is precisely backward for "if . . . then —"; the truth is that $B \to \bar{A}$ and $B \to . \bar{A}(A \vee B)$, both of which are theorems whenever $A \vee B$ is the Ddnf of a perfect tautology.

Before going on to see in the next subsection the conditions under which $\bar{A}(A \vee B) \to B$ *does* hold, we make two more observations.

First, though $\bar{A}(A \vee B) \to B$ holds for special cases, it does not hold in general, for (here comes the "independent proof" promised in §16.1), $\bar{A}(A \vee B) \to B$ iff $\bar{A}A \vee \bar{A}B \to B$, only if $\bar{A}A \to B$, which is absurd.

Second, we notice again that negation, which is at the bottom of all truth functional fallacies of relevance such as $\bar{A}A \to B$, plays a very weak role in entailments between truth functions: if $A_1 \vee \ldots \vee A_n \to B_1 \& \ldots \& B_m$ is such an entailment in normal form, then all negation signs occurring in the formula may be deleted without affecting validity. This feature of the situation again reinforces our claim (which stands in fact at the core of the tradition in formal logic), that the validity of a valid entailment depends never on the truth or falsity of antecedent or consequent alone.

We append just one more fact, to make good on a promise given in §11 relating to an understanding of the Lewis account of the necessity of A in terms of $\bar{A} \,\text{⥽}\, A$.

Fact 9. A formula A is Clavian — i.e., $\vdash \bar{A} \to A$, just in case, where the disjunctive normal form of $\bar{A}$ is $A_1 \vee \ldots \vee A_n$, for each i, j (including $i = j$), there is some variable p such that either p is in A_i and $\bar{p}$ in A_j, or $\bar{p}$ is in A_i and p in A_j.

§16.2.3. A special case of the disjunctive syllogism. The theorem we wish to prove, stated toward the beginning of §16.2, is this.

Theorem. Consider a perfect tautology in distinguished disjunctive normal form. Then the conjunction of this tautology with the denial of some of its disjuncts entails the disjunction of the remaining disjuncts just in case no pair among those denied differs in exactly one place.

Proof. We consider

$$3 \quad \overline{A_1 \vee \ldots \vee A_m} \& (A_1 \vee \ldots \vee A_m \vee A_{m+1} \vee \ldots \vee A_{2^n}) \to . A_{m+1} \vee \ldots \vee A_{2^n}$$

with $1 \leq m < 2^n$, where we know that

$$2 \quad A_1 \vee \ldots \vee A_{2^n}$$

is a distinguished disjunctive normal form of a perfect tautology; i.e., there are n variables $p_1, \ldots, p_n$ such that for each i ($1 \leq i \leq 2^n$), A_i is a state-description in $p_1, \ldots, p_n$; which is to say, $A_i = A_{i1}\& \ldots \& A_{in}$, where for $1 \leq i \leq n$, A_{ij} is either p_j or $\overline{p_j}$.

We want to show that 3 holds iff there is no pair $A_i, A_{i'}$ among $A_1, \ldots, A_m$ which differs in exactly one place. Let us set

4 $A = A_1 \vee \ldots \vee A_m$,

and

5 $B = A_{m+1} \vee \ldots \vee A_{2^n}$,

so that if 2 is associated properly, 3 comes to

6 $\bar{A}\&(A \vee B) \rightarrow B$.

We observe in the first place that by distribution and disjunction, 6 holds iff

7 $\bar{A}A \rightarrow B$,

and that by 4, 7 will hold iff

8 $\bar{A}A_i \rightarrow B$, each i ($1 \leq i \leq m$).

Further, by De Morgan and 4, 8 comes to

9 $\overline{A_1} \ldots \overline{A_m}A_i \rightarrow B$, each i ($1 \leq i \leq m$),

which by Fact 7 of §16.2.2 collapses into

10 $\overline{A_i}A_i \rightarrow B$, each i ($1 \leq i \leq m$).

So, to summarize what we have so far, 3 holds iff 10 does. Let

11 $A_i = A_{i1} \ldots A_{in}$,

so by 11 and De Morgan, 10 comes to

12 $(\overline{A_{i1}} \vee \ldots \vee \overline{A_{in}})(A_{i1} \ldots A_{in}) \rightarrow B$, each i ($1 \leq i \leq m$),

and hence

13 $\overline{A_{ij}}(A_{i1} \ldots A_{in}) \rightarrow B$, each i, j ($1 \leq i \leq m$, $1 \leq j \leq n$).

So 3 holds iff 13 does; i.e., 3 and 13 stand or fall together.

Now for half the theorem, suppose that no pair $A_i, A_{i'}$ in $A = A_1 \vee \ldots \vee A_m$ differs in exactly one place, and choose i ($1 \leq i \leq m$) and j ($1 \leq j \leq n$) arbitrarily.

Since by hypothesis the state-description which differs from A_i at exactly the jth place is not in A, and since $B = A_{m+1} \vee \ldots \vee A_{2^n}$ contains all the state-descriptions not in A, this state-description must be a disjunct of B. So

$\overline{A_{ij}}(A_{i1} \ldots A_{in})$ is a superconjunction of this disjunct of B; so 13 holds by Fact 5 of §16.2.2; so 3 holds.

For the converse, suppose $A = A_1 \vee \ldots \vee A_m$ contains a pair of state-descriptions A_i, $A_{i'}$ which differ in exactly the jth place. We claim 13 fails for that i and j. Invoking Fact 5 of §16.2.2, and 5, it suffices to show the failure of

$$14 \quad \overline{A_{ij}}(A_{i1} \ldots A_{in}) \rightarrow A_k,$$

i.e.

$$15 \quad \overline{A_{ij}}(A_{i1} \ldots A_{in}) \rightarrow A_{k1} \ldots A_{kn},$$

for each k ($m + 1 \leq k \leq 2^n$).

Fix k. We know that A_k differs from A_i (since $k \neq i$), but not at only the jth place — since by hypothesis the sole state-description $A_{i'}$ differing from A_i in only the jth place is in $A = A_1 \vee \ldots \vee A_m$, hence not in $B = A_{m+1} \vee \ldots \vee A_{2^n}$. Let A_i and A_k differ then in at least the j'th place, with $j' \neq j$; obviously the conjunct $A_{kj'}$ of A_k is not a conjunct of the antecedent of 15; so 15 fails. So 13, and thereby 3, fails to hold, as promised.

§16.3. A remark on intensional disjunction and subjunctive conditionals. As final evidence for our contention, we make the following observations:

The truth of A-or-B, with truth functional "or," is not a sufficient condition for the truth of "If it were not the case that A, then it would be the case that B." Example: It is true that either Napoleon was born in Corsica or else the number of the beast is perfect (with truth functional "or"); but it does not follow that had Napoleon *not* been born in Corsica, 666 would equal the sum of its factors. On the other hand the intensional varieties of "or" which *do* support the disjunctive syllogism *are* such as to support corresponding (possibly counterfactual) subjunctive conditionals. When one says "that is either *Drosophila melanogaster* or *D. virilis*, I'm not sure which," and on finding that it wasn't *D. melanogaster*, concludes that it was *D. virilis*, no fallacy is being committed. But this is precisely because "or" in this context *means* "if it isn't one, then it is the other." Of course there is no question here of a relation or logical *entailment* (which has been our principal interest); evidently some other sense of "if . . . then . . ." is involved, such as might be illuminated by $\mathbf{R}_{\rightarrow}$. But it should be equally clear that it is not simply the truth functional "or" either, from the fact that a speaker would naturally feel that if what he said was true, then if it *hadn't* been *D. virilis*, it *would* have been *D. melanogaster*. And in the sense of "or" involved, it does not follow from the fact that it is *D. virilis* that it is either *D. melanogaster* or *D. virilis* — any more than it follows solely from the fact that it was *D. virilis*, that if it hadn't been, it would have been *D. melanogaster*.

The logical differences we have been discussing are subtle, and we think it is difficult or impossible to give conclusive evidence favoring the distinctions among the various senses of "or" we have been considering. But whether or not the reader is in sympathy with our views, it might still be of interest to find a case (if such exists) where a person, other than a logician making jokes, *seriously* holds a proposition A-or-B, in a sense warranting inference of B with the additional premiss not-A, but is unwilling to admit *any* subjunctive conclusion from A-or-B. If no such examples exist, then we will feel we have made our case (and if examples do exist, we reserve the right to try to find something funny about them).

The connection between relevance logics and subjunctive conditionals is further discussed by Barker 1969, Bacon 1971, and Curley 1972. Some agree, and some disagree, but as we have indicated, we do *not* claim that our argument on this point, such as it is, is conclusive.

§17. Gentzen consecution calculuses. Some interest may attach to the availability of consecution calculus formulations of $\mathbf{E}_{\mathrm{fde}}$. Although the systems we are about to define, namely $\mathrm{L}\mathbf{E}_{\mathrm{fde1}}$ and $\mathrm{L}\mathbf{E}_{\mathrm{fde2}}$, lack the full force of the usual property of consecution calculuses according to which every formula occurring in the premisses of a rule also occurs as a part of the conclusion, nevertheless a somewhat weakened version of this property is available — and it is easy to show the equivalence of these systems to each other and to $\mathbf{E}_{\mathrm{fde}}$.

In the statement of the rules of the first system, $\mathrm{L}\mathbf{E}_{\mathrm{fde1}}$, we use a syntactic notion bound up with our two-valued conception of negation: given a formula A, define A' as the result of *adding* a sign of negation to A if the number of outer negation signs on A is *even* (or zero), or as the result of *removing* a sign of negation from A if the number is *odd*. Note that A'' is the same formula as A. α' is the sequence obtained from α by replacing each member A of α by A'.

POSTULATES FOR $\mathrm{L}\mathbf{E}_{\mathrm{fde1}}$.

Axioms.

$$A \vdash A$$

Structural rules.

$$\frac{\alpha \vdash \beta}{\alpha, A \vdash \beta}(\mathrm{K}\vdash) \qquad \frac{\alpha \vdash \beta}{\alpha \vdash A, \beta}(\vdash\mathrm{K})$$

$$\frac{\alpha_1, A, B, \alpha_2 \vdash \beta}{\alpha_1, B, A, \alpha_2 \vdash \beta}(\mathrm{C}\vdash) \qquad \frac{\alpha \vdash \beta_1, A, B, \beta_2}{\alpha \vdash \beta_1, B, A, \beta_2}(\vdash\mathrm{C})$$

$$\frac{\alpha,\ A,\ A \vdash \beta}{\alpha,\ A \vdash \beta}(\mathrm{W}\vdash) \qquad \frac{\alpha \vdash A,\ A,\ \beta}{\alpha \vdash A,\ \beta}(\vdash \mathrm{W})$$

$$\frac{\alpha \vdash \beta}{\beta' \vdash \alpha'}('\vdash')$$

Note that $('\vdash')$ is counted as a structural rule even though signs of negation are introduced — and eliminated — by its use. It is, however, like permutation ($\mathrm{C}\vdash$ and $\vdash\mathrm{C}$) in that — since $\alpha'' = \alpha$ — a second use of the rule brings one back to one's starting point.

Logical rules.

$$\frac{\alpha,\ A \vdash \beta}{\alpha,\ A\&B \vdash \beta}(\&\vdash) \qquad \frac{\alpha,\ B \vdash \beta}{\alpha,\ A\&B \vdash \beta}(\&\vdash)$$

$$\frac{\alpha \vdash A,\ \beta \qquad \alpha \vdash B,\ \beta}{\alpha \vdash A\&B,\ \beta}(\vdash\&)$$

$$\frac{\alpha \vdash A,\ \beta}{\alpha \vdash A\vee B,\ \beta}(\vdash\vee) \qquad \frac{\alpha \vdash B,\ \beta}{\alpha \vdash A\vee B,\ \beta}(\vdash\vee)$$

$$\frac{\alpha,\ A \vdash \beta \qquad \alpha,\ B \vdash \beta}{\alpha,\ A\vee B \vdash \beta}(\vee\vdash)$$

$$\frac{\alpha,\ A \vdash \beta}{\beta' \vdash \overline{A},\ \alpha'}(\vdash\sim) \qquad \frac{\alpha \vdash A,\ \beta}{\beta',\ \overline{A} \vdash \alpha'}(\sim\vdash)$$

One can assume in $(\vdash\sim)$ and $(\sim\vdash)$ that A has an odd number of outer signs of negation (so that A' is not the same as $\overline{A}$ and the application of the rule is not just an instance of $('\vdash')$).

$$\frac{A \vdash B}{\vdash A\rightarrow B}(\vdash\rightarrow)$$

There is no rule for implication on the left, since there are no nested arrows. In coming to understand the rules, the reader should interpret

$$A_1,\ \ldots,\ A_m \vdash B_1,\ \ldots,\ B_n$$

as

$$A_1\&\ \ldots\ \&A_m \rightarrow B_1\vee\ \ldots\ \vee B_n,$$

hence quite differently from the consecutions of §7 and §13. One should therefore not expect to have the usual Gentzen rule for negation,

$$\frac{\alpha,\ A \vdash \beta}{\alpha \vdash \overline{A},\ \beta}$$

for on this interpretation it would amount to the invalid rule of antilogism, from $C\&A \to B$ to infer $C \to \bar{A} \vee B$, hence yielding the abhorrent $C \to \bar{A} \vee A$ from the harmless $C\&A \to A$. (See §22.2.3.)

The consecution calculus $\mathbf{LE}_{fde1}$ above is defined in such a way as to remain as close as possible to the calculus of Gentzen 1934. A formulation in some respects more economical is given below as the calculus $\mathbf{LE}_{fde2}$.

POSTULATES FOR $\mathbf{LE}_{fde2}$.

Axioms. $\alpha_1, A, \alpha_2 \vdash \beta_1, A, \beta_2$
Structural rules. None.
Logical rules.

$$\frac{\alpha_1, A, B, \alpha_2 \vdash \beta}{\alpha_1, A\&B, \alpha_2 \vdash \beta}(\&\vdash)$$

$$\frac{\alpha \vdash \beta_1, A, \beta_2 \qquad \alpha \vdash \beta_1, B, \beta_2}{\alpha \vdash \beta_1, A\&B, \beta_2}(\vdash\&)$$

$$\frac{\alpha_1, A, \alpha_2 \vdash \beta \qquad \alpha_1, B, \alpha_2 \vdash \beta}{\alpha_1, A\vee B, \alpha_2 \vdash \beta}(\vee\vdash)$$

$$\frac{\alpha \vdash \beta_1, A, B, \beta_2}{\alpha \vdash \beta_1, A\vee B, \beta_2}(\vdash\vee)$$

$$\frac{\alpha_1, A, \alpha_2 \vdash \beta}{\alpha_1, \bar{\bar{A}}, \alpha_2 \vdash \beta}(\sim\sim\vdash)$$

$$\frac{\alpha \vdash \beta_1, A, \beta_2}{\alpha \vdash \beta_1, \bar{\bar{A}}, \beta_2}(\vdash\sim\sim)$$

$$\frac{\alpha_1, \bar{A}, \alpha_2 \vdash \beta \qquad \alpha_1, \bar{B}, \alpha_2 \vdash \beta}{\alpha_1, \overline{A\&B}, \alpha_2 \vdash \beta}(\sim\&\vdash)$$

$$\frac{\alpha \vdash \beta_1, \bar{A}, \bar{B}, \beta_2}{\alpha \vdash \beta_1, \overline{A\&B}, \beta_2}(\vdash\sim\&)$$

$$\frac{\alpha_1, \bar{A}, \bar{B}, \alpha_2 \vdash \beta}{\alpha_1, \overline{A\vee B}, \alpha_2 \vdash \beta}(\sim\vee\vdash)$$

$$\frac{\alpha \vdash \beta_1, \bar{A}, \beta_2 \qquad \alpha \vdash \beta_1, \bar{B}, \beta_2}{\alpha \vdash \beta_1, \overline{A\vee B}, \beta_2}(\vdash\sim\vee)$$

$$\frac{A \vdash B}{\vdash A\to B}(\vdash\to)$$

For each of these consecution calculuses the following Elimination Rules are readily shown admissible:

$$\frac{\alpha \vdash A, \beta \qquad \alpha, A \vdash \beta}{\alpha \vdash \beta}$$

$$\frac{\alpha_1 \vdash \beta_1 \qquad \alpha_2 \vdash \beta_2}{\alpha_3 \vdash \beta_3}$$

where every constituent of α_1 and every constituent of α_2 except A is a constituent of α_3, and where also every constituent of β_2 and every constituent of β_1 except A is a constituent of β_3.

The equivalence of these calculuses with $\mathbf{E}_{\text{fde}}$ is then straightforward, as is the fact that they lead to still another decision procedure for tautological entailmenthood. For although the full blown subformula theorem of Gentzen (§13.3) required for the decision procedures based on his consecution calculuses does not hold for $\mathbf{LE}_{\text{fde1}}$ and $\mathbf{LE}_{\text{fde2}}$ (sometimes a formula is lost as one passes down a proof), still a weakened version holds: if $\alpha \vdash \beta$ is provable in $\mathbf{LE}_{\text{fde1}}$ or $\mathbf{LE}_{\text{fde2}}$, then for every formula A occurring in a proof of $\alpha \vdash \beta$ in $\mathbf{LE}_{\text{fde1}}$ or $\mathbf{LE}_{\text{fde2}}$, either A itself, or the result of removing a single outer negation sign from A, is a subformula of $\alpha \vdash \beta$. In other words, every proof of $\alpha \vdash \beta$ is constructible wholly from subformulas and negations of subformulas of $\alpha \vdash \beta$.

§18. Intensional algebras (by J. Michael Dunn). It is well known that there are intimate connections between the two-valued propositional calculus and Boolean algebras. We shall develop in this section a similar intimate connection between the system $\mathbf{E}_{\text{fde}}$ of tautological entailments and a special algebraic structure called an *intensional lattice*.

We begin by recalling some of the high points in the development of algebraic logic, our aim being to provide a framework of established results concerning non-intensional logics with which our subsequent treatment of the algebra of intensional logics may be compared. (Those already familiar with "classical" algebraic logic may wish to skip to §18.1.) Although we shall chiefly be discussing the algebra of the classical propositional calculus, this discussion is intended to have a certain generality. We mean to emphasize the essential features of the relation of the classical propositional calculus to Boolean algebra, remarking from time to time what is special to this relation and what is generalizable to the algebra of other propositional calculuses. It should be mentioned that we here restrict ourselves to the algebra of propositional logics, despite the fact that profound results concerning the algebra of the classical predicate calculus have been obtained by Henkin and Tarski 1961, Halmos 1962, and others.

It should also be mentioned that we are not here concerned with setting down the history of algebraic logic, and that, much as in a historical novel, historical figures will be brought in mainly for the sake of dramatic emphasis. The interested reader may refer to Rasiowa and Sikorski 1963 for proofs and proper citations of most of the results we discuss.

About the middle of the last century, the fields of abstract algebra and symbolic logic came into being. Although algebra and logic had been around for some time, *abstract* algebra and *symbolic* logic were essentially new developments. Both these fields owe their origins to the insight that formal systems may be investigated without explicit recourse to their intended interpretations.

This insight led Boole 1847 to formulate at one and the same time perhaps the first example of a non-numerical algebra and the first example of a symbolic logic. He observed that the operation of conjoining two propositions had certain affinities with the operation of multiplying two numbers. He saw that by letting letters like "a" and "b" stand for propositions just as they stand for numbers in ordinary algebra, and that by letting juxtaposition of letters stand for the operation of conjunction just as it stands for multiplication in ordinary algebra, these affinities could be brought to the fore. Thus, for example, $ab = ba$ (or $a \wedge b = b \wedge a$) is a law of this *algebra of logic*, just as it is a law of the ordinary algebra of numbers. And a complete description of what it is to be a Boolean algebra is given by the equations

L1 $a \wedge a = a$, $a \vee a = a$
L2 $a \wedge b = b \wedge a$, $a \vee b = b \vee a$
L3 $a \wedge (b \wedge c) = (a \wedge b) \wedge c$, $a \vee (b \vee c) = (a \vee b) \vee c$
L4 $a \wedge (a \vee b) = a$, $a \vee (a \wedge b) = a$
L5 $a \wedge \bar{a} = \mathit{0}$, $a \vee \bar{a} = \mathit{1}$
L6 $a \wedge \mathit{0} = \mathit{0}$, $a \vee \mathit{1} = \mathit{1}$

At the same time, the algebra of logic has certain differences from the algebra of numbers since, for example, $aa = a$ (or $a \wedge a = a$). The differences are just as important as the similarities, for whereas the similarities suggested a truly *symbolic* logic, like the "symbolic arithmetic" that constitutes ordinary algebra, the differences suggested that algebraic methods could be extended far beyond the ordinary algebra of numbers.

Oddly enough, despite the fact that Boole's algebra was thus connected with the origins of both abstract algebra and symbolic logic, the two fields developed for some time thereafter in comparative isolation from one another. On the one hand, the notion of a Boolean algebra was perfected by Jevons 1871, Schröder 1890–1895, Huntington 1904, and others, and developed as a part of the growing field of abstract algebra. On the other hand, the notion of a symbolic logic was developed along subtly different lines from Boole's original algebraic formulation 1847, starting with Frege 1879 and receiving its classic statement in Whitehead and Russell 1910–1913. The divergence of the two fields was partly a matter of attitude. Thus Boole, following in the tradition of Leibniz, wanted to study the mathematics of logic, whereas the aim of Frege, Whitehead, and Russell was to study the logic of mathematics. The modern field of mathematical logic, of course, recognizes both approaches as methodologically legitimate, and indeed embraces them both under the very ambiguity of its name, "mathematical logic," but the Frege-Whitehead-Russell aim to reduce mathematics to logic obscured for some time the two-headedness of the mathematical-logical coin.

There is more than a difference in attitude, however, between Boole's algebraic approach to logic, and the Frege-Whitehead-Russell approach to logic, which for want of a better word we shall call *logistic*. We shall attempt to bring out this difference between the two approaches, which was either so profound or so subtle that the precise connection between the two ways of looking at logic was not discovered until the middle 1930's. (The difference we have in mind is essentially the distinction that Curry 1963, pp. 166–168, makes between a *relational* (algebraic) system and an *assertional* (logistic) system, though we shall have to be more informal than Curry since we do not have his nice formalist distinctions at hand.)

Let us begin by looking at a logistic presentation of the classical propositional calculus that is essentially the same as in **Principia Mathematica**, except that we use axiom schemata and thereby do without the rule of substitution, which was tacitly presupposed in **Principia**. This presentation begins by assuming that we have a certain stock of propositional variables p, q, r, etc., and then specifies that these are formulas and that further formulas may be constructed from them by the usual inductive insertion of logical connectives (and parentheses). The particular logical connectives assumed in this presentation are those of disjunction $\vee$ and negation $\overline{}$, although conjunction is assumed to be defined in terms of these so that $A\&B$ is an abbreviation for $\overline{\overline{A}\vee\overline{B}}$, and material implication is also assumed to be defined so that $A\supset B$ is an abbreviation for $\overline{A}\vee B$. A certain proper subset of these formulas are then singled out as axioms. These axioms are all instances of the following schemata:

1 $A\vee A\supset A$
2 $B\supset .A\vee B$
3 $A\vee B\supset .B\vee A$
4 $A\vee(B\vee C)\supset .B\vee(A\vee C)$
5 $B\supset C\supset .(A\vee B)\supset .A\vee C.$

These axioms are called *theorems*, and it is further specified that additional formulas are theorems in virtue of the following rule:

Modus ponens. If A is a theorem, and if $A \supset B$ is a theorem, then B is a theorem.

The point of this perhaps too tedious but not too careful rehearsal of elementary logic is to give us some common ground for a comparison of the classical propositional calculus with a Boolean algebra. There are certain surface similarities that are misleading. Thus, for example, a Boolean algebra has certain items called elements which are combined by certain operations to give other elements, just as the classical propositional calculus has certain items called formulas which are combined by the operation of inserting logical connectives to give other formulas. They are both then, from this point of view, abstract algebras. This fact might lead one to confuse the operation of disjoining two formulas A and B so as to obtain $A \vee B$, with the operation of joining two elements of a Boolean algebra a and b so as to obtain $\mathrm{a} \vee \mathrm{b}$. There are essential differences between these two binary operations. Consider, for example, that, where A is a formula, $A \vee A$ is yet another *distinct* formula since $A \vee A$ contains at least one more occurrence of the disjunction sign $\vee$ than does A. Yet in a Boolean algebra, where a is an element, $\mathrm{a} \vee \mathrm{a} = \mathrm{a}$, by L1. Further, in the algebra of formulas, where A and B are distinct formulas, the formula $A \vee B$ is distinct from the formula $B \vee A$ since although the two formulas are composed of the same signs, the signs occur in different orders. Yet in a Boolean algebra, $\mathrm{a} \vee \mathrm{b} = \mathrm{b} \vee \mathrm{a}$, by L2.

The trouble with the algebra of formulas is that, like the bore at a party, it makes too many distinctions to be interesting. Its detailed study might be of interest to the casual thrill-seeker who is satisfied with "something new every time," but the practiced seeker of identity in difference demands something more than mere newness. To such a seeker as Boole, the "identity" of two such different formulas as $A \vee A$ and A, or $A \vee B$ and $B \vee A$, lies in the fact that they express the "same proposition," but this was only understood at such an intuitive level until the 1930's, when Lindenbaum and Tarski made their explication of this insight.

Lindenbaum and Tarski observed that the logistic presentation of the classical propositional calculus could be made to reveal a deeper algebra than the algebra of formulas that it wore on its sleeve. Their trick was to introduce a relation of *logical equivalence* $\equiv$ upon the class of formulas by defining $A \equiv B$ iff both $A \supset B$ and $B \supset A$ are theorems. It is easy to show that the relation $\equiv$ is a genuine equivalence relation. Thus reflexivity follows because $A \supset A$ is a theorem, symmetry follows by definition, and transitivity follows from the fact that, whenever $A \supset B$ and $B \supset C$ are theorems, then

$A \supset C$ is a theorem (the rule form of transitivity). It is interesting to observe that since the classical propositional calculus has a "well-behaved" conjunction connective, i.e., $A \& B$ is a theorem iff both A and B are theorems, then the same effect may be obtained by defining $A \equiv B$ iff $(A \supset B) \& (B \supset A)$ is a theorem. It is natural to think of the class of all formulas logically equivalent to A, which we represent by $[A]$, as one of Boole's "propositions." Operations are then defined upon these equivalence classes, one corresponding to each logical connective, so that (using for convenience the same symbol for the algebraic operations on equivalence classes as we have already used for the logical connectives between formulas) $\overline{[A]} = [\bar{A}]$, $[A] \vee [B] = [A \vee B]$, $[A] \wedge [B] = [A \& B]$, and $[A] \supset [B] = [A \supset B]$. Since the replacement theorem (analogous to §8.9) holds for the classical propositional calculus, these operations may be shown to be genuine (single-valued) operations. The point of the replacement theorem is to ensure that the result of operating upon equivalence classes does not depend upon our choice of representatives for the classes. Thus, for example, if $A \equiv B$, then $[A] = [B]$. But then for the unary operation corresponding to negation to be single-valued, we must have $\overline{[A]} = \overline{[B]}$, i.e., $[\bar{A}] = [\bar{B}]$, i.e., $\bar{A} \equiv \bar{B}$, which is just what the replacement theorem guarantees us.

Let us call the algebra so defined the *Lindenbaum algebra* of the classical propositional calculus. It is simply a matter of axiom-chopping to see that this is a Boolean algebra. Thus, for example, it is easy to verify L1 $[A] \vee [A] = [A]$, even though $A \vee A$ and A are distinct formulas, for $(A \vee A) \supset A$ is an instance of axiom schema 1, and $A \supset (A \vee A)$ is an instance of axiom schema 2. Similarly, $[A] \vee [B] = [B] \vee [A]$ follows from two instances of axiom schema 3. The other laws of a Boolean algebra may be established analogously.

The essentials of the Lindenbaum-Tarski method of constructing an algebra out of the classical propositional calculus can be applied to most other well-motivated propositional calculuses, and, because of the intuitive properties of conjunction and disjunction, most of the resulting Lindenbaum algebras are lattices, indeed, distributive lattices. In particular, the Lindenbaum algebra of Lewis's modal logic **S4** is a closure algebra, and the Lindenbaum algebra of Heyting's intuitionistic logic **H** is a pseudo-Boolean algebra. (See McKinsey 1941, McKinsey and Tarski 1948, and Birkhoff 1948, pp. 195–196.) One of the most remarkable features of the reunion of logic and algebra that took place in the 1930's was this discovery that certain non-classical propositional calculuses that had captured the interest of logicians had such intimate connections with certain structures that had been developed by algebraists in the context of lattice theory — a generalization of the theory of Boolean algebras that by then stood on its own.

An even more striking example of the identification of notions and results that were of independent interest to both logicians and algebraists may be found in Tarski's 1930 theory of deductive systems, which was later seen to overlap the Boolean ideal theory of Stone 1936. Apparently Tarski did not realize the algebraic significance of his theory until he read Stone, and conversely, Stone did not realize the logical significance of his theory until he read Tarski. (See Kiss 1961, pp. 5–6.)

Intuitively, a *deductive system* is an extension of a logistic presentation of a propositional calculus (assumed not to have a rule of substitution) that has been obtained by adding additional formulas as axioms (however, Tarski defined the notion explicitly only for the classical propositional calculus). Stone defined a (lattice) ideal as we do in §18.1, and at the same time showed that Boolean algebras could be identified with idempotent rings (with identity), the so-called *Boolean rings*, and that upon this identification the (lattice) ideals were the ordinary ring ideals. This identification was of great importance since the value of ideals in ring theory was already well-established, the concept of an ideal having first been developed by Dedekind 1877 as an explication of Kummer's "ideal number," which arose in connection with certain rings of numbers (the algebraic integers). It is a tribute to the powers of abstract algebra that the abstract concept of an ideal can be shown to underlie both certain number theoretical concepts and certain logical concepts.

The connection between deductive systems and ideals becomes transparent upon the Lindenbaum identification of a formula with its logical equivalents. Then a deductive system is the dual of an ideal, namely, what in §18.1 we call a *filter;* and, conversely, a filter is a deductive system. Without going into the details of this connection, let us simply remark the analogy between a deductive system and a filter. Let us assume that F is a set of theorems of some extension of the classical propositional calculus, or of almost any well-known, well-motivated propositional calculus. Then both formal and intuitive considerations demand that if A, $B \in$ F, then $(A\&B) \in$ F, which corresponds to property F1 of our §18.1 definition of a filter, and that if $A \in$ F, then $A \vee B \in$ F, which corresponds to our property F2. It is interesting to observe that if we consider the set of *refutable* formulas, i.e., those formulas whose negations are theorems, then we get an ideal in the Lindenbaum algebra. The fact that theorems are more customary objects for logical study than refutables, while at the same time ideals are more customary objects for algebraic study than filters, has led Halmos 1962 to conjecture that the logician is the dual of the algebraist. By duality, we obtain as a corollary that the algebraist is the dual of the logician.

Upon the Lindenbaum identification of logically equivalent formulas, the filter of theorems of the classical propositional calculus has a particularly simple structure, namely, it is the trivial filter that contains just the *1* of the

Boolean algebra that so results. This fact depends upon one of the paradoxes of "implication," namely, that where B is a theorem, then $A \supset B$ is a theorem. This means that all theorems are logically equivalent and hence identified with each other in the same equivalence class, and that any theorem is logically implied by any formula and hence this equivalence class of theorems ends up at the top of the Boolean algebra. In short, A is a theorem iff $[A] = \mathit{1}$. This explicates a notion of Boole that a proposition a is a logical truth iff a $= \mathit{1}$. Since the same paradox of implication is shared with many other propositional calculuses, e.g., **S4** and the intuitionist logic **H**, this algebraically elegant characterization of theoremhood is widely applicable. But since in the intensional logics that we shall be studying it is not the case that all theorems are logically equivalent, we shall have to use a different algebraic analogue of theoremhood. Note that we can always resort to the inelegant characterization that A is a theorem iff $[A]$ is in the Lindenbaum analogue of the deductive system based on the logic. This means, in the case of the intensional logics that we shall be studying, that the algebraic analogue of the class of theorems is the filter generated by the elements that correspond to the axioms, although we shall find a more elegant way of putting this in §28.2. The same characterization actually holds for the Lindenbaum algebra of the classical propositional calculus, it being but a "lucky accident," so to speak, that this filter is the trivial filter that may hence be thought of as identical with the element $\mathit{1}$ that is its sole member. The algebra of intensional logics is thus demonstrably "non-trivial."

So far we have been discussing the algebra of the *syntactics* of a propositional logic, since the notions of *formula*, *theorem*, etc., by which the Lindenbaum algebra is defined, all ultimately depend only upon the syntactic structure of sequences of signs of the system. But there is another side to logic, namely, *semantics*, which studies the interpretations of logical systems. Thus, to use a well-known example, to say of the formula $A \vee \bar{A}$ that it is a theorem of the classical propositional calculus is to say something syntactical, whereas to say of $A \vee \bar{A}$ that it is a tautology is to say something semantical, since it is to say something about the formula's interpretations in the ordinary two-valued truth tables, namely, that its value is *true* under every valuation. Now we have already discussed an algebraic way of expressing the first fact, namely, we can say that $[A \vee \bar{A}] = \mathit{1}$. What we now want is an algebraic way of expressing the second fact. It is well-known that the ordinary truth tables may be looked at as the two-element Boolean algebra $\mathit{2}$ $(= \{\mathit{0}, \mathit{1}\}$, where *true* is $\mathit{1}$ and *false* is $\mathit{0}$). This allows us to define a *valuation* into $\mathit{2}$ (or any Boolean algebra) as a mapping of the formulas into the Boolean algebra that carries negation into complementation, disjunction into join, etc., all in the obvious way. We can then define a formula A as *valid* with respect to a class of Boolean algebras iff for every valuation v into a Boolean algebra in the class, $v(A) = 1$. We can define the classical propositional calculus as *consistent* with respect to a class of Boolean algebras iff every theorem is valid with respect to that class, and as *complete* with respect to the class iff every formula that is valid with respect to the class is a theorem. Observe that these definitions coincide with the usual definitions with respect to truth tables when the class of Boolean algebras in question consists of just the single Boolean algebra *2*. Observe also that similar definitions may be given for non-classical propositional calculuses once the appropriate algebraic analogue of theoremhood has been picked out.

It may easily be shown that the classical propositional calculus is both consistent and complete with respect to the class of all Boolean algebras. Thus consistency may be shown in the usual inductive fashion, showing first that the axioms are valid, and then that the rule (*modus ponens*) preserves validity. Completeness is even more trivial, since it may be immediately seen that if a formula A is not a theorem, then if we define for every formula B, $v(B) = [B]$, that under this valuation $v(A) \neq 1$. Of course, this completeness result is not as satisfying as the more familiar two-valued result since, among other things, it does not immediately lead to a decision procedure (the Lindenbaum algebra of the classical propositional calculus formulated with an infinite number of propositional variables not being finite). But it does form the basis for an algebraic proof of the two-valued result. We shall see this after a short digression concerning valuations and homomorphisms.

The notion of a homomorphism is the algebraic analogue of a valuation; the idea is that of a mapping from one algebraic structure into another which "respects" or "preserves" all the operations in which we are interested. In the Boolean case this means that for h to be a homomorphism from one algebra B with its operations $\wedge$, $\vee$, and $\sim$ into a second algebra B' with its operations $\wedge'$, $\vee'$, and $\sim'$, we would require, for all $a, b \in B$,

$$h(a \wedge b) = ha \wedge' hb,$$
$$h(a \vee b) = ha \vee' hb,$$

and

$$h(\sim a) = \sim' ha$$

From any valuation v of the classical propositional calculus into a Boolean algebra B we can define a homomorphism h of the Lindenbaum algebra into B as $h([A]) = v(A)$; and conversely, from any homomorphism h of the Lindenbaum algebra we can define a valuation v as $v(A) = h([A])$. The second fact is obvious, but the first fact requires a modicum of proof, which is not without intrinsic interest. What needs to be shown is that the function h is well-defined in the sense that its value for a given equivalence class as argument does not depend upon our choice of a formula as representative of that equivalence class, i.e., that if $A \equiv B$, then $v(A) = v(B)$. This amounts to a special case of the semantic consistency result, for what must be shown is that if $A \supset B$ is a theorem, then $v(A) \leq v(B)$, i.e., $v(A) \supset v(B) = v(A \supset B) = 1$. The fact that every valuation thus determines a homomorphism allows us to observe that the Lindenbaum algebra of the classical propositional calculus formulated with n propositional variables is free Boolean algebra with n free generators.

This latter concept is an important one for algebraic theory; for an algebra to be free in a family of algebras means that every element in the algebra can be generated by means of the operations from a privileged starting set called "free generators" having the following interesting property: no matter how these generators are mapped into another algebra of the given family, the mapping can be extended to *all* members of the first algebra in such a way as to constitute a homomorphism from the first free algebra into the second. For a logician, the paradigm case of a set of free generators is the set of equivalence classes of the propositional variables; indeed one might take the algebraic fact that they constitute a set of free generators as explicating the intuitive idea that one may "freely" assign any proposition to any propositional variable. Note that it is typical of algebraic logic that no artificial restrictions are placed upon the assumed cardinality of the stock of propositional variables. Although there may be very good metaphysical or scientific reasons for thinking that the number of actual or possible physical inscriptions of propositional variables is at most denumerable, still the proof we are about to sketch is not affected by questions of cardinality.

The proof that the set of equivalences classes of propositional variables (of any cardinality) forms a set of free generators begins by observing that distinct propositional variables determine distinct equivalence classes.

Let us suppose that the propositional variables are p_x, and that we have a mapping f of their equivalence classes $[p_x]$ into a Boolean algebra B. We can then define a new function s from the propositional variables into B by $s(p_x) = f([p_x])$. This function s then inductively determines a valuation v into B, and the valuation v in turn determines a homomorphism h of the Lindenbaum algebra into B, as we have just seen.

The situation we have described above is typical of the algebra of logic. We take a logic and form its Lindenbaum algebra (if possible). We then abstract the Lindenbaum algebra's logical structure and find a class of algebras such that the Lindenbaum algebra is free in the class. That the Lindenbaum algebra is *in* the class then amounts to the logic's completeness, and that it is *free* in the class amounts to the logic's consistency. The trick is to abstract the Lindenbaum algebra's logical structure in an interesting way. Thus, for example, it is interesting that the Lindenbaum algebra of **S4** is free in the class of closure algebras, and it is interesting that the Lindenbaum algebra of the intuitionist logic is free in the class of pseudo-Boolean algebras, because these algebras are rich enough in structure and in applications to be interesting in their own right. We remark that it is irrelevant whether the logic or the algebra comes first in the actual historical process of investigation.

Having thus picked an appropriate class of algebras with respect to which the logic may be shown consistent and complete, it is, of course, desirable to obtain a sharper completeness result with respect to some interesting subclass of the algebras. One perennially interesting subclass consists of the finite algebras, for then a completeness result leads to a decision procedure for the logic. McKinsey 1941 and McKinsey and Tarski 1948 have obtained such finite completeness results for **S4** with respect to closure algebras, and for the intuitionist logic **H** with respect to pseudo-Boolean algebras.

It might be appropriate to point out that due to the typical coincidence of valuations and homomorphisms, algebraic semantics may be looked at as a kind of algebraic *representation theory*, representation theory being the study of mappings, especially homomorphisms, between algebras. This being the case, one cannot expect to obtain deep completeness results from the mere hookup of a logic with an appropriate class of algebras unless that class of algebras has an already well-developed representation theory. Of course, the mere hookup can be a tremendous stimulus to the development of a representation theory, as we shall find when we begin our study of the algebra of intensional logics.

We close this section with an example of how a well-developed representation theory can lead to deep completeness results. We shall show how certain representation results for Boolean algebras of Stone 1936, dualized here for the sake of convenience from the way we report them in §18.1 to the way Stone actually stated them, lead to an elegant algebraic proof of the completeness of the classical propositional calculus with respect to *2*. Of course, in point of fact the completeness result (with respect to truth tables) was first obtained by Post 1921 by a non-algebraic proof using cumbersome normal form methods, but this is irrelevant to the point being made.

We shall show that a formula A is valid (in *2*) only if it is a theorem by proving the contrapositive. We thus suppose that A is not a theorem, i.e., that $[A] \neq 1$. By an algebraic result of Stone's we know that there is a maximal ideal (§18.1) M in the Lindenbaum algebra such that $[A] \in$ M. But by another purely algebraic result of Stone's we know that there is a homomorphism h of the Lindenbaum algebra that carries all members of M into $\mathbf{0}$. Thus h([A]) = $\mathbf{0}$. Now through the connection between homomorphisms and valuations, we can define a valuation v into *2* by v(A) = h([A]), and thus there is a valuation v such that v(A) = $\mathbf{0}$, i.e., v(A) ≠ $\mathbf{1}$, which completes the proof.

Even more remarkable connections between Stone's results and the completeness of the classical propositional calculus with respect to *2* have been obtained. Thus, for example, Henkin 1954 has essentially shown that Stone's representation theorem for Boolean algebras is directly equivalent to the completeness theorem stated in slightly stronger form than we have stated it (see also Łoś 1957 for critical modifications).

Let us remark several distinctive features of the "algebraic" proof of the completeness theorem we have given that make it algebraic. It uses not only the language of algebra but also the results of algebra. The only role the axioms and rules of the classical propositional calculus play in the proof is in showing that the Lindenbaum algebra is a Boolean algebra, and hence that the Boolean results may be applied. Further, the proof is wildly transfinite. By this we mean not only that no assumptions have been made regarding the cardinality of the propositional variables but also that a detailed examination of Stone's proof regarding the existence of the essential maximal ideal would reveal that he used the axiom of choice. The proof is at the same time wildly non-constructive, for we are given no way to construct the crucial valuation. A Lindenbaum algebra is thus treated by the same methods as any algebra. We note in closing that, although there may be philosophical objections to such methods of proof, these objections cannot be directed at just algebraic logic, but instead must be directed at almost the whole of modern algebra.

We return now to the system $\mathbf{E}_{\mathrm{fde}}$ of tautological entailments and to its relations with a special algebraic structure called an intensional lattice. Since this notion may best be looked at as a composite of familiar algebraic concepts, we shall begin by recalling the definitions of these underlying notions. These are all set forth in detail in Birkhoff 1948, and in Rasiowa and Sikorski 1963.

§18.1. Preliminary definitions. A set is said to be *partially ordered* by a relation $\leq$ if the relation is a binary relation on the set satisfying the following:

1	$a \leq a$	*reflexivity*
2	$a \leq b$ and $b \leq a$ imply $a = b$	*antisymmetry*
3	$a \leq b$ and $b \leq c$ imply $a \leq c$	*transitivity*

We may read "$a \leq b$" as "a is less than or equal to b," in analogy to the usual reading given in number theory, though we should emphasize that any relation whatsoever that satisfies the above three conditions is a partial ordering. The usual relation $\leq$ between numbers is a partial ordering but, of course, not the only one. The relation of $\subseteq$ of set inclusion is another example. And so is entailment between propositions.

An *upper bound* of a subset B of a partially ordered set A is an element u of A such that for every $b \in B$, $b \leq u$. The *least upper bound* (*l.u.b.*) of the subset B is an upper bound u which is such that if u′ is an upper bound of B as well, then $u \leq u'$. We define *lower bound* and *greatest lower bound* (*g.l.b.*) analogously. The l.u.b. of a set consisting of just two elements, a and b, is called a *join* and is denoted by "$a \vee b$." The g.l.b. of the set consisting of just a and b is called a *meet* and is denoted by "$a \wedge b$."

A *lattice* is defined as a non-empty partially ordered set such that any two of its elements have both a join and a meet. A trivial example of a lattice is provided by a set which consists of a single element a such that $a \leq a$. This is called a *degenerate lattice*, and unless we specify otherwise, we mean to exclude it when we talk of lattices. Clearly a partially ordered set A need not always be a lattice; i.e., there may be one or more pairs $\{a, b\}$ of A with neither a g.l.b. nor a l.u.b.

A sublattice of a lattice is a (non-empty) subset that is closed under the meet and join operations of the lattice. Two special kinds of sublattice are important. The first of these, an *ideal*, is a (non-empty) subset I such that

(I1) if $a, b \in I$, then $(a \vee b) \in I$, and
(I2) if $a \in I$, then $(a \wedge b) \in I$.

Equivalent definitions may be obtained by replacing (I2) with either,

(I2′) if $a \in I$ and $b \leq a$, then $b \in I$,

or by dropping (I2) altogether and strengthening (I1) to

(I1′) $(a \vee b) \in I$ iff $a \in I$ and $b \in I$.

The second of these two special kinds of sublattices is defined *dually*, i.e., by interchanging the roles of meet and join. A *filter* is accordingly a (non-empty) subset F such that

(F1) if $a, b \in F$, then $(a \wedge b) \in F$, and
(F2) if $a \in F$, then $(a \vee b) \in F$.

Equivalently, (F2) may be replaced by

(F2′) if $a \in F$ and $a \leq b$, then $b \in F$,

or (F2) may be dropped altogether in favor of the following strengthening of (F1):

(F1′) $(a \wedge b) \in F$ iff $a \in F$ and $b \in F$.

In the sequel we shall exploit these alternative definitions of *filter* and *ideal* without special mention.

An ideal {or filter} of a lattice A is said to be *maximal* if it is not identical with A and if no other ideal {or filter} other than A includes it. A *prime ideal* is an ideal I satisfying

(PI) if $(a \wedge b) \in I$, then $a \in I$ or $b \in I$,

and a *prime filter* is a filter F satisfying

(PF) if $(a \vee b) \in F$, then $a \in F$ or $b \in F$.

(I2) and (F2) allow the respective strengthening of (PI) and (PF) to

(PI′) $(a \wedge b) \in I$ iff $a \in I$ or $b \in I$, and
(PF′) $(a \vee b) \in F$ iff $a \in F$ or $b \in F$.

Observe that the notion of a prime ideal {or prime filter} can be attractively characterized by (I1′) and (PI′) (or by (F1′) and (PF′)) — we shall make use of this without explicit mention — and that the set theoretical complement of a prime ideal is a prime filter, and vice versa.

Since the intersection of ideals {or filters} is always an ideal {or filter}, we may define *the ideal* {or *filter*} *generated by a set* A as the least ideal {or filter} including A, and be sure that such exists. When A consists of just a single a, we speak of the *principal ideal* {or *filter*} *generated by* a, which is just the set of all x such that $x \leq a$ {or $a \leq x$}.

A *distributive lattice* is a lattice satisfying the following *distributive* laws:

(D1) $(a \wedge (b \vee c)) = ((a \wedge b) \vee (a \wedge c))$, and
(D2) $(a \vee (b \wedge c)) = ((a \vee b) \wedge (a \vee c))$.

(D1) and (D2) imply each other. The classic example of a distributive lattice is a *ring of sets*, i.e., a collection of sets that is closed under binary intersection (meet) and binary union (join).

Stone 1937 proved, using the axiom of choice, that for elements a and b of a distributive lattice, if $a \nleq b$, then there exists a prime filter P such that $a \in P$ and $b \notin P$.

A *Boolean algebra* may be defined as a *complemented distributive* lattice, where a lattice is said to be *complemented* if it has a least element $\mathit{0}$ and a greatest element $\mathit{1}$, and if it satisfies

(C) For any element a, there is an element $\bar{a}$ (called the *complement* of a) such that $(a \wedge \bar{a}) = \mathit{0}$, and $(a \vee \bar{a}) = \mathit{1}$.

The classic example of a Boolean algebra is a *field of sets*, i.e., a ring of sets that is closed under set-theoretical complementation.

Stone 1936 proved, again using the axiom of choice, that for any two elements a and b of a Boolean algebra, if $a \nleq b$, then there exists a maximal filter M such that $a \in M$ and $b \notin M$ (this is a special case of his 1937 result for distributive lattices, since in a Boolean algebra prime filters and maximal filters coincide). Stone further observed that a maximal filter M of a Boolean algebra contains for every element a, exactly one of a and $\bar{a}$.

§18.2. Intensional lattices. We define an *intensional lattice* as a (non-empty) set L, together with a relation $\leq$ on L, a unary function $\overline{}$ on L, and a subset T of L (all of which may be referred to as an ordered quadruple $\langle L, \leq, \overline{}, T\rangle$), that satisfies the following:

(DL) L is a distributive lattice under $\leq$,
(N1) for all $a \in L$, $\bar{\bar{a}} = a$,
(N2) for all $a, b \in L$, if $a \leq b$, then $\bar{b} \leq \bar{a}$, and
(T) T is a filter of L that is *consistent* in the sense that there is no $a \in L$ such that both $a \in T$ and $\bar{a} \in T$, and *exhaustive* in the sense that for all $a \in L$, either $a \in T$ or $\bar{a} \in T$. T is called a *truth filter* for reasons which will become clear.

The operation $\overline{}$ is called *intensional complementation.* (It is *not* in general a Boolean complementation.) It follows from (N1) and (N2) that $\overline{}$ is a one-to-one function of L onto itself that satisfies the following *De Morgan laws:*

(DeM) $\overline{a \wedge b} = \bar{a} \vee \bar{b}$, and
$\overline{a \vee b} = \bar{a} \wedge \bar{b}$.

In the sequel we shall refer to all variants of the De Morgan laws that involve implicit use of (N1), e.g., $\overline{\bar{a} \wedge b} = (a \vee \bar{b})$, simply by "(DeM)."

A one-to-one function of a lattice onto itself that satisfies the two De Morgan laws above is called a *dual automorphism* in Birkhoff 1948, and when such a function also satisfies (N1), i.e., is of a period two, Birkhoff calls it an *involution*. Thus, (N1) and (N2) are equivalent to the requirement that $\overline{}$ be an involution. Boolean complementation satisfies both (N1) and (N2).

Stone's result about the existence of maximal filters in all Boolean algebras (except the degenerate one), together with his observation that a maximal filter M of a Boolean algebra contains for every element a, exactly one of a and $\bar{a}$, implies that any Boolean algebra (except the degenerate one) has a subset T that satisfies (T), namely M. So an intensional lattice may be looked at as a generalization of a Boolean algebra, the filter playing the role of a designated maximal filter.

Anticipating the forthcoming applications of intensional lattices to tautological entailments, we remark that an intensional lattice L may be thought of as a set of propositions, $\leq$ as a relation of entailment, meet and join as propositional conjunction and disjunction, respectively, and $\overline{}$ as propositional negation. T is then thought of as the set of true propositions in L, which motivates the name "truth filter."

§18.3. The existence of truth filters. Separate study has been given to distributive lattices satisfying (N1) and (N2). These have been called *De Morgan lattices* in Monteiro 1960, *distributive involution lattices* (*i-lattices*) in Kalman 1958, and *quasi-Boolean algebras* in Białnicki-Birula and Rasiowa 1957. This naturally leads to the question of a necessary and sufficient condition for a De Morgan lattice to have a truth filter, and it turns out that a quite simple condition works (Belnap and Spencer 1966), namely that $\overline{}$ have no fixed point, i.e., that

(N3) $a \neq \bar{a}$.

The fact that this condition is both necessary and sufficient gains interest from the high intuitive plausibility of (N3) and (T) as conditions on propositional negation. Indeed, it is hard to think of anything more absurd than supposing a proposition could be its own negation, and it is hardly less absurd to suppose that it should be impossible exactly to divide the propositions into the False and the True, a proposition being true iff its negation is false, and a conjunction being true iff each conjunct is true. We now give the Belnap-Spencer proof. That (T) implies (N3) is obvious, since if for some a, $a = \bar{a}$, then every set containing a or $\bar{a}$ would be inconsistent. Hence to prove the equivalence it suffices to prove

THEOREM 1. If L is a De Morgan lattice such that $\overline{}$ has no fixed point, then some filter T of L is both consistent and exhaustive.

Observe that if L were a Boolean algebra, the theorem would amount to Stone's theorem about the existence of maximal filters. But in a De Morgan lattice it may be shown by example that not every truth filter is maximal, nor is every maximal filter a truth filter. Turning now to a proof of the theorem, let T_0 be the set of all elements q of L such that $q = (q \vee \bar{q})$, and let E be the set of all consistent filters in L containing T_0. We first show E non-empty by showing consistent the filter $F(T_0)$ generated by T_0. For suppose otherwise: then for some $p_1, \ldots, p_n \in T_0$ and some $b \in L$, we would have

$(p_1 \wedge \ldots \wedge p_i) \leq b$; and
$(p_{i+1} \wedge \ldots \wedge p_n) \leq \bar{b}$,

hence by (DL), for $p = (p_1 \wedge \ldots \wedge p_n)$, both $p \leq b$ (hence $\bar{b} \leq \bar{p}$) and $p \leq \bar{b}$, so that,

1 $\bar{p} = (p \vee \bar{p})$.

Now define a_n inductively as follows:

2 $a_1 = p_1;\ a_{i+1} = (p_{i+1} \wedge (\overline{p_{i+1}} \vee a_i))$.

We first prove inductively that

3 $a_i = ((p_1 \wedge \ldots \wedge p_i) \vee \overline{a_i})$, for all i $(1 \leq i \leq n)$.

$a_1 = p_1 \vee \overline{a_1}$ is immediate, and we now assume for induction that $a_i = (p_1 \wedge \ldots \wedge p_i) \vee \overline{a_i}$. Since $p_{i+1} \in T_0$, we have $p_{i+1} = (p_{i+1} \vee \overline{p_{i+1}})$, hence by (DL), $p_{i+1} \wedge (\overline{p_{i+1}} \vee a_i) = ((p_{i+1} \wedge a_i) \vee \overline{p_{i+1}})$; so that by 2,

4 $a_{i+1} = ((p_{i+1} \wedge a_i) \vee \overline{p_{i+1}})$.

By the hypothesis of the induction, 4 yields

5 $a_{i+1} = ((p_{i+1} \wedge ((p_1 \wedge \ldots \wedge p_i) \vee \overline{a_i})) \vee \overline{p_{i+1}})$
$= (p_1 \wedge \ldots \wedge p_{i+1}) \vee (\overline{p_{i+1}} \vee (p_{i+1} \wedge \overline{a_i}))$ by (DL)
$= (p_1 \wedge \ldots \wedge p_{i+1}) \vee \overline{p_{i+1} \wedge (\overline{p_{i+1}} \vee a_i)}$ by (DeM)
$= (p_1 \wedge \ldots \wedge p_{i+1}) \vee \overline{a_{i+1}}$ by 2

Hence 3 holds, which can be used to show that $a_n = \overline{a_n}$ as follows:

6 $a_n = (p \vee \overline{a_n})$ by 3
$= p \vee \overline{p \vee \overline{a_n}}$ by 3
$= p \vee (\bar{p} \wedge a_n)$ by (DeM)
$= (p \vee \bar{p}) \wedge (p \vee a_n)$ by (DL)
$= \bar{p} \wedge (p \vee a_n)$ by 1
$= \bar{p} \wedge (p \vee (p \vee \overline{a_n}))$ by 3
$= \bar{p} \wedge (p \vee \overline{a_n})$ by (DL)
$= \bar{p} \wedge a_n$ by 3
$= \overline{p \vee \overline{a_n}}$ by (DeM)
$= \overline{a_n}$ by 3

But this contradicts (N3), so that $F(T_0)$ must after all be consistent, and the set E of all consistent filters in A containing T_0 must be non-empty.

We are now ready to apply Zorn's Lemma, a well-known equivalent of the axiom of choice; but first recall some terminology necessary for its statement. Where E is a family of sets, C *is a chain of* E if (1) C is a subfamily of E and (2) for every pair of sets X and Y in C, either $X \subseteq Y$ or $Y \subseteq X$. The *union over* a family of sets C is that set $\cup C$ containing all and only members of members of C. A set T is *maximal* in a family of sets E if no member of E is a proper superset of T. Not all families of sets have maximal members, but *Zorn's Lemma* states

(ZL) if E is a non-empty family of sets, and if the union $\cup C$ over every nonempty chain C of E is itself a member of E, then E has at least one maximal member.

We have just shown that the family E of consistent filters in A containing T_0 is non-empty, so that, in order to apply (ZL) to E, we need to know that if G is a chain of consistent filters containing T_0, then $\cup G$ (the union over G) itself (i) is consistent, (ii) is a filter, and (iii) contains T_0.

(i) $\cup G$ is consistent, for if for some $a \in L$, both $a \in \cup G$ and $\bar{a} \in \cup G$, then there are some filters $F, F' \in G$ such that $a \in F$ and $\bar{a} \in F'$. But either $F \subseteq F'$ or $F' \subseteq F$, so both $a, \bar{a} \in F$, or both $a, \bar{a} \in F'$. But this contradicts the consistency of F and F′.

(ii) $\cup G$ is a filter. $\cup G$ satisfies (F1), for if $a, b \in \cup G$, then there are $F, F' \in G$ such that $a \in F$ and $b \in F'$. Now since either $F \subseteq F'$ or $F' \subseteq F$, either both $a, b \in F$ or both $a, b \in F'$. But then since both F and F′ are filters, either $(a \wedge b) \in F$ or $(a \wedge b) \in F'$, which means, since $F \subseteq \cup G$ and $F' \subseteq \cup G$, that $(a \wedge b) \in \cup G$. Also $\cup G$ satisfies (F2), since if $a \in \cup G$, then there is some filter $F \in G$ such that $a \in F$. But then $(a \vee b) \in F$, and hence $(a \vee b) \in \cup G$.

(iii) $\cup G$ includes T_0, since $G \subseteq E$, and every member of E includes T_0.

Now applying (ZL) we conclude that E has a maximal element T. T is obviously consistent, since E is that family of all *consistent* filters including T_0. We proceed to show that T is exhaustive as well, which will complete the proof of the theorem.

We observe to begin with that

7 $(a \vee \bar{a}) \in T$, for all $a \in L$,

since $a \vee \bar{a} = ((a \vee \bar{a}) \vee \overline{a \vee \bar{a}})$ by (DeM) and (DL); so that $(a \vee \bar{a}) \in T_0$. Furthermore, since T is a filter, it follows from 7 that $((c \vee \bar{c}) \wedge (d \vee \bar{d})) \in T$ for all $c, d \in L$; hence, since T is consistent,

8 $\overline{(c \vee \bar{c}) \wedge (d \vee \bar{d})} \notin T$, for all $c, d \in L$.

Now suppose for *reductio* that T is not exhaustive, so that for some b, neither $b \in T$ nor $\bar{b} \in T$. Then the filters $F(T, b)$ and $F(T, \bar{b})$ generated by T with b and T with $\bar{b}$, respectively, each have T as a proper subset; and so since T is maximal in E, neither $F(T, b)$ nor $F(T, \bar{b})$ can be consistent. Consequently, for some c, both c and $\bar{c}$ are in $F(T, b)$, and for some d, both d and $\bar{d}$ are in $F(T, \bar{b})$; so since $F(T, b)$ is the filter generated by T with b and since $F(T, \bar{b})$ is the filter generated by T with $\bar{b}$, it must be that for some $t_1, t_2, t_3, t_4 \in T$,

9 $(t_1 \wedge b) \leq c, (t_2 \wedge b) \leq \bar{c}, (t_3 \wedge \bar{b}) \leq d, (t_4 \wedge \bar{b}) \leq \bar{d}$,

which leads by (DL) to

10 $(t_1 \wedge t_2 \wedge t_3 \wedge t_4 \wedge (b \vee \bar{b})) \leq ((\bar{c} \wedge c) \vee (\bar{d} \wedge d))$.

Since t_1, t_2, t_3, $t_4 \in T$, and since also $(b \vee \bar{b}) \in T$ by 7, the fact that T is a filter implies that $(t_1 \wedge t_2 \wedge t_3 \wedge t_4 \wedge (b \vee \bar{b})) \in T$, and accordingly, with 10, that $((\bar{c} \wedge c) \vee (\bar{d} \wedge d)) \in T$. So by (DeM),

$$11 \quad \overline{(c \vee \bar{c}) \wedge (d \vee \bar{d})} \in T.$$

But 11 contradicts 8, rendering absurd the supposition that T fails to be exhaustive, which completes the proof.

§18.4. Homomorphisms of intensional lattices. We define a *homomorphism* of an intensional lattice L into an intensional lattice L′ as a function h of L into L′ such that for all a, b ∈ L

1 $h(a \wedge b) = (h(a) \wedge h(b))$,
2 $h(a \vee b) = (h(a) \vee h(b))$, and
3 $h(\bar{a}) = \overline{h(a)}$.

Following customary algebraic practice, we denote the "same" operations in different intensional lattices by the same symbols. Condition 2 follows from the other conditions *via* (DeM). An *isomorphism* is defined as a one-to-one homomorphism. We define a T-*preserving homomorphism*, or T-*homomorphism*, of an intensional lattice L with truth filter T into an intensional lattice L′ with truth filter T′ as a homomorphism h of L into L′ such that if a ∈ T, then h(a) ∈ T′. A T-*isomorphism* is a one-to-one T-homomorphism.

Homomorphisms and isomorphisms are "structure preserving" functions, which play a significant role in characterizing the structure of algebras. For example, Stone 1936 showed that a maximal filter of a Boolean algebra B determines a homomorphism of B into the particularly simple two-element Boolean algebra. We are about to prove an analogous theorem for intensional lattices, but first we must characterize the particularly simple intensional lattice, which we call M_0, that is used in our theorem.

We choose to do this by means of a Hasse diagram (see §22.1.3 for a matrix representation). Every lattice, indeed every partially ordered set, can be represented by a *Hasse diagram* (and the finite ones can be drawn on paper), in which distinct elements are represented by distinct points and $a \leq b$ is represented by a line or lines which ascend from a to b ($a \leq a$ is represented by a very short line consisting of just the single point a). The fact that a point a is below a point b in a Hasse diagram does not guarantee that $a \leq b$, since this is so just in case there is a steadily ascending sequence of lines connecting a to b; for instructive elementary examples see Dubisch 1964. Two lattices are isomorphic if they can be represented by the same Hasse diagram.

M_0 can be represented by the following Hasse diagram, where $\overline{\pm a} = \mp a$, and $T = \{+0, +1, +2, +3\}$:

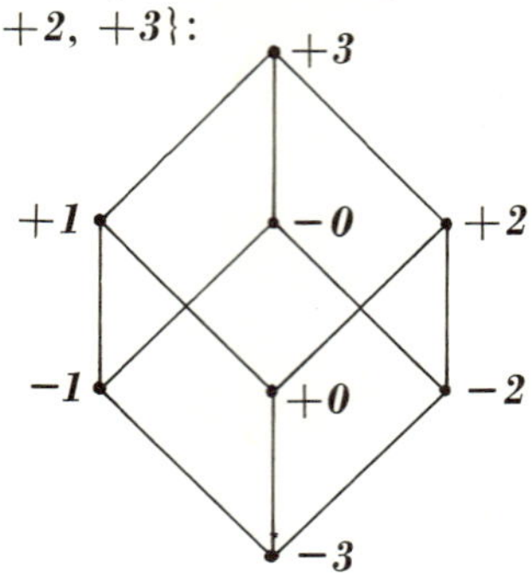

THEOREM 1. Every prime filter P of an intensional lattice $\langle L, \leq, \overline{}, T\rangle$ determines a T-homomorphism h of L into M_0 satisfying the following conditions (where F_i is the principal filter generated by i in M_0):

1 $h(a) \in F_{-1}$ iff $a \in P$,
2 $h(a) \in F_{-2}$ iff $\bar{a} \in -P$ (where $-P$ is the set-theoretical complement of P relative to L),
3 $h(a) \in F_{+0}$ iff $a \in T$.

These conditions uniquely determine a single-valued function h, for if we let μF_i be either F_i or $-F_i$, the set theoretical intersection of μF_{-1}, μF_{-2}, and μF_{+0} is always a unit set, for every such choice function μ.

We now prove that h is a T-homomorphism. We first show that $h(a \wedge b) = (h(a) \wedge h(b))$. In view of our remarks above concerning the unique determination of h, it suffices to show that $h(a \wedge b) \in F_i$ iff $(h(a) \wedge h(b)) \in F_i$ ($i = -1, -2, +0$). The cases $i = -1$ and $i = +0$ may be treated together. Thus $h(a \wedge b) \in F_{-1}$ $\{F_{+0}\}$, iff $(a \wedge b) \in P$ $\{T\}$, iff $a, b \in P$ $\{T\}$, iff $h(a), h(b) \in F_{-1}$ $\{F_{+0}\}$, iff $(h(a) \wedge h(b)) \in F_{-1}$ $\{F_{+0}\}$. We treat $i = -2$ by using the fact that $-P$ is a prime ideal (a well-known consequence of the fact that P is a prime filter), together with an immediate De Morgan property of $\overline{}$. Thus $h(a \wedge b) \in F_{-2}$, iff $\overline{a \wedge b} = (\bar{a} \vee \bar{b}) \in -P$, iff $\bar{a}, \bar{b} \in -P$, iff $h(a), h(b) \in F_{-2}$, iff $(h(a) \wedge h(b)) \in F_{-2}$.

We now show that $h(\bar{a}) = \overline{h(a)}$. Again it suffices to show that $h(\bar{a}) \in F_i$ iff $\overline{h(a)} \in F_i$ ($i = -1, -2, +0$). We first define $\overline{F_i}$ as the set of $\bar{b}$ such that $b \in F_i$, and observe that $F_{-1} = \overline{-F_{-2}}$, $F_{-2} = \overline{-F_{-1}}$, and $F_{+0} = \overline{-F_{+0}}$. Then $h(\bar{a}) \in F_{-1}$, iff $\bar{a} \in P$, iff $h(a) \in -F_{-2}$, iff $\overline{h(a)} \in \overline{-F_{-2}} = F_{-1}$. And finally, $h(\bar{a}) \in F_{+0}$, iff $\bar{a} \in T$, iff $a \in -T$, iff $h(a) \in -F_{+0}$, iff $\overline{h(a)} \in \overline{-F_{+0}} = F_{+0}$.

We complete our proof by noting that condition 3 insures that the homomorphism h is T-preserving.

We also have the converse

THEOREM 2. Every T-homomorphism h of an intensional lattice $\langle L, \leq, \ , T\rangle$ into M_0 determines a prime filter P of the intensional lattice in accord with conditions 1 and 2 of Theorem 1.

Given a T-homomorphism h, condition 1 by itself determines the set P. so it suffices to show that P is a prime filter and that it satisfies condition 2. That the set P is a prime filter follows immediately from the fact that if h is a homomorphism into a prime filter (in this case F_{-1}), then the inverse image of that prime filter under h (in this case P) is also a prime filter. We now demonstrate that P satisfies condition 2. Since $F_{-2} = \overline{-F_{-1}}$, we have that $h(a) \in F_{-2}$ iff $h(a) \in \overline{-F_{-1}}$, iff $\overline{h(a)} = h(\bar{a}) \in \overline{\overline{-F_{-1}}} = -F_{-1}$. But contraposing condition 1, we have $h(\bar{a}) \in -F_{-1}$ iff $\bar{a} \in -P$. So any set P which satisfies condition 1 also satisfies condition 2, and our proof is complete.

By combining Theorems 1 and 2 we obtain a natural one-to-one correspondence between prime filters and T-homomorphisms into M_0. That distinct filters never determine the same T-homomorphism is immediate, since they would differ at an element a, and hence the homomorphisms determined by them would differ in that one would send a into F_{-1} and the other would not. That distinct T-homomorphisms h and h′ never determine the same prime filter is shown as follows. Suppose for some element a, $h(a) \neq h'(a)$. We shall show that there is always some element b such that $h(b) \in F_{-1}$ but $h'(b) \notin F_{-1}$. The proof is by cases. Assume first that $h(a), h'(a) \in F_{-1}$. Make the further assumption that $a \in T$. Then since h and h′ preserve T, $h(a), h'(a) \in \{+1, +3\}$. Assume without loss of generality that $h(a) = +1$ and $h'(a) = +3$. Then $h(\bar{a}) = -1 \in F_{-1}$, but $h'(\bar{a}) = -3 \notin F_{-1}$. On the assumption that $a \notin T$, it follows immediately from the fact that h and h′ are T-preserving that $h(a), h'(a) \in \{-1, -0\}$. Assume without loss of generality that $h(a) = -1$ and $h'(a) = -0$. Then $h(\bar{a}) = +1 \in F_{-1}$, but $h'(\bar{a}) = +0 \notin F_{-1}$. A glance at the diagram of M_0 should convince the reader that the remaining case, in which $h(a), h'(a) \notin F_{-1}$, may also be divided according as to whether a is or is not in T, and treated analogously.

We obtain as an immediate consequence of Theorem 1, together with the theorem of Stone 1937 (see §18.1), the following

THEOREM 3. For any two elements a and b of an intensional lattice such that $a \not\leq b$, there exists a T-homomorphism h of the intensional lattice into M_0 such that $h(a) \in F_{-1}$ and $h(b) \notin -F_{-1}$.

Our proof of Theorem 3 implicitly relies upon the axiom of choice, since Stone's proof uses it. However, the following may be proved without appeal to the axiom of choice, even though the "only if" part is a weakening of Theorem 3.

THEOREM 4. A De Morgan lattice has a truth filter T iff it has a homomorphism into M_0.

The "only if" part is an immediate corollary of Theorem 1. This follows by considering the De Morgan lattice as an intensional lattice with truth filter T, and also by observing that the truth filter T is a prime filter. We can then apply Theorem 1, letting T play a double role, both as the truth filter and the prime filter, in determining the homomorphism.

To prove the converse, let L be an arbitrary De Morgan lattice, and h a homomorphism of L into M_0. Let T be the inverse image of F_{+0} under h. As has already been remarked in the proof of Theorem 2, the inverse image of a prime filter is a prime filter, so it only remains to show that T is both consistent and exhaustive. Suppose that T were inconsistent, i.e., that for some $a \in L$, both $a, \bar{a} \in T$. Then $h(a)$ and $h(\bar{a}) = \overline{h(a)}$ are both members of F_{+0}; but this is impossible since F_{+0} is consistent. Now suppose that T is not exhaustive, i.e., for some $a \in L$, neither $a \in T$ nor $\bar{a} \in T$. Then $h(a) \notin F_{+0}$ and $h(\bar{a}) = \overline{h(a)} \notin F_{+0}$ but this is impossible too since F_{+0} is exhaustive, which completes the proof.

§18.5. An embedding theorem. We are going to need Cartesian products of arbitrary cardinality, the definition of which requires the concept of indexing, which we first explain. Where A and X are sets and b a function, A is said to be *indexed by* X *under* b iff b is a function from X onto A; which is to say, iff X is the domain and A the range of b. The purpose of indexing is to provide a way of tagging the members of A analogous to, but more general than, the use of integers to sequence a set, and in order to enforce the analogy it is customary to write "b_x" for the value b(x) of the function b at the argument x. We can think of b_x as "the x-th element of A." The function b is said to be the *indexing function*, and its domain X is called the *indexing set*. The phrase "*indexed set*" is used ambiguously: sometimes it is used to refer to the range A of b, while even more often it is used to refer to the function b itself. It depends on whether one wants "indexed set" in the sense of "set which is indexed" (i.e., A) or in the sense "set with its indexing" (i.e., b, which is of course a subset of $X \times A$). In a corresponding way, the notation $\{b_x\}_{x \in X}$ is used sometimes (with propriety) to refer to the range of b, which is indexed by b, but more often (somewhat improperly) to refer to the indexing function b itself. The ambiguity is heightened by such sentences as "$b = \{b_x\}_{x \in X}$," which should be — but never is — written as "$b = \{\langle x, b_x \rangle\}_{x \in X}$." The use of the brace notation is illustrated in the second version of the definition of Cartesian product, which we give first without it. Let A be a collection of sets indexed by X under S (so that S_x is a set). Then the *Cartesian product*, $\times_{x \in X} S_x$, is defined as the set of all indexing functions b such that for each $x \in X$, $b_x \in S_x$. That is, b chooses one member from each set in the collection A of sets, and "b_x" gives us a name of the element chosen from "the x-th set" S_x. Now using the brace notation: given an indexed collection of sets $\{S_x\}_{x \in X}$, its *Cartesian product* $\times_{x \in X} S_x$ is the set of all indexed sets $\{s_x\}_{x \in X}$ such that $s_x \in S_x$.

Using these notations we may define the product of intensional lattices as follows: Let $\{\langle L_x, \leq_x, \overline{}^x, T_x \rangle\}_{x \in X}$ be an indexed set of intensional lattices, and let their *product* $\Pi_{x \in X} L_x$ be that quadruple $\langle L, \leq, \overline{}, T \rangle$ such that $T = \times_{x \in X} T_x$, $-T = \times_{x \in X} -T_x$, $L = T \cup -T$, and $\leq$ and $\overline{}$ are defined on elements $\{a_x\}_{x \in X}$ and $\{b_x\}_{x \in X}$ of L so that $\{a_x\}_{x \in X} \leq \{b_x\}_{x \in X}$ iff $a_x \leq_x b_x$ for all $x \in X$, and $\overline{\{a_x\}_{x \in X}} = \{\overline{a_x}^x\}_{x \in X}$. Because of the "component-wise" definition it is easily verified that the product of intensional lattices is itself an intensional lattice.

A product M^c, of M_0, is defined as $M^c = \Pi_{x \in X} L_x$, where $L_x = M_0$ for all $x \in X$, and where c is the cardinality of an indexing set X. We may now state the following embedding

THEOREM. Every intensional lattice $\langle L, \leq, \overline{}, T \rangle$ of cardinality d has a T-isomorphism into a product M^c of M_0 for some cardinal $c \leq d^2$.

The proof uses Theorem 3 of the previous section. We begin by considering as an indexing set, the set X of all pairs of elements $\langle a, b \rangle$ such that $a, b \in L$ and $a \nleq b$. Note that the cardinality of X, say c, is such that $c \leq d^2$, where d is the cardinality of L. We now construct a T-isomorphism h of L into M^c defined as above.

With each element $x = \langle a, b \rangle$ of X, we may by Theorem 3 of the previous section associate a T-homomorphism h_x of L into M_0 such that $h_x(a) \neq h_x(b)$. Then h, defined for $a \in L$ by $h(a) = \{h_x(a)\}_{x \in X}$ is the desired T-isomorphism. Thus h is obviously defined over all of L, since h_x is. And h is both into M^c and T-preserving, because each h_x is into M_0 and is T-preserving. And h is one-to-one, for suppose that $a \neq b$, and assume without loss of generality that $a \nleq b$. But then for $x = \langle a, b \rangle$, $h_x(a) \neq h_x(b)$, by the method of association given above, and so h(a) differs from h(b) in the x-th component.

It remains to show h a homomorphism; this follows immediately from the fact that operations in the product algebra were defined component wise: $h(a \wedge b) = \{h_x(a \wedge b)\}_{x \in X} = \{h_x(a) \wedge h_x(b)\}_{x \in X} = h(a) \wedge h(b)$, and $h(\bar{a}) = \{h_x(\bar{a})\}_{x \in X} = \{\overline{h_x(a)}\}_{x \in X} = \overline{h(a)}$.

The theorem is an analogue of the well-known result of Stone 1936 that every Boolean algebra is isomorphically embeddable in a product of the two-element Boolean algebra.

§18.6. Intensional lattices as models. By the *degree* of a formula we mean, as in §15, to indicate the depth of nesting of arrows within arrows. A *zero degree formula* (*zdf*) is purely truth functional, i.e., the only connectives which occur in it are those of negation, conjunction, and disjunction. In §19 first degree formulas will be dealt with at some length, but now we consider only *first degree entailments* (*fdes*), i.e., those formulas of the form $A \rightarrow B$, where both A and B are zdfs. The formulas of $\mathbf{E}_{\mathrm{fde}}$ (§15.2) are fdes.

We can employ intensional lattices as models for fdes. Keeping in mind the intuitive picture of an intensional lattice as a *propositional lattice*, i.e., a set of propositions closed under negation (as intensional complementation), conjunction (meet), and disjunction (join), and ordered by the relation of entailment, we define a *propositional model* (often just *model*) Q as a pair $\langle \mathrm{L}, \mathrm{s} \rangle$, where L is an intensional lattice and where s is an *assignment function*, assigning to each propositional variable an element of L. The terminology "propositional model" is further justified by the fact that in §40 we manage to pull off the metaphysical feat of *proving* from a few assumptions about propositions that every intensional lattice of at most denumerable cardinality is T-isomorphic to a propositional lattice.

Given a model $\mathrm{Q} = \langle \mathrm{L}, \mathrm{s} \rangle$, we define a (zero degree) valuation determined by Q as a function $\mathrm{v_Q}$ defined over all zdfs and having values in L as follows: for all zdfs A,

if A is a propositional variable, then $\mathrm{v_Q}(A) = \mathrm{s}(A)$;
if A has the form $\overline{B}$, then $\mathrm{v_Q}(A) = \overline{\mathrm{v_Q}(B)}$;
if A has the form $B \& C$, then $\mathrm{v_Q}(A) = \mathrm{v_Q}(B) \wedge \mathrm{v_Q}(C)$; and
if A has the form $B \vee C$, then $\mathrm{v_Q}(A) = \mathrm{v_Q}(B) \vee \mathrm{v_Q}(C)$.

We then say that a fde $A \rightarrow B$ is *true in a model* Q iff $\mathrm{v_Q}(A) \leq \mathrm{v_Q}(B)$, and otherwise *false in the model*. A fde is *valid in an intensional lattice* L iff it is true in every model $\mathrm{Q} = \langle \mathrm{L}, \mathrm{s} \rangle$, and otherwise is *falsifiable* in L. A fde is valid iff it is valid in every intensional lattice, and otherwise it is *falsifiable*.

The system $\mathbf{E}_{\mathrm{fde}}$ of tautological entailments is semantically *consistent* iff all fdes provable in $\mathbf{E}_{\mathrm{fde}}$ are valid; and the system $\mathbf{E}_{\mathrm{fde}}$ is semantically *complete* iff all valid fdes are provable in $\mathbf{E}_{\mathrm{fde}}$.

We shall eventually relate the provability of a fde in $\mathbf{E}_{\mathrm{fde}}$ with its validity, but first we take an algebraic detour.

§18.7. The Lindenbaum algebra of $\mathbf{E}_{\mathrm{fde}}$. The relation of *co-entailment* between two zdfs A and B holds just when $A \rightleftarrows B$ holds for $\mathbf{E}_{\mathrm{fde}}$ (see §10 for the notation "$\rightleftarrows$"). It is easily seen that the relation of co-entailment is an *equivalence relation*, i.e., it is *reflexive* ($A \rightleftarrows A$), *symmetric* ($A \rightleftarrows B$ implies $B \rightleftarrows A$), and *transitive* ($A \rightleftarrows B$ and $B \rightleftarrows C$ imply $A \rightleftarrows C$).

We may partition the set of zdfs into disjoint equivalence classes, letting $[A]$ be the equivalence class of all zdfs that co-entail A; i.e., "identify," so to speak, a zdf A with all the zdfs that are equivalent to it in the sense of co-entailment.

Then define operations upon these equivalence classes as follows:

$\overline{[A]} = [\bar{A}]$,
$[A] \wedge [B] = [A \& B]$, and
$[A] \vee [B] = [A \vee B]$.

It is important to note here that the replacement theorem holds for the zdfs of $\mathbf{E}_{fde}$ (see §15.2), for this ensures that the operations are well defined in the sense of not depending upon the particular representative zdfs that appear in the above definitions. For example, if $[A] = [B]$, then $\overline{[A]} = \overline{[B]}$, since $[A] = [B]$ implies $A \rightleftarrows B$ implies (by the replacement theorem) $\bar{A} \rightleftarrows \bar{B}$ implies $\overline{[A]} = \overline{[B]}$.

Now that we have operations defined upon the equivalence classes, we can introduce a relation by defining

$[A] \leq [B]$ iff $A \rightarrow B$ is provable in $\mathbf{E}_{fde}$.

When equivalence classes are available in a propositional logic which enables us to define operations and relations as above, we call the resulting algebra a *Lindenbaum algebra* (see Rasiowa and Sikorski 1963, pp. 244ff). It is now easily verified by browsing through the axioms and theorems of $\mathbf{E}_{fde}$ found in §15.2 that the Lindenbaum algebra of $\mathbf{E}_{fde}$, which we denote by "$\mathbf{E}_{fde}/\rightleftarrows$," satisfies postulates (DL), (N1), and (N2), of §18.2, which means that it is a De Morgan lattice. Further, it may be shown that $\mathbf{E}_{fde}/\rightleftarrows$ satisfies condition (N3), for if to the contrary there were some zdf such that $[A] = \overline{[A]}$, we would have $A \rightleftarrows \bar{A}$; but this is impossible since $\mathbf{E}_{fde}$ is readily seen to be a subsystem of the classical propositional calculus (taking the arrow to be a horseshoe), which is negation consistent. Then by Theorem 1 of §18.3 we have the following.

THEOREM. There is some subset T of the equivalence classes of zdfs in the Lindenbaum algebra $\mathbf{E}_{fde}/\rightleftarrows$ of $\mathbf{E}_{fde}$, such that $\langle \mathbf{E}_{fde}/\rightleftarrows, \leq, \overline{}, T \rangle$ ($\leq$ and $\overline{}$ defined as above) is an intensional lattice.

We record here, for future reference from other parts of this book, that if a calculus $\mathbf{S}$ has additional nesting connectives, e.g., $\rightarrow$ or $\circ$, by *the Lindenbaum algebra* $\mathbf{S}/\rightleftarrows$ of that calculus we always mean the result of adding clauses like the above, e.g. $[A] \rightarrow [B] = [A \rightarrow B]$ or $[A] \circ [B] = [A \circ B]$, one for each connective, to the fundamental definition $[A] = \{B : \vdash_{\mathbf{S}} A \rightleftarrows B\}$. And by

Lindenbauming a calculus **S** we always mean defining **S**/$\rightleftarrows$ with operations according to the above recipe.

§18.8. An algebraic completeness theorem for $\mathbf{E}_{\mathrm{fde}}$. The last theorem allows us to show that the system $\mathbf{E}_{\mathrm{fde}}$ of tautological entailments is complete in the sense of §18.6. We begin by defining the *canonical model* C for $\mathbf{E}_{\mathrm{fde}}$ as $\langle \mathbf{E}_{\mathrm{fde}}/\rightleftarrows, \mathrm{c}\rangle$, where $\mathbf{E}_{\mathrm{fde}}/\rightleftarrows$ is regarded as an intensional lattice outfitted with a truth filter T (as guaranteed by the last theorem), and c is the *canonical assignment function*, i.e., where A is a propositional variable, $\mathrm{c}(A) = [A]$. We then define the (zero degree) *canonical valuation* $\mathrm{v_c}$ as the (zero degree) valuation determined by c. It is easy to see that where A is a zdf, $\mathrm{v_c}(A) = [A]$.

It is now immediate that if a fde $A \rightarrow B$ is not provable in $\mathbf{E}_{\mathrm{fde}}$, then it is falsifiable in the canonical model under the canonical valuation, since then $[A] \nleq [B]$, and hence $\mathrm{v_c}(A) \nleq \mathrm{v_c}(B)$.

LEMMA. If a fde is not provable in $\mathbf{E}_{\mathrm{fde}}$, then it is falsifiable under the canonical valuation in the canonical model.

Although the above result establishes the completeness of $\mathbf{E}_{\mathrm{fde}}$, we call it a mere "lemma" because it establishes only a rather trivial sort of completeness, the sort that could be proved for any system on which can be defined a well-behaved equivalence relation so as to give rise to an algebra — with or without an attractive name. But the lemma, together with Theorem 1 of §18.4, gives rise to the following deeper completeness theorem.

THEOREM 1. If a fde is valid in M_0, then it is provable in $\mathbf{E}_{\mathrm{fde}}$.

We prove the contrapositive. On the assumption that the fde $A \rightarrow B$ is not provable in $\mathbf{E}_{\mathrm{fde}}$, then $[A] \nleq [B]$ in $\mathbf{E}_{\mathrm{fde}}/\rightleftarrows$. But since $\mathbf{E}_{\mathrm{fde}}/\rightleftarrows$ may be regarded as an intensional lattice by the theorem of the previous section, then by Theorem 1 of §18.4, there exists a homomorphism of $\mathbf{E}_{\mathrm{fde}}/\rightleftarrows$ into M_0 such that $\mathrm{h}([A]) \nleq \mathrm{h}([B])$. Now define hc as the composition of h and the canonical assignment function c, i.e., if D is a propositional variable, then $\mathrm{hc}(D) = \mathrm{h}([D])$. Obviously hc is an assignment function for M_0. Thus let $\mathrm{Q} = \langle \mathrm{M}_0, \mathrm{hc}\rangle$. It is easy to see that for all zdfs D, $\mathrm{v_Q}(D) = \mathrm{h}([D])$. Hence $\mathrm{v_Q}(A) \nleq \mathrm{v_Q}(B)$, and $A \rightarrow B$ is falsifiable in M_0.

COROLLARY. If $A \rightarrow B$ is not provable in $\mathbf{E}_{\mathrm{fde}}$, then there is a model $\mathrm{Q} = \langle \mathrm{M}_0, \mathrm{s}\rangle$ such that $\mathrm{v_Q}(A) \in \mathrm{F}_{-1}$.

For this we simply use Theorem 3 of §18.4 and proceed as in the proof of Theorem 1 above.

Turning now to the consistency of $\mathbf{E}_{fde}$, we have

THEOREM 2. If a fde is provable in $\mathbf{E}_{fde}$, then it is valid.

Inspection of the axioms of $\mathbf{E}_{fde}$ shows that they are valid. The conjunction axiom $A\&B \rightarrow A$, for example, is obviously valid, since $(a \wedge b) \leq a$ is true of all lattices. It is equally obvious that the rules preserve validity. Consider, for example, the entailment rule of transitivity, from $A \rightarrow B$ and $B \rightarrow C$ to infer $A \rightarrow C$; we have, if $v_Q(A) \leq v_Q(B)$ and $v_Q(B) \leq v_Q(C)$, then $v_Q(A) \leq v_Q(C)$, since the lattice ordering $\leq$ is transitive.

By combining Theorems 1 and 2, we obtain

THEOREM 3. A fde is provable in $\mathbf{E}_{fde}$ iff it is valid in M_0.

We remark in closing that Theorem 3 says *roughly* that M_0 is a characteristic matrix for tautological entailments. The only reason for hedging in this statement is that M_0 has no *operation* corresponding to entailment, but only the *relation* $\leq$. Of course, there was no reason for such an operation in dealing with fdes, since no nesting of arrows was involved. We have thus shown by algebraic means that M_0 stands in the same relation to tautological entailments as does Smiley's matrix of §15.3. Why then did we go the trouble of trotting in M_0? The first reason is merely to show off our algebraic methods. This reason is unimportant in the sense that M_0, fitted out with an operation corresponding to entailment, is already known to be characteristic for this fragment of E. We take the value of algebraic proofs to lie partly in the fact that they enable us to compare different systems. Thus our algebraic proof of Theorem 3 is analogous to the well-known algebraic proof of the completeness and consistency of the classical propositional calculus based upon Stone's 1936 representation theory for Boolean algebras. We could have, if we had chosen to do so, provided an algebraic proof of the fact that Smiley's matrix is characteristic, basing it upon results we have concerning De Morgan lattices that parallel those we have brought forth concerning intensional lattices. (Smiley's matrix is a De Morgan lattice, but not an intensional lattice, since, for example, when p is $\mathbf{2}$, $\bar{p}$ is $\mathbf{2}$ as well.)

The second reason is more basic to the entailment enterprise. In the study of semantics of formulas that do not have the form of an entailment $A \rightarrow B$ (where A and B are of zero degree), specific reference has to be made to truth values of their subformulas. Hence the need for intensional lattices with their truth filters. We close this section with two simple results bearing on this need, the second of which is used in the next section.

THEOREM 4. If a zdf A is not a two-valued tautology, then there is a truth-filter T for $\mathbf{E}_{\text{fde}}/\rightleftarrows$ such that $[A] \notin$ T.

PROOF. Since A is not a two-valued tautology, some assignment of ordinary truth values to the variables makes A come out false. Define T as the set of [B] such that B comes out true on this assignment. One needs to verify that representatives of an equivalence class are alike in truth value; then it is easily seen that T is a truth filter.

THEOREM 5. If A is not a two-valued tautology and $C \rightarrow D$ is not a theorem of $\mathbf{E}_{\text{fde}}$, there is an M_0-model Q = ⟨M_0, s⟩ such that $v_Q(A) \notin F_{+0}$, $v_Q(C) \notin F_{-1}$ and $v_Q(D) \notin F_{-1}$.

PROOF. This is an immediate consequence of Theorem 3 of §18.4, Theorem 4 above, and the idea of the proof of Theorem 1. This theorem illustrates incidentally the very considerable independence between truth values and entailment, since A on the one hand and C and D on the other are wholly unfettered in their relations.

§19. First degree formulas.

Our aim now is to extend some of our previous results to first degree formulas, which is as far as we can go without reintroducing the nesting of arrows we promised to eschew for the duration of this chapter. We remind the reader that a *zero degree formula* (*zdf*) is an ordinary truth functional expression writ in propositional variables, &, ∨, and ~, while a *first degree formula* (*fdf*) is any truth function of propositional variables and first degree entailments $A \rightarrow B$, with A and B both zdfs. In this section A, B, C, and D are zdfs, while F, G and H are fdfs.

§19.1. Semantics.

In §18.2 we introduced the concept of an intensional lattice and of a propositional lattice, each with its truth-filter T. In these terms we defined the notion of a propositional model Q = ⟨L, s⟩, with L an intensional lattice and s a function assigning to each propositional variable an element of L. Then the notion of a valuation v_Q determined by Q was defined as a function which assigns an element of L to each zdf, and finally we defined A→B as valid if always $v_Q(A) \leq v_Q(B)$. In these proceedings we made no reference to the truth filter T; but now having it available permits us to extend our semantic notions in such a way as to elicit plausible concepts of truth, falsity, and validity for all first degree formulas. After fixing the requisite definitions, we describe a Hilbert-style calculus of first degree formulas which is consistent and complete with respect to validity so defined.

As before (§18.6), let Q = ⟨L, S⟩ be a propositional model, with v_Q the valuation determined by Q; also let T be the truth filter of the lattice L. We can now capture the notion of truth for fdfs as follows:

A zdf A is true in Q iff $v_Q(A) \in T$;
$B \rightarrow C$ is true in Q iff $v_Q(B) \leq v_Q(C)$;
$\sim F$ is true in Q iff F is not;
$F \vee G$ is true in Q iff either F or G is;
$F \& G$ is true in Q iff both F and G are.

An fdf is said to be false in Q if it is not true; we leave it to the reader to determine that our terminology is well-chosen at least to the extent of satisfying the condition that an fdf is true exactly when its negation is false. We also define an fdf as *valid* if it is valid in every propositional model, and we call fdfs F and G *equivalid* if they are either both valid or both falsifiable.

§19.2. Axiomatization. We next offer a calculus designed to correspond to the semantics described above. Let us agree to call the calculus "$\mathbf{E}_{\text{fdf}}$," since it will turn out to be the fdf fragment of the system **E** of entailment. The axioms and rules of $\mathbf{E}_{\text{fdf}}$ are vastly redundant; the reader will have no difficulty in proving some from others. Our point here, however, is to give an intuitively convincing (rather than economical) set of axioms and rules.

We begin our formulation of $\mathbf{E}_{\text{fdf}}$ with exactly the same axioms as $\mathbf{E}_{\text{fde}}$:

Entailment Axioms. Each axiom of $\mathbf{E}_{\text{fde}}$ is also an axiom of $\mathbf{E}_{\text{fdf}}$.

Also, we assume excluded middle for arbitrary fdfs:

Truth functional axioms: $F \vee \overline{F}$.

The first rule allows us to manipulate truth functional connectives in sensible ways:

(TE) From $F_1, \ldots, F_n$ to infer G, where $F_1 \& \ldots \& F_n \rightarrow G$ is a substitution-instance of a tautological entailment.

Note that (1) since the F_i and G range over fdfs and hence can contain arrows, $F_1 \& \ldots \& F_n \rightarrow G$ will possibly not itself be an fdf, and (2) the rule (TE), though complex, is effective, since it is decidable whether or not a claimed inference really does fall under the rule. (Since the notions of effectiveness and decidability are important to the idea of a *logistic system*, as we understand the matter, and since we have never (and won't) talk about the topic, we refer the reader to Church 1956, §07.)

Now we all know about rules for transitivity, conjunction in the consequent, disjunction in the antecedent, contraposition, and the like; and it is obvious that if C follows from B, then $A \vee C$ follows from $A \vee B$. So the following rules suggest themselves immediately:

R1 *Transitivity.* $$\frac{F \vee ((A \to B) \& (B \to C))}{F \vee (A \to C)}$$

R2 *Conjunction.* $$\frac{F \vee ((A \to B) \& (A \to C))}{F \vee (A \to . B \& C)}$$

R3 *Disjunction.* $$\frac{F \vee ((A \to C) \& (B \to C))}{F \vee (A \vee B \to C)}$$

R4 *Contraposition.* $$\frac{F \vee (A \to B)}{F \vee (\bar{B} \to \bar{A})}$$

The foregoing rules tell us how to manipulate entailments; the next do an analogous job for negated entailments.

R5 *Transitivity.* $$\frac{F \vee (\overline{A \to B} \& (C \to B))}{F \vee \overline{A \to C}}$$

R6 *Transitivity.* $$\frac{F \vee (\overline{A \to B} \& (A \to C))}{F \vee \overline{C \to B}}$$

R7 *Conjunction.* $$\begin{array}{c} F \vee \overline{A \to . B \& C} \\ \hline\hline F \vee \overline{A \to B} \vee \overline{A \to C} \end{array}$$

R8 *Disjunction.* $$\begin{array}{c} F \vee \overline{A \vee B \to C} \\ \hline\hline F \vee \overline{A \to C} \vee \overline{B \to C} \end{array}$$

R9 *Contraposition.* $$\begin{array}{c} F \vee (\overline{A \to B}) \\ \hline\hline F \vee (\overline{\bar{B} \to \bar{A}}) \end{array}$$

Lastly, we need some rules relating entailments and negated entailments to truth functions:

R10 *Modus Ponens.* $$\frac{F \vee (A \& (A \to B))}{F \vee B}$$

R11 *Weakening.* $$\frac{F \vee (A \& \bar{B})}{F \vee \overline{A \to B}}$$

The calculus $\mathbf{E}_{\mathrm{fdf}}$ is then defined by the entailment axioms, the truth functional axioms, the rule (TE), and the rules R1–R11.

§19.3. Consistency.

THEOREM. $\mathbf{E}_{fdf}$ is consistent: all theorems are valid. Furthermore, its rules are not only validity-preserving but also truth-preserving.

PROOF. We need to show that all the axioms are valid, and that each rule preserves truth. Consider one of the entailment axioms, $A \to B$; we need to show that, for arbitrary Q = ⟨L, s⟩, $A \to B$ is true in Q, which requires that $v_Q(A) \leq v_Q(B)$. But this is so by Theorem 2 of §18.8. And that the truth functional axioms $F \vee \overline{F}$ are all true in Q follows immediately from the negation and disjunction clauses of our truth-definition. With respect to the rules (TE) and R1–R11, we must show that they preserve truth: if their premisses are true in Q, so are their conclusions. We illustrate the argument by an example, R6. Suppose for contraposition that the conclusion $F \vee \overline{C \to B}$ is not true in Q, so that neither F nor $\overline{C \to B}$ is true. Hence $C \to B$ is true; hence $v_Q(C) \leq v_Q(B)$. But then since $\leq$ is transitive, not both $v_Q(A) \not\leq v_q(B)$ and $v_Q(A) \leq v_Q(C)$, so not both $\overline{A \to B}$ and $A \to C$ are true. Hence $\overline{A \to B}\&(A \to C)$ is not true, so $F \vee (\overline{A \to B}\&(A \to C))$ is not true, as required. Other cases are similar, and $\mathbf{E}_{fdf}$ is therefore consistent with its intended interpretation.

§19.4. Facts. Ultimately we want to show that $\mathbf{E}_{fdf}$ is also complete, but first we take time out to see how the axioms and rules can be made to perform sensibly by exhibiting some facts about the system. We should add, we suppose, that the reason we have demoted these results from THEOREMS or LEMMAS to FACTS is that they don't seem to have much bearing on the philosophical issues we are worrying about. They are more like the kind of thing that would interest a plumber or a mathematician.

FACT 1. All rules of $\mathbf{E}_{fde}$ are derivable in $\mathbf{E}_{fdf}$.

We already have analogues of all the rules of $\mathbf{E}_{fde}$; for example, $\mathbf{E}_{fde}$ has the rule, from $A \to B$ and $B \to C$ to infer $A \to C$, while $\mathbf{E}_{fdf}$ has its analogue, from $F \vee ((A \to B)\&(B \to C))$ to infer $F \vee (A \to C)$. The only wrinkle needing ironing is that the $\mathbf{E}_{fde}$ rule of transitivity needs teasing out of its $\mathbf{E}_{fdf}$ analogue, as follows.

1	$A \to B$	Hypothesis
2	$B \to C$	Hypothesis
3	$(A \to B) \vee ((A \to B)\&(B \to C))$	1 (TE)
4	$(A \to B) \vee (A \to C)$	3 R1
5	$(A \to C) \vee ((A \to B)\&(B \to C))$	2 4 (TE)
6	$(A \to C) \vee (A \to C)$	5 R1
7	$A \to C$	6 (TE)

Other analogues are analogous, and proofs are left to the reader.

FACT 2. All tautological entailments (= theorems of $\mathbf{E}_{fde}$) are theorems of $\mathbf{E}_{fdf}$.

This follows immediately from the completeness of $\mathbf{E}_{fde}$ (§18.8) and Fact 1, since all the axioms of $\mathbf{E}_{fde}$ are axioms of $\mathbf{E}_{fdf}$.

FACT 3. $A \rightarrow B$ is derivable from

1 $D\&A \rightarrow B$ and
2 $A \rightarrow B \vee D$.

Notice that this means that if D appears on both the left and the right, it can (so to speak) be canceled. In view of the commutative, associative, and distributive properties of disjunction and conjunction (on both the left and the right of the arrow), the problem of getting to $A \rightarrow B$ from 1 and 2 can be solved as follows:

3	$B \rightarrow B$	Fact 2
4	$A \rightarrow A$	Fact 2
5	$A \rightarrow (B \vee D)\&A$	2 4 Fact 1
6	$(B \vee D)\&A \rightarrow B \vee (D\&A)$	Fact 2
7	$B \vee (D\&A) \rightarrow B$	1 3 Fact 1
8	$A \rightarrow B$	5 6 7 trans Fact 1

FACT 4. Each rule R1–R11 of $\mathbf{E}_{fdf}$ has an analogue in which the piece being worked on is any disjunct you like; i.e., if from $F \vee G$ to infer $F \vee H$ is one of R1–R11, then the rule,

from $F_1 \vee \ldots \vee F_{i-1} \vee G \vee F_{i+1} \vee \ldots \vee F_n$
to infer $F_1 \vee \ldots \vee F_{i-1} \vee H \vee F_{i+1} \vee \ldots \vee F_n$

is derivable in $\mathbf{E}_{fdf}$. The same holds for (TE).

The part for (TE) is trivial; and for R1–R11, obviously, we can move any disjunctive part to the extreme right, using commutation and association, i.e., (TE). Then, since all the rules R1–R11 deal only with the rightmost disjunct, we can use one of these rules, and then by (TE) move the result back to its original position.

FACT 5. If a rule

from G_1
G_2
$\vdots$
G_n

to infer H

is derivable in $\mathbf{E}_{fdf}$, then so is a rule

$$\begin{array}{ll} \text{from} & F_1 \vee \ldots \vee G_1 \vee \ldots \vee F_m \\ & F_1 \vee \ldots \vee G_2 \vee \ldots \vee F_m \\ & \vdots \\ & F_1 \vee \ldots \vee G_n \vee \ldots \vee F_m \\ \text{to infer} & F_1 \vee \ldots \vee H \vee \ldots \vee F_m. \end{array}$$

The proof is left to the reader. The only features of the system required for the proof are Fact 4 together with the fact that if G is an axiom, then by (TE), G flanked disjunctively by F's is a theorem.

FACT 6. If $A \rightleftarrows B$ is a tautological entailment, then $(\ldots A \ldots)$ and $(\ldots B \ldots)$ are interderivable in $\mathbf{E}_{fdf}$.

The course of the proof, details of which we omit, is as follows:

(i) If $A \rightleftarrows B$ is provable, then so is $(\ldots A \ldots) \rightleftarrows (\ldots B \ldots)$, where the dots represent truth functional contexts only.

(ii) Hence, the provability of $A \rightleftarrows B$ implies that corresponding members of each of the following pairs are interderivable:

$$\begin{array}{ll} (\ldots A \ldots) \rightarrow C & (\ldots B \ldots) \rightarrow C \\ C \rightarrow (\ldots A \ldots) & C \rightarrow (\ldots B \ldots) \\ \overline{(\ldots A \ldots) \rightarrow C} & \overline{(\ldots B \ldots) \rightarrow C} \\ \overline{C \rightarrow (\ldots A \ldots)} & \overline{C \rightarrow (\ldots B \ldots)} \end{array}$$

(iii) Hence, by Fact 5, the same holds when the members of the pairs are flanked by disjuncts.

(iv) Lastly, one can use a disjunctive-normal-form argument to show that the same holds for arbitrary truth functional contexts around the members of the pairs above, hence, also for any context $(\ldots A \ldots)$ legitimate in $\mathbf{E}_{fdf}$.

FACT 7. (Deduction theorem for material "implication"). If H is derivable from $G_1, \ldots, G_{n-1}, G_n$, then $\overline{G_n} \vee H$ is derivable from $G_1, \ldots, G_{n-1}$.

This follows immediately from Fact 5 together with the following: (1) each of $\overline{G_n} \vee G_1, \ldots, \overline{G_n} \vee G_{n-1}$ is derivable, by (TE), from $G_1, \ldots, G_{n-1}$; (ii) $\overline{G_n} \vee G_n$ is provable.

FACT 8. $C \rightarrow D$ is derivable from 1–4 below:

1 $\overline{B} \& C \rightarrow D \vee B$
2 $B \& \overline{B} \& C \rightarrow D$
3 $C \rightarrow D \vee \overline{B} \vee B$
4 $B \& C \rightarrow D \vee \overline{B}$

Note that 1–4 represent all possible ways of adding B or its negate either to the right or to the left of $C \rightarrow D$, short of sharing. For proof, first let 1 and 2 yield

5 $\bar{B} \& C \rightarrow D$. Fact 3

and 3 and 4 yield

6 $C \rightarrow D \vee \bar{B}$. Fact 3

Then 5 and 6 yield

7 $C \rightarrow D$

as desired, by a third use of Fact 3.

FACT 9. All tautologies are provable in $\mathbf{E}_{\text{fdf}}$.

Reduce the candidate to a conjunctive normal form by double negation, De Morgan, distribution, and Fact 6. Each conjunct will be an excluded middle with extra disjuncts added on, which can be proved by the truth functional axioms and (TE); and (TE) allows proof of the conjunction from its conjuncts.

§19.5. Completeness. That's enough facts. The thing to do now is to show that validity as defined above guarantees provability in $\mathbf{E}_{\text{fdf}}$. To this end we shall use a "normal form" style argument: we shall define for each fdf F an fdf F^* in "normal form" such that (1) F^* is interdeducible with F, hence by the consistency theorem, equivalid with F; and (2) F^* is so normal that it can be shown by direct argument to be invalid if unprovable.

The required normal form reduction is carried out in three stages.

In the first stage, F is reduced to a special sort of conjunctive normal form $F_1 \& \ldots \& F_q$, each conjunct F_i in turn having the form

$Z \vee E \vee N$,

where Z is a zdf, E is a disjunction $E_1 \vee \ldots \vee E_n$ of first degree entailments, and N is a disjunction $N_1 \vee \ldots \vee N_m$ of negations of first degree entailments. Such a formula will be said to be in *special conjunctive normal form*. (The only thing "special" is the left-to-right order: first truth functions (zdfs), then first degree entailments, then negated first degree entailments.)

To find the special conjunctive normal form $F_1 \& \ldots \& F_q$ of F, one merely applies De Morgan, double negation, and distribution to F in the way one usually does to find conjunctive normal forms, and then sorts each conjunct into order $Z \vee E \vee N$ by associativity and commutativity. The rule (TE) guarantees interderivability.

In the second stage, one insures that there is sufficient interaction between truth functional Z and each negated entailment in N as defined by the following notion of "representation": $\overline{A \rightarrow B}$ is said to be *represented in* Z if either A or $\bar{B}$ is a disjunctive part of Z.

To find a formula (still in special conjunctive normal form) such that each negated entailment in N is represented in Z, the appropriate tactic is to replace

$$Z \vee E \vee (\ldots \overline{A \rightarrow B} \ldots)$$

by

$$[(Z \vee A) \vee E \vee (\ldots \overline{A \rightarrow B} \ldots)] \& [(Z \vee \bar{B}) \vee E \vee (\ldots \overline{A \rightarrow B} \ldots)],$$

which are interdeducible in one direction by (TE) and in the other by R11, Fact 4, and (TE).

In the third and final stage we provide for interaction between entailments and negated entailments. (It is instructive to note that in disjunctive contexts there is *no* interaction between two entailments, nor between entailments and truth functions. There *can* be interplay between two negated entailments, as in $\overline{A \rightarrow \bar{A}} \vee \overline{\bar{A} \rightarrow A}$, but in our procedure such interaction is mediated by way of the truth functional segment.)

We say that $\overline{A \rightarrow B}$ is *represented in* $C \rightarrow D$ if any one of the following hold:

$\bar{A}$ and B are conjunctive parts of C.
A and $\bar{B}$ are disjunctive parts of D.
$\bar{A}$ is a conjunctive part of C, and A is a disjunctive part of D.
B is a conjunctive part of C, and $\bar{B}$ is a disjunctive part of D.

These conditions, on a little reflection, will make it clear how astonishingly close the logical tradition beginning with Boole got to the truth.

What needs doing, in order to be sure that every negated entailment in N is represented in every entailment in E, is to replace

$$Z \vee (\ldots (C \rightarrow D) \ldots) \vee (\ldots \overline{A \rightarrow B} \ldots)$$

by a conjunction of the four modes of "representation":

$$[Z \vee (\ldots (\bar{A} \& B \& C \rightarrow D) \ldots) \vee (\ldots \overline{A \rightarrow B} \ldots)] \&$$
$$[Z \vee (\ldots (C \rightarrow . D \vee A \vee \bar{B}) \ldots) \vee (\ldots \overline{A \rightarrow B} \ldots)] \&$$
$$[Z \vee (\ldots (\bar{A} \& C \rightarrow . D \vee A) \ldots) \vee (\ldots \overline{A \rightarrow B} \ldots)] \&$$
$$[Z \vee (\ldots (B \& C \rightarrow . D \vee \bar{B}) \ldots) \vee (\ldots \overline{A \rightarrow B} \ldots)].$$

We illustrate interdeducibility in one direction by deriving the third conjunct from the original formula.

1	$[Z\vee(\ldots\ (C\to D)\ \ldots)\vee(\ldots\overline{A\to B}\ldots)]$	Hypothesis
2	$[Z\vee(\ldots\ (D\to D\vee A)\ldots)\vee(\ldots\overline{A\to B}\ldots)]$	Fact 2 (TE)
3	$[Z\vee(\ldots\ (C\to D\vee A)\ldots)\vee(\ldots\overline{A\to B}\ldots)]$	1 2 (TE) R1 Fact 4
4	$[Z\vee(\ldots(\bar{A}\&C\to C)\ \ldots)\vee(\ldots\overline{A\to B}\ldots)]$	Fact 2 (TE)
5	$[Z\vee(\ldots(\bar{A}\&C\to D\vee A)\ldots)\vee(\ldots\overline{A\to B}\ldots)]$	3 4 (TE) R1 Fact 4

Interdeducibility in the other direction relies principally on Fact 8 to show that $C\to D$ is derivable, as follows, from the five listed hypotheses:

1	$\bar{A}\&B\&C\to D$	Hypothesis
2	$C\to D\vee A\vee\bar{B}$	Hypothesis
3	$\bar{A}\&C\to D\vee A$	Hypothesis
4	$B\&C\to D\vee\bar{B}$	Hypothesis
5	$A\to B$	Hypothesis
6	$D\vee A\to D\vee B$	5 Facts 1–2
7	$\bar{B}\&C\to\bar{A}\&C$	5 Facts 1–2
8	$\bar{B}\&C\to D\vee B$	3 6 7 trans Fact 1
9	$B\&\bar{B}\&C\to D$	1 as 8 from 3
10	$C\to D\vee\bar{B}\vee B$	2 as 8 from 3
11	$C\to D$	8 9 10 4 Fact 8

Therefore, by Fact 7,

$$\overline{A\to B}\vee C\to D$$

is derivable from 1–4. Now the desired conclusion follows by Fact 5 and (TE), which completes the proof of interdeducibility.

We have, as a consequence of these three stages, defined for each F a formula F^* such that (1) F^* is in special conjunctive normal form; (2) where the i-th conjunction F_i of F^* has the form $Z\vee E\vee N$ (Z truth functional, E a disjunction of entailments, N a disjunction of negated entailments), every negated entailment in N is represented in Z, and every negated entailment in N is represented in every entailment in E; and (3) F^* is interdeducible with (hence equivalid with) F. Let us say that F^* is in *first degree normal form*, and is *the first degree normal form of* F (ignoring fine points of uniqueness). What we now show is that every formula in first degree normal form is invalid if unprovable. Because a conjunction is provable iff its conjuncts are, and valid iff its conjuncts are, it suffices to prove for each conjunct $Z\vee E\vee N$ that it is invalid if unprovable, which we now proceed to do.

Let the disjunction of entailments E in $Z\vee E\vee N$ be $E_1\vee\ldots\vee E_n$. We find an M^n model (n = the number of entailments) $\mathrm{Q} = \langle\mathrm{M}^n, \mathrm{s}\rangle$ which makes each disjunct, hence the whole, false in Q. (M^n is defined in §18.5.)

Since the candidate is unprovable, (1) Z is not a tautology by Fact 9 and (TE); and (2) E_i is not provable in $\mathbf{E}_{\mathrm{fde}}$ by Fact 2 and (TE). Hence by Theorem 5 of §18.8, there is for each $E_i = (C_i\to D_i)$ an M_0-model $\mathrm{Q}_i = \langle\mathrm{M}_0, \mathrm{s}_i\rangle$

such that $v_{Q_i}(Z) \notin F_{+0}$, $v_{Q_i}(C_i) \in F_{-1}$, and $v_{Q_i}(D_i) \notin F_{-1}$. Now componentwise define the s of the desired $Q = \langle M^n, s\rangle$ so that for p a propositional variable,

$$s(p) = \langle s_1(p), \ldots, s_n(p)\rangle,$$

and accordingly for any formula A,

$$v_Q(A) = \langle v_{Q_1}(A), \ldots, v_{Q_n}(A)\rangle.$$

We may assume we use the *same* truth filter T not containing Z from Theorem 4 of §18.8 for each application of Theorem 5 of §18.8, so that $s_i(p)$ and $s_j(p)$ are alike in truth value, as required by the definition of M^n in §18.5. Evidently Z is false in Q, and with almost equal ease we see that E_i is false: it is precisely in the ith position that $v_Q(C_i) \leq v_Q(D_i)$ fails. The only work lies in showing that our efforts to falsify Z and the E_i have willy-nilly falsified N as well by making true each entailment $A \rightarrow B$ whose negation $\overline{A \rightarrow B}$ appears as a disjunct of N.

For $A \rightarrow B$ to be true in Q, we need for each i, $v_{Q_i}(A) \leq v_{Q_i}(B)$. That this indeed holds is due to the way $\overline{A \rightarrow B}$ is "represented" in Z and in E_i. Because there are two modes of representation in Z and four modes of representation in E_i, we shall have eight cases, of which we consider one just by way of example. Suppose A is a disjunctive part of Z, and $\overline{A}$ and B are both conjunctive parts of C_i. Then since $v_{Q_i}(Z) \notin F_{+0}$, this must also be true of its disjunctive part A: $v_{Q_i}(A) \notin F_{+0}$, hence $v_{Q_i}(A) \in \{-1, -2, -0, -3\}$. Also, since $v_{Q_i}(C_i) \notin F_{-1}$, this must also be true of its conjunctive parts $\overline{A}$ and B, so that $v_{Q_i}(A) \in \{+1, -1, +0, -3\}$ and $v_{Q_i}(B) \in \{-1, +1, -0, +3\}$. From the two known facts about A we may conclude $v_{Q_i}(A) \in \{-1, -3\}$. But then a glance at the picture of M_0 in §18.4 suffices to verify that $v_{Q_i}(A) \leq v_{Q_i}(B)$, as required.

This winds up the proof of the

COMPLETENESS THEOREM. E_{fdf} is complete: every valid fdf is provable.

§20. Miscellany. This section is analogous to §8; see the Analytical table of contents for its contents.

§20.1. The von Wright-Geach-Smiley criterion for entailment. We mentioned at the end of §5.1.1 that the valid entailment $A \rightarrow B \rightarrow .B \rightarrow A \rightarrow .B \rightarrow A$ appeared to violate a formal condition laid down by Smiley 1959, which is (as we see it) closely related to the informal suggestions of von Wright and Geach considered in §15.1. For reference we cite again the relevant passages:

> Von Wright 1957: "A entails B, if and only if, by means of logic, it is possible to come to know the truth of $A \supset B$ without coming to know the falsehood of A or the truth of B."

Geach 1958: "... A entails B if and only if there is an *a priori* way of getting to know that $A \supset B$ which is not a way of getting to know whether A or whether B."

Smiley 1959: "$A_1 \& \ldots \& A_n \supset B$ should not only be itself a tautology, but should also be a substitution instance of some more general implication [*sic*] $A_1' \& \ldots \& A_n' \supset B'$, where neither B' nor $\sim(A_1' \& \ldots \& A_n')$ are themselves tautologies."

As remarked in §15.1, it is hard to know exactly how to interpret such epistemological phrases as "coming to know" and "getting to know" in the context of formal logic, but even here some of the techniques previously developed *may* be of assistance.

If we use the analysis of a proof (i.e., the list of excuses written at the right, together with the arrangement thereof) as an analysis of "logical dependence," as we and Suppes 1957 (*malgré lui;* see §5.1) have suggested, then logical dependence might be taken as a clue as to how we "get to know." We do not wish either to elaborate or to defend this proposal, since we feel that it is fundamentally on the wrong track, for reasons to emerge later. But we give an illustration of how it might work in practice.

Consider the analysis

1	{1}	hyp
2	{2}	hyp
3	{3}	hyp
4	{1}	1 reit
5	{1,3}	3 4 →E
6	{2}	2 reit
7	{1,2,3}	5 6 →E
8	{1,2}	3–7 →I
9	{1}	2–8 →I
10		1–9 →I

Without knowing what the formulas 1 through 10 are, we can still tell from the analysis alone that each of the hypotheses was used in the course of arriving at the conclusion, and hence that we can "come to know" (von Wright) that step 1 entails step 9 without "getting to know" (Geach) the falsehood of A or the truth of B. In fact the proof analysis displayed above will work equally well for both $A \to B \to . B \to C \to . A \to C$ and $A \to A \to . A \to A \to . A \to A$. We "come to know" the truth of the consequent of the latter by noticing that it is a theorem, not by examining the analysis of the proof; theoremhood is in this sense independent of "coming to know" the truth of the consequent, since the analysis of the proof gives us no clues.

Contrast this situation with the following analysis of a proof in $\mathbf{FS4}_{\rightarrow}$:

1	├	hyp
2	│ ├	hyp
3	│ │	2 rep
4	│	2–3 →I
5		1–4 →I

We contend that in this case it is patent to the meanest intellect that step 4 is a theorem, regardless of how the formulas are (correctly) filled in. A child in his cradle would grasp the fact, even without knowing whether the analysis had been applied to $B\rightarrow.A\rightarrow A$, or to $A\rightarrow.A\rightarrow A$, or indeed to $A\rightarrow A\rightarrow .A\rightarrow A\rightarrow.A\rightarrow A$, or to any of a host of other examples; the theoremhood of 4 (whatever formula it may be) obtrudes itself on one's consciousness, simply on the basis of inspection of the analysis, in a way which is difficult to ignore.

We are not sure whether this suggestion helps (by providing the germ of a formal analysis of "coming to know") or hinders (by giving yet another guess, perhaps incompatible with others which are equally plausible, as to what might be involved in "coming to know"). We therefore abandon this line, and turn to what looks to us like a de-psychologized, or perhaps de-epistemologized, version of the proposal, namely Smiley's.

Here we are on firmer ground, since Smiley's criterion is stated in terms amenable to formal treatment, and also because he unmistakably takes $A\rightarrow B\rightarrow.B\rightarrow A\rightarrow.B\rightarrow A$ as violating his criterion (Smiley 1959, fn. 23). The criterion is susceptible to two quite distinct interpretations, which we shall discuss under two headings, both of which we refer to as "WGS" criteria, because of the affinities we find in the three quotations mentioned at the outset of this section.

§20.1.1. The intensional WGS criterion. The axiomatic system **S** proposed toward the end of Smiley 1959 has the following axioms and rules (where we write the arrow in place of the horseshoe, and take seriously Smiley's remark p. 251 that "$\rightarrow$, &, and $\sim$ must be taken as independent primitive connectives," though $\vee$ and $\equiv$ are definable as usual with the help of $\sim$ and &).

Axioms: S_1 $A \rightarrow A\&A$
S_2 $A\&B \rightarrow B\&A$
S_3 $A\&(B\&C) \rightarrow (A\&B)\&C$
S_4 $A\&B \rightarrow A$
S_5 $A\&(A\rightarrow B) \rightarrow B$
S_6 $A \rightarrow {\sim}{\sim}A$
S_7 ${\sim}{\sim}A \rightarrow A$

S_8 $\sim B\&(A\rightarrow B)\rightarrow\sim A$
S_9 $A\rightarrow B\rightarrow\sim(A\&\sim B)$
S_{10} $A\&(B\vee C)\rightarrow(A\&B)\vee(A\&C)$

Rules: From $A\rightarrow B$ and $B\rightarrow C$ to infer $A\rightarrow C$
From $A\rightarrow C$ and $B\rightarrow D$ to infer $A\&B\rightarrow C\&D$
From $A\rightarrow B$ to infer $\sim B\rightarrow\sim A$

Evidently **S** is a subsystem of **E**, and equally evidently the arrow is to be taken as a functor which permits nesting. Conspicuous by its absence is transitivity $A\rightarrow B\rightarrow.B\rightarrow C\rightarrow.A\rightarrow C$, which when added to **S** yields

$$A\rightarrow B\rightarrow.B\rightarrow A\rightarrow.B\rightarrow A,$$

which Smiley regards as objectionable, presumably on the grounds that it fails to meet his criterion; i.e., $A\rightarrow B\rightarrow.B\rightarrow A\rightarrow.B\rightarrow A$ is *not* a substitution instance of a tautology with a non-tautologous consequent. However, application of the criterion seems to require (a) interpreting the major connective as the independent functor required for the formulation of **S**, but (b) reading the first, third, fourth, and fifth occurrences of the arrow as material "implication," since we don't know how to apply "tautologous" to formulas with the independent primitive arrow.

No matter how we understand the criterion, however, it is clear that some of Smiley's words may be taken as *favorable* toward $A\rightarrow B\rightarrow.B\rightarrow A\rightarrow.B\rightarrow A$, at least if the proof is stated in the context of $\mathbf{FE}_{\rightarrow}$: ". . . inferences may be justified on more grounds than one, and the present theory requires not that there should be *no* cogent way of reaching the conclusion *without* using all the premisses, but only that there should be *some* cogent way of reaching it *with* all the premisses used" (p. 249).

(Actually, this condition on entailment won't do either, as is easily seen by examples. Proof of the uniqueness of the identity element in an Abelian group does not invoke the fact that the group operation is commutative; it nevertheless follows from the axioms for an Abelian group that the identity element is unique — or at any rate everyone says so. What *is* required is that there be some cogent way of reaching the conclusion with *some* of the (conjoined) premisses used.)

While we admit that what we refer to as the "intensional WGS criterion" may perhaps be patched up, we do submit that it has not been coherently stated (either by us or others). We therefore turn to a more plausible interpretation, which does have some salvageable parts.

§20.1.2. The extensional WGS criterion. Let us now, on the contrary, suppose that Smiley's formal system **S** of the previous section is to be interpreted as a theory of entailment between truth functions, so that the arrow when it appears other than as the major connective is to be thought of as material "implication." Then S_5 and S_8 become

$S_{5'}$ $A\&(\bar{A}\vee B) \rightarrow B$, and
$S_{8'}$ $\bar{B}\&(\bar{A}\vee B) \rightarrow \bar{A}$;

and the disputed formula becomes

$$\bar{A}\vee B \rightarrow \overline{\bar{B}\vee A}\vee\bar{B}\vee A \quad (\text{or } A\supset B\rightarrow .B\supset A\supset .B\supset A).$$

We notice first that from the point of view of §§15 and 16, **S** is too broad; neither $S_{5'}$ nor $S_{8'}$ is a tautological entailment. But on this interpretation the formula which fails to satisfy Smiley also fails to satisfy us. For it will be a tautological entailment only if all of

$\bar{A} \rightarrow B\vee\bar{B}\vee A$
$\bar{A} \rightarrow \bar{A}\vee\bar{B}\vee A$
$B \rightarrow B\vee\bar{B}\vee A$
$B \rightarrow \bar{A}\vee\bar{B}\vee A$

are explicitly tautological entailments, and the first and last of these fail to be such.

This coincidence leads us to ask whether the extensional WGS criterion, while not a sufficient condition for tautological entailmenthood, is at least a necessary condition; the answer proves to be affirmative.

THEOREM. If a first degree entailment $A\rightarrow B$ is provable in E_{fde}, then $A\rightarrow B$ meets the extensional WGS criterion.

PROOF. What the theorem means is that whenever a truth functional A entails a truth functional B, this fact has nothing to do with funny properties of one or the other (e.g., that A is contradictory or B tautologous); if A entails B, then that fact hangs on their meaning connections. There are several ways to show this, but the germ of all of them is in the fact that *explicitly tautological primitive entailments* (§15.1) have the feature that negation is irrelevant to their validity: $p\&\bar{q}\rightarrow\bar{q}\vee r$ is no better nor worse than $p\&q\rightarrow q\vee r$. For the proof we choose methods which are unfair in that they rely on facts not established until two sections hence; but such is life.

The theorem then relies on two observations.

(1) A necessary condition for A to be a contradiction $\{B$ to be a tautology$\}$ is that, in the language of §22 1.1, some variable p occur as both an antecedent and a consequent part of A $\{B\}$. (The condition is not, of course, sufficient: $p\&(\bar{p}\vee q)$ is no contradiction.)

(2) Whether or not $A\rightarrow B$ is a tautological entailment depends only on matching (or lack of matching) between antecedent parts of A and ante-

cedent parts of B, and between consequent parts of A with consequent parts of B.

So we proceed as follows. First make a list of all variables occurring at least once as an antecedent part of either A or B. Select for each of these variables p a brand new variable p' which is foreign to both A and B. Now replace all antecedent occurrences of p in A or B by p', but leave the consequent occurrences alone, yielding a formula $A' \rightarrow B'$ of which $A \rightarrow B$ is clearly a substitution instance. Then by the first observation A' cannot be a contradiction, nor B' a tautology, since in neither A' nor B' does any variable occur both antecedently and consequently; and since $A \rightarrow B$ is a tautological entailment, the second observation guarantees that $A' \rightarrow B'$ will be one as well. For any possible match between antecedent parts of A and B, or between consequent parts of A and B, will be preserved in corresponding matches between new antecedent parts, or old consequent parts, of A' and B'. So since $A' \rightarrow B'$ is a tautological entailment, *a forteriori* $A' \supset B'$ witnesses satisfaction by $A \rightarrow B$ of the extensional WGS criterion.

We now need add only that $A \rightarrow B$ is provable in $\mathbf{E}_{\text{fde}}$ iff $A \rightarrow B$ is a tautological entailment. Without needing to, we add further that this theorem seems to us to give a complete vindication of the WGS intuitions.

§20.2. A howler. We follow Maxwell 1959 in thinking of a howler as "an error which leads innocently to a correct result." Such we committed while trying to prove a theorem in Anderson and Belnap 1963, where (p. 317, line 6 from the bottom) we said ". . . we may be sure . . ." of something that is in fact false. The bad proof there given has (we hope) been rectified in §19, but it might be illuminating to see how the thing went sour.

The problem considered in that paper had to do with first degree entailments with individual quantifiers, a topic to which we shall return in Chapter VII. But the mistake in the proof can be illustrated at a more elementary level.

We were trying to show that first degree formulas of the form

$$1\ (A_1 \rightarrow B_1) \vee \ldots \vee (A_n \rightarrow B_n)$$

have the property that they are unprovable if each $A_i \rightarrow B_i$ is. Supposing that each disjunct is not provable, we would like to find a falsifying assignment for 1. The point then was to falsify each $A_i \rightarrow B_i$ by falsifying every formula in the bad branch of its "proof." To this end we suggested giving a value (say p_1) to each variable situated to the left of the first arrow (i.e., a part of A_1 in $A_1 \rightarrow B_1$) just in case neither that variable nor its denial occurred on the right of the arrow — the intent being that in the case of

$$(A \& \bar{A} \rightarrow B) \vee \ldots$$

we could then falsify the entailment by giving B a value which had nothing to do with the value given to A.

But the plan fizzled, as can be seen by consulting the obviously unprovable formula

$$(s\&\bar{s}\&t\&\bar{t}\rightarrow u)\vee(s\rightarrow t).$$

The assignment falsifies the left half all right; we can give p_1 to s and t, and p_2 to u. But in the same act we are giving Truth to the right half, thus counterexampling our putative proof.

Life is full of hazards, among which one must number making logical blunders in print; *nostra culpa* — we beg absolution, particularly since, though this proof won't work, the theorem is still correct.

§20.3. Facts and tautological entailments (by Bas van Fraassen). Not very long ago, facts (and their various relatives in the philosophical entourage) played a central role in the explication of logical relationships. But today the prevalent opinion seems to be that facts belong solely to the prehistory of semantics and either have no important use or are irredeemably metaphysical or both. In this section we shall explore first what kinds of facts must be countenanced if we are to take them seriously at all and, second, what we can do with them once we have them. We argue that there are several tenable positions concerning what kinds of facts there are, but we reach two main conclusions which are independent of these positions. The first is that facts can be represented within the framework of standard metalogic; the second is that facts provide us with a semantic explication of tautological entailment. These seem to us to be sufficient reasons to take facts seriously, and we shall argue in §20.3.1 that doing so involves no objectionable metaphysical commitment.

§20.3.1. Facts. "The first truism to which I wish to draw your attention" Bertrand Russell said in his 1918 lectures on Logical Atomism, ". . . is that the world contains facts . . . ." And he added, "When I speak of a fact . . . I mean the kind of thing that makes a proposition true or false" (Russell 1956, p. 182). But what facts are there? A most generous answer would consist in allowing that, if A is any sentence, then (*the fact*) *that A* names a fact. As is well known, Russell himself argued against this answer; but, to begin, let us consider the consequences of generosity.

The first consequence is clearly that some facts obtain (are the case), some facts do not obtain, some must obtain, and some cannot obtain. This notion of *obtaining*, or being the case, is somewhat like that of existence. Indeed, it may plausibly be held that "X is the case" means simply "X exists and X is a fact." Thus, Whitehead and Russell 1910–1913 (using "complex" where

Russell would later use "fact") say that an elementary judgment (i.e. an atomic statement) "is true when there is a corresponding complex, and false when there is no corresponding complex" (p. 44). We face here a difficulty (avoided by Russell) since in the terminology of many philosophers it makes no sense to say that some things do not exist, and, so, presumably, no sense to say that some facts do not, or cannot, obtain. But we have argued elsewhere (van Fraassen 1967) that sense can be made of it, and we shall use these expressions without further comment.

In what follows we shall everywhere accept that some facts obtain and others do not. But we shall not begin by committing ourselves to what we have called the "generous answer." Instead we wish to follow first Russell's procedure of admitting only such facts as we find ourselves forced to admit, given that we wish to have a viable theory of facts.

As the weakest possible principle of any theory of facts, we offer the following minimal explication of Russell's "first truism":

1. The truth value of a sentence is determined by the facts that are the case.

The question is now what kinds of facts there must be for this principle to hold. The minimal commitment would appear to be to "atomic facts," the *complexes* of **Principia Mathematica**: "We will give the name of 'a *complex*' to any such object as 'a in the relation R to b' or 'a having the quality q'...." The atomic fact, a's bearing relation R to b, is the case if and only if the atomic sentence *aRb* is true. It is important to see, however, that 1 does not require us to say that there is any other kind of fact. For with respect to more complex sentences, we can now give the usual truth definitions in terms of their components:

Not-*A* is true if *A* is not true, (*A* and *B*) is true if *A* and *B* are both true, $\forall xA$ is true if *A* is true for all values of *x*, and so on.

But this makes nonsense of any theory of facts that refuses to go beyond 1, or, correlatively, that seeks only to define truth conditions. For, of course, we do not need facts to define truth conditions for atomic sentences either: *bRc* is true if and only if b bears R to c. So we must look for a relation between sentences and facts more intimate than the relation defined by "*A* is true if and only if fact X obtains." And our first clue here is the remark added by Russell to his "first truism": a fact is the kind of thing that *makes* a sentence true.

As a less trivial explication of Russell's truism we therefore propose

2. A sentence *A* is true {false} if and only if some fact that makes *A* true {false} is the case.

If *A* is an atomic sentence, say *bRc*, and *A* is true, this still leads us only to the conclusion that the atomic fact of b's bearing R to c is the case. But what if *bRc* is false? Russell 1956, pp. 211–214, reports here on his (moderately famous) debate with Demos. The latter argued that one need not postulate "negative facts," for if *bRc* is false, this is because there is some (positive) fact that rules out that b bears R to c. Russell objected to this for various reasons.

Let us rephrase the question this way: suppose *A* is atomic, and not-*A* is true, made true by fact e. Is there then some other atomic sentence *B* that is made true by e? It appears that Demos answered "Always," and Russell "Never." But surely the answer depends on the structure of the language, specifically on the set of predicates. If some of these predicates have disjoint extensions or, better yet, have necessarily disjoint extensions, then the answer may sometimes be "Yes" and sometimes "No." Russell held, of course, that there is a unique "ideal language," of which the predicates express logically independent properties, from which point of view his answer is correct. But this atomism does not seem essential to the theory of facts, and it will suffice for us to say that the answer to this question depends on the structure of the language.

Russell argues secondly that there are no disjunctive facts; that is, that we need not postulate special facts whose function it is to make disjunctions true. For (*A* or *B*) will be made true by any fact that makes *A* true as well as by any fact that makes *B* true; and if no such facts obtain, then it is not true. It is not clear whether Russell sees any new problem arising from a false disjunction; he says that the truth value of (*A* or *B*) depends on two facts, one of which corresponds to *A* and one to *B*. We may take this to mean either that Russell vacillated between 1 and 2 or that he would have rejected 2. For if we accept 2, this does lead to a problem: if there are only atomic facts, there is no fact that makes (*A* or *B*) false.

Since (*A* or *B*) is false if and only if (not-*A* and not-*B*) is true, we may rephrase this as: if there are only atomic facts, then there is no fact that makes any conjunction true. Acceptance of 2, therefore, implies the acceptance of *conjunctive facts*. For every two facts e and e′ there is a conjunctive fact e·e′ that is the case if and only if both e and e′ obtain.

And we face the same problem, essentially, for the quantifiers. (The suspicion that Russell 1956 vacillated on his basic principles concerning facts is reinforced by his immediate postulation of both existential and universal facts (pp. 236–237) after having denied molecular facts; he admits that this may not be consistent (p. 237).) There is no reason (as yet) to admit existential facts, for if some fact makes *A* true (for some value of *x*), then it also makes $\exists xA$ true. But we must say that there are universal facts, if we accept 2, these universal facts being somewhat like infinite conjunctions.

Now it appears that we have reached (with Russell) a position where we can say, defensibly, that we have admitted into our ontology only such facts as we found it necessary to admit. But there are, of course, various possible ways to challenge that. It must be noted especially that Russell was not the only philosopher to attempt a semantic explication of logical and metalogical relationships through a theory of facts. For example, C. I. Lewis developed such a theory — for what Lewis calls "states of affairs" seems quite definitely to be the kind of thing that Russell called "facts." And Lewis's basic principles seem not to have been 1 and 2; indeed, he does not appear to address himself at all to the relation of *making true*.

Lewis 1943 says that a sentence *signifies* a state of affairs. He explains this to mean that the use of a sentence to make an assertion "*attributes* the state of affairs signified to the actual world" (p. 242). Besides the significance of a sentence, Lewis also discusses its "denotation," "comprehension," and "intension." From his discussion it appears that in standard logic we deal with denotations, and in modal logic with comprehensions and intensions. But Lewis did not develop the account of signification very far; he does not consider the question whether there are not some logical relationships for the explication of which signification is needed. It seems clear, however, that instead of the Russellian principle, 2, Lewis accepted

> 3. A sentence *A* is true if and only if every fact that *A* describes as being the case (or *signifies*) is the case, and false if and only if every fact that *A* describes as not being the case, is the case.

The question is now whether acceptance of 3 rather than 2 leads to an *essentially* different theory of facts.

For an atomic sentence *A*, the fact signified by *A* would seem to be exactly that fact which makes *A* true (if it obtains). But for molecular sentences one sees an obvious difference. (*A* and *B*) presumably describes as being the case whatever is so described by *A* and whatever is so described by *B*. There is, therefore, no need now to postulate conjunctive facts. On the other hand (*A* or *B*) cannot describe as being the case any fact so described by *A* or by *B* (in general); so now we must postulate disjunctive facts.

This is not surprising: signification is a relation "dual" to *making true*, and principle 3 is dual to 2; so the consequences are also dual to each other. But this is a bothersome problem for anyone who is seriously considering the question of which kinds of facts to admit into his ontology. For Russell's argument that we can do without disjunctive facts is good reason not to admit those, and the argument on the basis of Lewis's theory that we can do without conjunctive facts, is good reason not to admit those. But if we admit neither, our theory of facts is codified in principle 1 alone, and hence trivial; if we admit either to the exclusion of the other, we are arbitrary; and if we admit both, we are generous rather than parsimonious.

To cut this Gordian knot, we propose retaining our ontological neutrality, and will treat facts as we do possibles: that is, explicate "fact" discourse in such a way that engaging in such discourse does not involve ontological commitment. This means that we must represent facts, relations among facts, and relations between facts and sentences; this representation can serve to explicate fact discourse without requiring the claim that it also represents a reality. (Indeed, such a claim would, if unqualified, be necessarily false; for we wish to explicate discourse about non-existents and impossibles as well as about existents.) The nature of the representation is of course dictated by methodological considerations; unlike the ontologist, we cannot be embarrassed by achieving parsimony at the cost of being arbitrary. Purely for convenience our representation will be Russellian; and because we know that Lewis's approach is the dual of this, and the generous policy admits simply the sum of what is admitted by both Russell and Lewis, we can be assured that our results will be independent of this arbitrary choice.

We have so far talked about relations between facts and sentences, and, before we go on, we need to take a look at relations among facts. Suppose we say, with Whitehead and Russell, that to the atomic sentence *aRb* there corresponds the complex *that*-aRb. The first question is whether this determines entirely the class of complexes. To put it more clearly: are we to conceive of complexes as language-dependent entities, so that every complex corresponds to an atomic sentence? Or are we to say that there is a complex *that*-aRb whether or not the relation R and the individuals a, b are named by expressions of the language? Russell's debate with Demos suggests that he accepts the former (and we argued that, even granting that, he did not win the argument). For if we accept the latter, then for every complex *that*-aRb there is also a complex *that*-a$\overline{\text{R}}$b, which makes *aRb* false; and there will be no need to postulate any negative facts.

From here on we shall accept the latter course, partly for the convenience of not having to admit special negative facts, and partly because facts have traditionally been held to be independent of what anyone may think, or say, or be able to say, about them. The representation of the complex *that*-aRb may now conveniently be achieved by identifying it with the triple ⟨R, a, b⟩:

4. A *complex* is an $(n+1)$-tuple, of which the first member is a relation of degree n.

Various relations among complexes are now easy to define; for example, complexes e and e′ are incompatible (in the sense of Demos) if they differ only in that their first members are disjoint relations. For the explication of logical relationships, however, it is much more interesting to look at relations among molecular facts.

To represent the conjunctive fact whose components are complexes d and d′, we propose we use simply the set {d, d′}. This is convenient, for a universal fact can then be represented in the same way by just allowing an infinite set of complexes to be a fact also. (This would not be so easy if we had not decided that a complex need not correspond to a sentence; for, after all, all individuals might not be named in the language.) We shall make this precise in the next section, but in the meantime we adopt:

5. A *fact* is any non-empty set of complexes, and the fact e·e′ is the union of the two facts e, e′.

Clearly, the fact {d} plays the same role as the complex d, and so we have a certain amount of redundancy in our representation; from here on we shall find it convenient to say that the fact {⟨R, a, b⟩} rather than the complex ⟨R, a, b⟩ is what makes *aRb* true.

There is clearly an intimate relation between e·e′ and its components e, e′ both in the theory of facts and in our representation. We shall say that e·e′ *forces* e and e′, and this relationship is then represented by the relation of inclusion as subsets. (The term "forces" has already a use in metamathematics, but confusion seems hardly likely; and there are some analogies that make this adaption not too inappropriate.) Thus e forces e′ if and only if e′ is (represented by) a subset of e. Clearly e forces e, and we also have that if e forces both e′ and e″, then e forces e′·e″. We shall say that a sentence is made true {false} *in the wider sense* by any fact that forces some fact that makes it true {false}. After these remarks, the formal representation, to which we now turn, is straightforward. (In this we are much indebted to McKinsey 1948–49 and Schock 1962.)

§20.3.2. And tautological entailments. In the standard interpretation of a first order predicate language, truth is defined for sentences relative to a model and to an assignment of values to the variables. A model M comprises a domain D and relations R_1, R_2, . . . on that domain. Intuitively speaking, D is the set of existents involved in some (possible) situation, and R_i is the extension of the *i*th atomic predicate P_i in that situation. The values d(x) of the variables x are chosen from that domain, and the atomic sentence $P_i x_1 \ldots x_n$ is true (in M, relative to d) if ⟨d(x_1), . . . , d(x_n)⟩ belongs to R_i, and false otherwise. The truth values of complex sentences are defined in the familiar way, which we need not recount here. The question now is how we can represent in M the facts that constitute the situation which, intuitively speaking, M is meant to represent.

Following the intuitive remarks in the preceding section, we call a *complex in* M any ($n+1$)-tuple, whose first member is an n-ary relation on D and whose other members are members of D. We call a *fact in* M any non-empty set of complexes in M. We designate the union of facts e_1, . . . , e_n as $e_1 \cdots e_n$, and call this a conjunctive fact with components e_1, . . . , e_n. We say that e *forces* e′ in M if both e and e′ are facts in M, and e′ is a subset of e.

6. A complex $\langle \mathrm{R}, \mathrm{b}_1, \ldots, \mathrm{b}_n \rangle$ in M *is the case* (or obtains) if and only if $\langle \mathrm{b}_1, \ldots, \mathrm{b}_n \rangle \in \mathrm{R}$, and a fact e in M *is the case* (or obtains) if and only if all its members are the case (or obtain).

Turning now to the subject of truth and falsity, we shall define for every sentence A the set $\mathrm{T}(A)$ of facts that make A true in M and the set $\mathrm{F}(A)$ of facts that make A false in M (always relative to an assignment d, indicated by a subscript when necessary).

First, there is exactly one fact that makes $P_i x_1 \ldots x_n$ true in M (relative to d)—namely, $\langle \mathrm{R}_i, \mathrm{d}(x_1), \ldots, \mathrm{d}(x_n) \rangle$—and exactly one fact that makes it false in M—namely, $\langle \overline{\mathrm{R}}_i, \mathrm{d}(x_1), \ldots, \mathrm{d}(x_n) \rangle$, where we designate the complement of R in M as $\overline{\mathrm{R}}$. To define the sets $\mathrm{T}(A)$ and $\mathrm{F}(A)$ for a complex sentence A, some new notation may be helpful. When X and Y are two sets of facts, we shall call the *product* $\mathrm{X} \cdot \mathrm{Y}$ of X and Y the set of facts $\mathrm{e} \cdot \mathrm{e}'$ such that e is in X and e′ is in Y. The product of an infinite family of sets of facts is defined analogously. Note that such a product is still a set of facts. Second, we shall symbolize not-A as $\sim A$, (A and B) as $(A \& B)$, and (A or B) as $(A \vee B)$. The latter will be thought of as defined in terms of $\sim$ and &, and it will be seen that this yields the correct result for the sets $\mathrm{T}(A \vee B)$ and $\mathrm{F}(A \vee B)$.

7. $\mathrm{T}(\sim A) = \mathrm{F}(A)$; $\mathrm{F}(\sim A) = \mathrm{T}(A)$
$\mathrm{T}(A \& B) = \mathrm{T}(A) \cdot \mathrm{T}(B)$
$\mathrm{F}(A \& B) = \mathrm{F}(A) \cup \mathrm{F}(B)$
$\mathrm{T}_{\mathrm{d}}((x)A) =$ the product of the sets $\mathrm{T}_{\mathrm{d}'}(A)$ such that d′ is like d except perhaps at x
$\mathrm{F}_{\mathrm{d}}((x)A) =$ the union of the sets $\mathrm{F}_{\mathrm{d}'}(A)$ with d′ as above

We shall not have much occasion to use the definitions for the case of quantified sentences, and from here on we shall not treat this case in detail. The sets $\mathrm{T}(A)$ and $\mathrm{F}(A)$ are related to the sentence A as follows:

8. A is *true* (in model M) if and only if some fact in $\mathrm{T}(A)$ obtains, and *false* (in model M) if and only if some fact in $\mathrm{F}(A)$ obtains.

Recalling our notions of forcing, let us designate as $\mathrm{T}^*(A)$—respectively, $\mathrm{F}^*(A)$—the set of all facts that force some fact in $\mathrm{T}(A)$—respectively, in $\mathrm{F}(A)$. The members of $\mathrm{T}^*(A)$ make A true in the wider sense. It is easy to see that 8 still holds if we replace $\mathrm{T}(A)$ and $\mathrm{F}(A)$, respectively, by $\mathrm{T}^*(A)$ and $\mathrm{F}^*(A)$.

Facts will hardly be of interest if they serve only to redefine truth in a model; if facts are to have a use, they must serve to define interesting new semantic relations. And here the most promising avenue of approach would seem to be the replacement of the notion of "being true" by that of "being made true." Specifically, let us consider the relation of *semantic entailment* which is defined in terms of "being true." We say that A semantically entails

B ($A \Vdash B$) if, whenever A is true, so is B. (More precisely: if A is true in model M, then B is true in M.) To this corresponds then the tighter relationship:

9. Whatever makes A true, also makes B true.

We shall abbreviate this as $A \mid\mid\vdash B$. Were we to take 9 to mean that T(A) is included in T(B), this is a very tight relationship which even A&B does not bear to A, although A bears it to $A \vee B$. But if we explicate 9 as

10. $A \mid\mid\vdash B$ if and only if T*(A) is included in T*(B) in any model,

then this relationship turns out already to have a name: it is *tautological entailment*. This is a very encouraging result, for this relation did not previously have a semantic explication within the confines of standard model theory. (Dunn 1966 showed that tautological entailment could be explicated in terms of the topics that sentences are "about," but this relation of *being about* was not further explicated.) Before showing that 10 really does define tautological entailment, we shall make some intuitive remarks about this.

C. I. Lewis introduced strict implication to eliminate the paradoxes of material implication. For example, if A is true, then B materially implies A, but this does not hold for strict implication. However, if A is a tautology, then B strictly implies A. This, in effect, §16 argued to be a fallacy of relevance: the premise of an inference should be relevant to its conclusion, and the conclusion's being tautological does not make it so. So according to §16, A entails $A \vee \sim A$, and so does $\sim A$; but not just any sentence whatsoever. With this, our principle 10 agrees: what makes $A \vee \sim A$ true is exactly what makes A true if A is true, or what makes $\sim A$ true if $\sim A$ is true. But what makes B true has, in general, nothing to do with what makes A or $\sim A$ true.

We shall now prove that we have here an explication of tautological entailment, in two steps. In this proof we shall feel free to use the terminology explained in §15. Also, as in §15, we confine ourselves to propositional logic. It may be helpful, before we proceed with the proof, to point out that $T^*(A) \subseteq T^*(B)$ if and only if each member in T(A) forces some member in T(B). For if $T^*(A) \subseteq T^*(B)$ then $T(A) \subseteq T^*(B)$, so each member of T(A) forces some member of T(B); conversely, if the latter is the case, then, by the transitivity of forcing, the former follows. This will make the proofs shorter.

THEOREM 1. If A tautologically entails B, then $A \mid\mid\vdash B$.

PROOF. (a) If this is a primitive entailment with A being $A_1 \& \ldots \& A_n$ and B being $B_1 \vee \ldots \vee B_m$, then let T(A) be $\{e_1 \cdots e_n\}$ and let T(B) be $\{e'_1, \ldots, e'_m\}$. Since $A_i = B_j$ for some i and j, $e_i = e'_j$, so whatever forces $e_1 \cdots e_n$ also forces some member of T(B), namely e'_j. From this it follows at once that T*(A) is included in T*(B).

(b) If the entailment is in normal form, with A being $A_1 \vee \ldots \vee A_n$ and B being $B_1 \& \ldots \& B_m$, then each A_i tautologically entails each B_j. So we must validate the rules:

11. (i) if $A \Vdash B$ and $A \Vdash C$, then $A \Vdash B\&C$;
    (ii) if $A \Vdash C$ and $B \Vdash C$, then $A \vee B \Vdash C$.

Well, if T*(A) is included in both T*(B) and T*(C), then each member e of T(A) forces some member e_1 of T(B) and also some member e_2 of T(C). But then e forces $e_1 \cdot e_2$, which belongs to T($B\&C$). Second, let T*(A) and T*(B) both be included in T*(C). Then if e belongs to T($A \vee B$), it must belong to either T(A) or T(B), hence it forces some member of T(C). Thus both rules are valid.

(c) To allow every tautological entailment to be brought into normal form, we must validate the replacement rules: Commutation, Association, Distribution, Double Negation, and De Morgan. Of these, all but Distribution are immediate; we shall prove one part of Distribution as a representative example: $T(A\&(B\vee C)) = T(A)\cdot T(B\vee C) = T(A)\cdot(T(B)\cup T(C)) = (T(A)\cdot T(B))\cup(T(A)\cdot T(C)) = T((A\&B)\vee(A\&C))$. The third step is the important one; it follows from our definition of the product of two sets of facts.

THEOREM 2. If A does not tautologically entail B, then $A \Vdash B$ is not valid.

PROOF. Because of the replacement rules, we need to consider only entailments in normal form. If such an entailment is not a tautological entailment, its antecedent has a disjunct A and its consequent a conjunct B such that A is a primitive conjunction $A_1 \& \ldots \& A_n$ and B a primitive disjunction $B_1 \vee \ldots \vee B_m$ where $A_i \neq B_j$ for any i and j. It is easy to check that we can then choose facts in a model such that the single member of T(A_i) does not force the single member T(B_i), for any i and j.

In conclusion, we make two comments. The first is that our notion of fact can be extended to cover situations dealt with in non-classical logics. For example, in modal logic we would say that a complex is an $(n+2)$-tuple, where the first member is an n-ary relation, and the last member a possible world. (In many-valued logic, its last member could be an element of the logical matrix.) The facts that make Necessarily-A true would then be the conjunctions of facts that make A true in the various possible worlds; for Necessarily-A is true if and only if A is true in world a_1, *and* in world a_2, and so forth. Thus the relation of tautological entailment can be semantically extended to the case of a modal language; and it may be interesting to investigate its properties there. (For motivation, see the last section of van Fraassen

1969; for applications to the logic of adverbs and the construction of a theory of properties along corresponding lines, see van Fraassen 1973.)

Our second comment is that the question whether the theory of facts is philosophically interesting seems to us to hinge on the further question of what this theory can do regardless of which of the possible approaches to it we follow. For example, if we could explicate tautological entailment only by following Russell's rather than Lewis's approach, then this would not be interesting at all. But as we have seen, we can represent facts within the standard logical framework, and use the relations among them to explicate tautological entailment, regardless of the approach adopted.

CHAPTER IV

THE CALCULUS E OF ENTAILMENT

§21. $\mathbf{E} \simeq \mathbf{E}_{\overline{\rightarrow}} + \mathbf{E}_{fde}$. Thus far we have been concerned with fragments of the full calculus **E** to be considered in this chapter: the entailment fragment, the entailment-and-negation fragment, tautological entailments, and first degree formulas. The point of discussing these parts of the system **E** in such detail was twofold; on the one hand it was easier in that way to isolate and discuss various philosophical motivations for the construction of **E**, and second, we were able to provide solutions for those fragments to some standard questions, such as decidability (for $\mathbf{E}_{\rightarrow}$, $\mathbf{E}_{\overline{\rightarrow}}$, and first degree formulas) and completeness.

Our object in this chapter is to combine the calculus of entailment with negation of Chapter II with the calculus of tautological entailments of Chapter III to obtain a system which retains features of each — especially the avoidance of fallacies. A straightforward way to accomplish this suggests itself immediately: we could simply add to $\mathbf{E}_{\overline{\rightarrow}}$ axioms corresponding to the axioms and rules of the system $\mathbf{E}_{fde}$ to §15.2, thus making explicit the claim that the inferences used in reducing tautological entailments to explicitly tautological entailments are valid. In effect this is what we do, but it develops that combining $\mathbf{E}_{\overline{\rightarrow}}$ with $\mathbf{E}_{fde}$ proceeds most smoothly if certain minor changes are made in the course of consummating the marriage, so that **E** is only approximately equal ($\simeq$) to $\mathbf{E}_{\overline{\rightarrow}} + \mathbf{E}_{fde}$; these changes we discuss after describing **E** axiomatically.

NOTATION. Formulas are described as at the beginning of §15, as are our conventions for discussing them. In brief, we have a stock of propositional variables, and construct formulas by &, ∨, $\overline{}$, and →.

§21.1 Axiomatic formulation of E. The set of theorems can be characterized axiomatically in a number of ways, left to be considered under "axiom-chopping" in §26.1. For present purposes it will suffice to state our favorite set of axioms and rules.

POSTULATES FOR THE SYSTEM **E** OF ENTAILMENT

Entailment.

E1 $A \rightarrow A \rightarrow B \rightarrow B$

E2 $A \rightarrow B \rightarrow . B \rightarrow C \rightarrow . A \rightarrow C$

E3 $(A \rightarrow . A \rightarrow B) \rightarrow . A \rightarrow B$

Conjunction.

E4 $A\&B \to A$
E5 $A\&B \to B$
E6 $(A\to B)\&(A\to C)\to.A\to(B\&C)$

Distribution of necessity over conjunction.

E7 $\Box A\&\Box B \to \Box(A\&B)$ $[\Box A =_{\mathrm{df}} A\to A\to A.]$

Disjunction.

E8 $A \to A\vee B$
E9 $B \to A\vee B$
E10 $(A\to C)\&(B\to C)\to.(A\vee B)\to C$

Distribution of conjunction over disjunction.

E11 $A\&(B\vee C) \to (A\&B)\vee C$

Negation.

E12 $A\to\overline{A}\to\overline{A}$
E13 $A\to\overline{B}\to.B\to\overline{A}$
E14 $\overline{\overline{A}}\to A$

Rules:

$\to$E: given $A\to B$, from A to infer B.
&I: from A and B to infer $A\&B$.

We shall not bother to prove any theorems of **E** from the axiomatic formulation just given; proofs using only $\to$E and &I, while not altogether uninteresting, are difficult and time-consuming to construct, and tedious to check. We prefer to leave proofs to the reader after we establish connection between the axiomatic treatment here given and the equivalent Fitch-style formulation of §23, in which proofs are very easy. In the interim, however, we will occasionally mention certain formulas as being provable in **E**, hoping that the reader will believe us (or when in doubt check our claims with the techniques of §23.)

§21.2. Choice of axioms. As will emerge in §23, there is a sense in which the claim that $\mathbf{E} \simeq \mathbf{E}_{\overset{\sim}{\to}} + \mathbf{E}_{\mathrm{fde}}$, which provided the title for this section, can be strengthened to $\mathbf{E} = \mathbf{E}_{\overset{\sim}{\to}} + \mathbf{E}_{\mathrm{fde}}$, but appreciation of this fact depends on a natural deduction form of **E**, and from the point of view of the axiomatic formulation the equality looks only approximate, since e.g. neither E6 nor E7 appears in $\mathbf{E}_{\overset{\sim}{\to}}$ or $\mathbf{E}_{\mathrm{fde}}$. In this section we consider only those features of **E** which have not already been discussed in connection with subsystems of **E** in previous chapters. Since, as we shall leave it to the reader to check, De Morgan's laws hold for **E** in full generality, we may without loss consider disjunction to be defined, and let our remarks about conjunction deal, *mutatis mutandis*, with disjunction as well.

E1–E3 and E12–E14, together with →E, give us exactly $\mathbf{E}_{\rightarrow}$, and axioms E4–E5 (with the dual E8–E9), and E11 are familiar from $\mathbf{E}_{\mathrm{fde}}$. There remain only the axioms E6 (for conjunction on the right of the arrow), the rule of adjunction, and E7 for consideration in this section. We deal with these under two subheadings.

§21.2.1. Conjunction. The axioms E4 and E5 ($A\&B\rightarrow A$ and $A\&B\rightarrow B$) are of course exactly as one would expect, and require no comment. But the axiom E6, which has a conjunction in the antecedent, and which in a sense corresponds to the rule "from $A\rightarrow B$ and $A\rightarrow C$ to infer $A\rightarrow B\&C$" of $\mathbf{E}_{\mathrm{fde}}$, requires some comment, as does the primitive rule of adjunction (hereafter "conjunction introduction," or "&I").

It is frequently possible, in formal systems having an operator → intended to answer to "if . . . then —," for which *modus ponens* holds, to get the effect of a two-premiss rule by the following device: corresponding to a two-premiss rule, from A and B to infer C, we can write an axiom $A\rightarrow .B\rightarrow C$, and then use this together with two applications of →E to get C from premisses A and B whenever we wish. This device sometimes makes proof-theoretical studies easier, since for inductions on the length of proofs, the number of cases in the inductive step may be reduced to one. Why then don't we take as an axiom $A\rightarrow .B\rightarrow (A\&B)$, thus obviating the need for &I as a primitive rule (Feys 1956)?

The answer is straightforward. Prefixing B to the axiom E4 yields $B\rightarrow (A\&B)\rightarrow .B\rightarrow A$, which together with transitivity and $A\rightarrow .B\rightarrow A\&B$ would produce $A\rightarrow .B\rightarrow A$. So this course is open to two objections. (a) Addition of $A\rightarrow .B\rightarrow (A\&B)$ yields a system which contains theorems like $A\rightarrow .B\rightarrow A$ involving the arrow only, which can be proved *only* by taking a detour through conjunction; so the resulting system is not a conservative extension (§14.4) of $\mathbf{E}_{\rightarrow}$. (b) Addition of the formula produces $A\rightarrow .B\rightarrow A$, which is rubbish in its own right, independently of objection (a). So this won't do.

But one might still try the device on the rule, from $A\rightarrow B$ and $A\rightarrow C$ to infer $A\rightarrow B\&C$, of $\mathbf{E}_{\mathrm{fde}}$. Couldn't we take as an axiom $A\rightarrow B\rightarrow .A\rightarrow C\rightarrow .A\rightarrow (B\&C)$?

No. For again the extension would not be conservative. Taking B as A, identity would lead to $A\rightarrow C\rightarrow .A\rightarrow (A\&C)$, which, with the help of E4 etc. would give us $A\rightarrow C\rightarrow .A\rightarrow A$. The latter is bad, as is shown by the three-element matrix of §5.2.1: when $A = 1$ and $C = 2$, $A \rightarrow C \rightarrow . A \rightarrow A$ takes the undesignated value $\mathbf{0}$.

Worse (as was pointed out to us by John Bacon in correspondence), this formula abandons entailments to fallacies of relevance. For in $\mathbf{FE}_{\rightarrow}$ we can easily prove

$$C \rightarrow D \rightarrow . A \rightarrow B \rightarrow . C \rightarrow D \rightarrow A \rightarrow B,$$

and with the following special case of the bad formula $A \rightarrow C \rightarrow . A \rightarrow A$ above

$$(A \rightarrow B \rightarrow . C \rightarrow D \rightarrow A \rightarrow B) \rightarrow . A \rightarrow B \rightarrow . A \rightarrow B,$$

we get by transitivity,

$$C \rightarrow D \rightarrow . A \rightarrow B \rightarrow . A \rightarrow B,$$

and then by restricted permutation,

$$A \rightarrow B \rightarrow . C \rightarrow D \rightarrow . A \rightarrow B;$$

but both the latter embody fallacies of relevance. So this plan meets with the same two objections: it is radical (a) in the sense of being non-conservative, and (b) in the sense of being wild-eyed.

Examination of these two attempts to get the effect of &I with the help of $\rightarrow$E strongly suggests, though of course it does not prove, that &I must be taken as primitive, and that the axiom corresponding to the rule "from $A \rightarrow B$ and $A \rightarrow C$ to infer $A \rightarrow B \& C$" should simply mirror that rule as stated in English, which is what E6 does. We will have more to say about the situation from a formal point of view in §22.2.2 but a few remarks of a philosophical character are also in order here.

The situation is not altogether surprising. The system **E** is designed to encompass two branches of formal logic which are (as we have been arguing in the course of this entire treatment) radically distinct. The first of these, historically, is concerned with questions of relevance and necessity in entailments, both of which are at the root of logical studies from the earliest times. The second, extensional logic, is a more recent development to which attention was devoted partly in consequence (we believe) of the fact that the first was more recalcitrant — purely extensional logic *can* be developed in a mathematically interesting way simply by ignoring the problems of relevance and necessity, which got logic off the ground in the first place. Since **E** covers both kinds of territory, it is not surprising that two kinds of primitive rules are needed: the first, $\rightarrow$E, having to do with connections between meanings taken intensionally, and the second, &I, having to do with connections between truth values, where relevance is not an issue. And precisely the same distinction would have been apparent had our informal language contained natural ways of making it. We seriously mean to consider the rule of adjunction to be of the form "from A and B to infer A&B," but English constrains us to use the same sort of locution in the statement of →E though a more accurate reading of our intent might have been something like "if $A \rightarrow B$, then from A to infer B," or "given $A \rightarrow B$, from A to infer B," (as we *did* state the rule in §21.1), or some such. We realize that the latter locutions will tend to cause apoplexy among those on whom the use-mention distinction has a real hammerlock, a distinction entirely parochial to certain antique grammatical analyses of the language in which this book is written — on a par with "a noun is a person, place, or thing," of the nineteenth century — though somewhat more sophisticated. In any event, we promise to offend such delicate sensitivities only one more time in the remainder of this subsection, namely in pointing out that even the rule of transitivity for $\mathbf{E}_{fde}$ *should* be stated as "if $A \rightarrow B$, then from $B \rightarrow C$ to infer $A \rightarrow C$," since in this rule, unlike adjunction, relevance is preserved. The corresponding axiom, like the axiom corresponding to the single-premiss rule "from $A \rightarrow B$ to infer $\bar{B} \rightarrow \bar{A}$," is of course already present in $\mathbf{E}_{\rightarrow}$.

The radical intensional-extensional dichotomy does, however, admit of a certain Coherent Unity or Identity, since the intensional and extensional primitive rules of inference, so put, answer to the theorems $A \rightarrow B \rightarrow .A \rightarrow B$ (→E) and $A\&B \rightarrow .A\&B$ (&I), both of which are instances of $A \rightarrow A$. We find comfort in the thought that the law of identity, which in §1 we labeled "the archetypal form of inference, the trivial foundation of all reasoning," serves us in such an admirable way.

§21.2.2. Necessity. We offer two excuses for axiom E7.

In the first place we are led by a strong tradition to believe that the necessity of any theorem (of a formal system designed to handle the notion of logical necessity at all) should also be a theorem; unless this requirement is met, the system simply has no theory of its own logical necessities. For this reason we would like to have it be true that, whenever A is provable, the necessity of A is also provable. This condition could be satisfied by incoherent brute force, as it is for example in systems like **M** (Feys-von Wright), where a rule of necessitation is taken as primitive. It could equally well be satisfied by adding $A \rightarrow \Box A$ as an axiom. Both courses are equally odious, the latter because it destroys the notion of necessity, and the former because, if $A \rightarrow \Box A$ is neither true nor a theorem, then we ought not to have — in a coherent formal account of the matter — a primitive rule to the effect that $\Box A$ does after all follow from A. This constraint would not bother us if we were *simply* trying to define the set of theorems of **E** recursively in such a way that a digital computer, or some other equally intelligent being, could grind them out. But our ambitions are greater than this; we would like to have our theorems and our primitive rules dovetail in such a way that if **E** says or fails to say something, *we* don't contradict it, or violate its spirit. (Note that neither →E nor &I does so.)

Nevertheless it should be *true*, as a lucky accident, so to speak, that whenever A is a theorem, $\Box A$ is likewise. And this result can be secured by an induction on the length of formal proofs in **E**, provided we have E7 available: since all entailments are necessitives (§5.2), all the axioms are provably necessary; and with the help of $\Box(A \rightarrow B) \rightarrow . \Box A \rightarrow \Box B$ and $\Box A \& \Box B \rightarrow \Box(A \& B)$ we can get over the inductive steps for →E and &I. We dignify the upshot by calling it a

THEOREM. If $\vdash A$ in **E**, then $\vdash \Box A$ in **E**.

E7 also calls for a second remark, the point of which can be made clear now, even though we will be able to discuss it in detail only after propositional quantifiers are introduced. It develops that in the system $\mathbf{E}^{\forall\exists p}$ of Chapter VI, which consists of **E** together with appropriate axioms and rules for propositional quantifiers $\forall p$, $\exists p$, etc., we can define necessity more satisfactorily as follows:

$$\Box A =_{df} \forall p(p \rightarrow p) \rightarrow A,$$

which says that A is necessary just in case A is entailed by the law of identity. Then in $\mathbf{E}^{\forall\exists p}$ we can prove

$$A \rightarrow A \rightarrow A \rightleftarrows . \forall p(p \rightarrow p) \rightarrow A,$$

which means that the old definition exactly coincides with the new one, and, given the new definition, E7 emerges as a special case of E6 and can therefore be dispensed with. So the need for E7 in the present treatment depends only on the fact that **E** is a free-variable calculus; when propositional quantification is made explicit, as in Chapter VI, E7 is no longer required.

§22. Fallacies. Tradition has it that it is best to sort fallacies into two groups: logical or *formal* fallacies, and informal or *material fallacies*. The former are distinguished by their running afoul of rules having to do with the logical form of an argument, in abstraction from its content, and are to be detected by paying attention to formal features of the situation. For example, such plausibility as attaches to the argument from the premisses

All men are metal, and
All mortals are brittle,

to the conclusion

All men are brittle,

is dispelled, once we see that it commits the traditional *Formal Fallacy of Four Terms*. This we can recognize independently of the *matter* of the argument, by inspection of the form alone.

Material fallacies on the other hand depend on misinterpretations or misunderstandings of words and phrases. We cite the following venerable example of the *Material Fallacy of Accent:* "And he spake to his sons, saying, Saddle me the ass. And they saddled *him*" (Kings I, 13, 27).

We follow the tradition in making the division, but we modify it to suit present purposes, which differ from those for which the original dichotomy was invented.

Formal fallacies are exemplified by the widespread belief that relevance of antecedent to consequent is irrelevant to the validity of a conditional; here we have a formal criterion, and a proof that **E** meets it. The principal *material* fallacy, which concerns the interpretation of the formalism rather than proofs in or about it, is the contention that material "implication" is a kind of implication. We assimilate this case to the confusion as to where the sons put the saddle.

Since this section will contain our final frontal assault on fallacies, and since we now have available in **E** not only intensional logic but (as we shall see) extensional logic as well, we have made our catalogue as complete as we can: fallacies are formal (§22.1) and/or material (§22.2) and there are two subheads under each.

The remaining §22.3, by Robert K. Meyer, doesn't have directly to do with fallacies. On the other hand it does concern a topic on which we have touched from time to time (most recently in §21.2.2), namely the feeling that the motivating considerations for our formal systems, and the metalogical apparatus used in dealing with them, should somehow fit nicely together.

§22.1. Formal fallacies. The proof that **E** avoids these, here and in in what has gone before, rests on two main theorems: one (avoidance of fallacies of modality) to the effect that purely truth functional propositions never entail entailments (Ackermann 1956), the other (avoidance of fallacies of relevance) guaranteeing that antecedents and consequents of true entailments share propositional content (Belnap 1960, and independently Dončenko 1963). Both of these have been sharpened somewhat by Maksimova 1967, and we include here her improvements. We minimize philosophical discussion of the first two types of fallacies below, referring the reader to §5; here we simply present formal results. (Urquhart 1972c adds to our logical tool kit by characterizing, in addition to fallacies of relevance and modality, also fallacies of completeness and consistency.)

§22.1.1. Ackermann-Maksimova modal fallacies. Ackermann 1956 proves that for his system $\mathbf{\Pi'}$ (see §45), no theorems have the form $A\to.B\to C$ if A contains no arrow. We defer giving his proof until we discuss $\mathbf{\Pi'}$

in Chapter VIII since Maksimova's result is more general, and applies, as does Ackermann's, equally well to **E**.

Roughly speaking, Maksimova's generalization amounts to this. We know trivially from Ackermann's theorem that no theorems of **E** have the form

$$1 \quad A_1 \rightarrow . A_2 \rightarrow \ldots \rightarrow . A_k \rightarrow . A \rightarrow . B \rightarrow C,$$

where each A_i is a theorem, and A contains no arrow — since otherwise $A \rightarrow . B \rightarrow C$ would be a theorem, contrary to what Ackermann said. Maksimova shows that more relaxed conditions can be placed on the A_i and on A, and still no theorems of the form 1 can be proved. We have already seen in §12 that 1 remains unprovable if A_1 is the negation of any entailment, but Maksimova shows that entailment may enter into A or the A_i in many other ways as well, 1 remaining unprovable.

This is accomplished with the help of ten-element matrices derived from an intensional lattice (§18) with the following Hasse diagram.

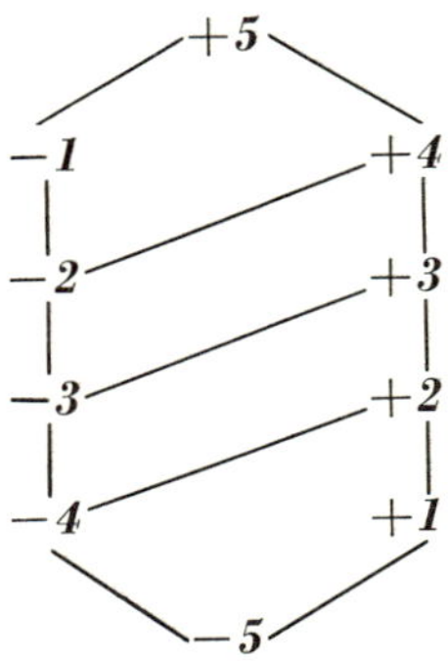

The tables for the extensional connectives are as follows.

&	−5	−4	−3	−2	−1	+1	+2	+3	+4	+5
−5	−5	−5	−5	−5	−5	−5	−5	−5	−5	−5
−4	−5	−4	−4	−4	−4	−5	−4	−4	−4	−4
−3	−5	−4	−3	−3	−3	−5	−4	−3	−3	−3
−2	−5	−4	−3	−2	−2	−5	−4	−3	−2	−2
−1	−5	−4	−3	−2	−1	−5	−4	−3	−2	−1
*+1	−5	−5	−5	−5	−5	+1	+1	+1	+1	+1
*+2	−5	−4	−4	−4	−4	+1	+2	+2	+2	+2
*+3	−5	−4	−3	−3	−3	+1	+2	+3	+3	+3
*+4	−5	−4	−3	−2	−2	+1	+2	+3	+4	+4
*+5	−5	−4	−3	−2	−1	+1	+2	+3	+4	+5

$\vee$	-5	-4	-3	-2	-1	$+1$	$+2$	$+3$	$+4$	$+5$
-5	-5	-4	-3	-3	-1	$+1$	$+2$	$+3$	$+4$	$+5$
-4	-4	-4	-3	-2	-1	$+2$	$+2$	$+3$	$+4$	$+5$
-3	-3	-3	-3	-2	-1	$+3$	$+3$	$+3$	$+4$	$+5$
-2	-2	-2	-2	-2	-1	$+4$	$+4$	$+4$	$+4$	$+5$
-1	-1	-1	-1	-1	-1	$+5$	$+5$	$+5$	$+5$	$+5$
$+1$	$+1$	$+2$	$+3$	$+4$	$+5$	$+1$	$+2$	$+3$	$+4$	$+5$
$+2$	$+2$	$+2$	$+3$	$+4$	$+5$	$+2$	$+2$	$+3$	$+4$	$+5$
$+3$	$+3$	$+3$	$+3$	$+4$	$+5$	$+3$	$+3$	$+3$	$+4$	$+5$
$+4$	$+4$	$+4$	$+4$	$+4$	$+5$	$+4$	$+4$	$+4$	$+4$	$+5$
$+5$	$+5$	$+5$	$+5$	$+5$	$+5$	$+5$	$+5$	$+5$	$+5$	$+5$

$\sim$	-5	-4	-3	-2	-1	$+1$	$+2$	$+3$	$+4$	$+5$
	$+5$	$+4$	$+3$	$+2$	$+1$	-1	-2	-3	-4	-5

For the intensional connective we have:

$\rightarrow$	-5	-4	-3	-2	-1	$+1$	$+2$	$+3$	$+4$	$+5$
-5	$+2$	$+2$	$+2$	$+2$	$+2$	$+2$	$+2$	$+2$	$+2$	$+2$
-4	-5	$+2$	$+2$	$+2$	$+2$	-5	$+2$	$+2$	$+2$	$+2$
-3	-5	-5	$+2$	$+2$	$+2$	-5	-5	$+2$	$+2$	$+2$
-2	-5	-5	-5	$+2$	$+2$	-5	-5	-5	$+2$	$+2$
-1	-5	-5	-5	-5	$+1$	-5	-5	-5	-5	$+2$
$+1$	-5	-4	-4	-4	-4	$+1$	$+2$	$+2$	$+2$	$+2$
$+2$	-5	-4	-4	-4	-4	-5	$+2$	$+2$	$+2$	$+2$
$+3$	-5	-5	-4	-4	-4	-5	-5	$+2$	$+2$	$+2$
$+4$	-5	-5	-5	-4	-4	-5	-5	-5	$+2$	$+2$
$+5$	-5	-5	-5	-5	-5	-5	-5	-5	-5	$+2$

We leave it to the reader to verify the following two lemmas (of which the first says simply that the matrices satisfy **E**).

LEMMA 1. If A is a theorem of **E**, then for every assignment of values to the variables in A, $\mathrm{v}(A) \geq +1$ (where $\mathrm{v}(A)$ is the value assumed by A for an assignment of values to its variables).

LEMMA 2. If A is any formula of **E**, and all variables of A are assigned the value $+3$, then $\mathrm{v}(A) \neq +1$.

To continue, we extend the definitions of *antecedent part* ("ap") and *consequent part* ("cp") *of A* given in §12. For entailment and negation the clauses are as before:

(a) A is a cp of A;
(b) if $\overline{B}$ is a cp {ap} of A, then B is an ap {cp} of A; and
(c) if $B \rightarrow C$ is a cp {ap} then B is an ap {cp} of A and C is a cp {ap} of A.

For conjunction and disjunction we add:

(d) if either $B \& C$ or $B \vee C$ is a cp {ap} of A, then both B and C are cps {aps} of A.

In stating her results, Maksimova uses *negative* and *positive* where we have used *antecedent* and *consequent*. We like the latter terminology partly because it seems to suggest more closely the connection with the intensional ideas carried by the arrow ("negative" and "positive" smack of truth functions), and partly because we wish to use "negative" and "positive" for a different purpose.

We first define *truth functional part of A: A* is a truth functional part of A, and if $\overline{B}$, $B \& C$, or $B \vee C$ are truth functional parts of A, so are B and C. Then we call a formula *negative* if it has no truth functional consequent parts of the form $B \rightarrow C$, and *positive* if it has no truth functional antecedent parts of the form $B \rightarrow C$.

Evidently some formulas (e.g. any purely truth functional formula) are both positive and negative, some are one and not the other (e.g., $(A \rightarrow B) \vee (C \rightarrow D)$ is positive but not negative; $\overline{(A \rightarrow B) \& (C \rightarrow D)}$ is negative but not positive), and some are neither (e.g. $\overline{A \rightarrow B} \vee C \rightarrow D$). Notice that always A is positive {negative} iff $\overline{A}$ is negative {positive}.

For ease in checking the next lemma we notice that the ten-element intensional lattice has two six-element sublattices ("sub" *qua* lattice, not *qua* intensional lattice, since they are not closed under complementation). They have the following Hasse diagrams:

top-6

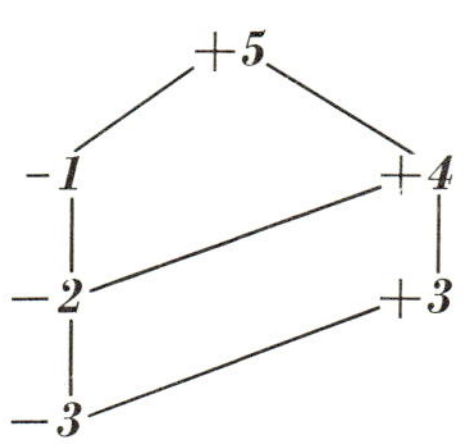

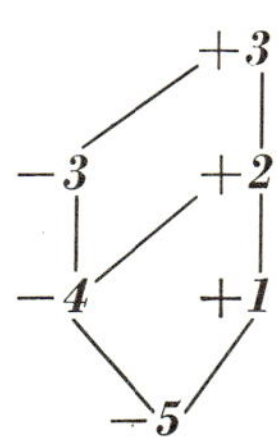

bot-6

(which we have drawn so as to suggest the way in which they were excised from the ten-element lattice); we call these "top-6" and "bot-6."

LEMMA 3. If A is a negative formula, and all variables in A are assigned the value $+3$, then (i) all consequent truth functional parts of A take values in top-6, and (ii) all antecedent truth functional parts of A take values in bot-6.

The proof is by induction on the length of parts B of A.

BASIS. If B is a variable, it receives $+3$, which is in both top-6 and bot-6, so it makes no difference whether B is a cp or an ap: (i) and (ii) both hold for B.

INDUCTIVE CASES. (1) B has the form $C \to D$. Then B is an antecedent part, since A is negative; and inspection of the table for the arrow shows that (ii) $v(B) \in$ bot-6. (2) B has the form $\overline{C}$. Then (i) if B is a cp, then C is an ap, hence by the inductive hypothesis $v(C) \in$ bot-6. But then inspection of the negation table reveals that $v(\overline{C}) \in$ top-6, i.e., $v(B) \in$ top-6, as required. The case (ii) where B is an ap is similar. (3)–(4). B has the form $C \& D$ or $C \vee D$. There are cases (i) and (ii) for each, and in each case the inductive hypothesis and inspection of the tables do the trick, as the reader can check.

The part of this lemma that interests us is part (i): every negative formula assumes a value in top-6 when all its variables are given the value $+3$. This gives us enough control to prove a similar fact about a set of formulas defined as follows:

We will say that A is a *Maksimova formula* if it is a member of the smallest class which is closed under *modus ponens* and adjunction, and contains (a) all axioms of **E**, and (b) all formulas $A \to B$ such that A is positive and B is negative.

The class of Maksimova formulas contains all theorems of **E**, but many other formulas as well: e.g., all formulas of the form $A \to B \to .\overline{C \to D}$, all formulas $A \to B$ where A and B contain only truth functions, etc., and con sequences of such. The next lemma deals with Maksimova formulas, invoking eight-element sublattices of the ten-element lattice, which we call "top-8" and "bot-8":

top-8

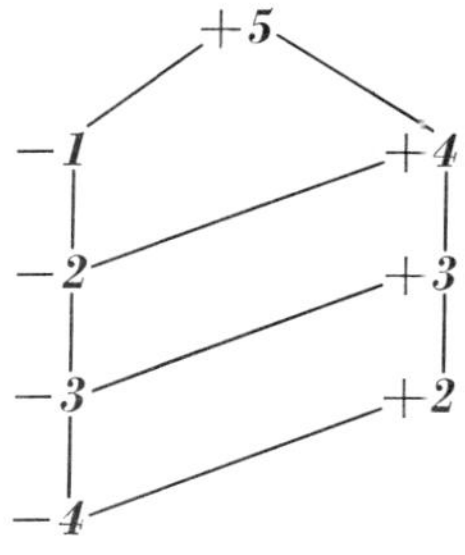

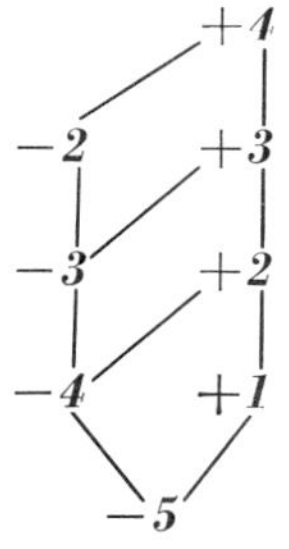

bot-8

LEMMA 4. If A is a Maksimova formula, and all variables in A are assigned the value $+3$, then A assumes a value in top-8.

The proof is by induction on the length of the proof that A is a Maksimova formula. We first assign the value $+3$ to every variable occurring in this proof, and then consider four cases, according as the ith step S_i in the proof is one of the initial formulas, or got by one of the two rules.

(1) S_i is an axiom of **E**. Then by Lemma 1, $v(S_i) \geq +1$, and by Lemma 2, $v(S_i) \neq +1$; hence $v(S_i) \in$ top-8.

(2) S_i is $A \to B$, where $\bar{A}$ and B are both negative formulas. Then since $\bar{A}$ and B are both consequent truth functional parts of themselves, we have by Lemma 3 both $v(\bar{A}) \in$ top-6 and $v(B) \in$ top-6. The negation table then says that $v(A) \in$ bot-6, and inspection of the entailment table shows that $v(A \to B) = +2$ or -4; hence $v(A \to B) \in$ top-8.

(3) S_i comes by *modus ponens* from S_j and $S_j \to S_i$, where the inductive hypothesis guarantees that both the latter formulas take values in top-8. Now we know from Lemma 2 that $v(S_i) \neq +1$, and it remains only to show that $v(S_i) \neq -5$, since then we will know that $v(S_i) \in$ top-8. To this end we note that among the values in top-8 only $+2$ and -4 occur inside the arrow table; so $v(S_j \to S_i)$ must be one of these. But if S_i takes -5, then for this assignment the table shows that $v(S_j \to S_i)$ must be $+2$, forcing $v(S_j)$ to be -5, contrary to the inductive hypothesis that $v(S_j) \in$ top-8. Hence $v(S_i)$ is neither $+1$ nor -5, so $v(S_i) \in$ top-8.

(4) S_i ($= S_j \& S_k$) comes by adjunction from S_j and S_k, where $v(S_j)$ and $v(S_k)$ are both in top-8. A glance at the Hasse diagram convinces us that top-8 is closed under conjunction; hence $v(S_i) \in$ top-8.

With Lemmas 2–4 we are now at long last in a position to prove the

THEOREM. If $A_1, \ldots, A_k$ are Maksimova formulas and A is a negative formula, then for no B and C is

$$1 \quad A_1 \to .A_2 \to \ldots A_k \to .A \to .B \to C$$

provable in **E**.

PROOF. Assign $+3$ to all the variables in 1. Now $v(B \to C) \neq +1$ by Lemma 2, so $v(B \to C) \in \{-5, -4, +2\}$, these being the only other values entailments take. By Lemma 3, $v(A) \in$ top-6 ($= \{-3, -2, -1, +3, +4, +5\}$). A case-by-case check then reveals that $v(A \to .B \to C) = -5$. Then Lemma 4 says that each $v(A_i) \in$ top-8 ($= \{-4, -3, -2, -1, +2, +3, +4, +5\}$), and if we consult the table we find that $v(A_i \to -5) = -5$. So $v(1) = -5$, and is therefore unprovable in **E**.

Ackermann's theorem for Π′ then emerges as a special case of Maksimova's: where $k = 0$ and A contains no arrows at all, $\overline{A\to.B\to C}$ is unprovable. The unprovability of every formula of the form $\overline{A\to B\to.C\to D}$ emerges as another special case, where again $k = 0$. The upshot is that not only are formulas such as these (which lead to modal fallacies) unprovable: they cannot even be made to follow from a wide variety of non-theorems.

Our reason for presenting Maksimova's result in this much detail is that the proof uses an intensional lattice, as does the proof of the principal theorem of the next section — and we have a fondness for intensional lattices. But just as some of the work for $\mathbf{E}_{\text{fde}}$ can be done with a lattice which has queer elements in it (i.e., elements which are their own negations; see §15.3), so we can use a funny unintensional lattice of a similar sort to prove Maksimova's generalization of Ackermann's theorem.

For this pleasant fact we are indebted to Meyer, who presented us in a letter of October, 1968, with the following five point chain:

1
P(roblematic)
A(ssertoric)
N(ecessary)
0

(Meyer's *A* faults itself, just like Smiley's *2* and *3*; skip quickly again to §15.3, which is very short.) In the interest of saving the reader trouble in checking the proof, we display the resulting matrices:

	~
*1	0
*P	N
*A	A
*N	P
0	1

∨	1	P	A	N	0
1	1	1	1	1	1
P	1	P	P	P	P
A	1	P	A	A	A
N	1	P	A	N	N
0	1	P	A	N	0

&	1	P	A	N	0
1	1	P	A	N	0
P	P	P	A	N	0
A	A	A	A	N	0
N	N	N	N	N	0
0	0	0	0	0	0

→	1	P	A	N	0
1	N	0	0	0	0
P	N	N	0	0	0
A	N	N	N	0	0
N	N	N	N	N	0
0	N	N	N	N	N

Meyer's proof then has a succinctness which we admire, and with his permission we conclude by quoting it:

> Then let $B_1, \ldots, B_n$ be Maksimova formulas, C negative, and D and E what you please. Falsify $B_1\to.B_2\to.\ \ldots\ C\to.D\to E$ by assigning all sentential variables *A*. C is *A* or above, $D\to E$ is *N* or below, and all the B_i are non-zero, which drives the whole thing to *0*.

§22.1.2. Fallacies of Modality (by J. Alberto Coffa). Except for a handful of contemporary foot-in-cheek epistemologists, the philosophical community seems to have been fairly unanimous in agreeing on what we will here refer to as the Platonic Principle (PP), the claim that necessary knowledge cannot derive from experience. Since, as Salmon has put it, "one man's *modus ponens* is another man's *modus tollens*," while some argued that PP established conclusively the existence of non-experiential sources of knowledge, others no less emphatically inferred from it the impossibility of (synthetic) necessary knowledge. Inferential discrepancies aside, the fact remains that PP seems to have been considered by members of otherwise hostile philosophical traditions, as a most probably correct restriction of the domain of allowable inferences.

As stated, the principle is far from perspicuous. Yet one part of its content seems clear enough: just as there is a "strong" modal property (necessity) that is preserved by valid arguments, the principle suggests that there is a "weak" modal property also preserved under entailment. Many philosophers have taken this weak property to be contingency, and so have argued that

(*) If A entails B and A is contingent, then so is B.

Routley and Routley 1969 have observed that (*) is untenable. After showing that belief in (*) is surprisingly widespread, they go on to concentrate their attack on the system **E** of entailment. One of the main philosophical arguments in favor of the adequacy of **E** as an explication of entailment is that it avoids fallacies of modality. But in a paragraph in Anderson and Belnap 1962a, a fallacy of modality is characterized as a violation of (*). Yet, as the Routleys remark, both $p\rightarrow(p\vee\sim p)$ and $(p\&(q\rightarrow q))\rightarrow.q\rightarrow q$ are obvious entailments (and theses of **E**) that violate (*). The Routleys argue that different formulations of (*) are either equally untenable or tautological, and therefore worthless as regulative principles. Hence, they conclude, "if we are right, a main ground for Anderson's and Belnap's choice of $\mathbf{E}_{\rightarrow}$ and **E** as the correct system of entailment is destroyed."

This implication happens to be even truth functionally false, since though the Routleys are clearly right in all of their objections against (*) and its variations, their arguments, as we will presently show, only establish the inadequacy of the earlier formulation of the essentially correct idea of what a modal fallacy is.

Contingency vs. the "Weak" Property. That people couldn't really have had in mind contingency while arguing for PP follows (cf. Salmon's Principle) from the fact that extremely trivial counterexamples like

1 $p\rightarrow(p\vee\sim p)$

are available. Surely, practically everyone would accept that 1 is a true entailment, since practically everyone would accept that

2 $p \rightarrow (p \vee q)$

is a true entailment, and that substitution of $\sim p$ for q in 2 preserves the validity of the inference.

Now, if one chooses to look at entailment as a relation of conclusive evidential support, one thing that immediately jumps to the eye is that what the (presumably contingent) proposition p establishes about $p \vee \sim p$ is precisely what it establishes about $p \vee q$, i.e., its truth. It also happens to be the case that (unbeknownst to p) $p \vee \sim p$ is not only true but necessary. Yet, we don't believe (and we don't have in **E**) that $p \rightarrow \Box(p \vee \sim p)$.

This suggests that what needs restriction in the formalization of PP is not the class of formulas that entail necessary statements but rather the class of those that entail statements to the effect that something is necessarily the case. Experience, we want to say, can be sufficient evidence for the fact that $p \vee \sim p$, but it cannot give us conclusive grounds to believe that $\Box(p \vee \sim p)$. Necessary knowledge, in the sense of PP, is not just knowledge of a proposition that happens to be necessary, but also of the fact that it *is* so. Inferences forbidden by (*) have necessary statements as conclusion, while those forbidden by PP terminâte in necessity claims.

But what about the premisses? Contemplate

3 $(p \& (q \rightarrow q)) \rightarrow . q \rightarrow q.$

The consequent of 3 is both necessary and (according to **E** and most other views about entailment) equivalent to a necessity claim. Its antecedent is clearly contingent, since its truth value is that of p. And, once again in violation of (*), 3 seems to qualify as a paradigm case of entailment. Is 3 also a counterexample to PP? It should be, if experiential statements are to be equated with contingent ones.

Yet, observe that the contingency of

4 $p \& (q \rightarrow q)$

is established via its falsehood conditions. 4 is contingent because it is conceivably false, and it is conceivably false because p is so. But the possible falsehood of 4 is as true as it is irrelevant to its empirical character. Try asking any Kantian whether the conjunction of Euclidean axioms with a contingent p is empirical. While contingency is established by showing that both truth and falsehood of the given statement are logically possible, empiricalness is decided by looking exclusively at its truth conditions. If verification of a statement requires verification of a necessity claim, then, quite independently of its possible falsehood conditions, the statement is not empirical. Thus, experiential knowledge in the sense of PP is not knowledge of contingent statements but of those that can be established as being true by "experience" alone, i.e., by means other than those required to verify necessity claims.

Traditionally, the validity of the putative modal principles asserting distributivity with respect to entailment of a strong and a weak modal property, has been both attacked and defended on the assumption that the weak property should be defined as the complement of the strong one. What we have been arguing so far, is that this common assumption is erroneous and that two essentially dissimilar partitions of the set of propositions is required in order adequately to interpret the content of these principles. The pair of concepts Necessity-Nonnecessity provides the proper partition for the principle concerning distributivity of the strong modal property. Let us now try to specify the partition associated with PP.

Definition of the "Weak" Property. We recall from §5.2 that a necessitive statement (by analogy with "conjunctive," "disjunctive," etc.) expresses a proposition to the effect that something is necessarily the case, or, technically, A is a necessitive iff there is a B such that A entails and is entailed by $\Box B$. Now, our previous discussion suggests that PP should be formulated in terms of the notion of Necessitivity rather than — as has been usually done — in terms of Necessity. But it also suggests that Necessitivity will not be quite enough. For example, 3 counterexemplifies the claim that necessitives are only entailed by necessitives. In fact, the way we have come to see it, PP says that necessitives *or propositions that say at least as much as they do* cannot be entailed by propositions whose content is less than that of a necessitive. Thus, for modalized statements, we have been led to a conception of content in terms of truth conditions rather than of falsehood conditions.

In order to formalize the idea suggested by our heuristic remarks for the particular case of propositional entailment, let us consider one such system in terms of $\Box$ — and truth functional connectives (e.g., $\vee$, & and $\sim$). Though we will presently allow for a certain kind of relativization of the modal notions to the systems in which they appear, we will assume throughout system-independent interpretations of the truth functional connectives. The notion that we now describe is that of a formula for which there is an assignment of truth values to its truth functional parts (see §22.1.1) that (under the system-independent interpretation of the truth functional connectives) makes the formula true while all of its truth functional necessitive parts (if any) are false. These will be the *weak* formulas, those that say "less" than a necessitive. In contrast a *strong* formula will be such that every assignment of truth values to its truth functional parts which makes the formula true also makes at least one of its necessitive parts true.

It is convenient to have a more botanical account of these matters.

For each formula F we define its tree $\mathrm{T}(F)$ as the result of performing the operations described by rules (i)–(vi) below, starting with $S_0 = F$. We say that at each application of a rule, the leftmost formula at stage S_n is being *activated.* An *atom* is a formula having none of the forms of the activated sentences in rules (i)–(v). Γ is a sequence (possibly empty) of formulas.

(i) $S_n = A\&B, \Gamma$
$S_{n+1} = \Gamma, A, B$

(ii) $S_n = A\vee B, \Gamma$
$S_{n+1} = \Gamma, A \; ; \; \Gamma, B$ (the semicolon indicates splitting of the tree)

(iii) $S_n = {\sim}(A\&B), \Gamma$
$S_{n+1} = {\sim}A\vee{\sim}B, \Gamma$

(iv) $S_n = {\sim}(A\vee B), \Gamma$
$S_{n+1} = {\sim}A\&{\sim}B, \Gamma$

(v) $S_n = {\sim}{\sim}A, \Gamma$
$S_{n+1} = A, \Gamma$

(vi) $S_n = A, \Gamma$ where A is an atom
$S_{n+1} = \Gamma, A.$

Branches and nodes are defined as usual. We stop the construction at each branch whenever we reach a node (*terminal node*) that contains only atoms. We now say that *F is at least as strong as a necessitive* iff $\mathrm{T}(F)$ contains a necessitive in each of its branches (not necessarily at the tip).

Given a class N of necessitives, we say that S(N) is the class of formulas at least as strong as formulas in N, and that W(N) is its complement. Now we can formulate precisely the portion of PP that we intend to use as a regulative principle for the identification of fallacies of modality:

(**) If $A\rightarrow B$ and $A \in \mathrm{W(N)}$, then $B \in \mathrm{W(N)}$.

But there is still an ambiguity in (**) that must now be eliminated.

Two kinds of modal fallacies. The question is, how are we going to identify N? Surely, if one is an **S5**-er, one will feel inclined to say that ${\sim}(p\rightarrow q)$, if true at all, is necessarily so, and in fact equivalent to its own necessity (i.e., ${\sim}(p\rightarrow q)\rightleftarrows\Box{\sim}(p\rightarrow q)$); hence a necessitive. Yet, in **E**, ${\sim}(p\rightarrow q)$ is not a necessitive. Hence, while the **S5**-er will claim that (**) is violated in **E** since it contains ${\sim}p\rightarrow{\sim}(q\rightarrow q\rightarrow p)$, the **E**-er will feel perfectly happy about that formula since, in his view, both antecedent and consequent are weaker than necessitives. If we want modal logicians to communicate on this issue without arguing across each other, we should try to find some way of objectivizing the problem without transforming it into an entirely uninteresting issue. We suggest the following way out of the predicament, based on a distinction between two sorts of modal fallacies: internal or absolute and external or relative.

Internal fallacies of modality (*ifm*) are characterized as follows. For each system **S** the class of necessitives (now we call it N_S) is defined to be the set of formulas that **S** recognizes as such (i.e., $N_S = \{A: \exists B(\vdash_S A \rightleftarrows \Box B)\}$, where the double arrow guarantees intersubstitutability). Given N_S, $W(N_S)$ and (**) are unambiguously determined, and it is now a technically well-defined problem whether **S** verifies (**) or not.

External fallacies of modality (*efm*) are defined by first identifying N using standards other than those available in the system under consideration. In this case one does not go to **S** in order to find out what things are to be counted as members of N. In the internal sense, one tries to compare the system with itself and see if it measures up to its own standards; in the external sense, one seeks to assess the adequacy of the system to one's own independent philosophical views. In this sense **S5**-ers are right in claiming that **E** violates (**), just as the principal authors of this book are right in finding fallacies in the intuitionist's $p \rightarrow .q \rightarrow p$ and in the **S5**-er's $\sim(p \rightarrow q) \rightarrow \Box \sim(p \rightarrow q)$. Whether a system has efm or not is a philosophical matter, and therefore one never to be solved to everybody's satisfaction. But whether it has ifm is a mathematical question that in principle allows an unambiguous answer.

Our version of (**) in its internal sense is most probably very close to what the principal authors had in mind while describing what they meant by fallacies of modality. At least three reasons suggest that this is the case: (i) that they almost said as much in their unpublished versions of this book — there, they state that "modal fallacies depend upon misdescribing the logical relations between necessitive and non-necessitive propositions" (cf. §5); (ii) that all theorems that have been considered partially to establish the absence of modal fallacies in **E**, do so under (**) (i.e., all statements thereby proved not to be in **E** are incompatible with (**); cf. §§22.1.1 and 22.1.3); and (iii) that they say so.

There is a fourth reason: **E** has no internal fallacies of modality.

In outline the proof is as follows. For each formula F in $W(N_E)$, rules (i)–(vi) describe an effective procedure to generate another formula in a certain normal form that entails F in **E**. By Maksimova's theorem, (§22.1.1), it turns out that formulas in this normal form cannot entail entailments. By transitivity of entailment in **E**, no formula in $W(N_E)$ can entail an entailment. But every formula in $S(N_E)$ entails an entailment. Hence the theorem follows. In more detail:

We start by defining an alternative characterization of the set $S(N_E)$ which will allow us to simplify one of the required proofs. For each formula

F in **E**, we define its value v(F) as follows: we assign to each propositional variable in F the value *3*, and then compute according to the following matrices; entailments are to receive the value *1* independently of the value of their components:

~	
1	*2*
2	*1*
3	*3*

∨	*1*	*2*	*3*
1	*1*	*2*	*3*
2	*2*	*2*	*2*
3	*3*	*2*	*3*

&	*1*	*2*	*3*
1	*1*	*1*	*1*
2	*1*	*2*	*3*
3	*1*	*3*	*3*

Intuitively: *1* is assigned to formulas at least as strong as necessitives; to those without this property we assign *2* if their negation has it and *3* if it does not. We say that a tree is *proper* iff all of its terminal nodes contain entailments. Henceforth F is a formula of **E**.

LEMMA 1. T(F) is proper iff v(F) = *1*.

The only atoms are propositional variables, their negations, entailments, and their negations. It is easy to check that if all formulas at a given node n at stage S_i have a value other than *1*, then so do all formulas at some node deriving from n at state S_{i-1}. Also, analysis of the matrices shows that if a given node n at stage S_i contains a formula with value *1*, then so do all nodes deriving from n at stage S_{i+1}. We also use the fact that the value of an atom is *1* iff it is an entailment.

Rules (i)–(vi) have the following helpful property.

LEMMA 2. For all nodes n′ immediately succeeding n, the conjunction of all formulas in n′ entails (in **E**) the conjunction of all formulas in n. Hence given any tip of the tree T(F), the conjunction of formulas at that tip entails F (in **E**).

This follows from inspection of the rules (i)–(vi), checking that **E** has all the required theorems.

LEMMA 3. If F entails an entailment in **E**, then T(F) is proper.

Suppose that for some A, B

4 $\vdash_{\mathbf{E}} F \rightarrow . A \rightarrow B$.

Suppose further that T(F) is not proper. Then one of its branches contains in its terminal node at most propositional variables, their negations, and negations of entailments. Call the conjunction of these atoms K. Then, by lemma 2 and the fact that entailment is transitive in **E**, we have

5 $\vdash_{\mathbf{E}} K \rightarrow F$

and from 4 and 5 we get

6 $\vdash_{\mathbf{E}} K \rightarrow .A \rightarrow B.$

Now K is always negative in the sense of Maksimova. Since, by Maksimova's theorem (§22.1.1) no negative formula can entail an entailment, 6 is impossible. Hence T(F) is proper.

REMARK. Since in **E** all necessitives entail entailments (see the definition of $\Box$), we have that if F is a necessitive in **E**, then T(F) is proper.

It is also worth observing that

LEMMA 4. If a formula A of **E** belongs to node n of T(F), and T(A) is proper, then all branches of T(F) above n contain entailments in their terminal nodes.

Consider rules (i)–(vi). When A (or any derivate of A) is in Γ, it reappears at all immediately succeeding nodes. When A (or any of its derivates) is activated, the operations on it in T(F) exactly parallel those performed in T(A), so that if some branch above n has no entailment at its terminal node, the same should be the case for T(A). (If the reader is unhappy about this intuitive argument, he can use Lemma 1 and observe that for each of the rules (i)–(vi), if at any node A is activated and v(A) = *1*, each of the formulas derived from A at the next stage has also value *1*. Hence, at each node above n and converging to it, there will be a formula with value *1*. The only such formulas at the terminal nodes are entailments.)

LEMMA 5. $F \in$ S($\mathrm{N_E}$) iff T(F) is proper.

From right to left the lemma is obvious since in **E** all entailments are necessitives. From left to right we apply the remark under Lemma 3, and Lemma 4.

Lemmas 1 and 5 show that the matrices are an alternative characterization of S($\mathrm{N_E}$).

THEOREM 1. For all A, B, C, if $\vdash_{\mathbf{E}} A \rightarrow .B \rightarrow C$, then $A \in$ S($\mathrm{N_E}$).

By lemmas 3 and 5.

LEMMA 6. If $F \in S(N_E)$ then, for some A, B, $\vdash_E F \rightarrow .A \rightarrow B$.

We prove by induction on the length of F that if $v(F) = 1$ then F entails an entailment, and if $v(F) = 2$ then $\overline{F}$ entails an entailment. (We are grateful to James Sterba for the correct formulation of this lemma.)

For example, suppose that $F = M \vee N$ and $v(M \vee N) = 1$. By inspection of the matrices we see that $v(M) = v(N) = 1$. By the inductive hypothesis there must be formulas K, L, K', L' such that

7 $\vdash_E M \rightarrow .K \rightarrow L$,
8 $\vdash_E N \rightarrow .K' \rightarrow L'$.

But since **E** contains $p \rightarrow p \vee q$ and $p \rightarrow q \vee p$, we infer

9 $\vdash_E M \rightarrow .(K \rightarrow L) \vee (K' \rightarrow L')$,
10 $\vdash_E N \rightarrow .(K \rightarrow L) \vee (K' \rightarrow L')$,

and using axiom E10, we have

11 $\vdash_E M \vee N \rightarrow .(K \rightarrow L) \vee (K' \rightarrow L')$.

But in **E** we also have

12 $\vdash_E (K \rightarrow L) \vee (K' \rightarrow L') \rightarrow .(K \& K') \rightarrow (L \vee L')$.

From 11 and 12 we get

13 $\vdash_E M \vee N \rightarrow .(K \& K') \rightarrow (L \vee L')$.

That is, if F is a disjunction, then the inductive hypothesis guarantees that F entails an entailment. The other cases are either trivial or solved by essentially the same method.

At last we have

THEOREM 2. **E** has no internal fallacies of modality; i.e., If $\vdash_E A \rightarrow B$, then, if $A \in W(N_E)$ then $B \in W(N_E)$.

Suppose that $\vdash_E A \rightarrow B$ and $B \in S(N_E)$. By Lemma 6, for some M, N

$\vdash_E B \rightarrow .M \rightarrow N$;

hence

$\vdash_E A \rightarrow .M \rightarrow N$.

But then, by Theorem 1, $A \in S(N_E)$.

Finally, we remark that since every negative formula in **E** belongs to

$W(N_E)$, but not conversely, our Theorem 2 is both more universal and more precise than Maksimova's corollary to the effect that no negative formula entails (in **E**) an entailment.

§22.1.3. Fallacies of relevance. The principal result here is that if $A \rightarrow B$ is provable in **E**, then A and B share intensional content, in the sense that they share a variable. Maksimova 1967 observes that the proof techniques of Belnap 1960b or Dončenko 1963 yield a slightly stronger theorem, to be stated after we prepare for the proof, which uses matrices derived from the intensional lattice M_0 of §18.

M_0

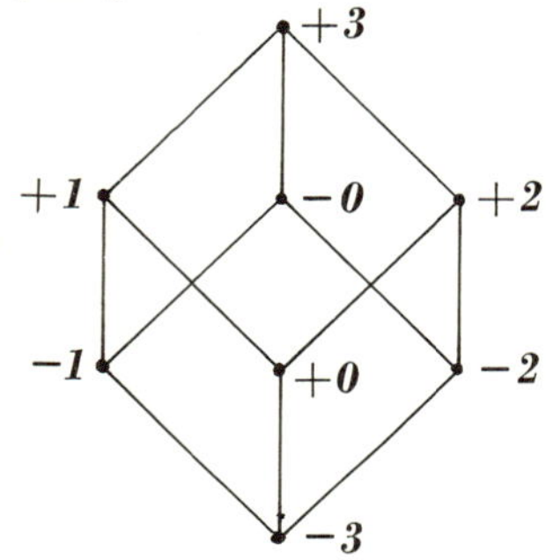

The tables are as follows:

&	−3	−2	−1	−0	+0	+1	+2	+3
−3	−3	−3	−3	−3	−3	−3	−3	−3
−2	−3	−2	−3	−2	−3	−3	−2	−2
−1	−3	−3	−1	−1	−3	−1	−3	−1
−0	−3	−2	−1	−0	−3	−1	−2	−0
+0	−3	−3	−3	−3	+0	+0	+0	+0
+1	−3	−3	−1	−1	+0	+1	+0	+1
+2	−3	−2	−3	−2	+0	+0	+2	+2
+3	−3	−2	−1	−0	+0	+1	+2	+3

∨	−3	−2	−1	−0	+0	+1	+2	+3
−3	−3	−2	−1	−0	+0	+1	+2	+3
−2	−2	−2	−0	−0	+2	+3	+2	+3
−1	−1	−0	−1	−0	+1	+1	+3	+3
−0	−0	−0	−0	−0	+3	+3	+3	+3
+0	+0	+2	+1	+3	+0	+1	+2	+3
+1	+1	+3	+1	+3	+1	+1	+3	+3
+2	+2	+2	+3	+3	+2	+3	+2	+3
+3	+3	+3	+3	+3	+3	+3	+3	+3

→	−3	−2	−1	−0	+0	+1	+2	+3
−3	+3	+3	+3	+3	+3	+3	+3	+3
−2	−3	+2	−3	+2	−3	−3	+2	+3
−1	−3	−3	+1	+1	−3	+1	−3	+3
−0	−3	−3	−3	+0	−3	−3	−3	+3
+0	−3	−2	−1	−0	+0	+1	+2	+3
+1	−3	−3	−1	−1	−3	+1	−3	+3
+2	−3	−2	−3	−2	−3	−3	+2	+3
+3	−3	−3	−3	−3	−3	−3	−3	+3

	~
−3	+3
−2	+2
−1	+1
−0	+0
+0	−0
+1	−1
+2	−2
+3	−3

THEOREM. If ⊢ *A*→*B*, then some variable occurs as an antecedent part of both *A* and *B*, or else as a consequent part of both *A* and *B*.

PROOF. We show that if *A*→*B* fails to satisfy the condition, we can find a falsifying assignment from the tables above. Any variable *p* in such a formula *A*→*B* will satisfy one of six conditions, for each of which we give *p* a different assignment, tabulated as follows:

	A	*B*	v(*p*)
p:	cp	—	−1
	ap	—	+1
	—	ap	−2
	—	cp	+2
	cp	ap	+3
	ap	cp	−3

The first row means, for example, that if the variable *p* occurs as a consequent part of *A*, and does not occur in *B*, then we give *p* the value −1; the other rows are read similarly. Clearly these six conditions exhaust the possibilities, for in any other case *A*→*B* would satisfy the condition of the theorem.

It develops that for this assignment, (i) every consequent part of *A* (including *A* itself) assumes a value in {±1, +3}, and (ii) every consequent part of *B* (including *B*) assumes a value in {±2, −3}; whence inspection of the arrow table leads us to believe that v(*A*→*B*) = −3, and that *A*→*B* is therefore unprovable in **E**. It remains only to prove (i) and (ii).

What we actually prove is something stronger for both *A* and *B*. As a sample, we state the lemma required for *A* and prove one case, leaving the remainder of the work to the reader.

LEMMA. Given the assignment described above, for every antecedent part C of A, $v(C) \in \{\pm 1, -3\}$, and for every consequent part C of A, $v(C) \in \{\pm 1, +3\}$.

Consider, e.g., the case where C has the form $D \rightarrow E$. If C is an ap, then D is a cp and E is an ap, whence by the obvious inductive hypothesis $v(D) \in \{\pm 1, +3\}$ and $v(E) \in \{\pm 1, -3\}$. Then the arrow table shows that $v(D \rightarrow E) \in \{\pm 1, -3\}$, which is what we wanted. If on the other hand C should chance to be a cp, then the inductive hypothesis and the table tell us that $v(D \rightarrow E) \in \{\pm 1, +3\}$.

With this we declare the theorem proved.

In §12 we proved that in theorems of $\mathbf{E}_{\rightarrow}$ *every* variable occurred at least once as an antecedent part and once as a consequent part (if it occurred at all), and observed that this was not true for **E**, which has as axioms

E4 $A\&B \rightarrow A$,
E5 $A\&B \rightarrow B$,
E8 $A \rightarrow A \vee B$, and
E9 $B \rightarrow A \vee B$,

in which a variable occurs but once. We notice that this situation does not occur elsewhere among the axioms for **E**; and moreover that in E4–5 we have conjunctions as antecedent parts, and in E8–9 disjunctions as consequent parts. All of this leads us naturally to ask whether E4–5 and E8–9 provide the *only* ways in which a variable can get into a theorem of **E** just once. The question can be answered affirmatively.

THEOREM. If A is a theorem of **E** containing no conjunctions as antecedent parts and no disjunctions as consequent parts, then every variable in A occurs at least once as an antecedent part *and* at least once as a consequent part (Maksimova 1967).

The proof uses the $\{\pm 1, \pm 3\}$ fragment of the matrices above, which is closed under $\sim$, $\&$, $\vee$, and $\rightarrow$. Suppose a variable B occurs in A *only* as consequent part; then we give B the value -3, and all other variables the value $+1$. Then by an induction of a type already too familiar we prove for any part C of A containing B, (i) if C is a consequent part of A then $v(C) = -3$, and (ii) if C is an antecedent part of A then $v(C) = +3$. As before, we give a sample case, after noticing first that always if C does *not* contain B, then $v(C) = \pm 1$.

Let C have the form $D \rightarrow E$.

(1) If C is a cp then the inductive hypothesis says that $v(D) = +3$ or ± 1

according as B is or is not in D, and $v(E) = -3$ or ± 1 according as B is or is not in E. Then consultation with the arrow table shows that if C contains B (so that one or both of D and E do also) then $v(C) = -3$, and otherwise $v(C) = \pm 1$.

All the other cases go the same way, and we consider the theorem proved.

We remarked near the end of §5.2 that theorems of $\mathbf{E}_{\rightarrow}$ had no loose pieces; the result just proved shows that the same holds true for $\mathbf{E}_{\overline{\rightarrow}}$, as claimed in §12, since the only way loose pieces can be introduced into a theorem is by weakening conjunctively in an antecedent part (E4–5) or disjunctively in a consequent part (E8–9). Otherwise **E** runs a very taut ship, demanding that in any other case, the arrangement of variables be like that of the archetypal foundation of all inference.

(Meyer has in correspondence (1968) pointed out the existence of a stowaway: ". . . running a taut ship *is* apparently compatible with the acceptability of items like $\overline{p\rightarrow p}\rightarrow .q\rightarrow q$, which satisfies the conditions of the Maksimova relevance theorem [above] but whose dependence upon the archetypal form of inference (in the sense of arranging variables the same way) is not such as to overwhelm the reader." We have, however, put this formula and his accomplices in irons on other charges.)

§22.2. Material fallacies. Material (perhaps unlike formal) fallacies meet an important traditional requirement of fallacies, namely, that they have a specious attractiveness which makes them convincing to the unwary. In this case the attractiveness is provided by truth: material fallacies arise from misinterpreting $\bar{A}\vee B$, among truth functional tautologies, as "if . . . then —," so that for example we are led from the truth $\bar{A}\vee(\bar{B}\vee A)$ to the falsehood "if A, then if B then A."

Though we are nearly alone in complaining about the identification of "if . . . then —" with "not . . . or —," still, every age has had its soothsayers. Hugh MacColl 1908, *vox clamantis in deserto*, wrote:

> For nearly thirty years I have been vainly trying to convince them [i.e., logicians] that this supposed invariable equivalence between a conditional (or implication) and a disjunction is in error.

We think that a large part of the attractiveness of the idea that "not . . . or —" means "if . . . then —" comes from the facts (i) that whenever A and $\bar{A}\vee B$ are theorems of **E** (or **R**) so is B (as we see in §25), which makes it look as if *modus ponens* is in the air, and (ii) that whenever B is, in a certain *outré* sense to be discussed, "deducible from" A, then $\bar{A}\vee B$ is a theorem, which makes it appear that a "deduction theorem" holds (in the sense of Tarski 1930). We shall, in what follows, consider the second case first.

But before proceeding to these cases we mention Rescher 1974, in which we found the quotation from MacColl cited above. Rescher argues, in ways we find convincing, that the enormous authority of Lord Russell was responsible for the fact that the voice of MacColl and others interested in modal logic went unheard. But having argued that Russell's anti-modal sentiments proved to have great influence, Rescher goes on to quote sentiments with which we heartily agree:

> Certainly nothing could be more wise and urbane than the pious sentiments of the concluding paragraph of Russell's [1906] review in *Mind* of MacColl's [1906] *Symbolic Logic and Its Applications:*
>
> > The present work . . . serves in any case to prevent the subject from getting into a groove. And since one never knows what will be the line of advance, it is always most rash to condemn what is not quite in the fashion of the moment.
>
> Anyone concerned for the health and welfare of modal logic as an intellectual discipline cannot but wish that Russell himself—and especially that majority among his followers who were perhaps even more royalist than their king — had seen fit to heed this eminently sound advice.

§22.2.1. The Official deduction theorem. As was suggested at the outset of Chapter I, this theorem is one of the principal targets for the philosophic polemics in this book, so we are anxious to make our objections as clear as possible. For this reason, and because of the prevailing Extensional Winds, we feel obliged to consider the subject at length — at greater length than we believe it deserves, since it appears to us that most of the views we are attacking should have been laughed out of court with only half a hearing in the first place. But duty calls, and we begin with an extended, malevolent discussion of what is meant Officially by a "deduction." We shall be speaking generally about finitary systems of logic of the usual sort, containing truth functions and perhaps some other gear, but we will always have in the backs of our minds the system **E**. We will also try from time to time, no doubt unsuccessfully, to temper the ill-humored tone of the discussion with a few bitter witticisms. As (what we think of as) a magnanimous conciliatory gesture, we conclude our tirade against the Official deduction theorem by showing that it holds for **E**.

By a "deduction" one means Officially a list L of formulas, each of which (relatively to the list in question) may be called either *adventitious* or *profectitious*. The adventitious formulas are chosen arbitrarily: they may be true, false, valid, invalid, or none of these, and they may or may not have any logical connection with any of the other formulas in the list.

Moreover they may appear anywhere in the list — top, bottom, middle, and as often as one wishes — without affecting the character of the list as a "deduction." In particular they may all be sent in a group to the top of the list, and *put before* all the other formulas. For this reason (presumably — it is difficult to think of any other) they are frequently called "premisses," from the Latin *praemittere:* to put (or send) before. This terminological choice is unfortunate, since it immediately invites confusion with a different sense of the word "premiss," namely the sense used when we try to sort out an argument, in mathematics, or morals, or whatever, with a view to arranging it in logical order for purposes, say, of evaluation. Here we are interested in what follows logically from what, and particularly in locating assumptions or premisses in the *logical* sense. But the Official sense has nothing (or has something only coincidentally) to do with that sense familiar in serious logical work. An Official "premiss" is simply a formula stuck into a "deduction," with the sole excuse that we could put it at the top if we felt like it — or at the bottom.

Profectitious formulas fall into one of two groups, again relatively to the list in question; they are treated either as axioms or as consequences of predecessors by one of the available rules of inference. Our own terminological choice may here also be deemed unfortunate, in that it includes axioms, whereas a Roman lawyer would give the term *profecticius* only to what was derived from one or more ancestors. But this defect, if it be one, can be remedied by pointing out that there is an extended sense of the word "ancestor" ("material ancestor") in which everyone is his own ancestor and may be derived from (or "materially derived from") himself by the *causa sui* rule — a stipulation which should sit equally well in both extensional logical and theological circles. Of course this sense of "ancestor" may not *exactly* coincide with the intuitions of naive, untutored folk, but it is quite adequate to Official needs, and for the rest of us who are reasonably sophisticated. If the reader is inclined to question the truth of, e.g., "everyone is his own ancestor," then this means that he has in mind some other use of "ancestor" than the material use.

A word should be said about the last or final formula in such a list, which is frequently called the "conclusion," presumably because it concludes the list; again, we can think of no better reason. This formula may be adventitious or profectitious, the latter in either of the two ways mentioned above, and hence may or may not have any logical connection with any of the other members of the list.

Before summarizing, we should add that the axioms are usually chosen with a view to expressing logical truths under some interpretation, and it is usually hoped that the rules represent valid modes of inference. Such at any rate was our intent in formulating **E**.

For the purpose of the theorem to follow, then, we consider finite lists L of formulas, some of which (call them $A_1, \ldots, A_k$) are adventitious and the rest materially profectitious, where (for notation) we let B be the final formula in L.

It remains only to point out with exasperation that as a matter of actual, historical, sociological *fact*, when an Officer gets hold of such a list, he writes $A_1, \ldots, A_k \vdash B$, and calls it *a deduction of B from the premisses* $A_1, \ldots, A_k$, or *a proof of B on* (or *under*) *the hypotheses* $A_1, \ldots, A_k$, or (when he is paying attention to the interpretation of the formalism) *a valid argument from* $A_1, \ldots, A_k$ *to B*, or *a demonstration that* $A_1, \ldots, A_k$ *have B as a consequence*, or the like. If the reader doubts this statement, we refer him to any standard textbook. Such a practice exists in spite of the fact that these "material" uses of terms like "deduction," "from," "premiss," "proof," "hypothesis," "valid," "demonstration," "consequence," and "argument" have virtually no connection with the same terms as understood outside the Official context — even by the Officer himself, when he takes off his uniform, as we see from examples in §§26.6.2 and 49.1.

Since we are bending over backward to be fair to the Official position, we say there is "virtually" no connection between the Official and the correct senses of the terms mentioned. For we must admit that in fact there is a tenuous, but quite precisely statable, connection between the Official terms (which we shall mention) and their standard correct uses (which we shall follow): the Official definition of "the list L is a deduction of B from the premisses $A_1, \ldots, A_k$" *does not absolutely preclude* the possibility that the list L is a deduction of B from the premisses $A_1, \ldots, A_k$.

That is the Official account of a "deduction."

Now if we let Σ be an arbitrary connective in a system **S** of the sort we have been considering, then with the help of these Official definitions we can state the

Official deduction theorem for Σ and **S**. If there is an Official deduction in **S** with adventitious formulas $A_1, \ldots, A_{k-1}, A_k$, and final formula B, then there is an Official deduction in **S** with adventitious formulas $A_1, \ldots, A_{k-1}$ and final formula $(A_k \Sigma B)$.

Our problem is now to consider the relations between the deduction theorem for Σ, and the intended interpretation of Σ in **S**, as an implicative connective, or conditional.

The first point is trivial: *no* kind of deduction theorem, either the Official one or any other, is sufficient to make us want to call the Σ in $(A\Sigma B)$ an implicative connective. For example, the Official deduction theorem holds for $(A\Sigma B) =_{df} (A\supset A)\vee B$ in **TV**, but no one thinks that Σ so considered is even remotely connected with "if . . . then —."

This point is helpful in interpreting the following

THEOREM. The Official deduction theorem holds when $(A\Sigma B)$ is material "implication," $\bar{A}\vee B$, in **E**.

PROOF. We borrow from subsequent sections the fact that $\bar{A}\vee A$, $B\rightarrow .\bar{A}\vee B$, $(\bar{A}\vee B)\&(\bar{A}\vee(B\rightarrow C))\rightarrow(\bar{A}\vee C)$, and $(\bar{A}\vee B)\&(\bar{A}\vee C)\rightarrow(\bar{A}\vee(B\&C))$, are all provable in **E**; then the theorem follows by an inductive argument in the standard Official way.

But of course $\bar{A}\vee B$ is no kind of conditional, since *modus ponens* fails for it, as we have remarked *ad nauseam* before. (To console the reader who thinks we have gone completely out of our minds we note that there is a connective, namely the enthymematic implication of §35, for which both the Official deduction theorem and *modus ponens* hold. The existence of such an "if . . . then —," one which we have often *used* in this book, and indeed one could hardly get along without (*think* of what it would be like *always* to state *all* the premisses for your argument), probably accounts, in part, anyway, for the attractiveness of the view that the Official deduction theorem has something to do with "if . . . then —.")

So provability of a deduction theorem for Σ in a system **S** is insufficient for interpretation of Σ as "if . . . then —." Is it necessary? Partly because of the way in which we have all learned logic, a student may get the idea that for *any* calculus with Σ as a connective, the Official deduction theorem should hold. It is, after all, the theorem on which we all cut our first teeth — our first inductive argument. But the Official definition of a "deduction," on which the theorem is based, leaves necessity out of account altogether; there is *no* suggestion that we are dealing with Relations of Ideas instead of Matters of Fact, and indeed Officially we allow that the adventitious formulas in an Official deduction might all be pure non-necessitives.

So suppose the connective Σ involves a necessity-claim; then it would be downright immodal to *require* that an Official deduction not involving the concept of necessity be provable. If Σ involves necessity, indeed, then not only does one want not to require that the Official deduction theorem hold, one wants to require that it not hold, since it yields $A \vdash (\mathrm{B}\Sigma A)$.

Did anyone ever think that a connective, such as material "implication," for which the Official deduction theorem holds, does involve some kind of logical claim? Perhaps not, but we do cite Peano 1889, who reads $\mathrm{A}\supset B$ as *ab A deducitur B*, and later on makes *on déduit* do the same work (van Heijenhoort 1967, p. 87). In any event, no one is confused today. For example, as Curry 1959 (p. 20) writes concerning a connective satisfying both *modus ponens* and the Official deduction theorem,

> The absoluteness of absolute [i.e., intuitionistic] implication does not depend on any claim to its being a definition of logical consequence. It does not pretend to be anything of the sort.

So a Σ involving a logical claim does not go together with the Official deduction theorem. On the other hand,

> . . . it is plausible to maintain that if strict implication is intended to systematize the familiar concept of deducibility or entailment, then some form of the deduction theorem should hold for it (Barcan Marcus 1953).

We are thereby led to the Official *modal* deduction theorem, first stated and proved for the stronger Lewis systems by Barcan Marcus 1946a, which we here give for the connective $\Box(\bar{A}\vee B)$ in **E**:

OFFICIAL MODAL DEDUCTION THEOREM. *If there is an Official deduction in* **E** *with adventitious formulas* $A_1, \ldots, A_{k-1}, A_k$, *and if each* A_i $(1 \leq i \leq k-1)$ *has the form* $\Box C$, *and if the final formula is* B, *then there is an Official deduction in* **E** *with adventitious formulas* $A_1, \ldots, A_{k-1}$ *and final formula* $\Box(\overline{A_k}\vee B)$.

But once more one should not suppose that this theorem makes a conditional out of $\Box(\overline{A_k}\vee B)$; *modus ponens* fails. And once more we point to a connective of §35, definable in **E** for which both this theorem and *modus ponens* hold.

Should one now suppose the Official or the Official modal deduction theorem should be required for entailment? Of course not; for where Σ is entailment it involves concepts of relevance wholly foreign to the Official ideas of a deduction. And furthermore one should require that these *fail* for entailment, or any relevant conditional, for even the Official modal deduction theorem leads to $\vdash(\Box B\Sigma(A\Sigma A))$, which wears its irrelevance on its sleeve. Nor is any of this surprising, since as our farfetched language in describing the Official concept of deduction was designed to emphasize, the Officers, fearing the rocky shoals of nonsense, refuse to hear the siren song of relevance.

Of course the leading idea of the deduction theorem and its intimate connection with *modus ponens* is absolutely essential to any sensible formal account of logic, as was pointed out clearly for the first time in Tarski 1930. And, to quote Barcan Marcus 1953 once more,

> It would be a curious explication of the concept of deducibility if, although B followed from the premiss A, B could not be said to be deducible from A.

Which is to say that an *appropriate* deduction theorem should be provable for the arrow of E. And such is the case, as we will see in §23.5.

(In fact, the search for a suitable deduction theorem for Ackermann's systems of Chapter VIII provided the initial impetus leading us to the research reported in this book.)

So there is a sense in which the Official deduction theorem and its modal counterpart are appropriate to material $\supset$ and strict $\prec$; but since these connectives do not bear the "if . . . then —" interpretation (*modus ponens* failing), there seems little point in calling them "deduction theorems." They are *theorems*, all right, but do not wear the clothes of deducibility. Aside from these objections, there is nothing wrong with them. The Official form of the "deduction" theorem is *of course* a useful methodological tool, and one we like to use ourselves. It is not such theorems, but their philosophical interpretations that we object to; not what they say, but how they are discussed informally.

It is as Wisdom 1936 said of puzzles connected with certain philosophical propositions: "It's not the stuff, it's the style that stupefies."

§22.2.2. Fallacies of exportation. We consider a particular formula $\overline{A\&B}\vee C\rightarrow\bar{A}\vee\bar{B}\vee C$, which while a valid entailment, on confusion of "not . . . or —" with "if . . . then —," becomes $A\&B\rightarrow C\rightarrow.A\rightarrow.B\rightarrow C$, which *might* be shown to be valid with the help of a deduction theorem as follows. Clearly one application of the extensional rule (&I) and one application of the intensional rule ($\rightarrow$E) suffices for a deduction of C from the premisses $(A\&B)\rightarrow C$, A, and B. Fallacious applications of the leading idea behind the deduction theorem might then take us mistakenly from this valid deduction to a "deduction" of $B\rightarrow C$ from premisses $(A\&B)\rightarrow C$ and A, thence to a "deduction" of $A\rightarrow.B\rightarrow C$ from $(A\&B)\rightarrow C$, and finally to the assertion of the "law" of exportation:

$$(A\&B)\rightarrow C\rightarrow.A\rightarrow.B\rightarrow C.$$

The plausibility of this celebrated truth functional fallacy of relevance (which is father to many, indeed perhaps all, others) arises also through a common maneuver involved in informal mathematical or logical expositions, which might, in a classroom, go something like this. "*We have shown* that S" (some sort of algebraic structure, say) "satisfies condition A, and it was proved in the last chapter of the textbook that if it satisfies both A and B, it satisfies C. So in order to prove condition C, *all we need to do* is to establish B." Then the instructor proves condition B, and adds, "Since we've just seen that *in this situation* B *is enough to give us* C, *we have* C."

Disregarding common abuses of language, like the confusion between conditions and propositions as exhibited in the quotation, the unwary student, in analyzing the form of the argument, might be led to something like

	$A\&B \rightarrow C$	("this situation")
	A,	("we have shown that")
therefore	$B \rightarrow C$.	("is enough to give us")
	B,	("all we need to do")
therefore	C.	("we have")

These perfectly innocuous informal locutions again lead by a natural, easy, and erroneous transition to the "law" of exportation as representing the form of the argument. The locution is innocent, in the sense that what the instructor did was to prove just exactly what was required to get the conclusion C in "this situation" (that $A\&B \rightarrow C$): he proved A, *and* he proved B. What is at fault is the misleading way of putting the matter, which suggests that somehow $B \rightarrow C$ follows from A, in "this situation."

This "law," as we pointed out, is canonized in the Official deduction theorem, and the fact that we are being misled by it becomes clear on reflecting that if we are told (ordinarily) that the premisses $A_1, \ldots, A_k$ imply B, we think of the premisses taken *conjointly* as implying B.

Think of examples. When it is said that the axioms of group theory imply that the identity element is unique, we understand that their *conjunction* implies this. No one would understand the statement as meaning that closure, associativity, and existence of *an* identity element conjointly imply that the-existence-of-an-inverse *entails* that the-identity-element-is-unique. How *could* we deduce anything about entailments from an incomplete set of axioms for groups, which *don't even mention* entailment? Similarly, no one has ever supposed he could *deduce* from the statement that Socrates is a man, that the fact that all men are mortal *entails* that Socrates is mortal. The statement that Socrates is a man has no consequences whatever which have to do with such essentially logical matters as entailment.

Officers generally would agree with all this, though no doubt resting their agreement on misguided grammatical arguments concerning their sharp distinction between the conditional and implication (see Appendix). But Lewis, who shares our grammatical carelessness, rejects *p*&*q*⥽*r*⥽.*p*⥽.*q*⥽*r*, on logical grounds. With sadness we note, however, that his view that "if . . . then —" can be captured by tacking a modal operator onto a truth function forces him to believe (at least with respect to his stronger systems) the equally pernicious □*p*&*q*⥽*r*⥽.*p*⥽.*q*⥽*r*.

§22.2.3. Christine Ladd-Franklin. In an article published in *Science* in 1901 (amended in 1913), Mrs. Ladd-Franklin simplified the rules of classical syllogistic reasoning, and coined the term *antilogism* to name her discovery. With her application of this principle to syllogistic reasoning (to which she confined it) we have no cavil. And we join her (1928) in deploring Mr. Johnson's pirating of her invention and her coinage: "I take it very ill of Mr. W. E. Johnson that he has robbed me, without acknowledgment, of my beautiful word 'antilogism.' . . . There is no good excuse for Mr. Johnson's having failed to recognize my claim to priority in the use of this word, for Keynes, with whom and with whose indispensable book, *Formal Logic* (1906), Mr. Johnson, by unusual good fortune seems to have been in constant touch, not only uses the word but gives me explicit credit for both the word and the thing."

To compound the crime, Mr. Johnson extended the use of the term beyond her own application, with the result that it has come to refer to the pretended equivalence of $A\&B \rightarrow C$ with $A\&\bar{C} \rightarrow \bar{B}$.

This confusion is no doubt in part attributable to the fallacy of exportation. For from

exp: $(A\&B)\rightarrow C \rightleftarrows . A\rightarrow . B\rightarrow C$

we get, through confusion with the valid co-entailment

$(A\rightarrow . B\rightarrow C) \rightleftarrows . A\rightarrow . \bar{C}\rightarrow \bar{B}$

the appropriately named

anti: $A\&B\rightarrow C\rightarrow . A\&\bar{C}\rightarrow \bar{B}$,

which brings ruination in its wake. Again with the help of E4 and E5 (and of weak jokes which turn us into monkeys' uncles) *anti* yields the disjunctive syllogism in its most virulently reduced form:

$A\&\bar{A} \rightarrow B$,

or contrapositively

$B\rightarrow A\vee\bar{A}$.

§22.3. On coherence in modal logics (by Robert K. Meyer). In this section we define a notion of coherence for modal logics and develop techniques which show that a wide class of logics are coherent; included in this class are not only familiar logics such as **S4**, but a number of logics such as the system $\mathbf{R}^{\Box}$ of §27.1.3, whose non-modal part is distinctly non-classical, and, by extension, the system **E** of entailment, and Ackerman's *strengen Implikation* (Chapter VIII). It will follow in particular that these logics have a number of interesting properties, including the **S4** property

$\vdash \Box A\vee\Box B$ iff $\vdash \Box A$ or $\vdash \Box B$.

§22.3.1. Coherence. Roughly, a logic is coherent if it can be plausibly interpreted in its own metalogic. Specifically, we presume a sentential logic **L** to be formulated with a necessity operator $\Box$, non-modal connectives $\rightarrow$, &, $\vee$, $\sim$ (and perhaps other connectives and constants which can be correlated with familiar truth functions), and formulas A, B, C, etc., built up as usual from sentential variables p, q, r, etc. Henceforth we identify **L** with its so-called *Lindenbaum matrix* — i.e., $\mathbf{L} = \langle \mathrm{F}, \mathrm{O}, \mathrm{T} \rangle$, where F is the set of formulas of **L**, T is the set of theorems, and O is a set of operations corresponding to connectives of **L**.

Let $2 = \langle 2, \mathrm{O}, \{1\} \rangle$ be the two-element Boolean algebra (considered as a matrix), where $2 = \{\mathbf{0}, 1\}$, and with operations in O corresponding to all non-modal connectives in **L** and defined as usual. A *metavaluation of* **L** shall be any function $\mathrm{v} : \mathrm{F} \rightarrow 2$ satisfying the following conditions, for all formulas A and B:

(i) $\mathrm{v}(\Box A) = 1$ iff $\vdash \Box A$ in **L**;

(ii) $\mathrm{v}(A \rightarrow B) = \mathrm{v}(A) \rightarrow \mathrm{v}(B)$, $\mathrm{v}(\sim A) = \sim \mathrm{v}(A)$, and similarly for other non-modal connectives.

A formula A of **L** is *true* on a metavaluation v iff $\mathrm{v}(A) = 1$; A is *metavalid* iff A is true on all metavaluations of **L**; **L** is *coherent* iff each theorem A of **L** is metavalid.

What makes the notion interesting is that not all logics are coherent; **S5**, for example, is incapable of interpretation in its own metalogic in our sense: **S5** is incoherent. For consider $\Box A \vee \Box \sim \Box A$ when neither $\vdash_{\mathbf{S5}} \Box A$ nor $\vdash_{\mathbf{S5}} \Box \sim \Box A$.

First an elementary consequence of coherence.

We call A and B *truth functionally equivalent* iff they are uniform substitution instances of formulas A_0 and B_0 such that (1) the sign $\Box$ does not occur in A_0 or B_0 and (2) $A_0 \equiv B_0$ is a classical tautology. For any formula C, let C' be a formula which results from C by replacement of truth functionally equivalent formulas.

The following theorem is trivial, but it generalizes well-known **S4** properties to all the logics that we shall prove coherent.

THEOREM 1. Let **L** be a coherent logic. Then

(i) $\vdash_{\mathbf{L}} (\Box A \vee \Box B)'$ only if $\vdash_{\mathbf{L}} \Box A$ or $\vdash_{\mathbf{L}} \Box B$;

(ii) $\vdash_{\mathbf{L}} (\Box A \& \Box B)'$ only if $\vdash_{\mathbf{L}} \Box A$ and $\vdash_{\mathbf{L}} \Box B$.

PROOF. Ad (1). Suppose neither $\Box A$ nor $\Box B$ are theorems of **L**. Then for an arbitrary metavaluation v, $\mathrm{v}(\Box A) = \mathbf{0}$ and $\mathrm{v}(\Box B) = \mathbf{0}$, whence $\mathrm{v}(\Box A \vee \Box B)' = \mathbf{0}$ on purely truth functional grounds. Since **L** is coherent, $\Box A \vee \Box B$ and all its truth functional equivalents are non-theorems.

Ad (ii). Similar.

§22.3.2. Regular modal logics. We shall prove coherent all modal logics which can be formulated with axioms and rules of certain kinds. In order to formulate our results in as general a way as possible, while keeping in mind those cases which are interesting in practice, we shall characterize the key notions rather sharply. $A[B_1, \ldots, B_n/p_1, \ldots, p_n]$ shall be the result of uniformly substituting the formulas $B_1, \ldots, B_n$, respectively, for the sentential variables $p_1, \ldots, p_n$ in the formula A; s(A) shall be the class of all uniform substitutions in A. Where $\langle A_0, \ldots, A_n\rangle$, $n > 0$, is a finite sequence of formulas, a uniform substitution $\langle A_0, \ldots, A_m\rangle$ $[B_1, \ldots, B_n/p_1, \ldots, p_n]$ shall be the sequence $\langle A_0[B_1, \ldots, B_n/p_1, \ldots, p_n], \ldots, A_m[B_1, \ldots, B_n/p_1, \ldots, p_n]\rangle$; s$(A_0, \ldots, A_n)$ shall be the class of all uniform substitutions in $\langle A_0, \ldots, A_n\rangle$.

A *scheme* shall be a pair $\langle A, \mathrm{s}(A)\rangle$, where A is called the *characteristic formula* of the scheme. A *rule* shall be a pair $\langle\langle A_0, \ldots, A_n\rangle, \mathrm{s}(A_0, \ldots, A_n)\rangle$, where the sequence of formulas $A_0, \ldots, A_n$, $n \geq 1$, is called the *characteristic sequence* of the rule, A_0 is called the *characteristic conclusion* of the rule, and $A_1, \ldots, A_n$ are called the *characteristic premisses* of the rule. A scheme is *tautologous* if its characteristic formula is a substitution instance of a truth functional tautology in which the sign $\Box$ does not occur; a rule is *truth functional* if the sign $\Box$ does not occur in its characteristic sequence and if the conditional whose antecedent is the conjunction of its characteristic premisses and whose consequent is its characteristic conclusion is a truth functional tautology.

Let $\mathbf{L} = \langle \mathrm{F, O, T}\rangle$ be a logic, let X be a set of schemes, and let R be a set of rules. $\langle$X, R$\rangle$ is a *formulation* of **L** provided that T is the smallest set which contains s(A) whenever $\langle A, \mathrm{s}(A)\rangle \in \mathrm{X}$ and of which $A_0[B_1, \ldots, B_n/p_1, \ldots, p_n]$ is a member whenever $\langle A_0, \ldots, A_m\rangle$ is a characteristic sequence of a member of R and each of $A_1[B_1, \ldots, B_n/p_1, \ldots, p_n], \ldots, A_m[B_1, \ldots, B_n/p_1, \ldots, p_n]$ belongs to T. If $\langle$X, R$\rangle$ is a formulation of **L**, we call members of X *axiom schemes* and members of R *primitive rules* of the formulation. Finally, we call a rule r *admissible* for a formulation $\langle$X, R$\rangle$ of **L** iff $\langle \mathrm{X, R} \cup \{\mathrm{r}\}\rangle$ is a formulation of **L** — i.e., following Lorenzen 1955 and Curry 1963, if taking r as a new primitive rule does not enlarge the class of theorems.

We shall call a modal logic *regular* only if it has a formulation $\langle$X, R$\rangle$ satisfying the following conditions:

(1) If $\langle A, \mathrm{s}(A)\rangle \in \mathrm{X}$, one of the following holds: (a) $\langle A, \mathrm{s}(A)\rangle$ is tautologous; (b) for some formula B, A is truth functionally equivalent to $\Box B \rightarrow B$; or (c–f) for some formulas B and C, A is truth functionally equivalent to

the entry in the left column below, and the corresponding entry in the right column is the characteristic sequence of an admissible rule of ⟨X, R⟩:

(c)	$\Box B \rightarrow \Box\Box B$	$\langle \Box\Box B, \Box B \rangle$
(d)	$\Box B \& \Box C \rightarrow \Box(B \& C)$	$\langle \Box(B \& C), \Box B, \Box C \rangle$
(e)	$\Box(B \rightarrow C) \rightarrow . \Box B \rightarrow \Box C$	$\langle \Box C, \Box(B \rightarrow C), \Box B \rangle$
(f)	$\Box(B \vee C) \rightarrow . \sim\Box\sim B \vee \Box C$	$\langle \Box C, \Box\sim B, \Box(B \vee C) \rangle$

(2) If $r \in$ R, one of the following holds: (a) r is truth functional; (b) the characteristic sequence of r is $\langle \Box B, B \rangle$ for some formula B. (c) the characteristic sequence of r is $\langle \Box B \rightarrow \Box C, \Box B \rightarrow C \rangle$ for some formulas B and C, and $\langle \Box C, \Box B \rightarrow \Box C, \Box B \rangle$ is the characteristic sequence of an admissible rule of ⟨X, R⟩.

It is readily observed that many familiar modal, deontic, and epistemic logics are regular, including the Lewis systems **S2**, **S3**, and **S4**, the Feys-Gödel-von Wright system **M**, the Lemmon system **S0.5**, and others. Of particular interest for present purposes is the fact that no conditions are placed on non-modal axioms and rules, save that they be classically valid; thus the Y-systems of Curry 1963 and the relevant modal logic $\mathbf{R}^{\Box}$ of §27.1.3 are regular.

We shall show that all regular modal logics are coherent by associating with each of them a special kind of structure. Let **L** be a regular modal logic. The *weak canonical matrix* W for **L** is the triple ⟨$2\times$F, O, D⟩, where $2\times$F is the set of pairs ⟨x, A⟩ such that x = $\mathit{0}$ or x = $\mathit{1}$ and A is a formula of **L**, ⟨x, A⟩ belongs to the set D of designated elements of $2\times$F iff x = $\mathit{1}$, and O is a set of operations corresponding to the connectives of **L** and defined as follows on all ⟨x, A⟩ and ⟨y, B⟩ in $2\times$F:

(I) $\langle x, A\rangle \rightarrow \langle y, B\rangle = \langle x \rightarrow y, A \rightarrow B\rangle$, $\sim\langle x, A\rangle = \langle \sim x, \sim A\rangle$, and similarly for other non-modal connectives and constants.

(II) $\Box\langle x, A\rangle = \langle \mathbf{\mathit{1}}, \Box A\rangle$ iff x = $\mathit{1}$ and $\Box A$ is a theorem of **L**; $\Box\langle x, A\rangle = \langle \mathbf{\mathit{0}}, \Box A\rangle$ otherwise.

A *canonical interpretation* of **L** in its weak canonical matrix W is any function f : F→$2\times$F satisfying the following conditions:

(a) If p is a sentential variable, $f(p) = \langle \mathbf{\mathit{0}}, p\rangle$ or $f(p) = \langle \mathbf{\mathit{1}}, p\rangle$;

(b) $f(A \rightarrow B) = f(A) \rightarrow f(B)$, $f(\Box A) = \Box f(A)$, and similarly for other connectives. A formula A of **L** is *weakly valid* in W iff $f(A) \in$ D for all canonical interpretations f of **L** in W. We now prove the key theorem.

THEOREM 2. Let **L** be a regular modal logic, and let W be its weak canonical matrix as defined above. Then for all formulas A of **L**, the following conditions hold.

(i) If A is a theorem of **L**, A is weakly valid in W;

(ii) $\Box A$ is a theorem of **L** iff $f(\Box A) = \langle 1, \Box A\rangle$ for all canonical interpretations of **L** in W;

(iii) $\Box A$ is a non-theorem of **L** iff $f(\Box A) = \langle 0, \Box A\rangle$ for all canonical interpretations f of **L** in W.

PROOF. (iii) follows directly from the definition of f and W. (ii) follows from (i) and the fact that if $f(\Box A) = \Box f(A) = \langle 1, \Box A\rangle$ for any canonical interpretation f, then by (II) (of the definition of W above), $\Box A$ is a theorem of **L**. We finish the proof of the theorem by proving (i).

Since **L** is regular, it has a formulation $\langle X, R\rangle$ satisfying the condition above. Hence if A is a theorem of **L**, there is a sequence of formulas $A_1, \ldots, A_n$ such that A_n is A and such that each A_i, $1 \leq i \leq n$, is either a substitution instance of the characteristic formula of a member of X or follows from predecessors by virtue of a rule in R. Given such a sequence, we assume on inductive hypothesis that A_h is weakly valid for all h less than arbitrary i, and we show that $f(A_i) = \langle 1, A_i\rangle$ for an arbitrary canonical interpretation f, and hence that A_i is weakly valid. There are two cases, with subcases corresponding to the conditions on regularity above.

CASE 1. $A_i \in s(B)$, where $\langle B, s(B)\rangle$ is an axiom scheme. (a) B is a truth functional tautology. Then B, and hence A_i, is a substitution instance of a classical tautology C in which $\Box$ does not occur. But C is weakly valid on purely truth functional considerations, whence so is A_i. (b) A_i is truth functionally equivalent to $\Box C \rightarrow C$, for some formula C. Then $f(A_i)$ is $\Box f(C) \rightarrow f(C)$, which is designated on truth functional grounds if $f(C) = \langle 1, C\rangle$; if $f(C) = \langle 0, C\rangle$, $\Box f(C) = \langle 0, \Box C\rangle$ by (II) above and so truth functionally $f(A_i) = \langle 1, A_i\rangle$. (c) A_i is truth functionally equivalent to $\Box C \rightarrow \Box\Box C$ for some formula C, and if $\Box C$ is a theorem of **L** so is $\Box\Box C$. By (II) unless it is the case that both $f(C) = \langle 1, C\rangle$ and $\Box C$ is a theorem of **L**, $f(A_i)$ is designated by falsity of antecedent; in the remaining case, it is designated by truth of consequent. (d) A_i is truth functionally equivalent to $\Box C \& \Box D \rightarrow \Box(C \& D)$, where if both $\Box C$ and $\Box D$ are theorems of **L** so also is $\Box(C\&D)$. By (II) unless it is the case that $f(C) = \langle 1, C\rangle$, $f(D) = \langle 1, D\rangle$, $\Box C$ is a theorem of **L**, and $\Box D$ is a theorem of **L**, $f(A_i)$ is designated by falsity of antecedent; in the remaining case, it is designated by truth of consequent. (e), (f): similar.

CASE 2. A_i follows from predecessors in virtue of a rule $r \in R$, where we may assume all predecessors weakly valid. (a) r is truth functional. Then on purely truth functional grounds, A_i is weakly valid. (b) A_i is $\Box C$, and for some $h < i$, A_h is C. On inductive hypothesis, $f(C) = \langle 1, C\rangle$ for an arbitrary canonical interpretation f, whence, since $\Box C$ is a theorem of **L**, $f(\Box C) = \langle \mathit{1}, \Box C\rangle$. (c) A_i is $\Box C \to \Box D$, and for some $h < i$, A_h is $\Box C \to D$; furthermore, if $\Box C$ and $\Box C \to \Box D$ are both theorems, so is $\Box D$. We may assume that $f(C) = \langle \mathit{1}, C\rangle$ and that $\Box C$ is a theorem of **L** (else $f(A_i)$ is designated by falsity of antecedent). Then $f(\Box C) = \langle \mathit{1}, \Box C\rangle$; furthermore, since A_h is weakly valid $f(D) = \langle \mathit{1}, D\rangle$ and, since $\Box D$ is a theorem of **L** on our assumptions, $f(A_i) = \langle \mathit{1}, A_i\rangle$ by truth of consequent. This completes the inductive argument and the proof of Theorem 2.

Theorem 2 has some interesting applications in addition to those with which we are primarily concerned here. If, for example, for a regular modal logic **L** we define a regular **L**-theory to be any set of formulas of **L** which contains all theorems of **L** and which is closed under the truth functionally valid rules of **L**, then for each such **L** there is a consistent and complete regular **L**-theory T such that $\Box A \in T$ iff $\Box A$ is a theorem of **L**, and hence, by consistency and completeness, such that $\sim\Box A \in T$ iff $\Box A$ is a non-theorem of **L**. For by Theorem 2, it is clear that the set of formulas which take designated values on any canonical interpretation in the weak canonical matrix will constitute such a theory.

We return to our main business with a corollary.

COROLLARY. Every regular modal logic **L** is coherent.

PROOF. We must show that each theorem A of **L** is true on an arbitrary metavaluation v. Define a canonical interpretation f of **L** in the weak canonical matrix W by letting $f(p) = \langle \mathit{0}, p\rangle$ if $v(p) = \mathit{0}$ and $f(p) = \langle \mathit{1}, p\rangle$ if $v(p) = \mathit{1}$ for each sentential variable p; clearly this suffices to determine the value of f on each formula of **L**.

We now show that $f(B) = \langle \mathit{1}, B\rangle$ if $v(B) = \mathit{1}$ and $f(B) = \langle \mathit{0}, B\rangle$ if $v(B) = \mathit{0}$, by induction on the length of B. This is true by specification for sentential variables, and it is trivial on inductive hypothesis if the principal connective of B is non-modal. Suppose finally that B is of the form $\Box C$. If $\Box C$ is a theorem of **L**, $v(\Box C) = \mathit{1}$ by definition of a metavaluation and $f(\Box C) = \langle \mathit{1}, \Box C\rangle$ by (ii) of the theorem; if $\Box C$ is a non-theorem of **L**, $v(\Box C) = \mathit{0}$ by definition and $f(\Box C) = \langle \mathit{0}, \Box C\rangle$ by (iii) of the theorem. This completes the inductive argument, and shows that $f(B)$ agrees with $v(B)$ for arbitrary B.

We complete the proof of the corollary by noting that, since by the theorem each theorem A of **L** is weakly valid, $v(A) = \mathit{1}$ for all metavaluations v. Hence if A is a theorem, A is metavalid, and so **L** is coherent.

§22.3.3. Regularity and relevance. We now apply Theorems 1 and 2 of §22.3.2 to the relevant logics $\mathbf{R}^{\Box}$ and **E**. That $\mathbf{R}^{\Box}$ is regular is simply a matter of checking the axioms and rules of §27.1.3 to see that they meet the conditions of regularity. This proves that $\mathbf{R}^{\Box}$ has by Theorem 1 the **S4** disjunction property; it also establishes that one cannot prove that two apodictic formulas of $\mathbf{R}^{\Box}$ are consistent unless one can prove both formulas. For let us introduce a co-tenability operator $\circ$ into $\mathbf{R}^{\Box}$ via the definition

$$A \circ B =_{\text{df}} \overline{A \rightarrow \bar{B}}.$$

Then since $A \& B$ is truth functionally equivalent to $A \circ B$, if one can prove in $\mathbf{R}^{\Box}$

$$\Box A \circ \Box B$$

then by Theorem 1 one can prove both $\Box A$ and $\Box B$. (The converse is trivial — if one can prove both $\Box A$ and $\Box B$, one can prove in $\mathbf{R}^{\Box}$ that they are co-tenable.) This solves for $\mathbf{R}^{\Box}$ a problem analogous to one raised in §26.7 for **E**. The problem is not quite analogous, for what is asked there is whether one can prove $\Box A$ consistent with $\Diamond B$, without being able to prove both in **E**, essentially. In this case it turns out there are formulas A and B of **E** such that one can; see §26.7.

$\mathbf{R}^{\Box}$ was introduced in Meyer 1968a because it putatively contained the system $\mathbf{E}^{\Box}$ exactly, on the definition (using $\Rightarrow$ for **E** and $\rightarrow$ for $\mathbf{R}^{\Box}$)

D1. $\mathrm{A} \Rightarrow \mathrm{B} =_{\text{df}} \Box(\mathrm{A} \rightarrow \mathrm{B})$.

This has remained conjecture, however, and so the results we have obtained for $\mathbf{R}^{\Box}$ do not automatically apply to **E**. [Added August 1973: the conjecture has turned out false; see end of §28.1.] Furthermore, in the §21.1 formulation of **E**, $\Box$ is defined by

D2. $\Box A =_{\text{df}} A \rightarrow A \rightarrow A$,

which, were it turned into a definitional axiom for a version of **E** with $\Box$ primitive, would not meet our conditions for regularity. The way out is to embed **E** in the system $\mathbf{E}^{\Box}$ of §27.1.3.

THEOREM 3. $\mathbf{E}^{\Box}$ (§27.1.3) is regular. Furthermore if A^* is the formula of $\mathbf{E}^{\Box}$ got by replacing in a formula A of **E** each subformula $B \rightarrow C$ with $\Box(B \rightarrow C)$, beginning with innermost parts, then A is a theorem of **E** iff A^* is a theorem of $\mathbf{E}^{\Box}$.

PROOF. That $\mathbf{E}^{\Box}$ is regular follows from the definition of regularity. To show that if A is a theorem of **E**, A^* is a theorem of $\mathbf{E}^{\Box}$, it suffices to show that if B is an axiom of **E**, B^* is a theorem of $\mathbf{E}^{\Box}$ and that *modus ponens* holds for $\Rightarrow$, as defined by D1, in $\mathbf{E}^{\Box}$; it follows that for each step $A_1, \ldots, A_n$ in a derivation of A in **E**, A_i^* is a theorem of $\mathbf{E}^{\Box}$. Actual verification of the axioms of **E** in $\mathbf{E}^{\Box}$ poses no problems and is left to the reader.

Conversely, suppose A^* is a theorem of $\mathbf{E}^{\Box}$. Replace each occurrence of the primitive sign $\Box$ in its proof with $\Box$ as defined by D2; it is easily seen that each step of the transformed derivation is a theorem of **E**; hence A^*, thus transformed, is a theorem of **E**. Finish the proof by showing that A^*, with $\Box$ defined by D2, entails A in **E**.

Theorem 3 suggests a new definition of coherence for a system in which entailment is taken as primitive. For **E** in particular, formulated with $\wedge$, $\vee$, $\sim$, and $\rightarrow$ primitive, we define a *metavaluation* to be any function defined on the set of formulas of **E** with values in $\{1, 0\}$ satisfying the following conditions, for all formulas A and B:

(i) $\mathrm{v}(A\rightarrow B) = 1$ iff $A\rightarrow B$ is a theorem of **E**;
(ii) $\mathrm{v}(A\vee B) = 1$ iff $\mathrm{v}(A) = 1$ or $\mathrm{v}(B) = 1$;
(iii) $\mathrm{v}(A\wedge B) = 1$ iff $\mathrm{v}(A) = 1$ and $\mathrm{v}(B) = 1$;
(iv) $\mathrm{v}(\sim A) = 1$ iff $\mathrm{v}(A) = 0$.

As before, we call a formula of **E** *metavalid* if it is true on all metavaluations; **E** is *coherent* if all its theorems are metavalid. We then have

COROLLARY. **E** is coherent. Furthermore a formula $A\rightarrow B$ is a theorem of **E** iff it is true on an arbitrary metavaluation. Accordingly, if the sign $\rightarrow$ does not occur in C, a formula $(A_1\rightarrow B_1)\vee \ldots \vee(A_n\rightarrow B_n)\vee C$ is a theorem of **E** iff either $(A_i\rightarrow B_i)$ is a theorem of **E** for some i or C is a truth functional tautology.

PROOF. Let A be a formula of **E**, and let A^* be the translation of A into $\mathbf{E}^{\Box}$ given by the theorem. Let v be any metavaluation of **E**, and let v* be the metavaluation of $\mathbf{E}^{\Box}$ which agrees with v on sentential variables. Use the theorem to show, for each subformula B of A, $\mathrm{v}(B) = \mathrm{v}^*(B^*)$. But if A is a theorem of **E**, A^* is a theorem of $\mathbf{E}^{\Box}$ and is hence, by the coherence of $\mathbf{E}^{\Box}$, true on all v*; so A is true on v. But v was arbitrary; hence **E** is coherent. This proves the first statement; the second is immediate from the definition of a metavaluation.

For the final part of the corollary, we know (§24.1.2) that all tautologies in which $\rightarrow$ does not occur are theorems of **E**. The sufficiency part of the last statement then follows by elementary properties of disjunction.

On the other hand, assume that none of $A_1\rightarrow B_1, \ldots, A_n\rightarrow B_n$ is a theorem of **E** and that C is not a tautology. Since $\rightarrow$ does not occur in C, there is an assignment of 0 or 1 to sentential variables which falsifies it. The extension of v to a metavaluation will falsify the disjunction, which is accordingly a non-theorem of **E**.

We remark in conclusion that of course Theorem 3 and its corollary are straightforwardly applicable to Ackermann's *strengen Implikation* of Chapter VIII, in view of the fact that it has the same theorems as **E**. They are also applicable, *mutatis mutandis*, to related systems straightforwardly translatable into regular modal logics. The reader is invited to convert this last into a proof, alternative to that of §28.4, that positive **R** (§27.1.1, no negation; call it $\mathbf{R}_+$) is prime: if $\vdash_{\mathbf{R}_+} A \vee B$ then either $\vdash_{\mathbf{R}+} A$ or $\vdash_{\mathbf{R}_+} B$. (For a start, try mapping p in **R** into $\Box p$ in $\mathrm{R}^{\Box}$.)

(After the article on which this section is based was published, I learned from Kit Fine, in correspondence, that he had hit independently on similar ideas. Meanwhile, the ideas themselves have proved capable of considerable generalization and simplification, with some remarkable consequences; in extended form, they serve as the backbone of §§61, 62 below.)

§23. Natural deduction. The Fitch-style formulations $\mathbf{FE}_{\rightarrow}$ of $\mathbf{E}_{\rightarrow}$ (§1.3) and $\mathbf{FE}_{\overline{\rightarrow}}$ of $\mathbf{E}_{\overline{\rightarrow}}$ (§9) generalize easily to the whole of **E**. The only new thing to consider concerning **FE** (the "natural deduction" form of **E**) is the role of the rule &I, since as remarked in §9, the lemma of §4 will continue to hold provided that all *axioms* added for truth functions are of the form $A \rightarrow B$. Since we are treating negation as classical we may let $A \vee B =_{\mathrm{df}} \overline{\overline{A} \& \overline{B}}$, and since the entailment-negation fragment $\mathbf{E}_{\overline{\rightarrow}}$ has already been fixed, we need worry only about adding conjunction.

§23.1. Conjunction. Our first problem here is how to handle the subscripts, which keep track of relevance. The solution is obvious, once we notice the equivalence

$$A \rightarrow (B \& C) \rightleftarrows .(A \rightarrow B) \& (A \rightarrow C),$$

which follows easily from the axiomatic formulation of §21.1. A is relevant to a conjunction of B and C just in case it is relevant to each. We take it as clear, therefore, that the rules for conjunction elimination (&E) should be

	⋮				⋮	
i	$A \& B_a$	?		i	$A \& B_a$	?
	⋮		and		⋮	
j	A_a	i &E		j	B_a	i &E
	⋮				⋮	

and the rule for conjunction introduction (&I) should be

	⋮	
i	A_a	?
	⋮	
j	B_a	?
	⋮	
k	$A\&B_a$	i j &I
	⋮	

the point being, for the introduction rule, that what is relevant to a conjunction is the same as what is relevant to *each* conjunct. The requirement that A and B have identical relevance-indices secures this and blocks such moves as that from A to $A\&(B\vee\overline{B})$, to which we objected in §15.1.

At first blush it may appear that this will suffice for conjunction, since from these rules we can produce the rules required for disjunction, as we now see.

§23.2. Disjunction. The introduction rule is obvious: from A_a to infer $A\vee B_a$, and from B_a to infer $A\vee B_a$; and the elimination rule is most conveniently put in the form: given $A\rightarrow C_a$ and $B\rightarrow C_a$, from $A\vee B_b$ we may infer $C_{a\cup b}$. These may be derived with $A\vee B$ defined as usual:

	⋮	
i	A_a	?
$i+1$	$\quad\overline{A}\&\overline{B}_{\{1\}}$	hyp
$i+2$	$\quad\overline{A}_{\{1\}}$	i+1 &E
$i+3$	$\overline{A}\&\overline{B}\rightarrow\overline{A}$	$(i+1)-(i+2)$ $\rightarrow$I
$i+4$	$\overline{\overline{A}\&\overline{B}}_a$ (i.e., $A\vee B$)	i $i+3$ contrap

and

i	$A\rightarrow C_a$	?
j	$B\rightarrow C_a$	?
k	$A\vee B_b$ [i.e., $\overline{\overline{A}\&\overline{B}}$]	?
$k+1$	$\quad\overline{C}_{\{n\}}$	hyp
$k+2$	$\quad A\rightarrow C_a$	i reit
$k+3$	$\quad\overline{A}_{a\cup\{n\}}$	$k+1$ $k+2$ contrap
$k+4$	$\quad B\rightarrow C_a$	j reit
$k+5$	$\quad\overline{B}_{a\cup\{n\}}$	$k+1$ $k+4$ contrap
$k+6$	$\quad\overline{A}\&\overline{B}_{a\cup\{n\}}$	$k+3$ $k+5$ &I
$k+7$	$\overline{C}\rightarrow\overline{A}\&\overline{B}_a$	$(k+i)-(k+6)$ $\rightarrow$I
$k+8$	$C_{a\cup b}$	k $k+7$ contrap
	⋮	

The fact that the set of relevance indices on the conclusion of the elimination rule for disjunction is the union of those on the premisses is not to be wondered at, once we reflect that the elimination rule takes the shape of a double-barreled *modus ponens*. The joining of sets in the conclusion reflects only the intensional part of the maneuver; the extensional step (namely, conjoining in effect the two entailments required as premisses) saves relevance by the device used for conjunction introduction: requiring subscripts to be the same.

We leave it to the reader to convince himself that these rules for conjunction and disjunction yield proofs of commutativity, associativity, and, with the help of negation rules from $FE_{\rightarrow}$, all sorts of De Morgan laws, pausing now only to give one example of how the rules interact.

1	$A\&B_{\{1\}}$	hyp
2	$A_{\{1\}}$	1 &E
3	$B_{\{1\}}$	1 &E
4	$B\vee C_{\{1\}}$	3 ∨I
5	$A\&(B\vee C)_{\{1\}}$	2 4 &I
6	$(A\&B)\rightarrow A\&(B\vee C)$	1–5 →I
⋮		
12	$(A\&C)\rightarrow A\&(B\vee C)$	(like 1–6)
13	$(A\&B)\vee(A\&C)_{\{1\}}$	hyp
14	$A\&B\rightarrow A\&(B\vee C)$	6 reit
15	$A\&C\rightarrow A\&(B\vee C)$	12 reit
16	$A\&(B\vee C)_{\{1\}}$	13 14 15 ∨E
17	$(A\&B)\vee(A\&C)\rightarrow A\&(B\vee C)$	13–16 →I

It would be nice if addition of the foregoing rules for conjunction and disjunction (two each) did the job for us; unfortunately another rule is required, namely, what amounts to the converse of 17 above.

§23.3. Distribution of conjunction over disjunction. In the course of proving the equivalence of FE with **E** we shall want to prove axiom E11 of E in FE. In a system in which relevance and necessity can be ignored the proof goes easily enough:

1	$A\&(B\vee C)$	hyp
2	A	1 &E
3	$B\vee C$	1 &E
4	B	hyp
5	A	2 reit (sic)
6	$A\&B$	4 5 &I (*sic!*)
7	$(A\&B)\vee C$	6 ∨I
8	$B\rightarrow(A\&B)\vee C$	4–7 →I **(sic!!)**
9	C	hyp
10	$(A\&B)\vee C$	9 ∨I
11	$C\rightarrow(A\&B)\vee C$	9–10 →I
12	$(A\&B)\vee C$	3 8 11 ∨E

But of course this is no good. Step 5 fails because A need not be an entailment, step 6 because the hypothesis B need not be relevant to A, and step 8 is a farce.

Nevertheless, the move from step 1 to step 12 is valid in the sense that it meets every criterion of relevance and necessity satisfied by the other theorems of **E**. Since respect for the arrow forbids us to write E11 as a consequence of 1–12 above. we are forced to add the move from 1 to 12 to our list of primitive rules. If we insisted on taking only $\rightarrow$, $\sim$, and & as primitive, the rule would lead to this mysterious-looking specimen:

$$A\&\overline{\overline{B}\&\overline{C}} \rightarrow \overline{\overline{A\&B}\&\overline{C}}.$$

But we prefer (suddenly) what is more conventional, and describe the rule thus:

	⋮	
i	$A\&(B\vee C)_{a}$	?
	⋮	
j	$(A\&B)\vee C_{a}$	i dist
	⋮	

This completes the rules required for **FE**, which incidentally take care also of axiom E7, as we now see.

§23.4. Necessity and conjunction. E7 now follows without an additional rule. We shall prove it at tedious length, so as to be able to make some remarks about details of the proof.

1	$(A\rightarrow A\rightarrow A)\&(B\rightarrow B\rightarrow B)_{\{1\}}$	hyp
2	$A\rightarrow A\rightarrow A_{\{1\}}$	1 &E
3	$B\rightarrow B\rightarrow B_{\{1\}}$	1 &E
4	$A\&B \rightarrow A\&B_{\{2\}}$	hyp
5	$A_{\{3\}}$	hyp
6	$A_{\{3\}}$	5 rep
7	$A\rightarrow A$	5 6 $\rightarrow$I
8	$A\rightarrow A\rightarrow A_{\{1\}}$	2 reit
9	$A_{\{1\}}$	7 8 $\rightarrow$E
⋮	⋮	
14	$B_{\{1\}}$	(like 5–9, for B)
15	$A\&B_{\{1\}}$	9 14 &I
16	$A\&B_{\{1,2\}}$	15 4 $\rightarrow$E
17	$A\&B\rightarrow.A\&B\rightarrow.A\&B_{\{1\}}$	4–16 $\rightarrow$I
E7	$\Box A\&\Box B \rightarrow \Box(A\&B)$	1–17 $\rightarrow$I

From the fact that E7 is thus provable in **FE**, it is apparent that *something* is hidden in **FE** that required explicit mention among the axioms of **E**, and it is instructive to see what this is. The quasi-concealed factor is brought into the open in steps 2, 3, and their reiterations at step 8 and step 13 (to be supplied). The rules of **FE** as currently in force allow us to reiterate *only* entailments, a condition which precludes reiteration of step 1. But the effect of the reiteration of a conjunction of entailments can always be obtained by two uses of &E, followed by two reiterations of *entailments*, followed by &I. We did not in fact use &I to conjoin steps 8 and 13, but we *did* use the fact that they *could* be conjoined, so to speak, in conjoining steps 9 and 14, which, because of step 1, have identical subscripts.

We argued in §2 that the restriction of reiteration to entailments reflected the intuitively plausible condition that we should be allowed to import into a hypothetical proof only propositions which were, if true at all, necessarily true. Allowing the *effect* of reiterating conjunctions of entailments amounts to allowing reiteration of *conjunctions* of propositions which are, if true at all, necessarily true, so that the conjunction itself is also regarded as necessary. If these last two dark sentences can be understood at all, they must be understood as saying precisely what E7 says, so by a somewhat circuitous route we arrive at a coherent understanding of E7 and reiteration of conjoined entailments.

Consideration of the formulation of **E** in §21.1 leads to the same conclusions. A check on the equivalence proof of $\mathbf{E}_{\rightarrow}$ with $\mathbf{FE}_{\rightarrow}$ in §4 reveals that the inductive case for reiteration calls for the use of

$$A\rightarrow B\rightarrow .A\rightarrow B\rightarrow C\rightarrow C,$$

or either of the formulas

$$(A\rightarrow .B\rightarrow C\rightarrow D)\rightarrow .B\rightarrow C\rightarrow .A\rightarrow D,$$

or (more simply)

$$A\rightarrow A\rightarrow B\rightarrow B,$$

all of which are equivalent in the presence of the other axioms of $\mathbf{E}_{\rightarrow}$.

For conjunctions of entailments, any of the following will do

$$((A\rightarrow B)\&(C\rightarrow D))\rightarrow .((A\rightarrow B)\&(C\rightarrow D))\rightarrow E\rightarrow E,$$
$$(A\rightarrow .((B\rightarrow C)\&(D\rightarrow E))\rightarrow F)\rightarrow .((B\rightarrow C)\&(D\rightarrow E))\rightarrow .A\rightarrow F,$$

or (more simply)

$$((A\rightarrow A)\&(B\rightarrow B))\rightarrow C\rightarrow C,$$

each of which may replace the other, or E7, in **E**. (Details will be found in §26.1.) But the intuitive content is, we believe, clearest as stated in E7.

§23.5. Equivalence of FE and E. In the foregoing sections we derived the disjunction rules for FE from those for conjunction, pretending that disjunction was defined. For the official versions of FE and E, however, we will take all four of the connectives $\rightarrow$, $\sim$, &, and $\vee$ as primitive, the reason being that from time to time we will want to isolate those axioms or rules which have to do just with some subset of the connectives, as for example $\mathbf{E}_{\stackrel{\sim}{\rightarrow}}$, or as when we come to consider $\mathbf{E}_{+}$ (§24.4), the positive part of **E**. For reference we summarize the rules, and state the equivalence theorem.

Hyp. A step may be introduced as the hypothesis of a new subproof, and each new hypothesis receives a unit class $\{k\}$ of numerical subscripts, where k is the rank (§8.1) of the new subproof.

Rep. A_{a} may be repeated, retaining the relevance indices a.

Reit. $A \rightarrow B_{\mathrm{a}}$ may be reiterated, retaining a.

$\rightarrow$I. From a proof of B_{a} on hypothesis $A_{\{k\}}$ to infer $A \rightarrow B_{\mathrm{a}-\{k\}}$ provided k is in a.

$\rightarrow$E. From A_{a} and $A \rightarrow B_{\mathrm{b}}$ to infer $B_{\mathrm{a} \cup \mathrm{b}}$.

&I. From A_{a} and B_{a} to infer $A \& B_{\mathrm{a}}$.

&E. From $A \& B_{\mathrm{a}}$ to infer A_{a}. From $A \& B_{\mathrm{a}}$ to infer B_{a}.

$\vee$I. From A_{a} to infer $A \vee B_{\mathrm{a}}$. From B_{a} to infer $A \vee B_{\mathrm{a}}$.

$\vee$E. From $A \vee B_{\mathrm{a}}$, $A \rightarrow C_{\mathrm{b}}$, and $B \rightarrow C_{\mathrm{b}}$, to infer $C_{\mathrm{a} \cup \mathrm{b}}$.

dist. From $A \& (B \vee C)_{\mathrm{a}}$ to infer $(A \& B) \vee C_{\mathrm{a}}$.

$\sim$I. From a proof of $\bar{A}_{\mathrm{a}}$ on the hypothesis $A_{\{k\}}$, to infer $\bar{A}_{\mathrm{a}-\{k\}}$, provided k is in a.

Contrap. From B_{a}, and a proof of $\bar{B}_{\mathrm{b}}$ on the hypothesis $A_{\{k\}}$, where k is in b, to infer $\bar{A}_{(\mathrm{a} \cup \mathrm{b})-\{k\}}$.

$\sim\sim$E. From $\bar{\bar{A}}_{\mathrm{a}}$ to infer A_{a}.

THEOREM. $\vdash A$ in **E** iff $\vdash A$ in FE.

The proof requires only an appropriate generalization of the lemma of §4, which in the context of **E** has the following intuitive content: since we may reiterate only necessitive propositions into subproofs, we can always prove $\Box A_{\mathrm{a}}$ in a subproof for every A_{a} such that a does *not* contain the relevance index on the hypothesis of that subproof (i.e., for every A_{a} got by reiteration or from reiterated formulas alone by applying rules). With this lemma in hand, reduction of innermost quasi-proofs goes just as in the case of $\mathrm{FE}_{\rightarrow}$ and $\mathrm{FE}_{\stackrel{\sim}{\rightarrow}}$, except that additional cases are required for &I, &E, $\vee$I, $\vee$E, and dist. Otherwise nothing new is involved, and we may leave the details for scratch-paper.

Observe that this theorem provides, in a sense, a completeness proof for **E**. We say "in a sense," since it is not a semantical completeness proof of the usual sort; it does not concern the assignment of truth values or other sorts of interpretations to the variables in such a way as to facilitate definitions of *true* or *valid* for formulas of the system. But equally "in a sense," the subscripting can be taken as giving us some sort of quasi-semantical understanding of what it means to preserve relevance. If, that is, we say that by "A is a logically sufficient and relevant condition for B" we mean that there exists a proof in FE with hypothesis $A_{\{1\}}$ and conclusion $B_{\{1\}}$, then the theorem above says that **E** is complete and consistent for this interpretation of theorems $A \rightarrow B$. But unlike the sort of thing usually put under the heading of "semantics," this definition is itself proof-theoretical, rather than set-theoretical in character. What we would like is a recognizably orthodox semantical completeness proof for **E**, such as has been provided for $\mathbf{E}_{\mathrm{fde}}$ (the first degree entailment fragment of **E**), $\mathbf{E}_{\mathrm{fdf}}$ (the first degree formula fragment of **E**), and will be given in §24 for $\mathbf{E}_{\mathrm{zdf}}$ (the zero degree formula fragment of **E**). This is the burden of Chapter IX.

§23.6. The Entailment theorem. We promised at the end of §22 to provide an "appropriate" deduction theorem for **E**, but before doing so we pause to point out a simplification of FE which makes clearer the connection between FE and $\mathbf{E}_{\mathrm{fde}}$, indicated in the title of §21. The rules 6–10, which deal with conjunction and disjunction, can be replaced by a single rule:

TAUTOLOGICAL ENTAILMENTS (TE): If $A_1 \& \ldots \& A_n \rightarrow B$ is a tautological entailment, then from $A_{1\mathrm{a}}, \ldots, A_{n\mathrm{a}}$ to infer B_{a}.

Proof of the possibility of this replacement is left to the reader (observe that it makes rule 13 redundant). If we think of (TE) as embodying the principles of $\mathbf{E}_{\mathrm{fde}}$ all at once, then we can say that FE is precisely $\mathrm{FE}_{\rightarrow} + \mathbf{E}_{\mathrm{fde}}$.

In order to state a sane deduction theorem for **E**, we must first repair the Official definition of a deduction; to this end we extend the star treatment of §3.

A *proof that* $A_1, \ldots, A_n$ *entail(s)* B consists of a list L of formulas $S_1, \ldots, S_m$, each of which is one of the premisses $A_1, \ldots, A_n$, or else an axiom of **E**, or else a consequence of predecessors by →E or &I, *such that* L *satisfies conditions* (i) and (ii):

(i) Stars (*) may be prefixed to the steps $S_1, \ldots, S_m$ of the proof so as to satisfy the following rules:

(a) if S_i is a premiss, then S_i is starred;

(b) if S_i is axiom which is not a premiss, then S_i is not starred;

(c) if S_i is a consequence of S_j and $S_j \rightarrow S_i$ by an application of →E, then S_i is starred if at least one of S_j and $S_j \rightarrow S_i$ is starred and otherwise not; and

(d) if S_i is a consequence of S_j and S_k by an application of &I, then if both S_j and S_k are starred, then S_i is starred, and if neither S_j nor S_k is starred, then S_i is not starred. (But no proofs of entailment are permitted in which an application of &I has just one starred premiss.)

(ii) In consequence of (i), the final step S_m $(=B)$ is starred.

Then as a corollary of the equivalence of **E** and **FE** we have the following

Entailment theorem. If there is a proof in **E** that $A_1, \ldots, A_n$ entail(s) B, then $(A_1 \& \ldots \& A_n) \rightarrow B$ is provable in **E**.

Corollary. If A is theorem of **E**, and B is a conjunction of all the axioms *used* in a proof of A, then $B \rightarrow A$ is provable.

Notice that the analogue of the corollary for systems of material and strict "implication" (if a rule of necessitation holds) are trivial, since in those systems if one has A as a theorem, then one always and trivially has $B \supset A$ (or $B \dashv A$) for arbitrary B. But for the theory of entailment the corollary is interesting and nontrivial, since it is not the case that every conjunction of axioms entails every theorem.

The Entailment theorem reflects a number of prejudices already insisted upon in §22.2. Among them (a) it is the *conjunction* of the premisses that entails the conclusion; and (b) the conclusion ought to follow *from* the premisses.

A "proof of B *under* hypotheses $A_1, \ldots, A_n$" ought not (as it may Officially) be a "proof of B *beneath* hypotheses $A_1, \ldots, A_n$"; we ought not let too much depend on notation alone, or on our habit of writing things in vertical lists. Of course the way formulas appear in a vertical list might be used (as the entailment theorem does) to keep some control on the notion of *from*, but the Officers don't even use it for that.

It is also to be noted that *some*, but not necessarily all, of the premisses must be *used* in arriving at B. This guarantees that the conjunction of the premisses is relevant to the conclusion, which is what is required of a sensible account of entailment, as we remarked before (see §22.2.2). When we say that $(A \rightarrow .B \rightarrow C) \rightarrow .A \rightarrow B \rightarrow .A \rightarrow C$ is a theorem of **E**, for example, we mean that it follows from axioms with the help of the rules. Of course we *don't* mean that *all* the axioms, or *all* the rules need to be used in its proof, but if B is a conjunction of axioms of **E**, including those used in the proof of A, then we would expect to have as a theorem $B \rightarrow A$; and this is precisely what the corollary above (with the help of E4–E5) guarantees.

The situation is to a certain extent clarified by the theorem at the end of §22.1.1, which gives us some understanding of formulas which admit irrelevant conjunctive parts into a formula: they must occur as antecedent parts. And as examples of the sort just discussed make clear, the entailment $A\&B \rightarrow A$ is not infected by the presence of the possibly irrelevant B: as was pointed out, this is *precisely* the kind of irrelevance that the *purely* extensional, truth functional notion of conjunction is designed to handle.

There is of course a stronger condition that we could put on the $A_1, \ldots, A_n$ in the Entailment theorem, namely that *all* the conjoined premisses are used in arriving at B. This relation is one which many of our readers have no doubt dealt with informally in the course of cleaning up a mathematical theorem for oral or written presentation; we all frequently make a final check to see just how the premisses enter into the proof, and to make sure that all the premisses have been used. (This common sort of activity, incidentally, gives the lie if anything does to the Official theory that relevance is simply too vague or mysterious or dubious to come clearly within the purview of our limited rational apparatus.) But if in fact a proof fails to use all the apparatus in the hypothesis, the argument is faulted on grounds of inelegance rather than logical incorrectness — and it is only the latter problem which is of overriding importance for **E**.

§24. Fragments of E. We have two topics to catch up with at this point. (1) We have built up to this chapter by discussing increasingly large fragments of the calculus **E**; one project remaining is to show, in as many cases as we can, that the motivation for the fragment is not lost when the fragments are combined into the full calculus. (2) A little reflection on the additions of Chapters II and III to the calculus $\mathbf{E}_\rightarrow$ of Chapter I, will convince the reader that we should expect to find plausible systems **R** of relevant implication (just like **E**, except that $\mathbf{R}_\rightarrow$ of §3 is taken as the underlying intensional "if . . . then —"), and **T** (just like **E**, except that $\mathbf{T}_\rightarrow$ of §6 is at the foundation of the intensional part). We hereby so christen **R** and **T**, and note that axioms for both can be found easily by making appropriate changes in §21.1, replacing the $\mathbf{E}_\rightarrow$ axioms by a group which yields $\mathbf{R}_\rightarrow$ or $\mathbf{T}_\rightarrow$. All of **R**, **E**, and **T** are defined again in §27.1.1, and **R** is discussed at length in Chapter V.

This takes care of (2), for the moment. We mention **R** and **T** now principally because the results concerning fragments of **E** are also available for **R** and **T**. Examples: the zero degree fragments of all three systems are identical; the first degree entailment fragments (those with only one arrow, and that the major connective) are identical; the first degree formula fragments (truth functions of those just mentioned) are identical. We will therefore adopt the following policy: fragments of **E** will be treated as fragments thereof, in honor of the fact that **E** is what this book is about. But where convenient, we will also point out by the way that some of the results hold for **R** and **T** as well.

So now we are back to (1), and we launch forthwith into a proof that the classical two valued calculus is in exactly the right sense contained in E. For philosophical reasons which we put aside at the moment (to be taken up seriously in §§30 and 33), "the right sense" involves reference (a) to the truth values *truth* ("***T***") and *falsehood* ("***F***"), which we take to be propositions of the sort discussed in §33, and (b) to *tautologyhood* in the classical sense. We defer consideration of (a), and expect the reader to know what is meant by (b).

§24.1. E and zero degree formulas. As before, the *degree* of a formula is a measure of its nesting of arrows; so zero degree formulas have no arrows at all. They are therefore formulas which involve only variables and extensional connectives. For present purposes we pick negation and disjunction as primitive (though any other functionally complete collection of truth functions could do as well), and we emphasize that, for the proof to follow, intensional idiosyncracies of our own or others are irrelevant: we give, quite independently of **E**, **R**, and **T**, a formulation of the two valued calculus which can be shown by altogether trivial proofs to be complete, consistent, decidable, and to have independent axioms. Then we show that it falls expectedly into place as a part of the intensional framework we have been discussing.

§24.1.1. The two valued calculus (TV). We want to think of formulas as made up of variables $p, q, r, \ldots$, negations $\overline{A}, \overline{B}, \overline{C}, \ldots$, and disjunctions $A \vee B$, $A \vee C, \ldots$. This gives us a recursive definition of *zero degree formula.* We also want to be able to select parts of a formula for computation. *Disjunctive parts* of a formula will fall under the same three headings, so that for example

$$p \vee \overline{q \vee r}$$

will have three disjunctive parts: (i) p (a variable), (ii) $\overline{q \vee r}$ (a negation), and (iii) $p \vee \overline{q \vee r}$ (a disjunction). More formally: A is a *disjunctive part* ("dp") of A, and if $B \vee C$ is a dp of A, so are B and C. Moreover, some dps are *atoms* (variables and their negations), and others are *non-atoms* (disjunctions and their negations, and double negations); we pursue the metaphor and call the latter *non-atomic molecules* (the terminology for use here and in §39) provided they are not disjunctions.

As notation for formulas and their disjunctive parts, we use $\varphi(A)$, with the understanding that this means

$$(\ldots \vee(A\vee \ldots)\vee \ldots)$$

or

$$(\ldots \vee(\ldots \vee A)\vee \ldots)$$

depending on how parentheses for the two-termed wedge are filled in (allowing that the φ part of $\varphi(A)$ might be void). Since it turns out that using the vinculum (overbar) for negation makes the way in which parentheses are deployed entirely immaterial, we may think of $\varphi(A)$ as

$$(\ldots \vee A\vee \ldots),$$

with the additional agreement that if $\varphi(A)$ means what was just said, then $\varphi(B)$ means

$$(\ldots \vee B\vee \ldots);$$

i.e., the context φ remains the same. Easy.

So much for notation. We are now confronted with the following problem. Someone hands us a formula A made out of variables, vincula, and wedges, and asks, "Is it the case that, whatever assignment of truth values $\boldsymbol{T}$ and $\boldsymbol{F}$ we give to the variables in A, it comes out $\boldsymbol{T}$ on the standard truth table evaluation; i.e., is A a tautology?" (If the formula is not made out of the parts just mentioned, fix it, using the usual definitions of &, $\supset$, $\equiv$, in terms of $\overline{}$ and $\vee$.)

The reply comes in two parts: a basis, and an inductive step.

1. Suppose A has no non-atomic molecules. Then A is a (perhaps many-termed) disjunction of atoms, and it will be a tautology just if there is some variable p such that both it and its negation crop up as disjunctive parts of A.

2. Suppose A has non-atomic molecules. Let B be the *leftmost* one, so that A is $\varphi(B)$, where B is either the negation of a negation or a negation of a disjunction (otherwise B is an atom). In either case we reduce the question concerning A to questions concerning shorter formulas, according to the *rules of tree construction:*

(i) if B has the form $\overline{\overline{C}}$, ask about $\varphi(C)$ instead;
(ii) if B has the form $\overline{C\vee D}$, ask about $\varphi(\overline{C})$ and $\varphi(\overline{D})$ instead.

Example. We are asked about $p\supset q\supset.q\supset r\supset.p\supset r$. First we fix it, getting $\overline{\bar{p}\vee q}\vee\overline{\bar{q}\vee r}\vee\bar{p}\vee r$. Then we construct the following tree, working from the bottom up:

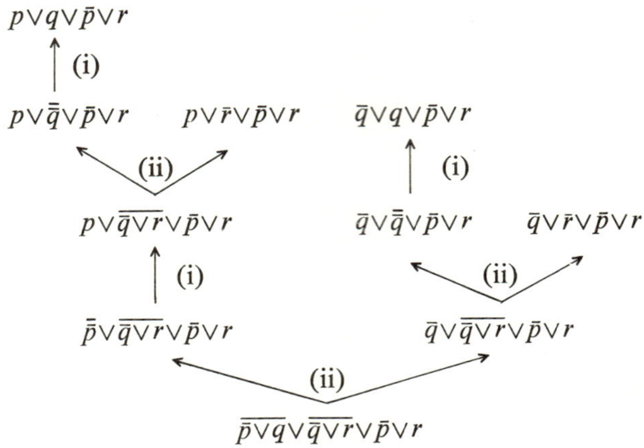

We notice that, as we go up the tree, formulas get shorter and shorter, until finally we arrive at formulas to which rules (i) and (ii) cannot apply, i.e. we are back to the basis case 1 above. We call formulas which are tautological according to 1 *axioms*, and we call the converses of the two rules for tree construction *rules of proof*. So the tree displayed above constitutes a proof of the formula at its base, since tips of all branches are axioms. We call the formal system so defined "**TV**" in honor of the truth values in which it receives its primary interpretation.

We clearly have a *decision procedure;* given a candidate, we construct its putative proof-tree mechanically, and when the procedure terminates, we examine the tips of the branches: if they are all axioms, we *have* a proof, and if they are not, there *is* none (the latter following from the fact that, since we are directed always to work on leftmost non-atomic molecules, there is at most one way to construct a proof).

We now want to see that if there is a proof, the formula at the base is a tautology, and otherwise not.

CONSISTENCY. Tautologyhood passes down each branch of a proof-tree; hence theorems of **TV** are tautologies.

PROOF. Obviously the tips are tautologies. Suppose $\varphi(C)$ is; then $\varphi(\bar{\bar{C}})$ is, since the values of C and $\bar{\bar{C}}$ are always the same. Suppose $\varphi(\bar{C})$ and $\varphi(\bar{D})$ both are. Then consider an assignment of truth values to variables in $\varphi(\bar{C})$ and $\varphi(\bar{D})$. If $\bar{C}$ (or $\bar{D}$) takes $\boldsymbol{F}$ for this assignment, then some other disjunctive part takes $\boldsymbol{T}$; so $\varphi(\overline{C\vee D})$ takes $\boldsymbol{T}$ for this assignment. Suppose on the other hand $\bar{C}$ is the *only* dp of $\varphi(\bar{C})$ which takes $\boldsymbol{T}$ for this assignment, then $\bar{D}$ is similarly the *only* dp of $\varphi(\bar{D})$ which takes T. Then $\bar{C} = \bar{D} = T$, whence $C = D = F$, so $C \vee D = F$ and $\overline{C \vee D} = T$; whence $\varphi(\overline{C \vee D}) = T$ for this assignment.

Hence all theorems are tautologies.

COMPLETENESS. All tautologies are theorems of **TV**.

PROOF. We prove the contrapositive. Suppose A is not a theorem; then in the tree with A at the base, there is a bad branch, i.e., a branch which terminates in a non-axiom. We simultaneously falsify every disjunctive part of every formula in this bad branch by assigning to a propositional variable p the value F if p occurs as a dp of a formula in the branch, and otherwise the value T. Clearly this falsifies every p occurring as a dp therein; and also every $\bar{p}$, for if $\bar{p}$ occurs as a dp, p does not, so $p = T$ and $\bar{p} = F$. Now argue inductively, with three cases. If $C \vee B$ is a dp, so are C and B; so $C = B = F$ by inductive hypothesis; so $(C \vee B) = F$. If $\bar{\bar{C}}$ is a dp, then by (i) so is C; so $C = F$, so $\bar{\bar{C}} = F$. If $\overline{C \vee B}$ is a dp, then by (ii) so is either $\bar{C}$ or $\bar{B}$. Suppose $\bar{C}$ is in the branch. Then $\bar{C} = F$, so $C = T$, so $\overline{C \vee B} = F$; similarly if $\bar{B}$ is in the bad branch. This assignment falsifies every disjunctive part of every formula of the bad branch; in particular it falsifies A, q.e.d.

This takes care of decidability, completeness, and consistency of **TV**, and for independence we simply note that no axiom can be the conclusion of a rule; hence each is independent of the others, as are (obviously) the rules.

§24.1.2. Two valued logic is a fragment of E. The extreme simplicity of the foregoing proofs of consistency and completeness of **TV** (due essentially to Schütte 1950; Anderson and Belnap 1959c was a later rediscovery) is connected partly with the fact that syntax and semantics are described in ways which makes the coincidence of provability with validity come as no surprise.

But in some mysterious way we do not know how to explain, the fact that the rules of the present syntactical formulation (unlike formulations with the primitive rule (γ): from A and $\bar{A} \vee B$ to infer B) are valid entailments, seems to us to contribute to the right-minded character of the metalogical proofs. For even in $\mathbf{E}_{\mathrm{fde}}$ we have

$$A \vee B \vee D \rightarrow A \vee \bar{\bar{B}} \vee D, \text{ and}$$
$$(A \vee \bar{B} \vee D) \& (A \vee \bar{C} \vee D) \rightarrow . A \vee \overline{B \vee C} \vee D,$$

with either or both of A and D void. That is, we have as theorems for any φ:

$$\varphi(B) \rightarrow \varphi(\overline{\overline{B}}), \text{ and}$$
$$\varphi(\overline{B})\&\varphi(\overline{C}) \rightarrow \varphi(\overline{B \vee C}),$$

and all the axioms (excluded middles with irrelevancy) are easy to prove in **E**, **R**, and **T**.

So the classical two valued calculus **TV** is included in **E**, **R**, and **T**. We note that though **E** is complete in this sense (relative to truth table tautologies), it is not "complete" in the sense of Post 1921. That is, it is not the case that addition to **E** of a (purely truth functional) non-tautology as an axiom necessarily renders the system incoherent by making *every* formula provable. And this is as it should be, since it is not the case that an arbitrary falsehood (or contradiction) implies everything. Of course if we add an object-language variable p as an axiom, with a rule of substitution, *then* the system is incoherent. But use of free variables in the presence of substitution is intended to have the effect of generality, i.e., universal quantification, so we are in effect adding $\forall pp$ as an axiom (to an extended propositional calculus based on **E**), and from the latter of course everything follows. But if we add (say) $p\&\bar{p}$, for some propositional variable p, then the only false propositions provable will be those involving p, since the principle of relevance is retained in **E**. Thus the principle of the "solidarity of contradictions," to the effect that any contradiction brings chaos in its wake, goes by the board, as it ought. (But see §29.11 for a mysterious occasion on which solidarity creeps back in.)

We must further show that **E** is conservative with respect to **TV**; i.e., that *only* tautologies among the zero degree formulas are provable in **E**. The way to show this is simply to reinterpret the arrow of **E** as plain material "implication," giving it the usual ***T-F*** table. Evidently all theorems of **E**, so construed, will turn out to be tautologies, so that in particular no non-tautologous zero degree formula can be found among the theorems of **E**. And similarly for **R** and **T**.

Hence the classical two valued calculus is exactly the extensional fragment of each of **E**, **R**, and **T**.

By taking a semantical detour, we can now prove for this fragment of **E** that if $\vdash A$ and $\vdash \bar{A} \vee B$, then $\vdash B$. For if both $\vdash A$ and $\vdash \bar{A} \vee B$, then both formulas are tautologies (consistency), whence B is a tautology (which requires a small argument), whence $\vdash B$ (completeness). This says that the rule (γ) of §25 is admissible for the zero degree fragment of **E** (and **T** and **R**). But notice that the proof of admissibility does *not* proceed by showing how to get from proofs of A and $\bar{A} \vee B$ to a proof of B; a proof-tree for the latter is constructible entirely on its own, and without reference to proofs of A and $\bar{A}\vee B$.

This is not altogether surprising, since in general A and $\bar{A}\vee B$ are not relevant to B, and we are inclined to think that the clumsiness of the usual proofs of elementary metalogical properties of the two valued calculus arises from the clumsiness of (γ) as a primitive rule.

§24.2. E and first degree entailments. We already know from §§21 and 23 that **E** contains $\mathbf{E}_{\text{fde}}$; i.e., if $A\rightarrow B$ is a tautological entailment, then $A\rightarrow B$ is a theorem of **E**. We also know that if $A\rightarrow B$ is of first degree (i.e., if A and B are both purely truth functional), then provability reduces for **E** (as for $\mathbf{E}_{\text{fde}}$) to provability of certain primitive entailments $A_i\rightarrow B_j$, as in §15.1. But application of the first theorem of §22.1.3 then tells us that for provability in **E**, each such A_i and B_j must share an atom, which tells us in turn that among formulas $A\rightarrow B$ of first degree, *only* tautological entailments are provable in **E**. Hence **E** is complete and consistent relative to the interpretation motivated in Chapter III; as are **R** and **T**. (The result also follows as a corollary of the next section, since $\mathbf{E}_{\text{fde}}$ is included in $\mathbf{E}_{\text{fdf}}$.)

§24.3. E and first degree formulas. We here establish the claim of §19.2 that $\mathbf{E}_{\text{fdf}}$ is exactly the first degree formula fragment of **E**.

It is easy to check that the axioms and the rules (TE), R1–R11 of $\mathbf{E}_{\text{fdf}}$ as formulated in §19.2 are theorems and rules of **E** (indeed for each of the rules the premisses entail the conclusion in **E**); hence **E** contains $\mathbf{E}_{\text{fdf}}$.

To prove that **E** contains no more first degree formulas than does $\mathbf{E}_{\text{fdf}}$, we suppose for reductio that F is a first degree formula provable in **E** but not in $\mathbf{E}_{\text{fdf}}$, and construct a model falsifying F, considered as a (putative) theorem of **E**. Now we know from §19.5, and the fact that $\mathbf{E}_{\text{fdf}}$ is a subsystem of **E**, that we can find in **E** a special conjunctive normal form F^* of F, and also an M^n-model of $\mathbf{E}_{\text{fdf}}$ and an assignment s of values in this M^n-model to the variables in F^*, such that F^* (hence F) comes out false in the model $\mathrm{Q} = \langle \mathrm{M}^n, \mathrm{s}\rangle$. But this is not yet sufficient for our purposes, since though this model Q satisfies $\mathbf{E}_{\text{fdf}}$, it has in it no *operation* corresponding to the arrow connective of **E**. Its *relation* $\leq$ suffices for unnested arrows, but cannot handle arrows within arrows.

What we must do is add an arrow operation to M^n having the following features:

(a) for a, b $\in \mathrm{M}^n$, a$\rightarrow$b $\in$ T (the truth filter) just in case a $\leq$ b;

(b) every theorem of **E** takes a value in T for every assignment of values to its variables.

For then if F is unprovable in $\mathbf{E}_{\mathrm{fdf}}$, the same assignment s which falsifies it in M^n will by (a) falsify it in M^n with $\rightarrow$, while by (b) every theorem of **E** is valid in M^n with $\rightarrow$. So F cannot be provable in **E**.

How now to define the arrow? The simplest definition that will work can be abstracted, as Meyer pointed out to us, from the Ackermann 1956 matrix mentioned in §5.2.1. Let L be any intensional lattice in the sense of §18.2 satisfying the following condition: there are elements $\boldsymbol{F}$, $\boldsymbol{t} \in \mathrm{L}$ such that $\boldsymbol{F}$ is the bottom of L ($\boldsymbol{F} \leq \mathrm{a}$, all $\mathrm{a} \in \mathrm{L}$) and $\boldsymbol{t}$ is the bottom of the truth filter ($\boldsymbol{t} \leq \mathrm{a}$ iff $\mathrm{a} \in \mathrm{T}$, all $\mathrm{a} \in \mathrm{L}$). Then add an arrow operation to the algebraic structure of L — let the result be $\langle \mathrm{L}, \rightarrow \rangle$ — with the following definition: $(\mathrm{a}\rightarrow\mathrm{b}) = \boldsymbol{t}$ if $\mathrm{a} \leq \mathrm{b}$, and $(\mathrm{a}\rightarrow\mathrm{b}) = \boldsymbol{F}$ if $\mathrm{a} \nleq \mathrm{b}$. The algebra is made a matrix by thinking of the elements of T as designated; and it is easy to show that all theorems of **E** are valid in $\langle \mathrm{L}, \rightarrow \rangle$. Since obviously $(\mathrm{a}\rightarrow\mathrm{b}) \in \mathrm{T}$ iff $\mathrm{a} \leq \mathrm{b}$, both (a) and (b) above hold, and we are home if we choose L as M^n.

This proof will work equally for the system **T**, since it is a subsystem of **E**. But it will not work for **R**: (b) fails for this definition of the arrow operation. In particular, $A\rightarrow.A\rightarrow A\rightarrow A$ fails for any $\mathrm{a} \in \mathrm{L}$ such that $\mathrm{a} \nleq \boldsymbol{t}$. To accommodate **R**, we must complicate the definition of $\rightarrow$ in the following way, due partly to Meyer, where $\boldsymbol{T} = \overline{\boldsymbol{F}}$ and $\boldsymbol{f} = \overline{\boldsymbol{t}}$: $(\mathrm{a}\rightarrow\boldsymbol{T}) = (\boldsymbol{F}\rightarrow\mathrm{b}) = \boldsymbol{T}$; $(\boldsymbol{t}\rightarrow\mathrm{a}) = \mathrm{a}$; $(\mathrm{a}\rightarrow\boldsymbol{f}) = \bar{\mathrm{a}}$; otherwise, $(\mathrm{a}\rightarrow\mathrm{b}) = \boldsymbol{t}$ if $\mathrm{a} \leq \mathrm{b}$ and $(\mathrm{a}\rightarrow\mathrm{b}) = \boldsymbol{F}$ if $\mathrm{a} \nleq \mathrm{b}$. Then if L is an intensional lattice with $\boldsymbol{F}$ and $\boldsymbol{t}$, and if *furthermore* $\boldsymbol{f} \nleq \boldsymbol{t}$, then all **R**-theorems are valid in $\langle \mathrm{L}, \rightarrow \rangle$; and (a) clearly stays true. (If we fail to add the *furthermore* condition, so that $\boldsymbol{f} \leq \boldsymbol{t}$, then the instance $\boldsymbol{f}\rightarrow\boldsymbol{t}\rightarrow.\boldsymbol{t}\rightarrow\mathrm{a}\rightarrow.\boldsymbol{f}\rightarrow\mathrm{a}$ of suffixing turns out badly for any a such that $\boldsymbol{t} \leq \mathrm{a}$, $\mathrm{a} \nleq \boldsymbol{t}$, $\mathrm{a} \neq \boldsymbol{T}$.) So choose L as M^n; for we know $\boldsymbol{f} \nleq \boldsymbol{t}$ in M^n.

§24.4. E and its positive fragment. The results of §§18–19 and §§24.2–3 specialize easily to a system $\mathbf{E}_+$ which we mentioned in passing in §23.5, where we pointed out that officially we take all four of the connectives of **E** and FE as primitive. This dodge enables us to consider separately the positive part $\mathbf{E}_+$ of **E**, defined by axioms E1–E11 of §21.1 (with of course $\rightarrow$E and &I), and the equivalent positive part FE_+ of FE, defined by rules 1–10 of §23.5. (Proof of the equivalence is trivial.)

As a calculus of propositions, $\mathbf{E}_+$ is not exactly a startling source of entertainment (yet it is not known to be decidable), but it will assume considerable interest in connection with the propositional quantifiers to be introduced in Chapter VI. This is true for several reasons to be discussed there, chief among them being the fact that in the system $\mathbf{E}_+^{\forall\exists p}$ (i.e., $\mathbf{E}_+$ with propositional quantification), intuitionistic propositional calculus may be precisely and naturally embedded with the help of the definitions

$$A\supset B =_{\mathrm{df}} \exists p(p\&(A\&p\rightarrow B))$$

of intuitionistic implication, and

$$\neg A =_{\mathrm{df}} A \supset \forall pp$$

of intuitionistic negation. We therefore include here a discussion of $\mathbf{E}_+$, and the first degree fragment $\mathbf{E}_{\mathrm{fde}+}$ thereof.

§24.4.1. $\mathbf{E}_+$: the positive fragment of E. Some of the results to be stated below can simply be taken over wholesale from previous discussions of $\mathbf{E}_{\mathrm{fde}}$ and $\mathbf{E}_{\mathrm{fdf}}$ (§§24.2–3), but in view of the absence of negation they may also be obtained in a much simpler direct way; in consequence some slight interest may attach to a few comments on the situation.

We observe first that $\mathbf{E}_+$ has an empty zero degree fragment: there are no tautologies in & and $\vee$ alone.

The positive fragment $\mathbf{E}_{\mathrm{fde}+}$ of $\mathbf{E}_{\mathrm{fde}}$ can be defined by the eight axioms and rules for $\rightarrow$, &, and $\vee$, of §15.2. It will be remembered from §15.3 that Smiley provided a four-element characteristic matrix for $\mathbf{E}_{\mathrm{fde}}$, the Hasse diagram for which is (with *1* designated):

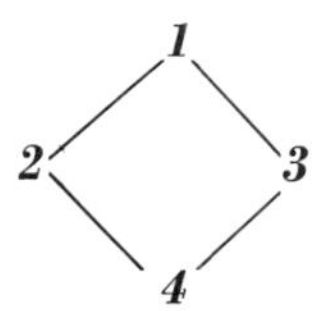

An even simpler lattice proves to be characteristic for $\mathbf{E}_{\mathrm{fde}+}$; it has the Hasse diagram (with T designated):

The truth tables for & and $\vee$ can be read off from the diagram, as can also, with a little imagination, the truth table for the arrow. Nor is it difficult to prove then that if A and B share no variables, then $A \rightarrow B$ is not provable in $\mathbf{E}_{\mathrm{fde}+}$. (We hasten to admit that these truth tables are not original with us.)

These hints should suffice to enable the reader to prove for himself that $\mathbf{E}_{\mathrm{fde}+} = \mathbf{E}_{+\mathrm{fde}}$, and also to characterize completely the first degree formula fragment $\mathbf{E}_{+\mathrm{fdf}}$ of $\mathbf{E}_+$, and prove it consistent and complete.

Truth functions do not entail entailments in $\mathbf{E}_+$ of course, since $\mathbf{E}_+$ is a subsystem of **E**. Unlike **E**, however, it is also the case that entailments do not entail truth functions in $\mathbf{E}_+$. Where A, B, and C are purely and positively truth functional, $(A \rightarrow B) \rightarrow C$ can be falsified with the help of the 8×8 matrix of §22.1.3, by giving all variables the value $-\mathbf{0}$.

Perhaps the most important fact about $\mathbf{E}_+$ is one due to Meyer, in the following section: of it **E** is a conservative extension.

§24.4.2. On conserving positive logics I (by Robert K. Meyer). Let $\mathbf{L}_+$ be a sentential logic without negation. One frequently wishes to know which classically valid negation axioms can be added *conservatively* to $\mathbf{L}_+$, in the sense that the negation-free fragment of the resulting logic **L** is precisely $\mathbf{L}_+$. This question becomes more urgent as the strength of the axioms to be added increases, for it frequently happens that one cannot add together axioms sufficient for the full classical principles of double negation, excluded middle, and contraposition conservatively. For example, the addition of plausible axioms expressing all these principles causes the negation-free fragment of intuitionism to collapse into classical logic, as is well-known.

In the present section we shall develop a general method which will enable us to prove, for several interesting systems, that their negation-free fragments are determined by their negation-free axioms. We take as the negation axioms to be added those of §21.1 and §27.1.1, namely (using names from the latter place)

A12 $A \rightarrow \overline{A} \rightarrow \overline{A}$

A13 $A \rightarrow \overline{B} \rightarrow . B \rightarrow \overline{A}$

A14 $\overline{\overline{A}} \rightarrow A$

We note in passing that these axioms lead in **E** (and in related systems) to the theoremhood of *all* forms of the double negation laws, the De Morgan laws, contraposition laws, and laws of the excluded middle and non-contradiction. In short they are strong axioms, raising non-trivial questions of conservative extension.

We assume the sentences of a positive logic $\mathbf{L}_+$ to be built up as usual from a denumerable set of sentential variables, binary connectives $\rightarrow$, &, $\vee$, and perhaps some additional connectives and constants. We assume in addition as rules of inference *modus ponens* ($\rightarrow$E) and *adjunction* (&I); the application of these rules to instances of some definite set of axiom schemes yields as usual the set T of theorems of $\mathbf{L}_+$. By the *negation completion* **L** of $\mathbf{L}_+$, we mean the result of adding negation to the formation apparatus, and taking as additional axioms all of the new sentences containing negation which are instances of the old axiom schemes together with all instances of A12–A14.

By a *possible matrix* for a logic **L**, we mean a triple $\mathrm{M} = \langle \mathrm{M}, \mathrm{O}, \mathrm{D} \rangle$, where M is a non-empty set, O is a set of operations on M exactly corresponding to the connectives, and D is a non-empty subset of M. An *interpretation* α of **L** in M is a homomorphism from the algebra of formulas (in the sense of §18.7) of **L** into M; a sentence *A* of **L** is *true* on the interpretation α if α maps *A* into a member of D, and a sentence *A* is *valid* just in case *A* is true on all interpretations α of *A* in M. Finally, M is an **L**-*matrix* iff all theorems of **L** are valid in M, $(a\&b) \in D$ whenever both $a \in D$ and $b \in D$, and $b \in D$ whenever both $a \in D$ and $(a\rightarrow b) \in D$.

Since we insist that the axioms of a logic **L** be given schematically, we have the familiar result that the set T of theorems of **L** is closed under the operation of substitution for sentential variables. Accordingly, for every logic **L** the *canonical* **L**-*matrix* $L = \langle F, O, T\rangle$ exists, where F is the set of sentences of L, O is the set of connectives (taken as operations on F), and T is the set of theorems. The *canonical interpretation* α_L of **L** in the canonical **L**-matrix L is simply the function which assigns, to each sentence of L, itself. (These ideas are simply generalizations to an arbitrary **L** of the definitions for **E** given in §25.2.2.)

We note the following truism.

LEMMA 1. Let **L** be a sentential logic. The following conditions are equivalent, for each sentence *A* of **L**:

1. *A* is a theorem of **L**;
2. *A* is valid in every **L**-matrix;
3. *A* is valid in the canonical **L**-matrix;
4. *A* is true on the canonical interpretation α_L.

PROOF immediate from definitions, in the manner of §25.2.2.

By Lemma 1, every non-theorem *A* of a positive logic $\mathbf{L}_+$ is invalid in some $\mathbf{L}_+$-matrix $M^+ = \langle M^+, O^+, D^+\rangle$; i.e., there exists some interpretation α of $\mathbf{L}_+$ in M^+ such that *A* is not true on α. To show that *A* remains a non-theorem of the negation completion **L** of $\mathbf{L}_+$, the expedient accordingly suggests itself of *enlarging* M^+ to a possible **L**-matrix $M = \langle M, O, D\rangle$, where

1. $M = M^+ \cup M^-$, where each element of M^- is the negation of some member of M^+;
2. O is the set of operations of M^+, extended to all of M, together with negation;
3. $D^+ \subset D$, and furthermore $M^+ \cap D = D^+$.

If we can work out the details of this plan in such a way that for every non-theorem *A* of $\mathbf{L}_+$ there exists in conformity to 1–3 an *enlargement* M of a matrix M^+ that rejects A, then $\mathbf{L}$ is a conservative extension of $\mathbf{L}_+$ provided that each such M is an $\mathbf{L}$-matrix — i.e., satisfies $\mathbf{L}_+$ together with the negation axioms A12–A14. The reason is that if these conditions are fulfilled, there was some interpretation α^+ of $\mathbf{L}_+$ in M^+ on which A is not true; letting α be the interpretation of $\mathbf{L}$ in M which agrees with α^+ on sentential variables (which uniquely determines α), we see by 2 that α and α^+ agree wherever the latter is defined and hence by 3 that A is not true on α; hence by Lemma 1, A is a non-theorem of $\mathbf{L}$.

So much for general strategy; how is it to be carried out in particular cases? Given a particular $\mathbf{L}_+$ and $\mathbf{L}_+$-matrix $\mathrm{M}^+ = \langle \mathrm{M}^+, \mathrm{O}^+, \mathrm{D}^+ \rangle$, the formation of a definite plan to enlarge M^+ to $\mathrm{M} = \langle \mathrm{M}, \mathrm{O}, \mathrm{D} \rangle$ requires specific answers to the following questions:

4. How is the set of new elements M^- to be determined?
5. How is negation to be defined on $M^+ \cup M^-$?
6. Given by 2 that the positive operations of O are to agree with corresponding operations of O^+ where the latter are defined, how are these operations to be defined when one or both of their arguments is in M^-?
7. Given that an element a in M^+ is to belong to D iff a belongs to D^+, which elements a in M^- shall belong to D?

Strong negation laws, though they are a burden in attempting to carry out syntactic proofs of conservative extension (since with strong laws new negation-free theorems could have come in lots of ways), have their uses in attempting to answer questions such as 4–7. In the presence of double negation, the correct answer to 4 and 5 is immediate:

I. M^- shall be a set disjoint from M^+ and there shall be a bijection * from M^+ onto M^- such that, for all $a \in M^+$, $\bar{a} = a^*$ and $\overline{a^*} = a$.

A partial answer can be given to 6 as well on the reflection that we want to preserve De Morgan laws and contraposition. Let $\rightarrow^+$, $\&^+$, $\vee^+$ be the operations whose existence we have required in O^+ (there may, of course, be others), and let $\rightarrow$, &, $\vee$ be the corresponding operations in O. Then, * being as in I,

II. For a, b in M^+,

a. $(a \rightarrow b) = (a \rightarrow^+ b)$;
b. $(a \& b) = (a \&^+ b)$;
c. $(a \vee b) = (a \vee^+ b)$;
d. $(a^* \rightarrow b^*) = (b \rightarrow a)$;
e. $(a^* \& b^*) = (a \vee b)^*$;
f. $(a^* \vee b^*) = (a \& b)^*$.

a–f defines the operations $\rightarrow$, &, and $\vee$ whenever both arguments are in M^+ or both are in M^-. The question now arises, how are M^+ and M^-

to be related? Of course this question is to be answered as simply as possible, and one simple answer is to allow everything in one of these sets intuitively to imply everything in the other; the felicitous choice turns out to be to let the M^+'s imply the M^-'s, and the desire to attend to elementary properties of conjunction and disjunction almost forces on us the following:

III. For all a, b in M^+,

a. $(a\&b^*) = (b^*\&a) = a$;
b. $(a\vee b^*) = (b^*\vee a) = b^*$;
c. $(a\rightarrow b^*) \in D$;
d. $(a^*\rightarrow b) \notin D$.

III takes leave of the self-evident principles that inspired I and II and by so much can be considered one among alternate strategies; our justification is that it often works and may be neatly pictured. For if, as is often the case, M^+ is a lattice in which $\&^+$ and $\vee^+$ deliver, respectively, greatest lower and least upper bounds, and the order $\leq^+$ is defined by setting $a \leq^+ b$ iff $(a\rightarrow b) \in D^+$, then the effect of I and II is to make of M^- a copy of M^+ (with the order relation reversed); III finishes the job of defining an extended order $\leq$ by making $a \leq b$ whenever $a \in M^+$ and $b \in M^-$; we complete the picture by observing that if M^+ is a {distributive} lattice, M is a {distributive} lattice.

c and d leave something to be desired, inasmuch as they do not yield specific values for the arrow operation. Forgetting this temporarily, we note that we automatically have an answer to our question 6; for since all $^+$'s imply all $^-$'s, we can only close D under *modus ponens* as follows:

IV. $D = D^+ \cup M^-$.

IV has the advantage of being simple, but also assures that, since D^+ was required to be non-empty, our matrix M will be inconsistent in the sense that, for some a, we have both $a \in D$ and $\bar{a} \in D$. This would have cost us some pain had we included the implicational paradox $A\&\overline{A} \rightarrow B$ among the negation axioms; as it is, all that is shown is that our simple-minded approach won't work for systems simple-minded enough to have implicational paradoxes of that sort.

I–IV already enables us to get conservative extension results for a number of positive logics $\mathbf{L}_+$. We shall not pause to derive them, however, since for logics such as **E** the vagueness of IIIc and IIId won't do. It turns out, however, that if the matrix M^+ from which we began has a certain simple property, we can specify IIIc and IIId satisfactorily for **E** also.

Let all of the operations of O be among $\{\rightarrow, \&, \vee, \overline{}, \circ, +\}$. (See §27.1.4 for $\circ$ and $+$.) We shall say that a matrix $M = \langle M, O, D\rangle$ is *rigorously compact* provided there exists an element $\boldsymbol{F} \in (M-D)$ and $\boldsymbol{T} \in D$ with the following properties:

V. For all $a \in M$, whenever the relevant operation is in O,

a. $(\boldsymbol{F}\rightarrow a) = (a\rightarrow\boldsymbol{T}) = (a+\boldsymbol{T}) = (\boldsymbol{T}+a) = (a\vee\boldsymbol{T}) = (\boldsymbol{T}\vee a) = \boldsymbol{T}$;
b. $(\boldsymbol{F}\&a) = (a\&\boldsymbol{F}) = (\boldsymbol{F}\circ a) = (a\circ\boldsymbol{F}) = \boldsymbol{F}$;
c. $(\boldsymbol{T}\&a) = (a\&\boldsymbol{T}) = (\boldsymbol{F}\vee a) = (a\vee\boldsymbol{F}) = a$;
d. If $a \neq \boldsymbol{F}$, $(a\rightarrow\boldsymbol{F}) = \boldsymbol{F}$ and $(a\circ\boldsymbol{T}) = (\boldsymbol{T}\circ a) = \boldsymbol{T}$;
e. If $a \neq \boldsymbol{T}$, $(\boldsymbol{T}\rightarrow a) = \boldsymbol{F}$ and $(\boldsymbol{F}+a) = (a+\boldsymbol{F}) = \boldsymbol{F}$;
f. $\overline{\boldsymbol{F}} = \boldsymbol{T}$ and $\overline{\boldsymbol{T}} = \boldsymbol{F}$.

a–f gives $\boldsymbol{T}$ and $\boldsymbol{F}$ many of the properties of two valued truth tables; in particular, if $M = \{\boldsymbol{F}, \boldsymbol{T}\}$, what a–f determine are the classical truth tables. A somewhat more interesting case arises when $M = \{\boldsymbol{F}, \boldsymbol{N}, \boldsymbol{T}\}$, $D = \{\boldsymbol{N}, \boldsymbol{T}\}$, where $\boldsymbol{N}$ is a (neuter) element distinct from $\boldsymbol{F}$ and $\boldsymbol{T}$. If we specify

VI. $\boldsymbol{N} = (\boldsymbol{N}\&\boldsymbol{N}) = (\boldsymbol{N}\vee\boldsymbol{N}) = (\boldsymbol{N}\rightarrow\boldsymbol{N}) = (\boldsymbol{N}\circ\boldsymbol{N}) = (\boldsymbol{N}+\boldsymbol{N}) = \boldsymbol{N}$,

and otherwise let operations on M be determined by a–f, the result is a matrix of some importance. It is, in fact, the first distinctive Sugihara matrix (§29.3); we shall call it M_3, as in §29.4, and note that it has a natural representation in the integers $\{-\mathit{1}, \mathit{0}, \mathit{1}\}$, with & going to *min*, ∨ to *max*, ‾ to inverse. (Its arrow-negation part is important at the end of §14.7.)

Apart from truth tables, most familiar matrices are not rigorously compact. The reason lies in Vd and Ve; in e.g., Boolean or pseudo-Boolean algebras (see Rasiowa and Sikorski 1963), for a such that $\boldsymbol{F} \neq a \neq \boldsymbol{T}$, $(\boldsymbol{T}\rightarrow a) = a$, violating Ve, and $(a\rightarrow\boldsymbol{F}) = \bar{a}$, violating Vd in the Boolean case. It is again those rigorously compact matrices that are lattices that are most easily pictured; it is obvious from definitions that $\boldsymbol{F}$ and $\boldsymbol{T}$ are then the lattice zero and unit; moreover $\boldsymbol{F}$ and $\boldsymbol{T}$ are *isolated* with respect to negation and the intensional operations →, ∘, + in the sense that if *one* argument to an intensional operation is $\boldsymbol{F}$ or $\boldsymbol{T}$, the value will be $\boldsymbol{F}$ or $\boldsymbol{T}$ whatever the other argument. Not surprisingly, some rigorously compact matrices have been useful for the semantics of relevant logics — e.g., in §§19 and 25; it turns out, as we prove below, that any matrix for a relevant logic may be trivially embedded in a rigorously compact matrix.

In fact, where $M = \langle M, O, D\rangle$ is an **L**-matrix, let the *rigorously compact extension of* M be the matrix $M^* = \langle M^*, O^*, D^*\rangle$, where M^* is got by adding $\boldsymbol{F}$ and $\boldsymbol{T}$ to M; D^*, by adding $\boldsymbol{T}$ to D; O^*, by extending the operations of O to M^* by Va–Vf. M^* is a possible **L**-matrix, though it is not necessarily an **L**-matrix; e.g., if **L** is classical sentential logic and M is truth tables, M^* is a 4-element matrix (the Sugihara matrix M_4, in fact) in which the classical theorem scheme $A\rightarrow(B\rightarrow A)$ is invalid; so M^* is not in this case an **L**-matrix.

It would be interesting, accordingly, to characterize the class of sentential logics **L** such that rigorously compact extensions of **L**-matrices are themselves invariably **L**-matrices. Our attempts to solve this problem have all foundered on counterexamples, but a necessary condition toward its solution is found in the following

LEMMA 2. Let **L** be a sentential logic whose connectives are among $\{\rightarrow, \&, \vee, \bar{\ }, \circ, +\}$. Suppose that for every **L**-matrix M the rigorously compact extension M* just defined for M is also an **L**-matrix. Then all theorems of **L** are valid in the Sugihara matrix M_3.

PROOF. It suffices to note that the 1-point matrix N = $\langle\{N\}, O, \{N\}\rangle$, with operations defined trivially by VI, is an **L**-matrix for every **L** with the connectives above, and that the rigorously compact extension of N is M_3. So trivially M_3 is by hypothesis an **L**-matrix, which was to be proved.

Lest the reader feel that we have cheated by dragging in the trivial matrix N, we present him with a Corollary: change "every" to "some" in the second sentence of Lemma 2; the lemma still holds. (Essentially the idea is that if the rigorously compact extension of any **L**-matrix M is an **L**-matrix, so is its image under the function that takes ***T*** into ***T***, ***F*** into ***F***, and everything else into ***N***. For that function is readily observed to be a *matrix homomorphism* from M* into M_3, whence the validity of the axioms of **L** in M_3 follows from their validity in M*.)

Every sublogic **L** of the classical sentential calculus admits a rigorously compact **L**-matrix — truth tables, as we have noted, will do. But what Lemma 2 and its Corollary show is that we cannot in general preserve **L**-matrixhood by tacking on ***T*** and ***F*** satisfying Va–Vf; in particular, such tacking on is never successful if any of $(p\&\bar{p}) \rightarrow q$, $p \rightarrow (q \rightarrow p)$, $p \rightarrow (q \vee \bar{q})$ and their ilk is a theorem of **L**, since all such are invalid in M_3. On the other hand, for the logics classified as relevant, such tacking on always works, an observation to be confirmed below where needed for present purposes.

That interlude being over, we can return to IIIc and IIId. Let $\mathbf{L}_+$ be a positive logic and let $M^+ = \langle M^+, O^+, D^+\rangle$ be a *rigorously compact* $\mathbf{L}_+$-matrix, where the operations in O^+ are $\rightarrow$, &, $\vee$. Then, * being as in I, let the *rigorous enlargement* M = $\langle$M, O, D$\rangle$ be defined by the specifications I–IV and VII, which by improving on IIIc–d will completely answer all our questions 4–7 for rigorously compact $\mathbf{L}_+$-matrices.

VII. For all a, b in M^+, where ***F*** and ***T*** are the particular elements of M^+ given by the conditions V determining rigorous compactness,

a. $(a \rightarrow b^*) = \boldsymbol{T}$;
b. $(a^* \rightarrow b) = \boldsymbol{F}$.

We shall now characterize certain logics as *rigorous*. By the *basic positive rigorous logic* $\mathbf{BR}_+$, we mean the logic formulated with $\rightarrow$, &, $\vee$ primitive, with rules of *modus ponens* for $\rightarrow$ and adjunction for &, and with the following axiom schemes (as in §27.1.1).

POSTULATES FOR $\mathbf{BR}_+$

A1	$A\rightarrow A$
A2	$A\rightarrow B\rightarrow .B\rightarrow C\rightarrow .A\rightarrow C$
A3	$A\rightarrow B\rightarrow .C\rightarrow A\rightarrow .C\rightarrow B$
A5	$A\&B \rightarrow A$
A6	$A\&B \rightarrow B$
A7	$(A\rightarrow B)\&(A\rightarrow C)\rightarrow .A\rightarrow B\&C$
A8	$A \rightarrow (A\vee B)$
A9	$B \rightarrow (A\vee B)$
A10	$(A\rightarrow C)\&(B\rightarrow C)\rightarrow .A\vee B\rightarrow C$

All of the axioms of $\mathbf{BR}_+$ are theorems of $\mathbf{E}_+$, but $\mathbf{BR}_+$ is a much weaker system, lacking in particular the **E**- and **T**-valid principles of distribution of & over $\vee$, contraction, and the **E**-theory of modality. Nevertheless, $\mathbf{BR}_+$ (and the negation-completion **BR** one gets by adding A12–14) is of some interest as a minimal relevant logic (see §8.11); it has a deduction theorem (of sorts) and familiar replacement properties hold for it; A5–A10, of course, are just lattice properties (§18.1). (For still more minimal minimal logics, see §§50 ff.)

A logic $\mathbf{L}_+$ is a *positive rigorous logic* if it can be formulated with the same connectives, axiom schemes, and rules of inferences as $\mathbf{BR}_+$, with perhaps one or more of the following taken as additional axiom schemes; **L** is a *rigorous logic* if it can be formulated as the negation completion, as defined at the beginning of this section, of a positive rigorous logic.

C1	$(A\rightarrow .A\rightarrow B)\rightarrow .A\rightarrow B$	(i.e., A4 of §27.1)
C2	$(A\rightarrow .B\rightarrow C\rightarrow D)\rightarrow .B\rightarrow C\rightarrow .A\rightarrow D$	(for which see §8.3.3)
C3	$A\rightarrow A\rightarrow B\rightarrow B$	(for which see §21.1)
C4	$\Box A\&\Box B \rightarrow \Box(A\&B)$	(i.e., A16 of §27.1.1)
C5	$A\&(B\vee C) \rightarrow (A\&B)\vee C$	(A11 of §27.1.1)
C6	$((A\&B)\rightarrow C)\&(A\rightarrow(B\vee C))\rightarrow .A\rightarrow C$	(see end of §27.1.1)
C7	$A\rightarrow B\rightarrow .A\rightarrow B\rightarrow .A\rightarrow B$	(A18)

We have chosen C1–C7 because their addition to $\mathbf{BR}_+$ produces logics in which people have taken an independent interest. For example, adding C1 and C5 produces the positive fragment $\mathbf{T}_+$ of the system **T** of ticket entailment (§6). Adding C2 and C4 to $\mathbf{T}_+$ (alternatively, adding C3 and

C4) produces $\mathbf{E}_+$. C6 is a strengthened version of C5, and adding C7 to **E** produces the Dunn-McCall system **EM** (§8.15, 27.1.1).

We can now prove our principal result.

THEOREM. Let $\mathbf{L}_+$ be a positive rigorous logic. Then the negation completion **L** of $\mathbf{L}_+$ which results from taking A12–14 as additional axiom schemes is a conservative extension of $\mathbf{L}_+$.

PROOF. We must show, for a given non-theorem A of $\mathbf{L}_+$, that A is a non-theorem of **L**. Suppose then that A is unprovable in L^+. Consider first the canonical $\mathbf{L}_+$-matrix L^+. By Lemma 1, A is not true in L^+ on the canonical interpretation $\alpha_{\mathbf{L}+}$.

Let $M^+ = \langle M^+, O^+, D^+\rangle$ be the rigorously compact extension of $\mathbf{L}_+$ whose definition was given above Lemma 2. We remark that by Lemma 2 and its informal Corollary at least a necessary condition that M^+ be an $\mathbf{L}_+$-matrix is fulfilled; the sufficient condition is that the axioms of $\mathbf{L}_+$ are valid in M^+ and that the rules are strongly satisfied; since M^+ is got from L^+ by adding $\boldsymbol{F}$ and $\boldsymbol{T}$, and since L^+ is known to be an $\mathbf{L}_+$-matrix, this is exhausting but easy; we do two cases and leave the rest to the reader.

Ad A5. Show for all a, b $\in M^+$, $(a\&b)\to a \in D^+$. Cases. (1) $a = \boldsymbol{F}$. By V, $(\boldsymbol{F}\&b)\to\boldsymbol{F} = (\boldsymbol{F}\to\boldsymbol{F}) = \boldsymbol{T} \in D^+$. (2) $a = \boldsymbol{T}$. By V, $(\boldsymbol{T}\&b)\to\boldsymbol{T} = \boldsymbol{T} \in D^+$. (3) $b = \boldsymbol{F}$. By V, $(a\&\boldsymbol{F})\to a = (\boldsymbol{F}\to a) = \boldsymbol{T} \in D^+$. (4) $a \neq \boldsymbol{F}$ and $a \neq \boldsymbol{T}$ and $b = \boldsymbol{T}$. By V, $((a\&\boldsymbol{T})\to a) = (a\to a)$, which is an element of L^+ and which is designated therein by the validity in L^+ of A1. (5) a and b are both distinct from $\boldsymbol{F}$ and $\boldsymbol{T}$. Then $(a\&b)\to a$ is by the validity of A5 in L^+ a designated element of L^+ and hence belongs to D^+. Cases exhausted.

Ad adjunction. Show for all a, b $\in D^+$, $a\&b \in D^+$. Cases. (1) $a = \boldsymbol{T}$. By Vc, $(\boldsymbol{T}\&b) = b \in D^+$ on assumption. (2) $b = \boldsymbol{T}$. Similar. (3) $a \neq \boldsymbol{T}$, $b \neq \boldsymbol{T}$. Adjunction holds on assumption for L^+. Cases closed.

So M^+ is an $\mathbf{L}_+$-matrix. Let $M = \langle M, O, D\rangle$ be the same rigorous enlargement of M^+ whose definition was completed by VII. Clearly A is not true on the interpretation α_M which agrees with $\alpha_{\mathbf{L}+}$ on sentential variables, for since A is negation-free, $\alpha_M(A) = \alpha_{\mathbf{L}+}(A) \notin D$ by definition.

Since as just noted A is invalid in M, A is by Lemma 1 a non-theorem of **L** provided that M is an **L**-matrix. We end the proof accordingly by showing M an **L**-matrix, given that M^+ is an $\mathbf{L}_+$-matrix. That D is closed under *modus ponens* and *adjunction* is clear from II and III. Show axioms valid by cases. Example — let * be as in I and let a, b, c belong to M^+; then as part of the verification of A2 note that by IIa $((a^*\to b^*)\to.(b^*\to c^*)\to(a^*\to c^*)) = ((b\to a)\to.(c\to b)\to(c\to a))$; but the latter belongs to D^+ because A3 is valid in M^+. Similar moves validate the negation axioms A12–A14, bearing in mind that I–IV and VII were chosen with the validation of those axioms in mind; example — for $\bar{\bar{a}} \in M^+$ note that $\bar{\bar{a}} \rightarrow a = (\overline{a^* \rightarrow a})$ (by I) $= (a \rightarrow a)$, which belongs to D^+ since **A1** is valid in M^+; this partially confirms the validity of A14 in M. So it goes, and the interested reader may amuse himself by checking all the computational possibilities in like manner.

§24.5. E and its pure entailment fragment. Using the semantic methods of Chapter IX, Meyer has shown that, indeed, **E** is a conservative extension of $\mathbf{E}_{\rightarrow}$; see §60.4.

§25. The disjunctive syllogism. We have already considered this "principle of inference" in several places, notably in §§16 and 22.2, and have given it deservedly bad marks. There are, however, several principles which have been put under this rubric, and in the first subsection we distinguish and discuss some of these. The remaining two subsections are devoted to a theorem of Meyer and Dunn 1969, which shows exactly where the truth lies.

§25.1. The Dog. The Dog plays a curious role in the history of the disjunctive syllogism. His thoughts on the topic were reported and discussed by such diverse authorities as Sextus Empiricus, the anonymous medieval Bestiarists, and Samuel Taylor Coleridge. Two of these reveal The Dog's proclivity for intensional senses of "or," and the third shows that The Dog has astounding logical acumen, despite The Man's attempt to muddy the analytical waters. We embark on these topics one by one.

Sextus Empiricus 200(?) wrote of the Real Dog as follows:

> Now it is allowed by the Dogmatists that this animal, The Dog, excels us in point of sensation: as to smell he is more sensitive than we are, since by this sense he tracks beasts that he cannot see; and with his eyes he sees them more quickly than we do; and with his ears he is keen of perception. . . . And according to Chrysippus, who shows special interest in irrational animals, The Dog even shares in the far-famed "Dialectic." This person, at any rate, declares that The Dog makes use of the fifth complex indemonstrable syllogism when, on arriving at a spot where three ways meet, after smelling at the two roads by which the quarry did not pass, he rushes off at once by the third without stopping to smell. For, says the old writer, The Dog implicitly reasons thus: "The creature went either by this road, or by that, or by the other: but it did not go by this road or by that: therefore it went by the other."

This passage *says* that Chrysippus's Dog was invoking the fifth indemonstrable syllogism of the Stoics: "Either the first or the second; but not the second; therefore the first." But if this is taken to mean that his dog endorsed $(A \vee B) \& \overline{B} \rightarrow A$, or the version we usually discuss, $A \& (\overline{A} \vee B) \rightarrow B$, then we will stoutly defend The Dog. Obviously this friend of Chrysippus could tell that the alternatives he was considering were relevant to one another, and that the "or" involved was not a simple truth function.

So could that equally splendid medieval specimen, whom the Bestiarist (according to T. H. White 1954) describes as follows:

> Now none is more sagacious than Dog, for he has more perception than other animals and he alone recognizes his own name. . . . When a dog comes across the track of a hare or a stag, and reaches the branching of the trail, or the criss-cross of the trail because it has split into more parts, then The Dog puzzles silently with himself, seeking along the beginnings of each different track. He shows his sagacity in following the scent, as if enunciating a syllogism. "Either it has gone this way," says he to himself, "or that way, or, indeed, it may have turned twisting in that other direction. But it has neither entered into this road, nor that road, obviously it must have taken the third one!" And so, by rejecting error, Dog finds the truth.

Finding truth in this way is perfectly all right with us, *provided* the "or" is either (a) truth functional in a *very* special case (as in the theorem of §16.2.3) or else (b) not truth functional at all, and hence not such as to allow the inference from A to A-or-B. Such senses of "or" we touched on at the end of §16, but having no satisfactory analysis of the topic, as the brief discussion there makes clear, we forthwith abandoned it.

It is at any rate apparent that we need charge neither The Dog of Chrysippus nor that of that Bestiarist with thinking that his "or" was truth functional; it remained for our human forebears to make this error.

Suppose now for the sake of argument that The Dog reasons as follows: "The arguments of §§16 and 22 make it clear that, when The Man accepts $A \& (\bar{A} \vee B) \rightarrow B$, he is making a simple inferential blunder. But surely The Man has *something* in mind, and we may charitably suppose him to have been believing that, whereas B clearly is not entailed by A and $\bar{A} \vee B$, B on the other hand *is* derivable from A and $\bar{A} \vee B$, in the sense that from A and $\bar{A} \vee B$ as premisses we can find a deduction of B."

This charity, though welcome, is misplaced, at least for a plausible understanding of what The Dog means by "derivable." For as the Entailment theorem of §23.6 teaches us, if there were a proof that A and $\bar{A} \vee B$ entailed B, then we would have $\vdash A \& (\bar{A} \vee B) \rightarrow B$, which we know is not so.

To be sure there are two other equally kind, relevant alternatives which The Dog may be attributing to The Man.

Maybe The Man meant that there were some axioms *Axioms* which when conjoined to the premisses, would produce the desired entailment: he may have supposed that $\vdash A\&(\bar{A}\vee B)\&Axioms{\rightarrow}B$. This could have happened if, being as confused as he is, he was thinking of the Official deduction theorem of §22.2.1; for in view of §23.6, the Entailment theorem, there would be an Official deduction with adventitious formulas A and $\bar{A}\vee B$ and final formula B just in case there is a conjunction of axioms *Axioms* such that $\vdash(A\&(\bar{A}\vee B)\&Axioms){\rightarrow}B$. But this thought of The Man's is not much better than his first, since any such formula can be falsified by the $8{\times}8$ matrix of §22.1.3 provided A and B are variables. Let A be $+\mathit{1}$, and let B and all other variables in *Axioms* have the value $-\mathit{3}$. Inspection of the $8{\times}8$ tables reveals that the $\{\pm\mathit{1},\ \pm\mathit{3}\}$ fragment is closed under all the connectives; hence the conjunction *Axioms* of the axioms takes either $+\mathit{1}$ or $+\mathit{3}$. Computation then shows that $(A\&(\bar{A}\vee B)\&Axioms)\rightarrow B$ takes the value $-\mathit{3}$. So B does not even derive "from" A and $\bar{A}\vee B$ in the Official sense (§22.2.1).

Perhaps, after all, what The Man meant was that inspection of *proofs* of A and $\bar{A}\vee B$ lead to a *proof* of B. Suppose that for arbitrary A and B such that A and $\bar{A}\vee B$ are provable we lay the proofs end-to-end, and that, *without using additional axioms*, we can then go on to derive B; i.e., we have

$$\left.\begin{array}{l}\left.\begin{array}{l}\left.\begin{array}{l}\vdots\\ A\end{array}\right\}\ \text{proof of } A\\ \left.\begin{array}{l}\vdots\\ \bar{A}\vee B\end{array}\right\}\ \text{proof of } \bar{A}\vee B\end{array}\right\}\ \text{end-to-end}\\ \vdots\\ B\end{array}\right\}\ \text{proof of } B$$

It is clear that, if there *were* some such uniform way of constructing a proof of B from proofs of A and $\bar{A}\vee B$, there would then be some psychological account, at least, of how The Man came to think that $A\&(\bar{A}\vee B)\rightarrow B$ in the first place. Just what we should mean by a "uniform way" is not altogether clear, but the question can be broadened a little by paying attention to the corollary to the Entailment theorem. We know that if $\vdash A$, then $\vdash Ax_A{\rightarrow}A$, where Ax_A is a conjunction of the axioms used in the proof of A; similarly if $\vdash\bar{A}\vee B$ then $\vdash Ax_{(\bar{A}\vee B)}{\rightarrow}(\bar{A}\vee B)$. So we have the question: is it then true that $\vdash Ax_A\&Ax_{(\bar{A}\vee B)}{\rightarrow}B$?

Answer: no. This simple counterexample was provided, as usual, by Meyer:

$\vdash A{\rightarrow}A$ without E11, distribution (§21.1).

$\vdash \overline{A{\rightarrow}A}\vee[((A{\rightarrow}A)\&B)\vee\bar{B}]$ without E11.

$\vdash((A{\rightarrow}A)\&B)\vee\bar{B}$ only *with* E11.

The reader can easily produce proofs in **E** of the first two formulas without using distribution. That the third cannot be proved using only the axioms invoked in the first two proofs follows from the fact that it cannot be proved in **E** without E11, distribution. Indeed, it cannot even be proved in **R** without E11. We know this because we know how to Gentzenize **R**-without-distribution in a decidable way, and the decision procedure rules out $((A\rightarrow A)\&B)\vee\overline{B}$. (Take the formulation **LK** of Gentzen 1934; delete the rules for the addition of arbitrary formulas which Gentzen's translators call "thinning," and which we follow Curry in calling "weakening." The result yields a formulation of **R** without distribution, as can be established via the usual elimination theorem.)

Summary of what is *not* true:

(i) It is not true that $A\&(\overline{A}\vee B)\rightarrow B$.

(ii) It is not true that there is a proof that A and $\overline{A}\vee B$ entail B.

(iii) It is not true that there is an Official deduction of B from premisses A and $\overline{A}\vee B$.

(iv) It is not true that the axioms yielding A together with those yielding $\overline{A}\vee B$ are always sufficient to give us B.

What *is* true is that whenever $\vdash A$ and $\vdash \overline{A}\vee B$, then $\vdash B$; i.e., the disjunctive syllogism, so construed, is an admissible rule (§7.2).

But this, like the rule of necessitation (§21.2.2) turns out to be, so to speak, another lucky accident — a much luckier accident, in view of the complication of the proof. The Meyer-Dunn argument below (§§25.2, 25.3) guarantees the existence of a proof of B, all right, but there is *no* guarantee that the proof of B is related in any sort of plausible, heartwarming way to the proofs of A and $\overline{A}\vee B$. For this reason we think that the correct analysis of The Dog's behavior is given in a passage from Coleridge 1863:

> . . . X is to be found either in A, or B, or C, or D: it is not found in A, B, or C; therefore it is to be found in D. — What can be simpler? Apply this to a brute animal. A dog misses his master where four roads meet; — he has come up one, smells to two of the others, and then with his head aloft darts forward to the fourth road without any examination. If this were done by a conclusion, the dog would have reason; — how comes it then, that he never shows it in his ordinary habits? Why does this story excite either wonder or incredulity? — If the story be a fact, and not a fiction, I should say — the breeze brought his master's scent down the fourth road to the dog's nose, and that therefore he did not put it down to the road, as in the two former instances.

The decision to take the fourth road, while correct, had little to do with believing that the first two were wrong, on Coleridge's account. His dog cleverly found Truth constructively, as it were; he did not *simply* avoid Error, as did the Bestiarist's beast, much less did he succumb to the pitfalls into which The Man tempted The Dog of Chrysippus. The proofs to be exhibited below make Coleridge's high constructive hopes look visionary. It must be remembered, however, that this Man was himself a Romantic Poet.

§25.2. The admissibility of (γ) in E; first proof (by Robert K. Meyer and J. Michael Dunn). By (γ) we mean that rule which Ackermann 1956 took as primitive for his systems Π′ and Π″ of Chapter VIII:

den Schluss von A und $\bar{A}\vee B$ auf B.

What we prove is that the addition of such a rule to **E** leads to nothing new; i.e., whenever A and $\bar{A}\vee B$ are both provable in **E**, then B is also, though our proof techniques give precious little information about *securing* a proof of B. If you tell us truly that A and $\bar{A}\vee B$ are provable in **E**, then probably someone could come up with a proof of B, were there world enough and time, but the only aid we get from Theorem 4 of §25.2.2 is assurance that a proof of B exists.

(γ), whose proof was at one time an achievement, has on further reflection (deeply into the nature of things, we add in proper immodesty) become so easy that, as van Fraassen puts it, any kid with truth tables can do it. But evidence for such assertions must be put off until further development of the semantic and coherence methods of Chapter IX and §§61–62, respectively.

§25.2.1. E-theories. By an **E**-*theory* we mean a triple ⟨L, O, T⟩, where (i) L is the set of formulas of **E**; (ii) O is the set of connectives of **E**, i.e., the members of O are the binary connectives →, &, ∨, and the singulary connective ‾ ; (iii) T is a subset of L such that, for all $A, B \in$ L,

(a) if $A \in$ T and $B \in$ T, then $(A\&B) \in$ T;
(b) if $\vdash_{\mathbf{E}} A\rightarrow B$ and $A \in$ T, then $B \in$ T.

We call an **E**-theory T *regular* if moreover

(c) if $\vdash_{\mathbf{E}} A$ then $A \in$ T.

As just above, we use T indifferently for the triple ⟨L, O, T⟩ and for the set T, the ambiguity being trivial and readily resolved in context. Moreover, we write $\vdash_{\mathrm{T}} A$ (and call A a *theorem* of T) when $A \in$ T. As usual, we let AB be $(A\&B)$.

When this section was originally written, as explained in Chapter IX, we did not expect *irregular* **E**-theories to play as important a role as they have come to play. Accordingly, when we speak *here* of **E**-theories, we mean always *regular* **E**-theories: i.e., T below always satisfies (a)–(c), not just (a)–(b). Importantly, regularity is built into the notion of *normality* below. Alternatively, we may characterize a regular **E**-theory T as any subset of L that is closed under &I and →E, and which contains all axioms of **E**.

It is clear that **E** itself is an **E**-theory. Furthermore, for every set S of formulas of L, there is a smallest **E**-theory T such that $S \subseteq T$ and such that if $\vdash_T A$, there exists a *derivation* of A from S, i.e., a finite sequence of formulas terminating in A each of whose members is either an axiom of **E**, a member of S, or the consequence of predecessors by →E or &I.

Let T and T′ be **E**-theories, $T \subseteq T'$; we call T′ an *extension* of T; if $S \cup T$ is a set of axioms for T′, we may call T′ the *axiomatic extension* of T by S, and members of S *proper axioms* of T′ (with respect to T). We write $S \vdash_T A$ if A is a theorem of the axiomatic extensions of T by S, abbreviating $\{A_1, \ldots, A_m\} \vdash_T A$ by $A_1, \ldots, A_m \vdash_T A$.

Let T be an **E**-theory. We call T a *prime theory* if, for all A and B in L, if $\vdash_T A \vee B$ then $\vdash_T A$ or $\vdash_T B$. We call T a *consistent theory* if there is no A such that both $\vdash_T A$ and $\vdash_T \bar{A}$. Finally, if T is both *prime* and *consistent*, we call T *normal*.

Normal **E**-theories will play an important role in our proof of (γ). For a normal **E**-theory contains, for every formula A of **E**, just one of A, $\bar{A}$. Suppose that the following is the case:

1. For every non-theorem B of **E**, there is a normal **E**-theory T such that $\vdash_T \bar{B}$.

On the hypothesis 1, (γ) is admissible for **E**. For suppose $\vdash_{\mathbf{E}} A$ and $\vdash_{\mathbf{E}} \bar{A} \vee B$. For *reductio*, assume also that B is a non-theorem of **E**. By 1, there is a normal **E**-theory T such that $\vdash_T \bar{B}$. Furthermore, since all **E**-theories are extensions of **E**, $\vdash_T A$ and $\vdash_T \bar{A} \vee B$. Since T is normal, it is prime; hence $\vdash_T \bar{A}$ or $\vdash_T B$. But in either case T then becomes inconsistent, contradicting the assumption of its normality and proving (γ) for **E** on 1.

1 is difficult to prove directly, however, and we shall in the end establish it by taking a semantical detour. After noting a few facts about **E**, however, we shall prove a weaker variant of 1 as our first lemma. The facts which we shall need are the following, for each **E**-theory T:

2. $A \vdash_T A \vee B$; $B \vdash_T A \vee B$;
3. if $A \vdash_T C$ and $B \vdash_T C$, then $A \vee B \vdash_T C$;
4. if $\vdash_T A$ and $A \vdash_T B$, then $\vdash_T B$.

PROOF of 2–4. 4 is an easy consequence of definitions, while 2 holds in virtue of →E and the axioms $A \to A \vee B$ and $B \to A \vee B$ of **E**.

To prove 3, assume (a) $A \vdash_T C$ and (b) $B \vdash_T C$. We show (a′) $A \vee B \vdash_T C \vee B$, (b′) $C \vee B \vdash_T C \vee C$, and (c) $C \vee C \vdash_T C$, from which it is clear that there is a derivation of C from $T \cup \{A \vee B\}$ and hence that $A \vee B \vdash_T C$.

To prove (a′), there is in virtue of (a) and the remarks about *derivations* above, a sequence $A_1, \ldots, A_m$ terminating in C each of whose members is either a theorem of T, A itself, or the consequence of predecessors by →E or &I. We assume as inductive hypothesis that $A_i \vee B$ is a theorem of the axiomatic extension of T by $\{A \vee B\}$ for $i < j$, and we show that $A \vee B \vdash_T A_j \vee B$, thus establishing (a′) in particular. There are four cases, according as A_i is a theorem of T, A itself, a consequence by →E, or a consequence by &I.

CASE 1. $\vdash_T A_j$. By 2 and 4, $\vdash_T A_j \vee B$; hence $A \vee B \vdash_T A_j \vee B$.

CASE 2. A_j is A. Trivially $A \vee B \vdash_T A_j \vee B$.

CASE 3. A_j is the consequence of predecessors A_h and $A_h \rightarrow A_j$. On inductive hypothesis, $A \vee B \vdash_T A_h \vee B$ and $A \vee B \vdash_T (A_h \rightarrow A_j) \vee B$. By &I, $A \vee B \vdash_T (A_h \vee B)((A_h \rightarrow A_j) \vee B)$. By distribution (in the appropriate form, provable in **E** from the axiom E11 of §21.1) and →E, $A \vee B \vdash_T A_h(A_h \rightarrow A_j) \vee B$. But

5. $\vdash_{\mathbf{E}} A_h(A_h \rightarrow A_j) \rightarrow .A_j \vee B$, and
6. $\vdash_{\mathbf{E}} B \rightarrow .A_j \vee B$,

whence by &I, axiom E10, and →E, we may conclude $A \vee B \vdash_T A_j \vee B$, disposing of this case.

CASE 4. A_j is the consequence of predecessors by &I. Similar to Case 3.

This completes the proof of (a′). The proof of (b′) is in like manner, using (b). And (c) follows immediately from the theorem

7. $\vdash_{\mathbf{E}} C \vee C \rightarrow C$,

concluding the proof of 3.

LEMMA 1. For every non-theorem B of **E**, there is a prime **E**-theory T such that $\vdash_T \overline{B}$ but not $\vdash_T B$.

PROOF. Enumerate the formulas of **E**, letting them be $B_1, \ldots, B_j, \ldots$. Form a succession of **E**-theories $T_0, T_1, \ldots, T_j, \ldots$, letting T_0 be **E** and forming T_i from T_{i-1} according to the following recipe:

If $B_i \vdash_{T_{i-1}} B$, let T_i be T_{i-1}. Otherwise let T_i be the axiomatic extension of T_{i-1} by $\{B_i\}$.

Let T be the union of all the T_i so formed. It is readily seen that T is an **E**-theory such that B is not a theorem of T; since $\vdash_{\mathbf{E}} B \vee \overline{B}$, we can conclude the proof of Lemma 1 by showing that T is prime.

Suppose rather that T is not prime. Then for some A and C, $\vdash_T A \vee C$, but neither $\vdash_T A$ nor $\vdash_T C$. By the instructions for the construction of T, $A \vdash_T B$ and $C \vdash_T B$; hence by 3, $A \vee C \vdash_T B$. Hence by 4, $\vdash_T B$, which is impossible. Accordingly T is prime, concluding our proof.

Lemma 1 may be viewed as a syntactical analogue of the Stone prime filter theorem for distributive lattices. As we seek to generalize our results, it will be important to note just which theorems and rules of **E** enter into its proof, a matter to which we shall return. We conclude this section with a theorem now rather trivial.

THEOREM 1. The following conditions are equivalent:

(i) $\vdash_E A$;
(ii) $\vdash_T A$, for all **E**-theories T;
(iii) $\vdash_T A$, for all prime **E**-theories T.

PROOF. Since all **E**-theories are extensions of **E**, (i) implies (ii). That (ii) implies (iii) is trivial. And that the denial of (i) implies the denial of (iii) is the content of Lemma 1; hence (iii) implies (i), completing the proof of Theorem 1.

§25.2.2. Semantics. By an **E**-matrix M, we mean a structure M = ⟨M, O, D⟩, where

(i) M is a non-empty set, and D is a subset of (*designated* elements of) M;
(ii) O is a set whose members are the binary operations →, &, ∨ and the singular operation ‾ on M;
(iii) for all elements a, b, and c of M, the following conditions (compare §21.1) are fulfilled:

E1′ (((a→a)→b)→b) ∈ D;
E2′ ((a→b)→((b→c)→(a→c))) ∈ D;
E3′ ((a→(a→b))→(a→b)) ∈ D;
E4′ (a&b→a) ∈ D;
E5′ (a&b→b) ∈ D;
E6′ ((a→b)&(a→c)→(a→.b&c)) ∈ D;
E7′ (□a&□b)→□(a&b)) ∈ D, where for all c ∈ M, □c is the element (c→c)→c;
E8′ (a→a∨b) ∈ D;
E9′ (b→a∨b) ∈ D;
E10′ ((a→c)&(b→c)→(a∨b→c)) ∈ D;
E11′ (a&(b∨c)→.(a&b)∨c) ∈ D;
E12′ ((a→ā)→ā) ∈ D;

E13′ $((\mathrm{a}\rightarrow\bar{\mathrm{b}})\rightarrow(\mathrm{b}\rightarrow\bar{\mathrm{a}})) \in \mathrm{D}$;
E14′ $(\bar{\bar{\mathrm{a}}}\rightarrow\mathrm{a}) \in \mathrm{D}$;
→E′ if $\mathrm{a} \in \mathrm{D}$ and $(\mathrm{a}\rightarrow\mathrm{b}) \in \mathrm{D}$, $\mathrm{b} \in \mathrm{D}$;
&I′ if $\mathrm{a} \in \mathrm{D}$ and $\mathrm{b} \in \mathrm{D}$, $(\mathrm{a}\&\mathrm{b}) \in \mathrm{D}$.

Let $\mathrm{M} = \langle \mathrm{M, O, D}\rangle$ be an E-matrix. M will be called a *prime matrix* if, for all a and b in M, if $(\mathrm{a}\vee\mathrm{b}) \in \mathrm{D}$ then $\mathrm{a} \in \mathrm{D}$ or $\mathrm{b} \in \mathrm{D}$. M will be called a *consistent* matrix provided that it is not the case that both $\mathrm{a} \in \mathrm{D}$ and $\bar{\mathrm{a}} \in \mathrm{D}$, for any $\mathrm{a} \in \mathrm{M}$. Finally, if M is both prime and consistent, we call M *normal.*

By an *interpretation* of **E** in an E-matrix M, we mean a function I defined on the propositional variables of **E** with values in M, and then recursively defined on all sentences of **E** in virtue of the following specifications:

(i) $\mathrm{I}(A\rightarrow B) = (\mathrm{I}(A)\rightarrow\mathrm{I}(B))$;
(ii) $\mathrm{I}(A\&B) = (\mathrm{I}(A)\&\mathrm{I}(B))$;
(iii) $\mathrm{I}(A\vee B) = (\mathrm{I}(A)\vee\mathrm{I}(B))$;
(iv) $\mathrm{I}(\bar{A}) = \overline{\mathrm{I}(A)}$.

A formula A of **E** is called *true* on an interpretation I in an E-matrix M provided that $\mathrm{I}(A) \in \mathrm{D}$, and otherwise *refuted* on I: if $\bar{A}$ is true on I, A is called *false* on I. If A is true on *all* interpretations I in an E-matrix M, A is called *M-valid;* if A is M-valid for all E-matrices M, A is called *valid*, and we write $\models_{\mathbf{E}}$A.

It is clear that each E-theory ⟨L, O, T⟩ may be considered an E-matrix M by taking the formulas of **E** as the elements of M, the connectives as the operations, and the theorems of T as the designated elements. Thus parsed, we note also that prime theories are prime matrices; consistent theories, consistent matrices; and hence normal theories, normal matrices. By the *canonical interpretation* $\mathrm{I_T}$ of **E** in an E-theory T (considered as an E-matrix), we mean the interpretation of **E** in T which is determined by setting $\mathrm{I}^*(p) = p$, for all propositional variables p. Clearly where $\mathrm{I_T}$ is a canonical interpretation of **E**, A is true on $\mathrm{I_T}$ iff $\vdash_{\mathrm{T}}A$; in particular, where T is **E** itself, A is true on $\mathrm{I_E}$ iff $\vdash_{\mathbf{E}}A$. The following result is then straightforward:

THEOREM 2. The following conditions are equivalent:

(i) $\vdash_{\mathbf{E}}A$;
(ii) $\models_{\mathbf{E}}A$;
(iii) A is M-valid, for all prime E-matrices M.

PROOF. To prove that (i) implies (ii), it suffices to show that all the axioms of **E** are true on an arbitrary interpretation I in any E-matrix M and that true-on-I is a property preserved under the rules of **E**. But the conditions

E1′–E14′, →E′, and &I′, were chosen for exactly this purpose, as comparison with the axioms and rules of §21.1 makes obvious. That (ii) implies (iii) is trivial. We finish the proof by assuming (iii) and proving (i). Suppose A is M-valid for all prime **E**-matrices M. Then in particular A is true on I_T, for each canonical interpretation I_T of **E** in a prime **E**-theory T. But then $\vdash_T A$ for all prime **E**-theories T. By Theorem 1, $\vdash_{\mathbf{E}} A$, concluding the proof of Theorem 2.

Theorem 2 is a completeness result of a sort, but it is insufficient for the proof of (γ). For we might have the following situation: both A and $\bar{A} \vee B$ are M-valid in all prime **E**-matrices, but there is nevertheless an interpretation I in a prime **E**-matrix M on which B is refuted. In that case, both $I(A)$ and $I(\bar{A})$ $(= \overline{I(A)})$ must be designated — i.e., M itself must be inconsistent. But there is nothing in the characterization of a prime matrix which prevents this possibility from being realized.

The moral to be drawn from this is that **E**-matrices in general, and even prime **E**-matrices in particular, are too wide a class. (On the other hand, inconsistent matrices are not wholly to be scorned, for they suggest the possibility of developing semantic machinery which can distinguish seriously inconsistent theories from theories which if inconsistent are so in a less serious sense, a point treated in some detail in the somewhat different context of Chapter IX.) For these still permit a situation in which the same sentence is *both* true and false on a given interpretation. On the other hand, *normal matrices* do not have this distressing property — on all interpretations I of **E** in normal matrices, it is clear that A is false on I iff A is refuted on I.

Accordingly, we lay down a recipe which associates with each prime **E**-matrix $\mathrm{M} = \langle \mathrm{M}, \mathrm{O}, \mathrm{D} \rangle$ a normal **E**-matrix $\mathrm{M}^* = \langle \mathrm{M}^*, \mathrm{O}^*, \mathrm{D} \rangle$. We do this by noting that there are three kinds of elements in M: (a) those which are designated and whose negations are designated; (b) those which are designated and whose negations are not designated; (c) those which are undesignated but those negations are designated. (By the **E**-theoremhood of the excluded middle and the primeness of M, a fourth possibility — that for some element a neither a nor $\bar{a}$ should be designated — goes unrealized.)

Now it is the elements which satisfy (a) that make M abnormal, if it is. Let us call the set of these elements N (for neuter), where $a \in N$ iff $a \in D$ and $\bar{a} \in D$. While we are at it, we affix the name T to the set of elements of M which satisfy (b) — i.e., $a \in D$ and $\bar{a} \notin D$ — and the name F to the set of elements of M which satisfy (c) — i.e., $\bar{a} \in D$ and $a \notin D$. And we note that these subsets of M obey a kind of three-valued truth table, which we reproduce: by putting F at the intersection of the row headed by T and the column headed by N in the arrow table, we mean that if $a \in T$ and $b \in N$, (a→b) ∈ F, and similarly in the other cases; we note also that the "truth table" is less than satisfactory where the principal connective is →.

&	T	N	F
T	T	N	F
N	N	N	F
F	F	F	F

∨	T	N	F
T	T	T	T
N	T	N	N
F	T	N	F

→	T	N	F
T	M	F	F
N	M	N∪F	F
F	M	M	M

	$\overline{}$
T	F
N	N
F	T

That the "truth table" is correct for $\overline{}$ follows from double negation and our definitions of T, N, and F; we proceed to show it correct for the other connectives, using whatever theorems of **E** are required for this purpose.

We begin with the entries for &. Suppose a ∈ F, and consider a&b; if (a&b) ∈ D, a ∈ D by E4′ and →E′ above; hence (a&b) ∈ F. Suppose a ∈ D and b ∈ D; (a&b) ∈ D by &I′ above; suppose moreover that a ∈ N; then $\bar{a}$ ∈ D; but (a&b→a) ∈ D, and hence ($\bar{a}\to\overline{a\&b}$) ∈ D by E13′, using also E2′, E14′, and →E′ to get the right form of contraposition; then by →E′, $\overline{a\&b}$ ∈ D; hence (a&b) ∈ N. Or suppose a ∈ T and b ∈ T: if a&b is nevertheless a member of N, $\overline{a\&b}$ ∈ D; but ($\overline{a\&b}\to\bar{a}\vee\bar{b}$) ∈ D in virtue of the De Morgan laws for **E**, whence ($\bar{a}\vee\bar{b}$) ∈ D; by the primeness of M, either $\bar{a}$ or $\bar{b}$ is then a member of D, contra hypothesis; so (a&b) ∈ T in this case, as claimed. The other entries for & hold by commutativity, ending the proof that this table is correct.

That the entries for ∨ are correct follows dually. Suppose a ∈ T, and consider a∨b; (a∨b) ∈ D by E7′ and →E′; if a∨b is nevertheless a member of N, then $\overline{a\vee b}$ ∈ D; but (a→.a∨b) ∈ D, whence by contraposition so is $\overline{a\vee b}\to\bar{a}$; by →E′ $\bar{a}$ ∈ D, contra hypothesis; so (a∨b) ∈ T in this case. Suppose then that a ∈ N and that b ∈ T; then $\bar{a}$ ∈ D and $\bar{b}$ ∈ D; hence by &I′, ($\bar{a}\&\bar{b}$) ∈ D; by the De Morgan laws for **E**, $\overline{a\vee b}$ ∈ D; hence since (a∨b) ∈ D, (a∨b) ∈ N. Finally suppose a ∈ F and b ∈ F; if (a∨b) ∈ D, a ∈ D or b ∈ D by the primeness of M; hence (a∨b) ∈ F in this case. Other cases for ∨ follow by commutativity, ending the proof that this table is correct.

We now examine the entries for →. Suppose a ∈ T and b ∉ T, but that a→b is nevertheless a member of D; then both $\bar{b}$ and $\bar{b}\to\bar{a}$ are members of D, the latter by contraposition, and so then by →E′ is $\bar{a}$, contra hypothesis. Or suppose that a ∈ N, b ∈ N, but (a→b) ∈ T; in fact, $\vdash_{\mathbf{E}} A\&\overline{B}\to\overline{A\to B}$, whence by Theorem 2 a→b is designated after all, proving the impossibility of this case. Or suppose that a ∈ N, b ∈ F, (a→b) ∈ D; then b ∈ D by →E′, contra hypothesis. This establishes that part of the table for → for which particular claims are made.

In order to normalize M — to find *the normalization* M* = ⟨M*, O*, D⟩ of M — let us enrich it with a set of items −N disjoint from M, such that for each (neuter) element a in N, there is exactly one corresponding element $-a$ in $-N$. Let M^* be $M \cup -N$.

Henceforth we shall think of the members of N as no longer neuter but true; the corresponding false items shall be the members of $-N$. Let a function h from M^* onto M be defined by cases as follows (where we write sometimes "ha" and sometimes "h(a)"):

(i) If $a \in M$, $ha = a$;
(ii) if $a \in -N$, and is hence an item $-b$, $ha = b$.

With the aid of h and the operations $\rightarrow$, &, $\vee$, and $\overline{}$ of M, we now define a set O^* of corresponding operations $\rightarrow^*$, &*, $\vee^*$, and $\overline{}^*$ on M^*, as follows:

(i) $(a\rightarrow^*b) = -(ha\rightarrow hb)$ provided that $a \in N$, $b \in -N$, and $(ha\rightarrow hb) \in N$; otherwise $a\rightarrow^*b = ha\rightarrow hb$.

(ii) $(a\&^*b) = -(ha\&hb)$ provided that $a \notin D$ or $b \notin D$, and moreover that $(ha\&hb) \in N$; otherwise $(a\&^*b) = (ha\&hb)$.

(iii) $(a\vee^*b) = -(ha\vee hb)$ provided that $ha\vee hb$ is a member of N and that $a \notin D$ and $b \notin D$; otherwise $a\vee^*b = ha\vee hb$.

(iv) $\bar{a}^* = -\bar{a} = -\overline{ha}$ provided that $a \in N$; otherwise $\bar{a}^* = \overline{ha}$.

We now make a number of observations:

8. The operations in O^* are well-defined. In particular, since $-$ is defined only on N, this requires that c be a member of N when $-c$ is listed as the result of an operation. But this is explicit in (i) and (iv) just above, and may be read off the "truth tables" in the other cases.

9. We now give "truth tables" for the operations of O^*.

&*	T	N	−N	F
T	T	N	−N	F
N	N	N	−N	F
−N	−N	−N	−N	F
F	F	F	F	F

∨*	T	N	−N	F
T	T	T	T	T
N	T	N	N	N
−N	T	N	−N	−N
F	T	N	−N	−F

	$\overline{}^*$
T	F
N	−N
−N	N
F	T

→*	T	N	−N	F
T	M	F	F	F
N	M	N∪F	−N∪F	F
−N	M	N∪F	N∪F	F
F	M	M	M	M

Verification of these "truth tables" is straightforward given the earlier ones and conditions (i)–(iv) above. We note in particular that $a\rightarrow^*b$ is a member of the original set M unless $a \in N$ and $b \in -N$.

10. The function h defined above is in fact a homomorphism from the matrix ⟨M*, O*, D⟩ to the prime **E**-matrix ⟨M, O, D⟩, in the sense that h preserves operations — i.e., h(a→*b) = (ha→hb), h(a&*b) = (ha&hb), h(a∨*b) = (ha∨hb), and h(ā*) = $\overline{\text{ha}}$, as one easily verifies on conditions (i)–(ii) of the definition of h and conditions (i)–(iv) defining the operations of O*; furthermore h preserves D, for if a ∈ D, ha = a ∈ D by definition.

11. ⟨M*, O*, D⟩ is in fact a normal **E**-matrix. To show this, we must show (a) that ⟨M*, O*, D⟩ is an **E**-matrix, (b) that ⟨M*, O*, D⟩ is prime, and (c) that ⟨M*, O*, D⟩ is consistent. To prove (b), suppose (a∨*b) ∈ D = (T∪N); consulting the "truth table," we note that a ∈ D or b ∈ D. To prove (c), we note by the table for ‾* that a ∈ D iff ā* ∉ D; hence for no a are both a and ā* members of D. We turn to the proof of (a).

To show that ⟨M*, O*, D⟩ is a normal **E**-matrix, we need now show only that conditions E1′–E14′, →E′, and &I′ are satisfied by ⟨M*, O*, D⟩. We note first that conditions E1′–E14′ state in this context that items x→*y are members of D, where x and y are of a special form. Now there are two cases under the definition (i) of →*; in what we shall call the *special case*, (x→*y) = −(hx→hy); in what we shall call the *otherwise case*, (x→*y) = (hx→hy). But we note that if x→*y is defined in accordance with the otherwise case for an instance of E1′–E14′, x→*y is a member of D, for since h is a homomorphism hx→hy must already be an instance for ⟨M, O, D⟩ of one of E1′–E14′; since ⟨M, O, D⟩ is on assumption an **E**-matrix, (hx→hy) ∈ D.

On the other hand, if x→*y is defined according to the special case, x→*y cannot be a member of D, since it is of form −a. So to show that each of E1′–E14′ hold for ⟨M*, O*, D⟩, it is sufficient to show that →* is defined for all of their instances according to the specifications of the otherwise case. Accordingly, for each of E1′–E14′ we assume for *reductio* that an instance x→*y is defined according to the special case, and we derive a contradiction therefrom. Since we assume that we are in the special case, our assumptions in each instance will be that x ∈ N, y ∈ −N, and (hx→hy) ∈ N, in accordance with the restrictions of the special case.

E1′. ((a→*a)→*b) ∈ N, b ∈ −N. But (a→*a) = (ha→ha) ∈ D, since *A*→*A* holds in **E** and since ⟨M, O, D⟩ is an **E**-matrix. But then by table ((a→*a)→*b) ∈ −N∪F, contradicting the initial assumption.

E2′. (a→*b) ∈ N, ((b→*c)→*(a→*c)) ∈ −N. By table (b→*c) ∈ N, (a→*c) ∈ −N; again by table, a ∈ N and c ∈ −N. Where then is b? If b ∈ D, b→*c is not, by table, a member of D, contradicting its being in N; if b ∈ −N∪F, a→*b is not a member of D, contradicting its being in N. Conclusion: b is not anywhere, a contradiction.

E3′. (a→*(a→*b)) ∈ N, (a→*b) ∈ −N. Then a ∈ N. But then by table, a→*(a→*b) is not in N, a contradiction.

E4′. (a&*b) ∈ N, a ∈ −N. Impossible by table.

E5′. $(a\&^*b) \in N$, $b \in -N$. Ditto.

E6′. $((a\rightarrow^*b)\&^*(a\rightarrow^*c)) \in N$, $(a\rightarrow^*(b\&^*c)) \in -N$. Then $a \in N$, $(b\&^*c) \in -N$. Then $b \in -N$ or $c \in -N$, by table; in either case, the table confirms that $(a\rightarrow^*b)\&^*(a\rightarrow^*c)$ is not in D, a contradiction.

E7′. $(((a\rightarrow^*a)\rightarrow^*a)\&^*((b\rightarrow^*b)\rightarrow^*b)) \in N$, $((a\&^*b\rightarrow^*.a\&^*b)\rightarrow.a\&^*b) \in -N$. Then $(a\&^*b) \in -N$, and so is one of a, b, by table: arguing as for E1′ above, the tables confirm that $((a\rightarrow^*a)\rightarrow^*a)\&^*((b\rightarrow^*b)\rightarrow^*b)$ is not designated, a contradiction.

E8′. $a \in N$, $(a\vee^*b) \in -N$. Impossible by table.

E9′. $b \in N$, $(a\vee^*b) \in -N$. Ditto.

E10′. $((a\rightarrow^*c)\&^*(b\rightarrow^*c)) \in N$, $(a\vee^*b\rightarrow^*c) \in -N$. Then $(a\vee^*b) \in N$, $c \in -N$. Then by table $a \in N$ or $b \in N$; the initial assumption that $((a\rightarrow^*c)\&^*(b\rightarrow^*c)) \in N$ then fails by the tables, a contradiction.

E11′. $(a\&^*(b\vee^*c)) \in N$, $((a\&^*b)\vee^*c) \in -N$. Apply tables to get a contradiction.

E12′. $(a\rightarrow^*\bar{a}^*) \in N$, $\bar{a}^* \in -N$. By table $a \in N$; but then $a\rightarrow^*\bar{a}^*$ is undesignated, a contradiction.

E13′. $(a\rightarrow^*\bar{b}^*) \in N$, $(b\rightarrow^*\bar{a}^*) \in -N$. $b \in N$, $\bar{a}^* \in -N$; but then $a \in N$, $\bar{b}^* \in -N$, and $a\rightarrow^*\bar{b}^*$ is undesignated, a contradiction.

E14. $\bar{\bar{a}}^* \in N$, $a \in -N$. Impossible by table.

Since each instance of E1′–E14′ already falls under the otherwise case, as we have seen, $\langle M^*, O^*, D\rangle$ meets these conditions on an **E**-matrix. We now show that it meets conditions →E′ and &I′ also. But &I′ holds by table considerations; to prove →E′, suppose $a \in D$ and $(a\rightarrow^*b) \in D$. Since $a \in D$, if $b \in -N\cup F$ we note by table that so also is $a\rightarrow^*b$, contra hypothesis; so $b \in D$ also, concluding the proof that the matrix $\langle M^*, O^*, D\rangle$ which we have associated with the prime **E**-matrix $\langle M, O, D\rangle$ is itself a normal **E**-matrix.

The next lemma is the key one.

Lemma 2. Suppose *A* is not a theorem of **E**. Then there is a normal **E**-matrix $\langle M^*, O^*, D\rangle$ and an interpretation I^* such that *A* is refuted on I^*.

Proof. Since on assumption *A* is not a theorem of **E**, there exists by Theorem 2 a prime **E**-matrix $\langle M, O, D\rangle$ and an interpretation I such that *A* is refuted on I, for otherwise *A* would by Theorem 2 be provable in **E**. Let $\langle M^*, O^*, D\rangle$ be the normal **E**-matrix associated with $\langle M, O, D\rangle$ by the construction outlined above.

Let I^* be the interpretation of **E** in $\langle M^*, O^*, D\rangle$ which agrees with I on sentential variables, and let h be the function from M^* to M defined above. We now show, by induction on length of sentences, that for every sentence *B* of **E**, $h(I^*(B)) = I(B)$. This is clear for the basis case, since I and I^* agree on sentential variables and since h is the identity on M. There are four cases in the induction step, according as the main connective of B is $\rightarrow$, &, $\vee$, or $\overline{}$. We omit three.

CASE 1. B is of the form $C \rightarrow D$. Then $I^*(B)$ is $I^*(C) \rightarrow^* I^*(D)$, and $h(I^*(B))$ is (since h is by 10 a homomorphism) $h(I^*(C) \rightarrow h(I^*(D))$. But on the inductive hypothesis $h(I^*(C)) = I(C)$ and $h(I^*(D)) = I(D)$, whence $h(I^*(C \rightarrow D)) = I(C \rightarrow D)$, as required.

Other cases being similar, $h(I^*(A)) = I(A)$ for our selected non-theorem A of **E**. But $I(A)$ is not in D; since by 10, h preserves D, $I^*(A)$ is not in D either. Accordingly A is refuted on the interpretation I^* in the normal **E**-matrix $\langle M^*, O^*, D\rangle$, which was to be proved.

We now state and prove the resulting theorems.

THEOREM 3. The following conditions are equivalent to each other and to those stated in Theorems 1 and 2:

(i) $\vdash_E A$;
(ii) A is M-valid in all normal **E**-matrices M;
(iii) $\vdash_T A$, for all normal **E**-theories T.

PROOF. That (i) implies (ii) is a consequence of Theorem 2. We prove that (ii) implies (iii) by proving the contrapositive. Suppose that there is a normal **E**-theory T of which A is not a theorem; $\langle L, O, T\rangle$ is a normal **E**-matrix, as we have observed above, and A is refuted on the characteristic interpretation I_T of **E** in $\langle L, O, T\rangle$; hence A is not M-valid for some normal **E**-matrix M. So if A is M-valid for all normal M, $\vdash_E A$. We conclude the proof of Theorem 3 by showing that (iii) implies (i).

Suppose that A is not a theorem of **E**. Then there is by Lemma 2 an interpretation I^* of **E** in a normal **E**-matrix $\langle M^*, O^*, D\rangle$ such that A is false on I^*. Let T be the set of sentences of **E** true on I^*; it is readily established that T is a normal **E**-theory of which A is not a theorem. So if $\vdash_T A$ for all normal **E**-theories T, $\vdash_E A$.

The next theorem is (γ).

THEOREM 4. Suppose $\vdash_E A$ and $\vdash_E \bar{A} \vee B$. Then $\vdash_E B$.

PROOF. 1 of §25.2.1 is true by Theorem 3, so we already have a proof. For a proof with a semantic flavor, we note that by Theorem 3, $I(A) \in D$

and $I(\bar{A}\vee B) = (\overline{I(A)}\vee I(B)) \in D$, for all interpretations I in normal **E**-matrices ⟨M, O, D⟩. By normality, $I(B) \in D$ for all such I; by Theorem 3, $\vdash_{\mathbf{E}} B$.

§25.2.3. Generalizations. A trivial generalization results if we change the cardinality of the set of sentential variables of **E**, assumed denumerable in the formulation we are using. Since every theorem is deducible from axioms of **E** in a finite number of steps, clearly Theorem 4 continues to hold under alphabetic inflation.

We turn now to a general method by which (γ) may be proved for **E**-logics, where by an **E**-*logic* we mean an **E**-theory ⟨L, O, T⟩ whose set T of theorems is closed under substitution for propositional variables. Clearly, we may define, for each **E**-logic T, T-matrices and T-theories as we defined the analogous notions for **E** above. We note then the following analogue of Theorems 1 and 2.

THEOREM 5. Let T be an **E**-logic. Then the following conditions are equivalent:

(i) $\vdash_{T} A$;
(ii) $\vdash_{T'} A$, for all T-theories T′;
(iii) $\vdash_{T'} A$, for all prime T-theories T′;
(iv) $\vdash_{T} A$;
(v) A is M-valid, for all prime T-matrices M.

PROOF. The only non-trivial matter which must be attended to in carrying out analogous proofs is the proof of Lemma 1. But for the proof of Lemma 1, we required only certain theorems of **E**; since these are automatically theorems of extensions of **E**, that proof goes through for all **E**-logics T, establishing Theorem 5.

Since the proof of Theorem 4 depends only on finding, for every prime **E**-matrix M, a corresponding normal **E**-matrix M* such that if A can be refuted in M it can be refuted in M*, we can generalize Theorem 4 to any **E**-logic T which has the property that, if ⟨M, O, D⟩ is a prime T-matrix, the associated ⟨M*, O*, D⟩ is a normal T-matrix. Since ⟨M*, O*, D⟩ is at any rate a normal **E**-matrix by 10 above, we need only show that all non-theorems of **E** provable in T are M*-valid. In particular, if T is got from **E** by adding a finite number of new axiom schemata, it suffices to show that conditions on T-matrices which correspond to the axiom schemata of T as E1′–E14′ above correspond to the axiom schemata of **E** hold in ⟨M*, O*, D⟩ if they hold in ⟨M, O, D⟩. We illustrate by proving (γ) for three illustrative systems.

Among the **E**-logics are those which one gets by adding one or both of the following schemata:

A1 $A \to ((A \to A) \to A)$
A2 $(A \to B) \to ((A \to B) \to (A \to B))$

By enriching **E** with the scheme A1, one gets the system **R** (§§3, 27.1.1, 28). By adding A2 to **E**, one gets **EM** (§§8.15, 27.1.1). If one adds both A1 and A2 to **E**, one gets **RM** (§§8.15, 27.1.1, 29.3–5).

For all of these systems, we shall now show that (γ) holds. Since the problem has been reduced to showing that, for each system T, if $\langle M, O, D\rangle$ is a prime T-matrix then $\langle M^*, O^*, D\rangle$ is a normal T-matrix, it will suffice to show the following:

A1′. If for all elements a of a prime **E**-matrix $\langle M, O, D\rangle$, $(a \to ((a \to a) \to a)) \in D$, then for all elements b of the associated normal **E**-matrix $\langle M^*, O^*, D\rangle$, $(b \to^* ((b \to^* b) \to^* b)) \in D$;

A2′. if for all elements a and b of a prime **E**-matrix $\langle M, O, D\rangle$, $((a \to b) \to ((a \to b) \to (a \to b))) \in D$, then for all elements c and d of the associated normal **E**-matrix $\langle M^*, O^*, D\rangle$, $((c \to^* d) \to^* ((c \to^* d) \to^* (c \to^* d))) \in D$.

The method of proof is that of 11 of the previous section; we may assume, for *reductio*, that we are in the special case.

A1′. Suppose $a \in N$, $((a \to^* a) \to^* a) \in -N$. Then $a \in -N$ by "truth table," a contradiction.

A2′. Suppose $(a \to^* b) \in N$, $((a \to^* b) \to^* (a \to^* b)) \in -N$. Then $(a \to^* b) \in -N$ by table, a contradiction.

We sum up.

THEOREM 6. Let **T** be one of the systems **R**, **RM**, **EM**, and suppose $\vdash_T A$ and $\vdash_T \bar{A} \vee B$. Then $\vdash_T B$.

PROOF. As indicated.

In fact, that A1 and A2 can be added to **E** as new axioms without disturbing the admissibility of (γ) is a particular case in each instance of a more general phenomenon. For take any scheme $A \to X$, where X is a (schematic) theorem of **E**, as a new axiom scheme; there will be no interpretation in the matrix $\langle M^*, O^*, D\rangle$ associated with a prime T-matrix of the resulting logic T which refutes $A \to X$, since $\langle M^*, O^*, D\rangle$ is at any rate an **E**-matrix and hence X cannot be given the undesignated value $-N$ therein; thus in particular A2 leaves (γ) provable. Similarly, take any scheme $A \to (X \to A)$ as a new axiom scheme and apply the argument of A1′ above.

We can continue in this way, proving (γ) after meddling with **E** and **R** in various respects. Accordingly, the question comes up whether (γ) holds for all **E**-logics: this question has been settled negatively by showing that there are certain **RM**-logics (i.e., **RM**-theories, and hence **E**-theories, closed under substitution for sentential variables) for which (γ) fails (see §29.4).

So much for logics stronger than **E**. Can the method of proof used here be employed to show (γ) admissible for logics weaker than **E**, or weaker in some respects and stronger in others? The answer again is "Yes," though we shall be even more sparing of details than in our investigation of logics stronger than **E**. Clearly, however, **E** can be stripped of those theorem schemata which do not appear in an essential way in our proofs. In particular, we note that the axiom scheme $((A\rightarrow A)\rightarrow B)\rightarrow B$ is not required for our results and that it figures in them only in that we put the corresponding condition E1′ on an **E**-matrix. Neither is E7′ (distribution of necessity over conjunction) required. Accordingly, (γ) remains admissible when these axiom schemes are dropped; this enables us to prove in particular that (γ) holds when the truth functional axioms E4–6, E8–14 of §21.1 are added to the systems of ticket entailment of §6, yielding the system **T** (§27.1.1). The transitivity axiom E2 and the contraction axiom E3 of **E** can be weakened in various respects without losing the admissibility of (γ).

So much for the superfluous axioms. Which axioms are necessary that our proof go through? The most conspicuous of these is the distribution axiom, which, as was pointed out in §25.1, seems to figure heavily in proving theorems that one would expect to hold by (γ). Since the distribution axiom is in some other respects a headache for **E** and related systems, it is interesting to observe here that it plays a key role. This is obvious from our proof of Lemma 1, and from the proof of the Official deduction theorem of §22.2.1. It is also the case that, if the distribution axiom is dropped, counterexamples to (γ) appear; e.g., $p\&(q\rightarrow q)\vee\bar{p}$ in that case is a non-theorem, although it would follow by (γ) from the E-theorems $q\rightarrow q$ and $p\&(q\rightarrow q)\vee\bar{p}\vee\overline{q\rightarrow q}$, for which distribution is not required. Similarly, the other axioms governing & and $\vee$ (E4–E6 and E8–10) figure in essential ways both in the proof of Lemma 1 and in the construction of our "truth tables." Likewise, although the transitivity and contraction axioms can be weakened, our Lemma 1 uses the theorem $A\&(A\rightarrow B)\rightarrow B$, which is proved in **E** using the contraction axiom E3: $(A\rightarrow .A\rightarrow B)\rightarrow .A\rightarrow B$.

Finally we turn to the negation axioms. The extent to which these may be weakened is perhaps of special interest, for although many interesting systems contain all of the negation-free theorems of **R** (and hence of **E**), **R** and **E** have quite strong negation axioms. For example, all negation-free theorems of **R** are provable in the absolute system **HA** of Curry 1963. (This follows from results noted in §36.2.) But **HA** is simultaneously the negation-free fragment of the intuitionistic system **HJ**, the minimal system **HM**, and the complete system **HD** of Curry 1963, all of which lack the strong double negation principle $\bar{\bar{A}}\rightarrow A$ of the relevant logics. Nevertheless, (γ) holds for the three systems just mentioned (though not for the Kripke system **HE** (no relation) also presented in Curry 1963); for **HM** and **HD** in particular, however, some argument is required to show the admissibility of (γ). Accordingly, it is interesting to generalize the above argument to situations where the underlying negation axioms are weaker, but this is not done here (where we rely on De Morgan laws, excluded middle, contraposition, and double negation). See §25.3 and, for further developments, the relevant sections of Chapter IX and X cited above.

We close with observations on the significance of our proof of (γ) for **E** and for **R**. The first observation to be made is that **E** turns out in fact to be equivalent to the system Π' of Ackermann 1956, in the sense that **E** and Π' have the same stock of theorems. Since §45.1 shows that Π' contains the Ackermann system Π'' in a straightforward way, it follows that the varying considerations which motivate the system of entailment and Ackermann's *strengen Implikation* wind up in essentially the same place, which is itself an argument for the stability of the system.

Secondly, however, there remains an important difference between Π' and and **E**, which in our opinion is to be counted in favor of the latter. For on the usual understanding that the primitive rules of a system **L** of logic remain rules of the various regular **L**-theories that the system engenders on the addition of proper axioms, there is associated with **E** a much richer class of theories than is associated with Π'. In particular, rejection of the paradoxes means to us not only rejection of particular theorems $A\&\bar{A}\rightarrow B$ but also rejection of the notion that every sentence B is to be asserted in any inconsistent theory T. Yet, if (γ) holds not only for a system of pure logic but also for all its extensions, then the contention that every contradiction (we might say, every mistake) is equally and totally disastrous remains in force. So though we welcome the admissibility of (γ) for systems of *pure* logic — these *ought* to be semantically stable — we think it likely that interesting applications will result when the requirement of such stability is not imposed on *arbitrary* theories. In fact, to postulate (γ) for a theory would seem reasonable only in the presence of, or at least faith in, the consistency of that theory; though we should not wish to knock such faith, past experience would seem to indicate that it be indulged in with some discretion.

§25.3. Meyer-Dunn theorem; second proof. Such is our enthusiasm for the theorem just stated, and the argument supporting it, that we forthwith undertake the job again: in this section we give a second argument that our philosophically motivated deletion from Ackermann's Π' (Chapter VIII) of the rule (γ), from A and $\bar{A}\vee B$ to infer B, causes a minimum of mathematical disruption. Which is to say that (γ) is admissible in **E**. The proof just given emphasized the connections with **E**; that to follow is designed to warm the cockles of an algebraist's heart.

The proof we sketch below is based squarely on that of §25.2, although we have rearranged it and given it an algebraic flavor. We begin with some definitions, then give some abstract properties of a theory T pertinent to the proof of (γ), then list the lemmas required in the proof, and finally state and prove the main theorems.

§25.3.1. Definitions. Although we have tried to make this section self-contained, a look at the early parts of Rasiowa and Sikorski 1963 is recommended.

V is a set of *propositional variables*.

L is the algebra of formulas in $\vee$, &, $\overline{}$, $\rightarrow$, with V as free generators. *A*, *B*, *C* range over L.

$\langle$L, T$\rangle$ (we sometimes just use "T") is a *theory* if T $\subseteq$ L. (We omit the O of §25.2.1 since it is fixed as $\{\&, \vee, \overline{}, \rightarrow\}$, and we do not require the closure conditions there imposed.)

M = $\langle$M, D$\rangle$ ranges over *matrices* similar to L, i.e., matrices with grammatically similar operations, and with D $\subseteq$ M the designated elements. (O omitted as above.)

v is an M-*assignment* iff v is defined for all $p \in$ V, and always v(p) $\in$ M.

Given an algebra M and an M-assignment v, v_M is the (clearly unique) homomorphic extension of v; e.g., $v_M(A \rightarrow B) = v_M(A) \rightarrow v_M(B)$. v_M, a function from L into M, is a *valuation*.

$\langle$M, D$\rangle$ is a T-*matrix* iff for each $A \in$ T, $v_M(A) \in$ D for every M-assignment v.

D is a *filter* in M iff for a, b $\in$ M, (a$\wedge$b) $\in$ D iff a $\in$ D and b $\in$ D.

D is *prime* in M iff for a, b $\in$ M, (a$\vee$b) $\in$ D iff a $\in$ D or b $\in$ D.

D is *exhaustive* in M iff for a $\in$ M, either a $\in$ D or $\bar{a} \in$ D.

D is *consistent* in M iff for a $\in$ M, not both a $\in$ D and $\bar{a} \in$ D.

D is a *truth-filter* in M iff D is a consistent and exhaustive prime filter.

If $\rightleftarrows$ is an equivalence relation (reflexive, transitive, symmetrical) on L, [A] is defined as the equivalence class of A.

$A \Rightarrow B$ iff there is a C such that $C \in$ T and $(A \wedge C) \rightleftarrows (A \wedge C \wedge B)$.

If $\rightleftarrows$ is a congruence on L with respect to &, $\vee$, $\rightarrow$, $\overline{}$, and T (see T1 of the next section), L/T is defined as the algebra $\{[A] : A \in L\}$, with operations defined in the usual way; e.g., $([A] \rightarrow [B]) = [A \rightarrow B]$.

D_T is the set $\{[A] : A \in T\}$.

For $\langle$M, D$\rangle$ a matrix, $M_{[D]}$ is the algebra $\{\mathit{0}, \mathit{1}\} \times$M with operations defined as follows:

$$(\langle s_1, a\rangle \wedge \langle s_2, b\rangle) = \langle s_1 \wedge s_2, a \wedge b\rangle$$
$$((\langle s_1, a\rangle) \vee \langle s_2, b\rangle) = \langle s_1 \vee s_2, a \vee b\rangle$$
$$\overline{\langle s, a\rangle} = \langle \bar{s}, \bar{a}\rangle$$
$$(\langle s_1, a\rangle \rightarrow \langle s_2, b\rangle) = \langle sg(a \rightarrow b) \wedge (s_1 \rightarrow s_2), a \rightarrow b\rangle,$$

where the following conventions obtain: (1) operations on the left of pairs are the usual ones on $\{\mathit{0}, \mathit{1}\}$ *qua* Boolean algebra, and (2) *sg*(a) is defined in terms of D by: $sg(a) = \mathit{1}$ if $a \in D$, $sg(a) = \mathit{0}$ if $a \notin D$. (Read "*sg*(a)" as "the sign of a.")

For $\langle M, D\rangle$ a matrix:

$\overline{D} = \{a : \bar{a} \in D\}$. Note: $\overline{D}$ is the *inverse* image of D under $^{-}$.

$M_{(D)} = ((\{\mathit{1}\}\times D)\cup(\{\mathit{0}\}\times\overline{D}))$. Note: a set, not an algebra.

M_D = the smallest subalgebra of $M_{[D]}$ which contains $M_{(D)}$.

(In the interesting cases, $M_D = M_{(D)}$; i.e., $M_{(D)}$ is closed under the operations as defined on $M_{[D]}$.)

$$D_M = \{\langle \mathit{1}, a\rangle : \langle \mathit{1}, a\rangle \in M_D\} = ((\{\mathit{1}\}\times M)\cap M_D)$$

(In the interesting cases, i.e., when $M_D = M_{(D)}$, it obviously turns out that $D_M = \{\mathit{1}\}\times D$.)

$$\langle M, D\rangle^* = \langle M_D, D_M\rangle, \text{ which is a matrix if } \langle M, D\rangle \text{ is.}$$

A formula A is T-*-*surviving* iff it survives the *-transformation: for every T-matrix $\langle M, D\rangle$ with D a filter, and for every M_D-assignment, $v_{M_D}(A) = \langle sg(a), a\rangle$, some $a \in M$. ($v_{M_D}(A) \leq \langle sg(a), a\rangle$ would suffice, but we have the more precise property in our cases.)

Let a rule R be reified as a relation between an n-tuple of premisses and a conclusion, hence a subset of $L^n\times L$.

R is T–*–*surviving* iff for each of its members (instances) $\langle\langle A_1, \ldots, A_n\rangle, A_{n+1}\rangle$, for each T-matrix $\langle M, D\rangle$ with D a filter, and for each M_D-assignment v, setting $v_{M_D}(A_i) = \langle s_i, a_i\rangle$, we have

$$s_1\wedge \ldots \wedge s_n \leq s_{n+1}.$$

That is, if $s_1, \ldots, s_n$ each $= \mathit{1}$, then $s_{n+1} = \mathit{1}$.

(Note that a rule need not preserve T–*–survivingness of formulas in order itself to be T–*–surviving. In fact adjunction *does* preserve T–*–survivingness of formulas under the already granted assumption that D is a filter, but *modus ponens*, though itself T–*–surviving, does not preserve T–*–survivingness of formulas. Thus, one might well have a theory T with axioms T_0 and rules R_0 all of which are T–*–surviving, and still not all theorems of T would be T–*–surviving. (Hence the proof to follow depends on the particular axiomatization of T — an oddity.)

§25.3.2. Abstract properties. We codify some properties of a theory T which together are sufficient for a theory to have (γ) as an admissible rule. The point of each property is stated first, and then the technical definition.

T1. T has enough properties so that it can be algebraized. That is, there is a congruence relation $\rightleftarrows$ on the set of formulas L with respect to &, $\vee$, $\rightarrow$, $^{-}$, and T such that

$A \rightleftarrows A$,
if $A \rightleftarrows B$ then $B \rightleftarrows A$, and
if $A \rightleftarrows B$ and $B \rightleftarrows C$ then $A \rightleftarrows C$;

and if $A \rightleftarrows A'$ and $B \rightleftarrows B'$ then each of

$\overline{A} \rightleftarrows \overline{A'}$,
$A\&B \rightleftarrows A'\&B'$,
$A \vee B \rightleftarrows A' \vee B'$, and
$A \rightarrow B \rightleftarrows A' \rightarrow B'$,

holds, and finally also

$A \in \mathrm{T}$ if and only if $A' \in \mathrm{T}$.

The most familiar way of defining such a congruence is in terms of T and a two-argument function σ from L^2 (i.e., the Cartesian product $\mathrm{L} \times \mathrm{L}$) into L:

$A \rightleftarrows B$ iff $\sigma(A, B) \in \mathrm{T}$.

And the most familiar σ is

$\sigma(A, B) = ((A \rightarrow B)\&(B \rightarrow A))$.

T2. T has enough properties beyond T1 so that L/T is a distributive lattice:

$A \vee B \rightleftarrows B \vee A$	$A\&B \rightleftarrows B\&A$
$A \vee (B \vee C) \rightleftarrows (A \vee B) \vee C$	$A\&(B\&C) \rightleftarrows (A\&B)\&C$
$(A\&B) \vee B \rightleftarrows B$	$A\&(A \vee B) \rightleftarrows A$
$A\&(B \vee C) \rightleftarrows (A\&B) \vee (A\&C)$	$A \vee (B\&C) \rightleftarrows (A \vee B)\&(A \vee C)$

T3. T has enough properties beyond T1 so that $\mathrm{D_T}$ is a filter in L/T:

$(A\&B) \in \mathrm{T}$ iff $A \in \mathrm{T}$ and $B \in \mathrm{T}$.

T4. T has enough properties beyond T1 so that $\langle \mathrm{L/T, D} \rangle$ is a T-matrix for all D such that $\mathrm{D_T} \subseteq \mathrm{D} \subseteq (\mathrm{L/T})$:

T is closed under substitution, which is to say that T is a logic in the sense of §25.2.3.

T5. T has enough properties beyond T1 to insure that if $\mathrm{D_T} \subseteq \mathrm{D} \subseteq (\mathrm{L/T})$, and if D is a prime filter in L/T, then $(\mathrm{L/T})_\mathrm{D} = (\mathrm{L/T})_{(\mathrm{D})}$; i.e., $(\mathrm{L/T})_{(\mathrm{D})}$ is closed under the operations as defined on $(\mathrm{L/T})_{[\mathrm{D}]}$:

T5.1. Closure under complementation requires that if $\mathrm{a} \in \mathrm{D}$ then $\bar{\mathrm{a}} \in \overline{\mathrm{D}}$, for which one needs

$A \Rightarrow \overline{\overline{A}}$.

T5.2. Closure under $\wedge$ requires that if $(\bar{a}\vee\bar{b}) \in D$, then $(a\wedge b) \in \overline{D}$, for which one needs

$$\bar{A}\vee\bar{B} \Rightarrow \overline{A\&B}.$$

T5.3. Closure under $\vee$ requires that if $(\bar{a}\wedge\bar{b}) \in D$ then $(a\vee b) \in \overline{D}$, for which one needs

$$\bar{A}\&\bar{B} \Rightarrow \overline{A\vee B}.$$

T5.4. Closure under $\rightarrow$ requires two properties: (1) if either $\bar{a} \in D$ or $b \in D$, and if also $(a\rightarrow b) \notin D$, then $(a\rightarrow b) \in \overline{D}$; and (2) if $a \in D$, $b \in \overline{D}$, then $(a\rightarrow b) \in \overline{D}$. The following suffice:

$$\bar{A}\vee B \Rightarrow (A\rightarrow B)\vee\overline{A\rightarrow B}$$
$$A\&\bar{B} \Rightarrow \overline{A\rightarrow B}$$

T6. T has enough properties beyond T1 so that if D is a prime filter in L/T containing D_T then $\langle sg[p], [p]\rangle \in (L/T)_D$; or at least $\langle s, [p]\rangle \in (L/T)$ for some $s \in \{\boldsymbol{0}, \boldsymbol{1}\}$; i.e.,

$$(A\vee\bar{A}) \in T.$$

Note: This is the *only* place that a non-intuitionistic property seems required. It implies the first part of T5.4.

T7. T has enough properties beyond T1 so that $\langle M, D\rangle^*$ is a T-matrix if $\langle M, D\rangle$ is a T-matrix with D a filter:

> T is the closure of a set of T–*–surviving axioms under T–*–surviving rules.

The following are used in verifying the T–*–survivingness of the axioms of certain systems:

T7.1. $(A\rightarrow A) \in T$. (This is guaranteed by T1 and T3 if $\sigma(A, B) \Rightarrow (A\rightarrow B)$.)

T7.2. $A\&(A\rightarrow B) \Rightarrow B$. (This is the *only* place the derivability of $\rightarrow$E is used.)

T7.3. If $\langle M, D\rangle$ is a T-matrix and D a filter, then if $a \in D$ and $b \in D$, $(a\wedge b) \in D$. (This is true by definition.)

§25.3.3. Facts. In this section we list with neither comment nor proof a series of facts required to make the next section short.

FACT 1. If T1–T3, and if $A \notin T$, then (1) L/T is a distributive lattice; (2) D_T is a filter; and (3) $[A] \notin D_T$.

FACT 2. (Prime filter theorem; Stone 1937.) If M is a distributive lattice, if D′ is a filter, and if $a \notin D'$, then there is a D such that (1) D is a prime filter in M, (2) $D' \subseteq D$, and (3) $a \notin D$.

FACT 3. If T1 and T4, and if $D_T \subseteq D$, then $\langle L/T, D\rangle$ is a T-matrix.

FACT 4. If T1 and T5, and if $D_T \subseteq D$, and if D is a prime filter in L/T, then (1) if $a \in D$ then $\bar{a} \in \overline{D}$; (2) if either $a \in \overline{D}$ or $b \in \overline{D}$ then $(a \wedge b) \in \overline{D}$; (3) if both $a \in \overline{D}$ and $b \in \overline{D}$ then $(a \vee b) \in \overline{D}$; (4) if $(\bar{a} \vee b) \in D$, and if $(a \rightarrow b) \notin D$, then $(a \rightarrow b) \in \overline{D}$; (5) if $a \in D$ and $b \in \overline{D}$, then $(a \rightarrow b) \in \overline{D}$.

FACT 5. If T1 and T6, and if $D_T \subseteq D$, and if D is prime in L/T, then D is exhaustive in L/T.

FACT 6. If T7, and if $\langle M, D\rangle$ is a T-matrix, then $\langle M, D\rangle^* = \langle M_D, D_M\rangle$ is a T-matrix.

FACT 7. D_M is a truth filter in M_D. Note: No assumptions are needed other than that $\langle M, D\rangle$ is a matrix.

FACT 8. If $(D \cup \overline{D}) = M$, and if v′ is an M-assignment, then v defined by $v(p) = \langle sg(v'(p)), v'(p)\rangle$ is an M_D-assignment; i.e.,

$$\langle sg(v'(p)), v'(p)\rangle \in M_D.$$

FACT 9. If T1 and if v is any $(L/T)_D$-assignment such that $v(p) = \langle s, [p]\rangle$ for some $s \in \{\boldsymbol{0}, \boldsymbol{1}\}$, then for all $A \in L$,

$$v_{(L/T)D}(A) = \langle s, [A]\rangle$$

for some $s \in \{\boldsymbol{0}, \boldsymbol{1}\}$.

FACT 10. If (1)–(5) as in Fact 4 hold of $\langle M, D\rangle$, and if D is a prime filter in M, then

$$M_D = M_{(D)};$$

i.e., M_D is closed under the operations defined on $M_{[D]}$.

FACT 11. If $M_D = M_{(D)}$, and if $a \notin D$, then $\langle s, a\rangle \notin D_M$, all $s \in \{\boldsymbol{0}, \boldsymbol{1}\}$.

§25.3.4. Punch line. Theorems and proofs go like this.

THEOREM 1. If T1–T7 hold for a theory T, then for any $A \notin T$ there is a $\langle M, D'\rangle$ such that (1) $\langle M, D'\rangle$ is a T-matrix, (2) D′ is a truth filter in M, and (3) $v_M(A) \notin D'$, for some M-assignment v.

PROOF. Suppose T1–T7 and $A \notin T$. Then by T1,

(a) $\langle L/T, D_T\rangle$ is a matrix,

and by Fact 1 with T1–T3, using also Fact 2, there is a D such that

(b) $D_T \subseteq D$,
(c) D is a prime filter in L/T, and
(d) $[A] \notin D$.

By Fact 3 with (b), T1, and T4,

(e) $\langle L/T, D\rangle$ is a T-matrix;

and by Fact 4 with (b), (c), T1, and T5, for a, b $\in$ (L/T),

(f) if $a \in D$ then $\bar{a} \in \overline{D}$,
(g) if either $a \in \overline{D}$ or $b \in \overline{D}$ then $(a \wedge b) \in \overline{D}$,
(h) if both $a \in \overline{D}$ and $b \in \overline{D}$ then $(a \vee b) \in \overline{D}$,
(i) if $(a \rightarrow b) \notin D$ then $(a \rightarrow b) \in \overline{D}$, and
(j) if $a \in D$ and $b \in \overline{D}$ then $(a \rightarrow b) \in \overline{D}$.

In fact, by T1, T6, (b), and (c) the condition (i) can be strengthened by Fact 5 to

(i′) if $a \notin D$ then $a \in \overline{D}$.

Now consider $\langle L/T, D\rangle^* = \langle (L/T)_D, D_{(L/T)}\rangle$. We need to show that (1)–(3) of Theorem 1 hold. By Fact 6 with (c) and (T6),

(1) $\langle L/T, D\rangle^* = \langle (L/T)_D, D_{(L/T)}\rangle$ is a T-matrix.

By Fact 7 with (e),

(2) $D_{(L/T)}$ is a truth filter in $(L/T)_D$.

For $p \in V$ define

(3.1) $v(p) = \langle sg[p], [p]\rangle$.

where *sg* depends on D as in the definition of $M_{[D]}$. By Fact 8 with (i′),

(3.2) v is an $(L/T)_D$-assignment,

and by Fact 9, with (3.2) and (T1),

(3.3) $v_{(L/T)_D}(A) = \langle s, [A]\rangle$, some $s \in \{\mathit{0}, \mathit{1}\}$.

By Fact 10 with (c), (f), (g), (h), (i), and (j),

(3.4) $(L/T)_D = (L/T)_{(D)}$,

so by Fact 11 with (d), (3.3), and (3.4),

(3.5) $v_{(L/T)_D}(A) \notin D_{(L/T)}$.

Hence, by (3.2) and (3.5),

(3) $v_{(L/T)D}(A) \notin D_{(L/T)}$, some $(L/T)_D$-assignment v.

Theorem 2. T1–T7 hold for **E**, **R**, and many others.

Proof. We mention only the status of T7 for the standard axiomatization of **E** as in §21.1: **E** is the closure of a set of **E**–*–surviving axioms under **E**–*–surviving rules.

1. Axioms E4, E5, E8, E9, E11, E12, and E14, are T–*–surviving, for arbitrary T, hence are **E**–*–surviving.
2. Axiom E1 is **E**–*–surviving since T7.1 holds for **E**.
3. Axioms E2, E3, and E13, are **E**–*–surviving since T7.2 holds for **E**.
4. Axioms E6 and E10 are **E**–*–surviving since T7.2 and T7.3 hold for **E**.
5. Axiom E7 is **E**–*–surviving since all of T7.1, T7.2, and T7.3 hold for **E**.
6. The rules →E and &I are T–*–surviving for arbitrary T, hence for **E**.

Theorem 3. The rule (γ) is admissible in **E**, **R**, and many others.

Proof (for **E**). Suppose for *reductio* that A and $\bar{A}\vee B$ are theorems of **E**, but that B is not. By Theorems 1 and 2, there is an **E**-matrix ⟨M, D⟩ with D a truth filter such that for some M-assignment, v, $v_M(B) \notin D$. Since ⟨M, D⟩ is an **E**-matrix, $v_M(A) \in D$, so by the consistency of D, $\overline{v_M(A)} = v_M(\bar{A}) \notin D$. But again by the fact that ⟨M, D⟩ is an **E**-matrix, $v_M(\bar{A}\vee B) \in D$, so by the primeness of D, $v_M(\bar{A}) \in D$ or $v_M(B) \in D$; which contradicts the supposition that $v_M(\bar{A}) \notin D$ and $v_M(B) \notin D$.

§26. Miscellany. This section is analogous to §8; see the Analytical table of contents for a list of topics discussed.

§26.1. Axiom-chopping. We are getting a little bored with this topic, but there are still probably one or two things worth pointing out.

E7 (§21.1) may be replaced by any of several axioms, all having the effect of getting us over one inductive step (that involving &I) required for the lemma used in proving that **E** contains **FE**. All of these have the fault of being somewhat unnatural, and of messing up various independence results and separation theorems; they are therefore aesthetically displeasing. Of these the simplest is

1 $((A\to A)\&(B\to B))\to C\to C$,

which we notice makes E1 redundant. Take E4 in the form

2 $(A\to A)\&(B\to B)\to(A\to A)$;

then suffix C twice, and use 1 to get E1 by $\rightarrow$E.

To see that E7 also follows from 1, we suffix A to 2, getting

3 $(A\rightarrow A\rightarrow A)\rightarrow.((A\rightarrow A)\&(B\rightarrow B))\rightarrow A.$

Acting similarly on E5 we get

4 $(B\rightarrow B\rightarrow B)\rightarrow.((A\rightarrow A)\&(B\rightarrow B))\rightarrow B,$

and from the last two we get by easy maneuvers with E4–6 and the definition of $\Box$ to

5 $\Box A\&\Box B\rightarrow.((A\rightarrow A)\&(B\rightarrow B))\rightarrow.A\&B.$

The consequent of 5 is a necessitive, so we have

6 $\Box A\&\Box B\rightarrow\Box[((A\rightarrow A)\&(B\rightarrow B))\rightarrow.A\&B].$

Distribution of $\Box$ over 1 (with $A\&B$ for C) gives

7 $\Box[((A\rightarrow A)\&(B\rightarrow B))\rightarrow.A\&B]\rightarrow\Box(A\&B),$

and 6 and 7 yield

E7 $\Box A\&\Box B\rightarrow\Box(A\&B),$

thus fulfilling one of the prophecies made at the end of §23.4. The other two will be left to the reader.

§26.2. Independence (by John R. Chidgey). We offer complete independence results for the formulation of **E** of §21.1; other formulations are treated in Chidgey 1974. The matrices to which we refer will be found in §29.9. The results are presented in approximately the style of §8.4.1; e.g.

E1 $A\rightarrow A\rightarrow B\rightarrow B$ M VII: $A = \mathit{1}$ or $A = \mathit{4}$; $B = \mathit{2}$

means that axiom E1 is independent in the set {E1–E14, $\rightarrow$E, &I} in virtue of Matrix Set M VII of §29.9; an undesignated value for E1 being produced by assigning A either *1* or *4*, and assigning B the value *2*.

Independence of the axioms and rules of **E** (formulation of §21.1).

E1	$A\rightarrow A\rightarrow B\rightarrow B$	M VII:	$A = \mathit{1}$ or $A = \mathit{4}$; $B = \mathit{2}$.
E2	$A\rightarrow B\rightarrow.B\rightarrow C\rightarrow.A\rightarrow C$	M VIII.1:	$A = \mathit{2}, B = \mathit{1}; C = \mathit{0}$.
E3	$(A\rightarrow.A\rightarrow B)\rightarrow.A\rightarrow B$	M IX:	$A = \mathit{1}; B = \mathit{0}$.
E4	$A\&B\rightarrow A$	M X.1:	$A = \mathit{0}; B = \mathit{1}$.
E5	$A\&B\rightarrow B$	M X.2:	$A = \mathit{1}; B = \mathit{0}$.
E6	$(A\rightarrow B)\&(A\rightarrow C)\rightarrow.A\rightarrow(B\&C)$	M I.5:	$A = \mathit{1}; B = \mathit{1}; C = \mathit{1}$.
E7	$\Box A\&\Box B\rightarrow\Box(A\&B)$	M XXI:	$A = \mathit{2}; B = \mathit{3}$; or
	$[\Box A =_{df} A\rightarrow A\rightarrow A]$		$A = \mathit{3}; B = \mathit{2}$.

E8	$A \to A \vee B$	M XI.1:	$A = \mathit{2}$; $B = \mathit{1}$.
E9	$B \to A \vee B$	M XI.2:	$A = \mathit{1}$; $B = \mathit{0}$.
E10	$(A \to C)\&(B \to C) \to .(A \vee B) \to C$	M XI.3:	$A = \mathit{0}$; $B = \mathit{0}$; $C = \mathit{0}$ or $\mathit{1}$.
E11	$A\&(B \vee C) \to (A\&B) \vee C$	M XII:	$A = \mathit{2}$; $B = \mathit{3}$ or $\mathit{4}$; $C = \mathit{1}$ or $A = \mathit{4}$; $B = \mathit{1}$ or $\mathit{2}$; $C = \mathit{3}$.
E12	$A \to \bar{A} \to \bar{A}$	M XIII.1:	$A = \mathit{1}$.
E13	$A \to \bar{B} \to .B \to \bar{A}$	M X.4:	$A = \mathit{2}$; $B = \mathit{1}$.
E14	$\bar{\bar{A}} \to A$	M X.5:	$A = \mathit{0}$.
$\to$E	From A and $A \to B$ to infer B	M XIV	$A = \mathit{2}$; $B = \mathit{0}$.
&I	From A and B to infer $A\&B$	M XI.4	$A = \mathit{2}$; $B = \mathit{2}$.

§26.3. Intensional conjunctive and disjunctive normal forms. In §22.1.1 we defined *truth functional part* (*tfp*) of a formula A in the obvious way, so that the tfps of A are just those well-formed parts which occur in A otherwise than within the scope of an arrow. We can think of a formula A, therefore, as a truth function of its tfps, and if a tfp B of A is a propositional variable, or has the form $C \to D$, then we will call B an *intensional variable* of A.

(The terminology is unhappy, though it does suggest what we mean, i.e., an atomic piece of notation which may assume propositions in intension as values; but it is the best we have been able to do. The point is that formulas of **E** may be written in a kind of conjunctive or disjunctive normal form, where both propositional variables and formulas of the form $A \to B$ play the role of atomic expressions in the usual extensional treatments, in the sense that neither is susceptible of further truth functional decomposition. "Intensional atoms" would do, except that we have already used "atoms" in §24 to include denials of propositional variables, and for reasons to emerge presently we do not want to construe formulas of the form (e.g.) $\overline{A \to B}$ as among the "building blocks" of our normal forms. "Intensional unit" is close, but it has an unsavory, septic sound to it. Happily the use of the term "intensional variable" won't last long.)

We will say that A' is an *intensional conjunctive normal form* of A if it arises from A by making the following replacements as often as possible. Replace a tfp of the form:

(i)	$\bar{\bar{B}}$	by	B;
(ii)	$\overline{B\&C}$	by	$\bar{B} \vee \bar{C}$;
(iii)	$\overline{B \vee C}$	by	$\bar{B}\&\bar{C}$;
(iva)	$A \vee (B\&C)$	by	$(A \vee B)\&(A \vee C)$;
(ivb)	$(B\&C) \vee A$	by	$(B \vee A)\&(C \vee A)$.

Any result A' of applying such replacements as often as possible to A is an intensional conjunctive normal form of A, and in view of the fact that for each of (i)–(iv) the left-hand side co-entails the right, we have that $\vdash A \rightleftarrows A'$. We do not bother to make these forms unique, though of course this could be done by adding associativity and commutativity for & and $\vee$, contractions $A \rightleftarrows (A \vee A)$ and $A \rightleftarrows (A \& A)$, which we may always allow, if we wish, for convenience in finding an A', and requiring some sort of order for conjuncts and disjuncts. We will in fact assume generalized associativity, so that we may write e.g. $B_1 \& B_2 \& \ldots \& B_m$; justification for this practice is trivial. As will become clear below, pairwise grouping of such conjunctions (or disjunctions) is immaterial to the use to which such forms are put, and in general when we need an intensional conjunctive form A' of A, any old such A' will do. *Mutatis mutandis*, the same remarks apply to

Intensional disjunctive normal forms, which are defined similarly, just replacing (iva) and (ivb) by:

(iva′)	$A \& (B \vee C)$	by	$(A \& B) \vee (A \& C)$;
(ivb′)	$(B \vee C) \& A$	by	$(B \& A) \vee (C \& A)$.

The fact that every formula A *has* at least one of each of the forms, each co-entailing A, is easy to show, and will be left without proof.

Our reasons for introducing intensional normal forms at this juncture is to establish a little fact about their relations to positive and negative formulas (§22.1.1) which will be of use to us later on.

LEMMA. Let A be a formula of **E** with an intensional conjunctive normal form $B_1 \& B_2 \& \ldots \& B_m$ (where $A \rightleftarrows B_1 \& B_2 \& \ldots \& B_m$, each B_i is a disjunction $C_1 \vee \ldots \vee C_n$, parentheses restored *ad lib.*, and each C_j is an intensional variable, or the denial of such). If D is an intensional variable, then (1) D is an ap (§22.1.1, after Lemma 2) of A iff for some B_i, D is an ap of B_i, and (2) D is a cp of A iff for some B_j, D is a cp of B_j.

The proof is obvious from the fact that if D is a cp {ap} of a replaced formula on the left of (i)–(iv) above, then D is also a cp {ap} of the replacing formula on the right (by the definition of cp and ap), (iva) and (ivb) guaranteeing that any part of the left (with the exception of $B \& C$ itself) is part of at least one of the conjuncts on the right; hence the "only if" of (1) and (2). "If" follows from the fact that all replacements are co-entailments, and recovering A from an intensional conjunctive normal form A' of A simply requires retracing the steps leading from A to A'. A careful and detailed proof will be left to the reader.

The parallel lemma for an intensional disjunctive normal form $B_1 \vee B_2 \vee \ldots \vee B_m$ of A states that for any intensional variable D, (1) D is an ap of A iff for some B_i, D is an ap of B_i, and (2) D is a cp of A iff for some B_i, D is a cp of B_i.

As an example we rewrite the formula

$$(\bar{B} \vee \overline{A \vee (B \rightarrow C \& A)})\&(B \rightarrow C)$$

in one of its intensional conjunctive normal forms:

$$(B \vee \overline{B \rightarrow C})\&(\bar{A} \vee \overline{B \rightarrow C})\&(B \vee A)\&(\bar{A} \vee A)\&(B \rightarrow C),$$

noticing that each of the intensional variables which occurs as a *positive* part of the example (namely B, A, $B \rightarrow C$) crops up also as a *positive* part in one (or more) of the conjuncts, and each of the intensional variables occurring *negatively* in the example (namely: A, $B \rightarrow C$) crops up *negated* in one of the conjuncts — as the lemma would lead us to expect. Similar (dual) remarks hold for an intensional disjunctive normal form of the same formula:

$$(B \& \bar{A} \& (B \rightarrow C)) \vee (\overline{B \rightarrow C} \& A \& (B \rightarrow C)).$$

COROLLARY. If A is a positive {negative} formula, and B is either an intensional disjunctive or conjunctive normal form of A, then B is a positive {negative} formula.

§26.4. Negative formulas; decision procedure. It turns out that, if A is a negative formula of **E**, then it is decidable whether $\vdash_{\mathbf{E}} A$. The lemma and corollary of the previous section help by reducing the question to that of the provability of an intensional conjunctive normal form $B_1 \& \ldots \& B_m$ of A, where each B_i has the form $C_1 \vee \ldots \vee C_n$, each C_j being either a propositional variable, the denial thereof, or of the form $\overline{D \rightarrow E}$. (None can have the form $D \rightarrow E$, by the corollary, since A is negative.) So to check provability of A, we need only check provability of all the B_i, each of which is negative.

The decision procedure is as follows. Construct a tree for B_i as in §24.1.1, except that we use three rules for tree construction:

(i) $\dfrac{\varphi(A)}{\varphi(\bar{\bar{A}})}$ (ii) $\dfrac{\varphi(\bar{A}) \qquad \varphi(\bar{B})}{\varphi(\overline{A \vee B})}$ (iii) $\dfrac{\varphi(A) \qquad \varphi(\bar{B})}{\varphi(\overline{A \rightarrow B})}$

We first verify that if the conclusion of a rule by (i)–(iii) is a negative formula then so are the premisses. (Notice that this is not altogether trivial, since for (iii), $A \rightarrow B$ may be of *any* degree.) We consider case (iii), leaving the others to the reader. Suppose that $\varphi(\overline{A \rightarrow B})$ is negative, but $\varphi(A)$ is not;

then $\varphi(A)$ has a consequent tfp $C\to D$. This part cannot occur as a consequent part in the context around A, else it would appear in $\varphi(\overline{A\to B})$, contrary to hypothesis. Hence $C\to D$ must occur as a consequent part of A. Then $C\to D$ is an antecedent part of $A\to B$, hence a consequent part of $\overline{A\to B}$, contrary to hypothesis. (For this we require two obvious lemmas which are provable from the definitions of antecedent and consequent parts in §22.1.1; they will be left to the reader.) The proof for the other premiss of (iii) is similar.

Since the candidates B_i above are negative formulas, the preceding observation guarantees that when constructing trees in accordance with (i)–(iii), we never run into a formula $\varphi(C\to D)$; hence (i)–(iii) suffice to drive provability of the candidate B_i back to primitive disjunctions at the tips of branches. As before (§24.1), if the primitive disjunctions at the tips each contain a propositional variable and its denial as disjunctive parts, we pronounce the candidate provable in **E** — otherwise not. What remains is to see that this pronouncement is correct; i.e., that $\vdash_{\mathbf{E}} B_i$ just in case the decision procedure just outlined yields a satisfactory answer.

Evidently all the tips in a good tree are provable in **E**, and the following theorems (left to the reader) guarantee that provability in **E** passes down the tree to B_i:

$$\vdash_{\mathbf{E}} \varphi(A) \to \varphi(\overline{\overline{A}}),$$
$$\vdash_{\mathbf{E}} \varphi(\overline{A})\&\varphi(\overline{B}) \to \varphi(\overline{A\vee B}), \text{ and}$$
$$\vdash_{\mathbf{E}} \varphi(A)\&\varphi(\overline{B}) \to \varphi(\overline{A\to B}).$$

And if some branch terminates in a non-axiom, we can falsify every disjunctive part of every formula in the bad branch (in particular B_i) by means of the two valued matrix, which satisfies all of **E** if the arrow is taken as material "implication."

Hence a negative formula A is provable in **E** iff it passes the test. (This fragment of **E** is, incidentally, the only fragment for which we have a decision procedure depending neither on degree nor absence of certain connectives.)

§26.5. Negative implication formulas. All apologists for *any* formal theory of "if . . . then —" have been unanimously adamant on one aspect of their theory: "if A then B" must be construed as false when A is true and B is false. Frege and his followers wanted "if A then B" to be true in any other case; Frege indeed even wanted to count "if 2 then 6" as true. (That is, he says that

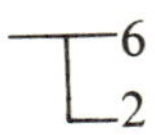

is the same as

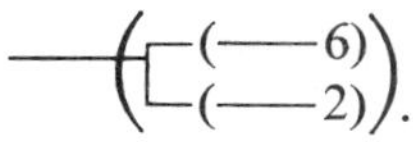

the value of which is the True, since the values of both ——6 and ——2 are the False. See Frege 1893, §12.)

As we have argued at length, this truth condition is implausibly liberal; but for any sane analysis of the conditional, indeed for *any* analysis, sane or not, A-and-not-B is surely a sufficient condition for the *falsity* of "if A then B."

This suggests that if some candidate for an analysis of "if . . . then —" figures in a formula A in such a way that only the falsehood conditions for the "if . . . then —" are relevant to the validity of A, then in such contexts the various proposed analyses ought to coincide. Somewhat more precisely, if a formula A involves $B \rightarrow C$ (in the sense of **E**), or $B \supset C$ $(=_{\mathrm{df}} \overline{B} \vee C)$, or $B \dashv C$ $(=_{\mathrm{df}} \Box(B \supset C))$ only in an essentially negative way (a notion we make clear in a moment), then $B \supset C$, $B \dashv C$, and $B \rightarrow C$ ought in such contexts to coincide. And, plausibly understood, this turns out to be true.

We begin by noting that if either $\varphi(\overline{A \supset B})$ or $\varphi(\overline{A \dashv B})$ is a negative formula, then so are $\varphi(A)$ and $\varphi(\overline{B})$. Hence we may generalize the lemma and corollary of §26.3 in the following way. For notation we use $\supset$ and $\dashv$ as well as $\rightarrow$, $\sim$, and $\vee$, with $\supset$ and $\dashv$ defined but, for present notational purposes, left in their defined form, as above. Then a *negative* {*positive*} *implication formula* is one containing no consequent {antecedent} occurrences of either $A \rightarrow B$, $A \supset B$, or $A \dashv B$. We leave it to the reader to reconstruct the appropriate generalizations of the lemma and corollary of §26.3, so as eventually to get the following result:

THEOREM. If A is a negative implication formula, then $\vdash_{\mathbf{E}} A$ iff for certain formulas $B_1, \ldots, B_m$ of **E**, (a) A co-entails the conjunction of the B_i, and (b) each B_i has the form $C_1 \vee \ldots \vee C_n$, where each C_j is either a propositional variable, the denial thereof, or else has the form $\overline{A \rightarrow B}$, $\overline{A \supset B}$, or $\overline{A \dashv B}$.

The decision problem for negative implication formulas then reduces to that for the B_i above, and solution to the problem amounts then simply to adding two more rules for tree construction:

$$\text{(iii}')\quad \frac{\varphi(A) \qquad \varphi(\overline{B})}{\varphi(\overline{A \supset B})}, \quad \text{and} \quad \text{(iii}'')\quad \frac{\varphi(A) \qquad \varphi(\overline{B})}{\varphi(\overline{A \dashv B})}.$$

This procedure yields a decision procedure for negative implication formulas, a fact which we will let the reader verify.

But it yields a little more, namely an account of the fact that *everybody* is clear on falsehood conditions for "if . . . then — ." For clearly $A\supset B$, $A \mathrel{-\!\!\!3} B$, and $A\to B$ are inter-replaceable in negative implication formulas. And if we take the phrase above:

> "if . . . then —" figures in a formula A in such a way that only falsehood conditions for the "if . . . then —" are relevant to the validity of A,

as explicated by "A is a negative implication formula," then we now have some formal account of the matter.

We illustrate with the negative implication formula $\overline{A\to B}\vee\overline{A}\vee B$, which is provable in **E**. The theorem above then tells us that $\overline{A\supset B}\vee\overline{A}\vee B$ and $\overline{A \mathrel{-\!\!\!3} B}\vee\overline{A}\vee B$ are equally provable in **E**. As another example, the negative formula $\overline{\overline{A\to B}}\to .C\to D$ considered in §12 is provable in **E** just in case $\overline{\overline{A\supset B}}\supset .C\supset D$ is provable, which it clearly isn't. (Notice however that this does not preclude adding $\overline{A\to B}\to .C\to D$ to **E** as an axiom — but the resulting system is *not* **E**, and most of **E**'s nice properties are not available for it.)

So for a well-defined collection of formulas of **E**, $A\supset B$, $A \mathrel{-\!\!\!3} B$, and $A\to B$ are interchangeable. Nor is it unusual to find cases where A and B may be interchangeable *salva veritate* in certain contexts, even though it is not the case that $A\rightleftarrows B$ is a theorem. For example $\vdash A\equiv B$ (material "equivalence") is sufficient to go from $\vdash(\ldots A\ldots)$ to $\vdash(\ldots B\ldots)$ in **E**, provided the context $(\ldots\quad\ldots)$ is purely truth functional (this holds even if A and B contain arrows, as is easily shown). And even in contexts involving arrows, for example, we have the following case. Let F be any truth functionally contradictory formula; then, though of course $B\&\overline{B}\rightleftarrows F$ generally fails in **E**, we may still replace $B\&\overline{B}$ by any such F in the theorem $(A\to .B\&\overline{B})\to A$, as the reader may check.

As Meyer has pointed out to us (conversation), connections between $A\circ B$ and $\overline{A\to\overline{B}}$ discussed in §27.1.4 (they are equivalent in **R**) give us some leverage for expanding the results just stated for positive **R** with $\circ$ as primitive — and perhaps elsewhere. We leave the problem for any interested spectators.

§26.6. Further philosophical ruminations on implications. In arguing that the classical truth functional analysis is an accurate mathematical reconstruction of certain uses of conjunction and disjunction, one commonly finds authors explaining that just as the truth tables for $A\&B$ and $A\vee B$ have the right properties, so the De Morgan properties show similarly that denial of conjunctions and disjunctions have acceptable features. But *no one* goes on to extol the classical equivalence of "it is not the case that if A then B" with $A\&\overline{B}$ — and with good reason: nothing could be further from the truth.

There are of course Officers who have persuaded themselves that "although most conditional statements assert more than a merely material implication between antecedent and consequent," it is still not unreasonable "to symbolize *any* occurrence of 'if . . . then —' by the truth functional connective '$\supset$'." Some have even convinced themselves that though they sometimes may (very informally and vaguely, and out of lofty condescension to the man in the street) use "if . . . then —" in a different sense, still when they are being hardheaded and careful, they themselves really mean "if . . . then —" in the alleged material sense. Such logicians of course disagree with our claim that "all" should replace "most" in the first remark quoted just above, and though it is hard to dispute another's claim that *he* means $\bar{A} \vee B$ when *he* says "if A then B," the following debater's ploy has worked often enough to convince us that pretending to believe that one means $\bar{A} \vee B$ by "if A then B" is a full-time job, and, like so many false fronts, difficult to keep up. The ploy has lots of variants, but one form goes like this.

§26.6.1. Facetious. Present to an Officer the following remark: "If every signal has a maximum velocity, then there is a maximum velocity which no signal exceeds." The conversation is likely to go something like this:

Officer: But surely not; whoever said that was confused about his quantifiers.

Eristic: But look here, I thought you knew about all this. According to Einstein's theory of relativity, the velocity of no signal can exceed that of light.

O: What's that got to do with it?

E: Well, according to *your theory*, it guarantees the truth of the proposition you just rejected.

O: What? Oh . . . , well, . . . , heh-heh, . . . , I didn't mean *that;* what I meant was that there was no way of arguing correctly from the antecedent to the consequent.

E: Well, presumably if there *were* such a way of arguing correctly from the antecedent to the consequent, we would be able to use other relevant facts about the world and about logic, wouldn't we?

O: I suppose so.

E: Well, why can't we use Einstein's theory in the way indicated? (Note, please, this is *your* theory of "if . . . then —," not mine.)

O: But you see, . . .

E (interrupting): If you call the conditional false, you are committed to a repudiation of Einstein's theory, according to your analysis; so you're obliged to say that the conditional is true. I really can't see what you're objecting to.

O: You'll have to excuse me; I think I hear my mother calling me.

There are probably subtler ways of escaping the situation in which we have put O, but even so E, if malicious enough, has it in his power to discredit O by going around whispering to O's friends: "You know what? O claims that if every signal has a maximum velocity, then there is a maximum velocity which no signal exceeds." This will no doubt lead to the view that O has been badly trained, though of course he really understands what he is saying; the trouble is that he doesn't realize that it is false (or at least, true only in a sense of "if . . . then —" that doesn't mean if . . . then —).

Continuing in the same vein, we report an example communicated to us in correspondence by Charles L. Stevenson. He considers the enviable position of a man who believes that the concept of a benevolent and omnipotent Deity leaves no room for malice in His analogical heart, and consequently rejects as false the statement "If God exists then there is evil in the world." That is (?), he claims that $\sim(G \supset E)$, from which it follows that God exists and there is no evil in the world, thus (as Stevenson remarks) "dismissing atheism and pessimism at one fell swoop." (See also Stevenson 1970.)

No one argues this way, but with a little perversity, the theory that "if . . . then —" means "not . . . or —" can be made to look as if it sanctions such nonsense.

§26.6.2. Serious. We would now like to discuss a seriously meant case in the literature, where the author is convinced that he means by "if A then B" that A intuitionistically "implies" B. We would also like the reader to shift gears with us at this point, since we want to drop completely the flippant tone of the preceding subsection. The reason for this abrupt about-face is that the passage to be discussed comes from a profound and justifiably influential work by one of the foremost members of the logical fraternity, and we do not want to impute any sort of confusion *at all* to the author. Indeed the whole volume in question discusses a very intricate topic with monumental clarity; our only intent is to suggest that there are alternative locutions, which fit better with the philosophical tenor of this book, and which should not (we hope) be found offensive by the author we cite. (Note: we will try to make the discussion self-contained; knowledge of recursive functions will not be required for the points at issue.)

We begin by recalling that

$$\vdash_{\mathbf{E}} A \to B \to . \bar{A} \vee B,$$

since negation is classical, but

$$\dashv_{\mathbf{E}} \bar{A} \vee B \to . A \to B,$$

since the latter leads to paradox. Intuitionistically, however, we have

$$\dashv_{\mathbf{H}} A \to B \to . \neg A \vee B,$$

(where **H** is Heyting's 1930 intuitionism) since intuitionistic negation ($\neg$) and implication are nonclassical, and

$$\dashv_{\mathbf{H}} \neg A \vee B \rightarrow . A \rightarrow B,$$

which leads to $\vdash_{\mathbf{H}} \neg A \rightarrow . A \rightarrow B$ and $\vdash_{\mathbf{H}} B \rightarrow . A \rightarrow B$, both paradoxical.

Now in the light of this, consider the following, from Kleene 1952, p. 273:

> EXAMPLE 2. To settle a point raised in §45, suppose φ is 3-recursive but not 2-recursive, and ψ is 2-recursive but not 1-recursive, i.e. not primitive recursive. Then "if ψ is primitive recursive, then φ is primitive recursive" is vacuously true, but "φ is primitive recursive in ψ" is false, since that would make φ 2-recursive.

Our reading of this passage involves two points.

(1) Concerning the truth of "if ψ is primitive recursive, then φ is primitive recursive," Kleene feels it necessary to qualify the claim with "vacuously," so as to obviate the confusion which might easily arise if one were to take the claim at face value, and go on to suppose that involved in the claim was the notion that the primitive recursiveness of ψ had something to do with the primitive recursiveness of φ; this would (as we believe) be the natural and obvious understanding of the "if . . . then —" locution in this context. But Kleene has antecedently given allegiance to the belief that when *he* uses "if . . . then —" he uses it in such a way that "if $\neg A$, then if A then B" is true of the conditional, and consequently the natural understanding of the quoted sentence at issue is not available to him.

(2) He is therefore led to define the notion "φ is primitive recursive in ψ," as follows (p. 224):

> A function φ is *primitive recursive in* $\psi_1, \ldots, \psi_l$ (briefly Ψ), if there is a finite sequence $\varphi_1, \ldots, \varphi_k$ of (occurrences of) functions (called a *primitive recursive derivation of* φ from Ψ) such that each function of the sequence is either one of the functions Ψ (the *assumed functions*), or an initial function, or an immediate dependent of preceding functions, and the last function φ_k is φ.
>
> Since this definition has the same form as that of deducibility, to each of the general properties of $\vdash$ (§20) a corresponding principle can be stated now. For example, if φ is primitive recursive in Ψ, and some of the functions Ψ are primitive recursive, then φ is primitive recursive in the rest of the functions Ψ. An example (with $l = 1$) will be given later in which "if $\psi_1, \ldots, \psi_l$ are primitive recursive, then φ is primitive recursive" is true, but "φ is primitive recursive in $\psi_1, \ldots, \psi_l$" is false (Example 2 §55).

(The example is that displayed above).

The point is that, in the Example 2 above, there is no *primitive recursive derivation* of φ from ψ. "From" is still not taken as seriously as we would like, of course, since Kleene would no doubt allow that if φ is primitive recursive, then it is *primitive recursive in* (vacuously) any functions $\psi_1, \ldots, \psi_l$ one likes. But "from" is taken seriously at least in the sense that if neither φ nor ψ is primitive recursive and still φ is *primitive recursive in* ψ, then we must *use* equations for ψ in the course showing the latter fact; i.e., a *connection* must be established in the course of establishing the claim. (Compare the fact that if B is false, and $A \supset B$ holds intuitionistically, features of A must be *used* in getting intuitionistically to B — even if the only feature is that A is false.)

In spite of our protestations, some readers will no doubt believe us to be accusing Kleene of some error, which, as we want to emphasize strongly, we are *not* doing; if he wishes to speak in the way he does, who are we to complain? We don't want to complain, but we will still allow ourselves to enter a minor cavil.

It seems to us more "natural" to say, not that "if ψ is primitive recursive, then φ is primitive recursive" is true (on the grounds that the antecedent is false), but rather that it is false (on the grounds that φ is not *primitive recursive in* ψ, which is what is most easily taken as a sensible understanding of the "if . . . then —" proposition in the first place).

We recognize that all this is captious, and we certainly don't recommend that any changes be made in Kleene's eminently clear exposition. We simply leave it to the reader to ponder the merits of the case if so inclined.

We conclude by observing that this is the *only* case we know of in the literature where funny properties of a writer's theory about "if . . . then —" are thus taken seriously — which is the reason we have discussed it at such length. (We promptly dishonor our observation to conclude by quoting the following from Kleene 1952, p. 532:

> QUINE, WILLARD VAN ORMAN
>
> 1940. **Mathematical logic.** New York (Norton), xiii+348 pp. See Rosser 1942 and Quine 1941, concerning the fact that the Burali-Forti paradox arises in the system of this book (although Cantor's paradox apparently is avoided), as was discovered by Rosser and by Roger C. Lyndon. Revised ed., Harvard University Press, 1951, xii+346 pp.

The reader is invited to speculate on the question as to *how*, given the classical or intuitionistic view that *any* contradiction leads to what Meyer calls a "psychotic break," we can get the Burali-Forti paradox while avoiding Cantor's paradox.)

§26.7. $A\to B$, $\overline{C\to D}$, **and** $\overline{A\to B\to .C\to D}$. Evidently the first of these entails that the second entails the third, as a result of contraposing the consequent of

$$A\to B\to .(A\to B\to .C\to D)\to .C\to D,$$

a special case of *restricted assertion*. For material "implication" we also have both

$$\overline{A\supset B\supset .C\supset D}\supset .A\supset B,$$

and

$$\overline{A\supset B\supset .C\supset D}\supset .\overline{C\supset D},$$

each of which obviously fails for the arrow. However, it is not altogether implausible to guess that if

$$\vdash_{\mathbf{E}_{\overset{\sim}{\to}}} \overline{A\to B\to .C\to D}$$

then both $\vdash_{\mathbf{E}_{\overset{\sim}{\to}}} A\to B$ and $\vdash_{\mathbf{E}_{\overset{\sim}{\to}}} \overline{C\to D}$; i.e., it is hard to see how to get a theorem of the form displayed just above, except by way of

$$\vdash_{\mathbf{E}_{\overset{\sim}{\to}}} A\to B\to .\overline{C\to D}\to .\overline{A\to B\to .C\to D},$$

and the two antecedents thereof. At any rate it was hard for Anderson 1963, who conjectured that we could prove $\overline{A\to B\to .C\to D}$ only if we had already proved the two antecedent formulas.

Meyer 1966, not sharing Anderson's myopia, presents a counterexample to the conjecture. As a first step, he shows that for any theorem T of **E**, we have

$$\vdash_{\mathbf{E}} {\sim}(T\to \Box((\bar T\to T)(\overline{T\to .\bar T\to T}))).$$

(Notice that the formula has the form ${\sim}(B\to\Box C)$, where C is an arcane compound built up out of B. Meyer genially admits that this is not the sort of counterexample which should be expected to leap to the mind, and accordingly absolves Anderson of guilt for not having seen it himself.)

We prove the formula with the help of techniques of FE.

1	$\vert\_\ (\bar A\to .A\to \bar A)((\bar A\to .A\to \bar A)\to A)_{\{1\}}$	hyp
2	$\vert\ \ A_{\{1\}}$	1 &E (twice), →E
3	$(\bar A\to .A\to \bar A)((\bar A\to .A\to \bar A)\to A)\to A$	1–2 →I

Then contraposing the first occurrence of $\bar A\to .A\to \bar A$ yields

4 $(\overline{A\to \bar A}\to A)((\bar A\to .A\to \bar A)\to A)\to A,$

where the antecedent now has the form $(B\to D)(C\to D)$. This being equivalent to $(B\vee C)\to D$, we write

5 $(\overline{A\to \bar A}\vee(\bar A\to .A\to \bar A))\to A\to A.$

Contraposition (once on the antecedent; once on 5) and De Morgan equivalences then give us

$$6\quad \overline{A}\to.\sim(\overline{A}\to.(A\to\overline{A})\overline{(\overline{A}\to.A\to\overline{A})}).$$

We temporarily abbreviate $((A\to\overline{A})\overline{(\overline{A}\to.A\to\overline{A})})$ as D, in order to observe succinctly that since $\Box D\to D$ is a theorem of **E**, prefixing and contraposition give us

$$7\quad \sim(\overline{A}\to D)\to\sim(\overline{A}\to\Box D).$$

The antecedent of 7 is the consequent of 6, so we now have by transitivity

$$8\quad \overline{A}\to.\sim(\overline{A}\to\Box((A\to\overline{A})\overline{(\overline{A}\to.A\to\overline{A})})).$$

Finally, choosing A as $\overline{T}$, where T is any theorem of **E**, we have by double negation and →E

$$9\quad \sim(T\to\Box((\overline{T}\to T)\overline{(T\to.\overline{T}\to T)})),$$

as a required step toward finding Meyer's counterexample. Were Anderson's conjecture correct, 9 would be provable only if both T and $\sim\Box((\overline{T}\to T)\overline{(T\to.\overline{T}\to T)})$ were provable; the former is of course a theorem (having been chosen to be such), but for a special case of T we can use the following matrix, which satisfies **E**, to show that $\sim\Box(\overline{T}\to T)\overline{(T\to.\overline{T}\to T)})$ is not provable. (This arrow matrix of Meyer's is an adaptation of the matrix in §22.1.3; matrices for truth functions are the same as there.)

→	−3	−2	−1	−0	+0	+1	+2	+3
−3	+3	+3	+3	+3	+3	+3	+3	+3
−2	−3	+0	−3	+2	−3	−3	+0	+3
−1	−3	−3	+0	+1	−3	+0	−3	+3
−0	−3	−3	−3	+0	−3	−3	−3	+3
+0	−3	−2	−1	−0	+0	+1	+2	+3
+1	−3	−3	−3	−1	−3	+0	−3	+3
+2	−3	−3	−3	−2	−3	−3	+0	+3
+3	−3	−3	−3	−3	−3	−3	−3	+3

Choosing T as $BC\to B$, we note that if $B = +1$ and $C = +2$, then $+1\&+2\to+1 = +0\to+1 = +1$, according to the matrices; i.e., $T = +1$ for this assignment to B and C. The reader can check that $\sim\Box((\overline{T}\to T)\overline{(T\to.\overline{T}\to T)})$ then takes the value -0, and is hence unprovable.

§26.8. Material "implication" is sometimes implication. This is immediate from §14.6, where we showed that there exist formulas A and B such that both $\vdash_{\mathbf{E}} \bar{A}\rightarrow.A\rightarrow B$ and $\vdash_{\mathbf{E}} B\rightarrow.A\rightarrow B$. From these we have $\vdash_{\mathbf{E}} \bar{A}\vee B\rightleftarrows.A\rightarrow B$, which shows, amusingly enough, that in at least one carefully contrived, altogether artificial case, "not . . . or —" coincides with "if . . . then —."

§26.9. Sugihara's characterization of paradox, his system, and his matrix. Sugihara 1955 provided the first *general* characterization of implicational paradoxes of which we know, and the first matrix usable for showing the unprovability of an entire family of irrelevant Bad Guys; see §5.1.2. Here we present Sugihara's account of paradox, his own little-studied paradox-free system **SA**, and his interesting matrix.

Sugihara on paradox. Sugihara 1955 suggests characterizing a system as paradoxical in terms of the formulas $\exists p\forall q(p\rightarrow q)$ and $\exists p\forall q(q\rightarrow p)$. Since he wishes to apply his characterization to systems without propositional quantifiers, we probably catch his intentions in the following definition: relative to a given connective intended as implicational, we shall say that a system **S** is *paradoxical in the sense of Sugihara* just in case it has either a weakest or a strongest formula, and that a system is *paradox free in the sense of Sugihara* if it is not paradoxical in that sense. Here we use "strongest" and "weakest" in the sense of §8.14: relative to a given connective, →, intended as implicational, a formula A is *strongest* if one can prove $A\rightarrow B$ for every formula B, and *weakest* if $B\rightarrow A$ is provable for all B. As Sugihara points out, even the most minimal Lewis systems are (doubly) paradoxical in virtue of the formulas $p\&\bar{p}$ (strongest) and $p\vee\bar{p}$ (weakest).

Sugihara's system. Sugihara calls by the name "**SA**" the system he proves paradox free. Taking as primitives $\bar{\ }$ (negation), & (conjunction), and → (implication), and with the four rules *modus ponens* for the arrow, *adjunction* for &, *substitution*, and *replacement* of A by B whenever $(A\rightarrow B)\&(B\rightarrow A)$ is provable, he gives the following axioms:

$$A\&B \rightarrow B\&A;\quad A\&B \rightarrow A;\quad A \rightarrow A\&A;\quad (A\&B)\&C \rightarrow A\&(B\&C);$$
$$A\rightarrow\bar{\bar{A}};\quad (A\rightarrow B)\&(B\rightarrow C)\rightarrow.A\rightarrow C;\quad (A\&(A\rightarrow B))\rightarrow B;\quad \bar{A}\rightarrow B\rightarrow.\bar{B}\rightarrow A.$$

He also observes that adding any of the following would leave the system paradox free, where $A\vee B =_{\text{df}} \overline{\bar{A}\&\bar{B}}$ and $\Diamond A =_{\text{df}} \overline{A\rightarrow\bar{A}}$:

$$(A\rightarrow B)\&(B\&C\rightarrow D)\rightarrow.A\&C\rightarrow D;\quad A\rightarrow B\rightarrow\overline{A\&\bar{B}};\quad (A\rightarrow.A\&B)\rightarrow.A\rightarrow B;$$
$$A\&(B\vee C)\rightarrow.(A\&B)\vee(A\&C);\quad (A\rightarrow B)\&\Diamond A\rightarrow.\Diamond B;\quad \Diamond(A\&B)\rightarrow\Diamond A;$$
$$A\rightarrow B\rightarrow.\Diamond A\rightarrow\Diamond B;\quad \Diamond\Diamond A \rightarrow \Diamond A;\quad \Diamond A \rightarrow {\sim}\Diamond{\sim}\Diamond A.$$

The proof of freedom from paradox is with the help of a matrix.

Sugihara's matrix. We quote:

Consider the set of s_i's and t_j's under the following conditions as a model: 1) i and j are integers, positive, 0 or negative. 2) $\mathbf{s}_i < \mathbf{s}_j$ and $\mathbf{t}_i < \mathbf{t}_j$ where $i < j$, and $\mathbf{t}_i < \mathbf{s}_j$ for any i and j. 3) Every $\mathbf{s}_i$ is a designated value, and every $\mathbf{t}_i$ is an undesignated value. 4) $\overline{\mathbf{s}_i} = \mathbf{t}_{-i}$ and $\overline{\mathbf{t}_i} = \mathbf{s}_{-i}$. 5) a&b $=$ min(a, b). [5′) a$\vee$b $=$ max(a, b).] 6) a$\rightarrow$b $= \bar{\text{a}}\vee$b when a $\leq$ b, and a$\rightarrow$b $= \bar{\text{a}}$&b when a $>$ b.

A Hasse diagram would look as follows:

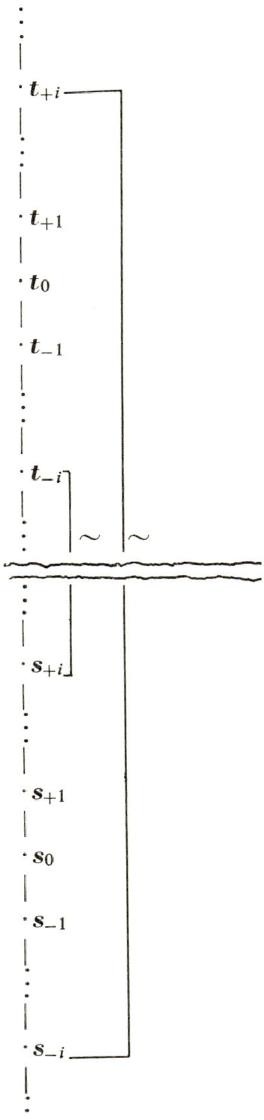

This matrix can be used to show that the system **SA** is paradox free in the sense of Sugihara; and more generally: every system all of whose axioms and rules are satisfied by the matrix is in the sense of Sugihara paradox free. For suppose for *reductio* that A is a strongest formula in a system **S** satisfied by the matrix. Pick out some — any — assignment of values, and let A on this assignment take the value a. Let p be a variable not occurring in A; and assign p some value less than a. Combine the two valuations; then the value of $A \rightarrow p$ is undesignated, so A is not strongest. Similarly A is not weakest.

Analysis of this simple argument shows that if any system is satisfied by a matrix without a bottom element (the ordering being determined by whether or not a→b is designated) then the system has no strongest formula; and if it is satisfied by a matrix without a top element, then it has no weakest formula. So with only a little mud in your eye we can say that a system is paradox free in the sense of Sugihara just in case it is satisfied by at least one matrix with neither a top nor a bottom element.

Evidently Sugihara's own matrix is not the simplest such; the following will do exactly the same work as his (we show only the Hasse diagram and negation, keeping the same account of implication, + values designated):

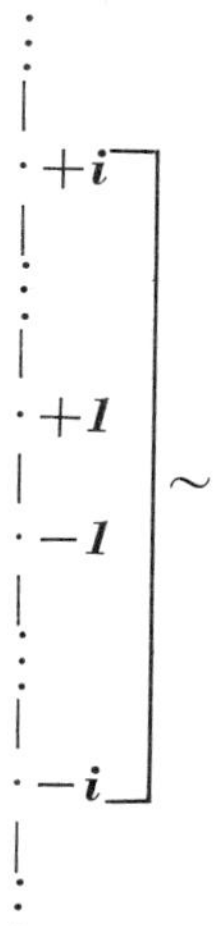

(Or one might insert *0* between -1 and $+1$, counting it designated.) We have accordingly come to think of the matrix answering to this Hasse diagram as *the* Sugihara matrix, certainly pardonable since the matrix so exactly catches Sugihara's ideas. Consequently, in spite of historical inaccuracy, we shall in the sequel permit ourselves — and our friends — to so refer.

At some point Meyer suspected that this matrix might be characteristic for the system **RM** (R-mingle), a system deriving not from Sugihara's **SA** but from combining an idea of Onishi and Matsumoto 1962 with the ideas of **R**. (Further information about **RM**, which is defined in §27.1.1, is to be found in §§8.15, 29.3–5). Meyer's success in establishing this unobvious connection is reported in §29.3, and there is further elaboration due to Dunn in §29.4, including a revealing algebraic characterization of the underlying ideas of Sugihara's matrix and its cousins.

CHAPTER V

NEIGHBORS OF E

§27. A survey of neighbors of E. One of the hazards associated with setting forth a system of formal logic at length, is that it invites tinkering. Much of this fooling around is of course well-intentioned: we get a better sense of the formal structure we are dealing with if we can answer lots of questions of the form, "what happens if we add (drop) certain axioms?" And finding out what happens gives us new leads for philosophical reflection. In this chapter we consider a bunch of addings and droppings, some of which have a certain attractiveness, and some not. The loveliest of the former — the calculus **R** of relevant implication — we examine in some detail in §28, while about one of the latter we growl a bit in §29.5. This section itself is constituted by an axiomatic survey of some immediate neighbors (§27.1), a natural deduction survey (§27.2), and a brief mention of some more distant neighbors (§27.3).

§27.1. Axiomatic survey. This is the place in this book where we bring together axiomatic formulations of **E** and its neighbors.

§27.1.1. Neighbors with same vocabulary: T, E, R, EM, and RM. The idea for defining **T, E, R, EM,** and **RM** is simple-minded: just add the truth functional axioms of **E** to the already discussed pure implicational systems $\mathbf{T}_{\rightarrow}$, $\mathbf{E}_{\rightarrow}$, $\mathbf{R}_{\rightarrow}$, $\mathbf{EM}_{\rightarrow}$ and $\mathbf{RM}_{\rightarrow}$ of Chapter I. We remind the reader of the leading ideas of these systems by the following table:

- **T:** ticket-entailment. Even stricter than **E**. See §6.
- **E:** relevance and necessity. See §4.
- **R:** relevant implication; but no modal notions. See §3.
- **EM:** E-mingle. See §8.15.
- **RM:** R-mingle. See §8.15.

One formulation of these systems can be given via the following list of axioms; for all systems the rules are just

→E: from $A \rightarrow B$ and A to infer B

and

&I: from A and B to infer $A\&B$.

Axiom list:

A1 $A\to A$
A2 $A\to B\to.B\to C\to.A\to C$
A3 $A\to B\to.C\to A\to.C\to B$
A4 $(A\to.A\to B)\to.A\to B$
A5 $A\&B\to A$
A6 $A\&B\to B$
A7 $(A\to B)\&(A\to C)\to.A\to.B\&C$
A8 $A\to A\vee B$
A9 $B\to A\vee B$
A10 $(A\to C)\&(B\to C)\to.(A\vee B)\to C$
A11 $A\&(B\vee C)\to(A\&B)\vee C$
A12 $A\to\bar{A}\to\bar{A}$
A13 $A\to\bar{B}\to.B\to\bar{A}$
A14 $\bar{\bar{A}}\to A$
A15 $A\to B\to.A\to B\to C\to C$
A16 $\Box A\&\Box B\to.\Box(A\&B)$, where $\Box A =_{df} A\to A\to A$
A17 $A\to.A\to A\to A.$
A18 $A\to B\to.A\to B\to.A\to B$

Now the systems:

T: A1–A14.
E: **T**+(A15, A16) = A1–A16.
R: **E**+A17 = A1–A17.
EM: **E**+A18 = A1–A16, A18.
RM: **R**+A18 = **EM**+A17 = A1–A18.

This formulation of **T** is perhaps as economical as any; but with respect to **E**, A3 is redundant. (The §21.1 formulation of **E** is different only by collapsing the work of A1 and A15 into $A\to A\to B\to B$.) The same redundancy of course obtains with respect to **EM**.

The formulations of **R** and **RM** are, however, clumsy, inasmuch as they are based on an awkward combination of modality-preserving axioms (A15, A16, A18) with a modality-destroying axiom (A17). This is a disadvantage when arguing, as we do later, that **R** or **RM** axioms have certain properties; therefore, in spite of the fact that we are tired of axiom-chopping, we present a separate economical set of axioms for **R** and **RM**. We use the following

AXIOMS FOR **R**

R1	$A{\rightarrow}A$	(= A1)
R2	$A{\rightarrow}B{\rightarrow}.B{\rightarrow}C{\rightarrow}.A{\rightarrow}C$	(= A2)
R3	$A{\rightarrow}.A{\rightarrow}B{\rightarrow}B$	(compare A15)
R4	$(A{\rightarrow}.A{\rightarrow}B){\rightarrow}.A{\rightarrow}B$	(= A4)
R5	$A\&B \rightarrow A$	(= A5)
R6	$A\&B \rightarrow B$	(= A6)
R7	$(A{\rightarrow}B)\&(A{\rightarrow}C){\rightarrow}.A{\rightarrow}.B\&C$	(= A7)
R8	$A \rightarrow A\vee B$	(= A8)
R9	$B \rightarrow A\vee B$	(= A9)
R10	$(A{\rightarrow}C)\&(B{\rightarrow}C){\rightarrow}.(A\vee B){\rightarrow}C$	(= A10)
R11	$A\&(B\vee C) \rightarrow (A\&B)\vee C$	(= A11)
R12	$A{\rightarrow}\overline{B}{\rightarrow}.B{\rightarrow}\overline{A}$	(= A13)
R13	$\overline{\overline{A}}{\rightarrow}A$	(= A14)

Then for **RM** we add

R14	$A{\rightarrow}.A{\rightarrow}A$	(compare A18)

For both systems the rules are $\rightarrow$E and &I. Note that adding A3, A12, A15, A16, or A17 would be redundant in **R**. With respect to A12, the *reductio* axiom, the fact is due to Meyer; who also points out that **R** can be axiomatized by substituting A12 for R4 (contraction) in the above formulation (see §14.1.3).

That **R** is well axiomatized by our formulation is shown in §28.3.2. **RM** is not hereby well-axiomatized; see §8.15.

While we're at it, we mention one more series of systems of recent vintage. They arise by adding to any of the systems so far discussed one of the following:

$(A{\rightarrow}B)\&(B\&C{\rightarrow}D){\rightarrow}.A\&C{\rightarrow}D$ (*replacement of conjoined antecedent*)
$(A{\rightarrow}.B\vee C)\&(C{\rightarrow}D){\rightarrow}.A{\rightarrow}.B\vee D$ (*replacement of disjoined consequent*)
$(A{\rightarrow}.C\vee B)\&(B\&A{\rightarrow}C){\rightarrow}.A{\rightarrow}C$ (*strong distribution;* or maybe *cut*)

The second axiom answers to an interesting subproof rule, indicated in §27.2. And some revealing semantic information is given in §47. Other than that we have little information about these systems.

A final oddity answering to a tinker with subscripts (see end of §27.2): $(A{\rightarrow}.A\&(C{\rightarrow}C){\rightarrow}B){\rightarrow}.A{\rightarrow}B$.

We include the following picture, where **S3**–**S5** are the Lewis systems and **TV** is the two valued calculus.

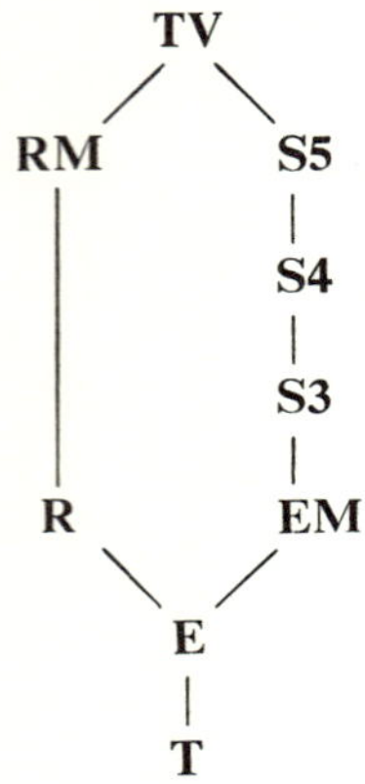

The picture, as Storrs McCall points out, correctly places **RM** and **S5** next to **TV**; for between these systems and **TV** there lie only systems having a *finite* characteristic matrix; see §29.4.

§27.1.2. Neighbors with propositional constants: R and E with t, f, w, w', T, F. One family of neighbors arises by adding as new primitive notation one or more of the following propositional constants, whose intuitive interpretation we tabulate as follows:

t: conjunction of all logical truths.
f: disjunction of all logical falsehoods.
w: conjunction of all truths ("the world").
w': disjunction of all falsehoods.
T: disjunction of all propositions.
F: conjunction of all propositions.

By a conjunction of a set of propositions or sentences we mean a weakest proposition or sentence which implies every member of the set; and dually for disjunction.

We draw on the following axiom list.

$t1$	$\Box t$; i.e., $t \to t \to t$
$t1'$	t
$t2$	$t \to . A \to A$
$f1$	$f \rightleftarrows \bar{t}$
$f2$	$A \to f \to f \to A$
$w1$	$A \equiv (w \to A)$; i.e., $(\bar{A} \vee (w \to A)) \& (A \vee \overline{w \to A})$
$w'1$	$w' \rightleftarrows \bar{w}$
$T1$	$A \to T$
$F1$	$F \to A$

Adding $\boldsymbol{t}$. Add $\boldsymbol{t}$ to **R** or **E** via $\boldsymbol{t}1$ and $\boldsymbol{t}2$; $\boldsymbol{t}1'$ and $\boldsymbol{t}2$ may be used for **R**. In E one then has $\vdash \Box A \rightleftarrows (\boldsymbol{t} \rightarrow A)$, (where $\Box A = (A \rightarrow A \rightarrow A)$), whence in **R** but not **E**, one will quickly obtain $\vdash A \rightleftarrows (\boldsymbol{t} \rightarrow A)$; which constitutes another way to add $\boldsymbol{t}$ to **R**. (Note that *no* propositional constant can be introduced into **E** in this way, on pain of destroying the distinction between necessitives and non-necessitives and thus reducing **E** to **R**.)

Adding f. f can be added always via $f1$. To **R** but not **E**, as pointed out in §14.3, one can also add f to *positive* **R** via $f2$, whence given $\bar{A} = (A \rightarrow f)$ as a definition, one gets **R** back again.

Adding $\boldsymbol{w}$ and $\boldsymbol{w}'$. **R**, because of $\vdash A \rightleftarrows (\boldsymbol{t} \rightarrow A)$ and hence $\vdash A \equiv (\boldsymbol{t} \rightarrow A)$, cannot distinguish logical from contingent truth, i.e., cannot distinguish $\boldsymbol{t}$ from $\boldsymbol{w}$. But **E** can, which suggests adding $\boldsymbol{w}$ to **E**. We are by no means clear how to do this. $\boldsymbol{w}1$ should certainly turn out to be a theorem, but *not* $A \rightarrow . \boldsymbol{w} \rightarrow A$ or $\boldsymbol{w} \rightarrow A \rightarrow A$. $\boldsymbol{w}$ should also turn out to be a theorem, but a contingent one; i.e., one should *not* have $\Box \boldsymbol{w}$. (Nor should one have the necessity of $\boldsymbol{w}1$.) Of course $\vdash \boldsymbol{w}$ would follow from the instance $\boldsymbol{w} \equiv (\boldsymbol{w} \rightarrow \boldsymbol{w})$ of $\boldsymbol{w}1$, provided one had detachment for $\equiv$; which boils down to our old friend (γ), the disjunctive syllogism. One should also have $\vdash \boldsymbol{w} \rightarrow (A \vee B) \equiv . (\boldsymbol{w} \rightarrow A) \vee (\boldsymbol{w} \rightarrow B)$, and could obtain it similarly. And perhaps in the presence of (γ) $\boldsymbol{w}1$ suffices for *everything* one wants. But the question is: is there a formulation of **E** with $\boldsymbol{w}$ in which $\rightarrow$E and &I are the sole rules? We do not know. Nor have we any information about $\boldsymbol{w}'$, except the obvious $\boldsymbol{w}'1$.

Adding $\boldsymbol{T}$ *and* $\boldsymbol{F}$. To any system, $\boldsymbol{T}$ could be added via $\boldsymbol{T}1$, and $\boldsymbol{F}$ via $\boldsymbol{F}1$.

§27.1.3. Neighbors with necessity as primitive: $R^{\Box}$ and $E^{\Box}$.

For book-keeping purposes we list the following axioms together, where now $\Box$ is to be thought of as primitive.

$\Box 1$ $\quad \Box A \rightarrow A$
$\Box 2$ $\quad \Box(A \rightarrow B) \rightarrow . \Box A \rightarrow \Box B$
$\Box 3$ $\quad \Box A \& \Box B \rightarrow \Box(A \& B)$
$\Box 4$ $\quad \Box A \rightarrow \Box\Box A$
$\Box 5$ $\quad \Box A \rightarrow . \Box A \rightarrow \Box A \rightarrow \Box A$

We also consider the rule of necessitation.

$\Box$I: if A is a theorem, so is $\Box A$.

In each case below, instead of the rule of necessitation one could specify that if A is an axiom, then so is $\Box A$, and then prove $\Box$I inductively — a preferable course; see §21.2.2. This is to be understood whenever we mention the rule of necessitation.

Adding necessity to **R**: $\mathbf{R}^{\Box}$. This addition is of some philosophical importance. Meyer 1968a defines the system $\mathbf{R}^{\Box}$ by adding the rule of necessitation and the axioms $\square 1$–$\square 4$ to those defining **R**; i.e., to R1–R13. (Bacon 1966, while not discussing $\mathbf{R}^{\square}$ in detail, points out that if we *start* with **R**, then we may add e.g. $\square 4$, as we wish, or not, thus suggesting further similarities between Lewis' $\supset$ and $\square\supset$ on the one hand, and our $\rightarrow$ and $\square\rightarrow$ on the other.) The interest of $\mathrm{R}^{\square}$ is described in §28.1.

Adding necessity to **E**: $\mathbf{E}^{\square}$. One could add necessity to **E** by way of the simple equivalence $\square A \rightleftarrows . A\rightarrow A\rightarrow A$. However, if one wants a sense of necessity which might be stronger than $A\rightarrow A\rightarrow A$ — which is not committed to collapsing to $A\rightarrow A\rightarrow A$ — then one can proceed as suggested by Meyer: define $\mathbf{E}^{\square}$ as the result of adding to **E** (i.e., to A1–A16) the axioms $\square 1$–$\square 5$ together with the usual rule of necessitation. ($\mathbf{E}^{\square}$ is used in §23.2.2.)

§27.1.4. R with intensional disjunction and co-tenability as primitive. We use "co-tenability" for the operation $\circ$ which may be added to **R** by way of the axioms

$\circ 1$ $A\rightarrow .B\rightarrow .A\circ B$

$\circ 2$ $(A\rightarrow .B\rightarrow C)\rightarrow .A\circ B\rightarrow C$

There are several remarks to be made about this connective, for some of which we are indebted to Dunn 1966 (p. 143), Woodruff, and Meyer (we'll not try to distinguish the contributions).

1. In §16.1 we suggested that (as C. I. Lewis clearly foresaw in a number of articles about sixty years ago) there were intensional senses of disjunction. Some of these can no doubt be grabbed by defining "A or B" as "if not-A then B," where the "if . . . then —" is intensional. A minimal amount of informal muddling convinces us immediately that this sort of intensional "or" is, for example, commutative, given classical views about negation. So it seems sane to define

$$A+B =_{\mathrm{df}} \bar{A}\rightarrow B,$$

where the arrow is that of $\mathbf{T}_{\rightarrow}$, $\mathbf{E}_{\rightarrow}$, or $\mathbf{R}_{\rightarrow}$, with a view to trying on the idea for size.

2. Intensional disjunction could be added to positive **R**, with suitable axioms, but without pursuing this suggestion, we leap off to an obvious association: what about an "intensional conjunction" to go (via De Morgan) with it? Such a connective in **R** can be defined as follows

$$A\circ B =_{\mathrm{df}} \overline{A\rightarrow\bar{B}},$$

yielding $\circ 1$–$\circ 2$. But now it turns out, as the reader may verify, that if we add $\circ 1$–$\circ 2$ to **R** with $\circ$ as a new primitive, we can prove the relevant equivalence answering to the definition just above. That is, $\circ$, looked at as a "positive" connective, can be added independently of negation to **R**.

Furthermore, in **R** it has such memorable and delightful properties as $(A\circ B)\circ C\rightleftarrows A\circ(B\circ C)$, $A\circ B\rightleftarrows B\circ A$, and $A\circ B\rightarrow .A\circ C\rightarrow B\circ C$, which makes it very easy to handle. This becomes of first importance for Dunn's algebraic treatment of **R** (§28.2), for his consecution formulation of positive **R** (§28.5), and for the Meyer-Routley semantics of **R** (§48). (Meyer 1973b puts it as follows. "$\circ$ and $+$ can be interpreted as intensional analogues of conjunction and disjunction, respectively; they were introduced into the relevant logics in Belnap 1960a and were studied in a number of dissertations, including Belnap's, mine, and Dunn's ($\circ$ and $+$ are the *sort* of connectives one studies in dissertations, though we shall give reasons below for the post-graduate utility of $\circ$, as Fisk 1964 gave reasons for the utility of $+$.)")

3. How then to interpret $\circ$? We confess puzzlement.

In some ways $\circ$ looks like conjunction (viewed from the classical material standpoint). For from $\circ$1 by suffixing in the consequent we get

$$A\rightarrow .A\circ B\rightarrow C\rightarrow .B\rightarrow C,$$

whence permutation leads to the converse of $\circ$2, so we have

$$(A\rightarrow .B\rightarrow C)\rightleftarrows A\circ B\rightarrow C.$$

This together with $\circ$1 makes $\circ$ have some of the features classically attributed to &, which we discussed in §21.2.2. It also shares the property that if $\vdash A$ and $\vdash B$, then $\vdash A\circ B$.

But $\circ$ fails to have the property $A\circ B\rightarrow A$; so it isn't conjunction.

4. Lewis and Langford 1932 defined a consistency operation reminiscent (in temporal reverse) of $\circ$:

$$A\circ B =_{\text{df}} \sim(A \prec\!\!3 \sim B),$$

and discussed (in their Chapter VI, section 4) a number of its properties. One of the more important of these is that for the Lewis theory, the equivalence of $\Diamond A$ with $A\circ A$ holds (on which point see §11, from which our attitude toward this equivalence can be constructed). It follows, in theories of strict implication, that $\overline{A\circ(B\&\overline{B})}$, which is as it should be if $\circ$ is interpreted as consistency *simpliciter;* but this property fails for the notion we have defined above, so the notion expressed by $\circ$ doesn't look exactly like consistency either.

5. Goodman 1955, in the course of discussing counterfactual conditionals, defines *A is co-tenable with S* (he writes "cotenable") as "it is not the case that *S* would not be true if *A* were," or (as we shall try to put it a little more luminously) "*A* does not preclude *S*." Bacon 1971 (and also in correspondence, for which we are grateful) suggests co-opting Goodman's terminology, and reading $A\circ B$ as "that *A* and that *B* are co-tenable." This is probably the best ploy, since "co-tenable" is an intelligible neologism which

hasn't yet had any formal work to do; it also answers nicely to $\overline{A\to\overline{B}}$, on practically anyone's account of the arrow. In earlier discussions of the matter, we and others have used "relevant consistency," and Meyer 1970c calls it "intensional conjunction"; no doubt these less satisfactory terms will crop up frequently in the sequel. If so, the reader is urged to bear in mind that relevant consistency is the same as co-tenability, which is just like intensional conjunction (actually they are all the same as $\circ$).

§27.2. Natural deduction survey: FR, FE, FT, FRM, and FEM. These Fitch-style formulations represent, as we think, the most perspicuous way of understanding these systems; they are certainly the easiest to use in actually constructing proofs for the corresponding Hilbert-style systems of the last section.

In order to facilitate comparison of the systems, we regroup the rules so as to exhibit more clearly the way in which various restrictions function. Conceptually the rules fall into four groups: (i) structural rules, (ii) rules having to do with the intensional connective, (iii) mixed rules, connecting intensional and extensional notions, and (iv) rules for purely extensional uses of connectives (which are the same for all three systems). The mingle rules come under a separate heading, as does the rule corresponding to the second of the axioms mentioned at the end of §27.1.1.

As is immediately obvious, this grouping cuts across the grouping by connectives, since for example *modus tollens* (it's a long time since we have used this phrase: it means the inference from $A\to B$ and $\overline{B}$ to $\overline{A}$) and $\vee$E share with *modus ponens* the feature that implication is required among the premisses for application of the rule: all three amount to elimination rules for the arrow.

We shall furthermore state the rules as candidly as possible, by which we mean we will avoid cute economies which don't advance one's understanding. Example: the rule of repetition is redundant. We could, if we like, prove $A\to A$ thus:

1	$\lfloor\ A_{\{1\}}$	1 hyp
2	$A\to A$	1–1 $\to$I

explaining that the second step consists of the hypothesis, followed by an arrow, followed by the last step of the hypothetical subproof (where, notice, $1 \in \{1\}$). Similarly one can prove

1	$\vert\ \underline{A_{\{1\}}}$	1 hyp
2	$\vert\ A\& A_{\{1\}}$	1 1 &I
3	$A\to A\& A$	1–2 $\to$I

But not much seems to be gained, and we think of our earlier proof of $A\to A$ in §1 as more forthright.

With this understanding we begin by stating the structural rules, remarking first that the rule of reiteration is the only one with any punch. Restrictions on reiteration have exactly the effect of restrictions on permutation of antecedents for **FR** and **FE**, an effect which is amplified in **FT** by restricting $\rightarrow$E, using subscripts for control.

The rules as stated below hold for all three systems, unless otherwise indicated.

Structural rules

Hyp. A step may be introduced as the hypothesis of a new subproof, and each new hypothesis receives a unit class $\{k\}$ of numerical subscripts, where k is the rank (§8.1) of the new subproof.

Rep. A_a may be repeated, retaining the relevance indices a.

Reit. A_a may be reiterated (retaining subscripts) into hypothetical subproofs in **FR** with no proviso, and in **FE** and **FT** provided A has the form $B \rightarrow C$.

Intensional rules

$\rightarrow$I. From a proof of B_a on the hypothesis $A_{\{k\}}$ to infer $A \rightarrow B_{a-\{k\}}$, provided k is in a.

$\rightarrow$E. From A_a and $A \rightarrow B_b$ to infer $B_{a \cup b}$, where for **FT**, max(b) $\leq$ max(a) (unrestricted for **FR** and **FE**).

Mixed rules (all unrestricted for **FR** and **FE**)

$\sim$I. From $A \rightarrow \bar{A}_a$ to infer $\bar{A}_a$.

$\sim$E. From $\bar{B}_a$ and $A \rightarrow B_b$ to infer $\bar{A}_{a \cup b}$, where for **FT**, max(b) $\leq$ max(a).

$\vee$E. From $A \vee B_a$, $A \rightarrow C_b$, and $B \rightarrow C_b$, to infer $C_{a \cup b}$, where for **FT**, max(b) $\leq$ max(a).

Extensional rules

$\sim\sim$I. From A_a to infer $\sim\sim A_a$.
$\sim\sim$E. From $\sim\sim A_a$ to infer A_a.
&I. From A_a and B_a to infer $A \& B_a$.
&E. From $A \& B_a$ to infer A_a. From $A \& B_a$ to infer B_a.
$\vee$I. From A_a to infer $A \vee B_a$. From B_a to infer $A \vee B_a$.
&$\vee$. From $A \& (B \vee C)_a$ to infer $(A \& B) \vee C_a$.

The reader will note that we have rechristened *contrap* as $\sim$E, the "mixed rules" sharing the feature that there is a mixture of extensional and intensional connectives. (We have also rechristened the rule "Dist" of §23.3 as &$\vee$, because of analogies with quantifiers which will appear later.)

Mingle rules

From A_a and A_b to infer $A_{a \cup b}$ (for **FRM**; for **FEM** it is required that A have the form $B \rightarrow C$).

Rule $\vee\mathrm{E}^{\mathrm{s(trong)}}$

This involves replacement of $\vee$E and &$\vee$ by a rule $\vee\mathrm{E}^{\mathrm{s}}$, which is most easily simply displayed schematically:

⋮	
$A\vee B_a$	
$\vert$ A_a	hyp
$\vert$ ⋮	
$\vert$ C_b	
$\vert$ B_a	hyp
$\vert$ ⋮	
$\vert$ C_b	
C_b	$\vee\mathrm{E}^{\mathrm{s}}$ (provided $a\subseteq b$)

Note that the rule $\vee\mathrm{E}^{\mathrm{s}}$ involves subscripting hypotheses with a whole class of subscripts — not just a singleton as heretofore. (The idea has been suggested independently by Dunn.)

It is easy to verify then that both $\vee$E and &$\vee$ hold in the presence of $\vee\mathrm{E}^{\mathrm{s}}$. So this has the interesting feature of giving us distribution for free. But of course more besides.

Proofs of the equivalence of the systems heading §§27.1.1 and 27.2 will be left to the reader; they are straightforward, and involve nothing more than extensions of familiar techniques. For those systems mentioned at the end of §27.1.1, involving strong distribution principles, the situation is at present writing fluid; we leave the questions for further research.

A final oddity: suppose we allow subscript deletion, i.e., from $A_{a\cup b}$ to infer A_a. This turns out to cause somewhat less trouble than one might suppose. Nevertheless, we suppress details except to mention that it does not add any pure implicational theorems to **R**, that it allows proof of $(A\rightarrow .A\&\boldsymbol{t}\rightarrow B)\rightarrow .A\rightarrow B$ for $\boldsymbol{t}$ a theorem, and that the latter, when added to the arrow-conjunction axioms of **R**, is complete with respect to the arrow-conjunction part of **FR** with subscript deletion — Chidgey has shown this. And Pottinger has obtained further results akin to those of §8.20, but with additional connectives involved.

§27.3. More distant neighbors. We continue the survey initiated in §27.1 by mentioning a selection of systems which appear to overlap with **E** with respect to their motivating considerations. But for these systems we shall not do as much as list axioms, hoping to satisfy the reader with citations to places wherein he may find motivation, axiomatics, and semantics.

Parry's system of analytic implication, and Dunn's modification thereof, are discussed in §29.6.

McCall's connexive implication: see §29.8.

The propositional logic with subjunctive conditionals of Angell 1962 is also discussed in §29.8.

Routley's **I** and its cousins involving $A{\rightarrow}B{\rightarrow}.A\&C{\rightarrow}.B\&C$ are treated in Routley 1972 and Routley and Routley 1972; in the latter paper Routley's **I** is shown to have the same first degree fragment as **E**.

The system **SIR** ("system of intensional relations") of Barker 1969, motivated by problems of subjunctive conditionals, law, and causality, is shown by Barker to be equivalent to **R**.

Smiley 1959 contains an entailment calculus, discussed in §20.1.

The system of "strict implication free from implicational paradoxes" of Sugihara 1955 is discussed in §26.9.

The Stalnaker and Thomason system **CS** of conditional sentence logic is presented in Stalnaker 1968, and further discussed in Stalnaker and Thomason 1970, and Stalnaker 1970. Close cousins are discussed in David Lewis 1971.

§28. Relevant Implication: R. This section is devoted to **E**'s most important, delightful, and beloved neighbor: the system **R** of relevant implication, whose essential nature was adumbrated in §3. We begin with the reasons for **R**'s especial interest and proceed to a series of studies by Dunn and by Meyer of such of its more important properties as have not as yet been touched upon.

We remark that among the earlier sections, additional information about **R** is to be found in §§3, 7, 8.3.4, 8.4.4, 8.16, 8.18, 14.1.3, 14.2, 14.3, 14.4, Chapter III, §§22.1.3, 24, and 25.

§28.1. Why R is interesting. There are a number of reasons why one should find **R** (axiomatized in §27.1.1) and its modalized enrichment $\mathbf{R}^{\Box}$ (axiomatized in §27.1.3) of interest. Most or all of these have been observed by Meyer or by Dunn, but we shall not pause for attribution.

Age. The pure implicational fragment $\mathbf{R}_{\rightarrow}$ of **R** is the oldest of the relevance logics, having been formulated independently in Moh 1950 and Church 1951; both papers contain appropriate deduction theorems, as remarked in §3. Neither Moh nor Church considered the possibility of obtaining **R** by adding axioms for truth functions to $\mathbf{R}_{\rightarrow}$ in the straightforward way suggested by Ackermann's addition of truth functional axioms to $\mathbf{E}_{\rightarrow}$. However, the heart of relevance in **R** lies in the aged Moh-Church implicational fragment, which is one reason why **R** deserves at least respect if not outright veneration.

Isolating relevance. In Chapter I we motivated the calculus $\mathbf{E}_{\rightarrow}$, which from one point of view is the kernel of **E**, by adversion to two sets of considerations, one involving necessity and the other relevance. Now it is perfectly clear from the way Chapter I goes that these two sets of considerations are separable if not independent, the former issuing as it does in the calculus $\mathbf{S4}_{\rightarrow}$, whereas the latter leads us to $\mathbf{R}_{\rightarrow}$. Furthermore, when viewed from the historical perspective, it is obvious that $\mathbf{E}_{\rightarrow}$ says nothing new about necessity except insofar as it interacts with relevance — a sizable exception, of course (see especially Coffa §22.1.2). Accordingly, one would expect that much of what is of fundamental importance in the line of thought beginning with Ackermann 1956 should be available in the system **R** to which one is led by minding relevance while ignoring necessity. In fact it is not too much to say that in **R** one has an even clearer view of relevance than in **E**, just because of the absence of modal complications. Another advantage of studying relevance when separated from necessity is that one can be perfectly sure that one is not pigheadedly studying *logical* implication from what some would think a perverse point of view; for the connective of **R**, like that of intuitionism, makes no claim to logical force.

Stability. All systems invite tinkering, as we have said before, but **R** has proved remarkably stable in something like a mathematical sense: small changes by any of addition, subtraction, or substitution of either axiomatic or semantic postulates tend to lead us rapidly away from the initial motivating criteria. Of course such a claim must be taken with salt, since it is based only on our own experiences; but with this caution given, we report that regardless of whether we try fiddling axiomatically or semantically, the result, if the fiddle is small, is always a system or semantics discernibly more ugly than the one with which we began. The most spectacular instance of the phenomenon is the fiddle defined by adding the harmless looking $A\rightarrow.A\rightarrow A$, which yields the hopelessly irrelevant calculus **RM** (§§29.3–5).

Richness. The most important sense in which **R** is rich is that to this day it has defied every effort to provide it with a decision procedure, a mechanical test as to whether or not a given candidate is a theorem. It is also rich in containing, in plausible and intellectually satisfying ways, both positive intuitionism and two valued logic. Lastly, it is rich in that it or its fragments have multiple connections with various aspects of logic: Kripke-like semantics, Gentzen formulations, natural deduction formulations, perspicuous axiomatics, algebraic semantics, etc.

Easy proof theory. In spite of the fact that **R** is not at present known to be decidable, one soon develops an accurate feel for what is and is not provable, and for how proofs go, especially in the natural deduction formulation. Of course it's not as easy as truth tables; rather, we think that **R** is on a par with intuitionism and the simpler Lewis modal systems in respect of its ability to generate a sense of rightness about how arguments hang together.

Fragments. **R** has all the proper relations to its fragments: it admits the same implications between truth functions, and the same first degree formulas as does **E**, so that all of the motivation and results of Chapter III apply. Furthermore, all the desirable conservative extension results hold for **R**; for example, **R** is in fact a conservative extension of the pure implicational calculus $\mathbf{R}_{\rightarrow}$.

Applicability. E lacks a relevant, *contingent* sense of "if-then"; accordingly, **R** promises to have application wherever what is wanted is a conditional the meaning of which, while non-logical, involves relevance of antecedent to consequent. Since certain it is that conditional locutions in English usually do not have logical force, one would expect the arrow of **R** more often than that of **E** to be a suitable target for translation from the vernacular. We hasten to add that these remarks are programmatic, and though we have given them a little thought, we are not prepared to be as fierce about the topic as we are about material implication.

The most sustained investigations of the applicability of **R** are to be found in Barker 1969 and Bacon 1971.

Extensibility. One may extend **R** in natural ways with individual or propositional quantifiers, as is done in subsequent chapters.

Of more immediate importance is the fact that **R** can be naturally and conveniently extended by the addition of modalities, issuing in the calculus $\mathbf{R}^{\Box}$ due to Meyer 1968a (§27.1.3). The importance of $\mathbf{R}^{\Box}$ in the context of our investigation is considerable. In the first place, if the separate motivating talk about necessity in §2 and relevance in §3 holds water, then the most natural account of entailment should derive from $\mathbf{R}^{\Box}$, which itself keeps necessity and relevance separate; i.e., $\mathbf{R}^{\Box}$ can define entailment as the necessitation of its non-modal but relevant implication: $\Box(A\rightarrow B)$. And if **E** is really the system which correctly arises out of the considerations of §2 and §3, then the translation of **E** into $\mathbf{R}^{\Box}$ obtained by mapping the arrow of **E** into $\Box(A\rightarrow B)$ in $\mathbf{R}^{\Box}$ should constitute an exact translation: provable if and only if. Meyer 1968a, who so conjectures, observes that this is half trivial: if a formula is provable in **E**, its translation is provable in $\mathbf{R}^{\Box}$. The converse is at this writing not yet established; it is, however, a matter of current research — and we add that it just *must* be true (Method of authority); but though it is difficult to speak confidently, we predict that if in fact it is found that $\mathbf{R}^{\Box}$ and **E** diverge, then we shall, with many a bitter tear, abandon **E**.

(Since this manuscript was submitted for publication, we have obtained evidence that the Method used above is not to be trusted. In fact Maksimova 1973 shows by the techniques of Chapter IX that $\Box(\mathrm{A}\rightarrow\Box(\mathrm{B}\rightarrow\mathrm{C}))\&\Box(\mathrm{B}\rightarrow.\mathrm{A}\vee\mathrm{C})\rightarrow\Box(\mathrm{B}\rightarrow\mathrm{C})$ is a theorem of $\mathbf{R}^{\Box}$, though $[(\mathrm{A}\rightarrow(\mathrm{B}\rightarrow\mathrm{C}))\&(\mathrm{B}\rightarrow.\mathrm{A}\vee\mathrm{C})]\rightarrow.\mathrm{B}\rightarrow\mathrm{C}$ is not a theorem of **E**. Pending a more satisfactory understanding of the matter, we postpone, at least to Volume II, the decision as to whether to laugh or cry. But without being either risible or lachrymose we can include here a matrix proof due to Chidgey 1974 of the non-E-theoremhood of the Maksimova formula above, which drives a wedge between **E** and $\mathbf{R}^{\Box}$. It is rejected by the following →-matrix (when $A = 4$, $B = 3$, $C = 3$) which, when adjoined to the & and ∨ matrices of §22.1.3, satisfies all the *positive* axioms of **E**:

→	*0*	*1*	*2*	*3*	*4*	*5*	*6*	*7*
0	7	7	7	7	7	7	7	7
1	0	6	0	7	0	0	6	7
2	0	0	6	6	0	7	0	7
3	0	0	0	6	0	0	0	7
**4*	0	0	0	0	7	7	7	7
**5*	0	0	0	0	0	7	0	7
**6*	0	0	0	0	0	0	6	7
**7*	0	0	0	0	0	0	0	7

So Maksimova's formula is not a theorem of $\mathbf{E}_+$; nor a theorem of **E** itself since **E** is by §§28.3.2 and 60.4 a conservative extension of $\mathbf{E}_+$.)

Second, $\mathbf{R}^{\Box}$ has some of the nice properties of **E**; in particular, Meyer 1968a establishes that $A \to \Box B$ is never a theorem of $\mathbf{R}^{\Box}$ when A is $\Box$-free, and that **E** and $\mathbf{R}^{\Box}$ have exactly the same first degree fragment (no nesting of arrows, no necessities). Third, one could tinker with the modal structure of $\mathbf{R}^{\Box}$, leaving relevance untouched, a point made in Meyer 1968a and in Bacon 1966; see also §29.1.

Last, $\mathbf{R}^{\Box}$ is the appropriate arena in which not only to observe the separate natures of relevance and modality but to ponder their delicate interconnections. That this is so becomes most evident in the revealing semantics of Chapter IX.

§28.2. The algebra of R (by J. Michael Dunn). See §27.1.1 for an axiomatization of **R** via axioms R1–R13 and rules →E and &I. It follows from our preceding algebraic work in §18 (which we presuppose) together with more of the same upcoming in §40, that the Lindenbaum algebra (see §18.7) of **R** is an intensional lattice, and indeed a free one in the sense of §40.1. (To make a Lindenbaum algebra out of **R**, we suppose added to §18.7 a clause $[A] \to [B] = [A \to B]$ defining the arrow operation on equivalence classes in **R**.) But these observations contribute little to elucidating the algebraic structure of the whole of **R**. The difficulty is that there is no *operation* in intensional lattices that corresponds to relevant implication. Relevant implication can be represented in intensional lattices only as a *relation*, so that an axiom of **R**

like $A \rightarrow B \rightarrow . B \rightarrow C \rightarrow . A \rightarrow C$ can be only imperfectly represented in intensional lattices. These lattices do represent the transitivity of the implication relation that is a consequence of this axiom, since the rule of transitivity — from $A \rightarrow B$ and $B \rightarrow C$ to infer $A \rightarrow C$ — is represented by the transitivity of the partial-ordering relation. But for an algebraic structure to represent this axiom directly it would have to have an "arrow" operation such that for elements a, b, c, $a \rightarrow b \leq (b \rightarrow c) \rightarrow (a \rightarrow c)$, just as in the Lindenbaum algebra of **R**, $[A] \rightarrow [B] \leq ([B] \rightarrow [C]) \rightarrow ([A] \rightarrow [C])$.

It turns out that a certain kind of residuated lattice-ordered monoid has such an "arrow" operation to represent relevant implication. (The connection between residuation and relevant implication was first suggested by Meyer.) The reader is warned that no "deep" theorems concerning either **R** or the related special kind of monoids will be forthcoming in the sequel. There is, at least in the present stage of investigation, no "spin off" of the sort associated with the McKinsey-Tarski (1944, 1946) identification of the study of **S4** with the study of closure algebras. Thus the identification of the study of **R** with the study of the special kind of monoid does not magically solve any of the unsolved problems concerning **R**; nor does this identification yield any significant theorems concerning the special kind of semi-groups. But "the algebraic point of view" has been fruitful in a variety of ways, as illustrated in §§28.3 and 42, Chapter IX (especially §59), and elsewhere.

The subsequent theorems then should be of no interest to the logician *qua* logician, nor to the mathematician *qua* mathematician; but they should be of interest to the logician *qua* mathematician or his dual (both of which fall under the common abstraction of the mathematical logician). Needless to say, we hope ultimately that the connection between **R** and the special kind of monoid will prove of real use in the study of both since, in mathematics, two points of view are often better than one.

§28.2.1. Preliminaries on lattice-ordered semi-groups. A *semi-group* $\langle S, \circ \rangle$ is a non-empty set S that is closed under a binary operation $\circ$ that is associative, i.e., $a \circ (b \circ c) = (a \circ b) \circ c$. We say of a semi-group that it is commutative if $ab = ba$. (Note that we permit ourselves to drop the "$\circ$" notation for "multiplication" in favor of juxtaposition when convenient.) A semi-group is *with identity* and is called a *monoid* if it has an element $\boldsymbol{t}$ (*the identity*) such $\boldsymbol{t}a = a\boldsymbol{t} = a$ (recall that $\boldsymbol{t}$ is provably unique). A monoid is a *group* if for every element *a* there exists an element a^{-1} (*the inverse* of a) such that $aa^{-1} = a^{-1}a = \boldsymbol{t}$ (recall that inverses are provably unique).

We say of a semi-group that it is *lattice-ordered* (briefly, that it is an *l-semi-group*) if it is a lattice as well (though neither the meet nor the join need be the same as the multiplication operation $\circ$), and $a(b \vee c) = (ab) \vee (ac)$, and

$(b \vee c)a = (ba) \vee (ca)$. Note that it follows from this that $a \leq b$ implies $ca \leq cb$ and $ac \leq bc$ (see Certaine 1943, p. 39). Note in particular that the l-semi-group may be a group, in which case we call it an *l-group*.

The concept of an l-semi-group should probably be credited to Ward and Dilworth 1939 (although they assume commutativity and the existence of an identity). Birkhoff 1942 drops the commutativity assumption (although he still retains the identity assumption) and calls the resulting structure a "groupoid." Certaine 1943, in his classic study, follows Birkhoff's usage. The reader should be warned that many modern writers mean by a groupoid simply a set with a binary operation (with no assumptions even about associativity, let alone about the existence of an identity), as in Fuchs 1963. The reader should be further warned that Birkhoff 1948 means by a "semi-group" what we have called a monoid and what he called in 1942 a "groupoid."

We say of an l-semi-group that it is *right residuated* if for every pair of elements a, b there exists an element a:b (called the *right residual* of a by b) such that $x \leq a{:}b$ iff $xb \leq a$. Clearly a:b is the greatest element x such that $xb \leq a$, and indeed is the join of all such elements x. Similarly, *left residuation* is defined so that for every pair of elements a, b there exists an element a::b such that $x \leq a{::}b$ iff $bx \leq a$. If an l-semi-group is commutative, then it is right residuated iff it is left residuated, and $a{:}b = a{::}b$. If an l-semi-group is both right and left residuated, then we simply say that it is *residuated*. (The concept of residuation, although as old as ideal quotients and division, was probably first abstractly formulated by Ward and Dilworth 1939.)

Residuation is a common abstraction of the ideal quotient in ring theory and of division in number theory. For examples of the former see Fuchs 1963, p. 190. For an instructive example of the latter consider the multiplicative group of the positive rationals Q^+ ordered so that $a \leq b$ iff a integrally divides b, i.e., b/a is a positive integer. This is a residuated l-group in which $a{:}b = a/b$ (see Birkhoff 1948, p. 202). Indeed Certaine 1943 has shown that any l-group is residuated and that $a{:}b = ab^{-1}$ and $a{::}b = b^{-1}a$ (p. 61).

An especially instructive example of a residuated l-semi-group, from a logical point of view, is a Boolean algebra in which meet is taken as the multiplicative operation and $a{:}b = \bar{b} \vee a$. Similarly instructive is a pseudo-Boolean algebra in which meet is taken as the multiplicative operation and a:b is the pseudo-complement of b relative to a. Both of these examples can be found in Certaine 1943, p. 61.

We now gather some useful properties of commutative l-semi-groups. Since we are dealing with the commutative case we need not distinguish between left and right residuals and shall denote both by a:b.

P1 $a \leq b$ implies $a{:}c \leq b{:}c$
P2 $b{:}a \leq (b{:}c){:}({:}a{:}c)$
P3 $a \leq b$ implies $c{:}b \leq c{:}a$
P4 $b{:}a \leq (c{:}a){:}(c{:}b)$
P5 $ab \leq c$ iff $a \leq c{:}b$ iff $b \leq c{:}a$
P6 $c{:}(ab) = (c{:}b){:}a = (c{:}a){:}b$
P7 $a(b{:}a) \leq b$
P8 $a \leq b{:}(b{:}a)$
P9 $a \leq (ab){:}b$
P10 If $\wedge_{x \in X} a_x$ exists, then so does $\wedge_{x \in X}(a_x{:}b)$ and $\wedge_{x \in X}(a_x{:}b) = (\wedge_{x \in X} a_x){:}b$.
P11 If $\vee_{x \in X} a_x$ exists, then so does $\wedge_{x \in X}(b{:}a_x)$ and $\wedge_{x \in X}(b{:}a_x) = b{:}(\vee_{x \in X} a_x)$.
P12 $(a \wedge b)(a \vee b) \leq ab$

(§40.1 explains the notation of P10–P11.) Proofs of properties P1–11 may be found in Certaine 1943, pp. 68–69. P12, in the presence of commutativity, follows immediately from Certaine's 3) on p. 39. We also have

P13 $a \leq c$ and $b \leq d$ imply $ab \leq cd$

PROOF. $a \leq c$ implies $ab \leq cb$, but $b \leq d$ implies $cb \leq cd$, and hence by transitivity $ab \leq cd$.

If we assume that the l-semi-group has an identity $\boldsymbol{t}$, we get the following further properties (proven in Certaine 1943, pp. 69–70) for what we might call *l-monoids:*

P14 $a{:}\boldsymbol{t} = a$.
P15 $\boldsymbol{t} \leq a{:}a$.
P16 $a \leq b$ implies $\boldsymbol{t} \leq b{:}a$, and conversely.

If we further assume that the l-semi-group is *upper semi-idempotent* or *square increasing*, i.e., that $a \leq aa$, we get the following:

P17 $a \wedge b \leq ab$

PROOF.

1	$(a \wedge b)(a \vee b) \leq ab$	P12
2	$a \wedge b \leq a \wedge b$ and $a \wedge b \leq a \vee b$	Lattice properties
3	$(a \wedge b)(a \wedge b) \leq ab$	From 1 and 2 *via* P13
4	$a \wedge b \leq ab$.	From 3 by upper semi-idempotency

P18 $(b{:}a){:}a \leq b{:}a$

PROOF:

1	a((b:a):a) $\leq$ b:a	P7
2	aa((b:a):a) $\leq$ b	From 1 by P5
3	a((b:a):a) $\leq$ b	From 2 by P13 since a $\leq$ aa
4	(b:a):a $\leq$ b:a	From 3 by P5

It might be anticipated that we are about to define a *De Morgan lattice-ordered semi-group* $\langle S, \circ, \leq, \overline{}\rangle$. We shall, of course, want $\langle S, \circ, \leq\rangle$ to be a lattice-ordered semi-group and $\langle S, \leq, \overline{}\rangle$ to be a De Morgan lattice (§18.3). But we should further like $\overline{}$ to have some direct interaction with $\circ$. Let us then require the following:

$$\text{(A)}\quad a\circ b \leq c \ \text{ iff } \ b\circ\bar{c} \leq \bar{a}, \quad \text{and} \quad a\circ b \leq c \ \text{ iff } \ \bar{c}\circ a \leq \bar{b}.$$

Property (A) will ultimately be motivated by the applications to **R** that are forthcoming in the next section. But even at this stage we can attach some algebraic meaning to it by observing that it holds both in Boolean algebras treated as l-semi-groups (where $\overline{}$ is Boolean complementation) and in l-groups (where $\overline{}$ is taken as inversion). Indeed, under these interpretations of $\overline{}$ both Boolean algebras and l-groups turn out to be De Morgan lattice-ordered semi-groups. It was observed in Birkhoff 1942 that inversion is a dual automorphism of period two on the underlying lattice of an l-group, and that the underlying lattice is necessarily distributive (see pp. 301 and 306). Kalman 1958 offered De Morgan lattices that are "normal" in his sense as a "common abstraction of Boolean algebras and l-groups," thus seemingly presenting them as an answer to problem 105 of Birkhoff 1948: "Is there a common abstraction which includes Boolean algebras (rings) and l-groups as special cases?" Unfortunately, De Morgan lattices by themselves take no account of the multiplicative operation of the group. We therefore suggest that our De Morgan lattice-ordered semi-groups, in this respect at least, provide a better answer to Birkhoff's problem; this answer could be further improved by adding the postulate that the De Morgan lattice is "normal" in the sense of Kalman, as well as the postulate that the semi-group has an identity.

It is interesting to note the strength of postulate (A). Thus it gives us

THEOREM 1. Every De Morgan lattice-ordered semi-group is residuated, with $a{:}b = \overline{b\circ\bar{a}}$ and $a{::}b = \overline{\bar{a}\circ b}$.

We show the first identity by using the first part of (A). Thus $x\circ b \leq a$ iff $b\circ\bar{a} \leq \bar{x}$. But since $\overline{}$ contraposes, this last holds iff $x \leq \overline{b\circ\bar{a}}$. So $x\circ b \leq a$ iff $x \leq \overline{b\circ\bar{a}}$, and the first identity follows immediately from the definition of a:b. The second identity follows similarly from the second part of (A).

It is interesting to note that an operation $+$ dual to $\circ$ can be defined *via* $a+b = \overline{\bar{a} \circ \bar{b}}$. This operation is associative, but distributes over meet (instead of join) and satisfies the dual of (A), namely, $a+b \geq c$ iff $b+\bar{c} \geq \bar{a}$, and $a+b \geq c$ iff $\bar{c}+a \geq \bar{b}$. We may then define operations of right-difference ($\dot{-}$) and left difference ($\ddot{-}$) by $x \geq a \dot{-} b$ iff $a \leq b+x$, and $x \geq a \ddot{-} b$ iff $a \leq x+b$, and it may be shown that $a \dot{-} b = \overline{b{:}a} = a \circ \bar{b} = \overline{\bar{a}+b}$, and similarly for $a \ddot{-} b$. All of this suggests an obvious equivalent but dual formulation of De Morgan lattice-ordered semi-groups, in which $+$ is taken as primitive and $\circ$ is defined in terms of it.

Let us now define a special kind of De Morgan lattice-ordered semi-group that will be useful in the sequel. Since we shall no longer be talking of De Morgan lattice-ordered semi-groups in general, we shall not risk confusion if we call this special kind a De Morgan semi-group for the sake of brevity. A *De Morgan semi-group* then is a quadruple $\langle S, \circ, \leq, \overline{} \rangle$ that is a commutative, upper semi-idempotent, De Morgan lattice-ordered semi-group. When the underlying De Morgan lattice is an intensional lattice as well, we shall speak of an *intensional semi-group*. Similarly, when we hereafter speak of *De Morgan monoids* and *intensional monoids*, we intend that they have commutativity and upper semi-idempotence as built-in features. Note that a De Morgan semi-group is *not* a generalization of a l-group, not even a commutative l-group, because of upper semi-idempotence. And an intensional semi-group is even less a generalization, since in any group $\boldsymbol{t} = \boldsymbol{t}^{-1}$. But De Morgan semi-groups and intensional semi-groups are still generalizations of Boolean algebras.

A few more definitions, and then we shall be prepared to discuss the algebraic structure of **R**. Let us define a *homomorphism* between two De Morgan semi-groups, $\langle S, \circ, \leq, \overline{} \rangle$ and $\langle S', \circ', \leq', \overline{}' \rangle$, as a homomorphism h between $\langle S, \leq, \overline{} \rangle$ and $\langle S', \leq', \overline{}' \rangle$ (see §18.4) that is also a semi-group homomorphism between $\langle S, \circ \rangle$ and $\langle S', \circ' \rangle$, i.e., $h(a \circ b) = h(a) \circ' h(b)$. Observe that, in virtue of the interdefinability of multiplication and residuation, the condition that h is a semi-group homomorphism is equivalent to the condition that h preserves residuation, i.e., $h(a{:}b) = h(a){:}'h(b)$. A one-to-one homomorphism is an *isomorphism*.

If the two De Morgan semi-groups S and S′ have identities $\boldsymbol{t}$ and $\boldsymbol{t}'$, respectively, i.e., if they are *De Morgan monoids*, then if h is a homomorphism between them such that $h(\boldsymbol{t}) = \boldsymbol{t}'$, we shall call h an *identity-preserving homomorphism* (or, *$\boldsymbol{t}$-homomorphism*). If h is one-to-one, it is a *$\boldsymbol{t}$-isomorphism*.

In accord with these definitions, we shall define *the free De Morgan semi-group with n free generators $\boldsymbol{a}_i$, FDMSG(n)*, as a De Morgan semi-group such that any mapping of the $\boldsymbol{a}_i$ into an arbitrary De Morgan semi-group S can be extended to a homomorphism of *FDMSG*(*n*) into S. We shall similarly define *the free De Morgan monoid with n free generators* $\boldsymbol{a}_i$, *FDMM*(n), as a De Morgan monoid such that any mapping of the $\boldsymbol{a}_i$ into an arbitrary De Morgan monoid can be extended to a $\boldsymbol{t}$-homomorphism. We may define *free intensional semi-groups* and *free intensional monoids* similarly.

We here summarize for reference the various sorts of algebraic structures we have defined in this section and in §18, using for this purpose the following series of definitions governing a non-empty set S, a binary relation $\leq$, a binary operation $\circ$, and a unary operation $\overline{}$. We omit further consideration of the binary operations $\dot{-}$, $\ddot{-}$, and $+$, and we bring in the residuation operations a:b and a::b, the identity $\boldsymbol{t}$, and the truth filter T existentially. The concept of normality, due to Meyer and not used until §42.2, is included here for convenience.

$\langle S, \leq \rangle$ is a *partial ordering* iff for all a, b, c $\in$ S,

$a \leq a$	(reflexivity)
$a \leq b$ and $b \leq c$ imply $a \leq c$	(transitivity)
$a \leq b$ and $b \leq a$ imply $a = b$	(antisymmetry)

x is a *join* {*meet*}, or *least upper bound* {*greatest lower bound*} of a and b in S with respect to $\leq$ iff $a \leq x$ and $b \leq x$ {$x \leq a$ and $x \leq b$}, and for all $z \in S$, $a \leq z$ and $b \leq z$ jointly imply $x \leq z$ {$z \leq a$ and $z \leq b$ jointly imply $z \leq x$}.

$\langle S, \leq \rangle$ is a *lattice* iff it is a partial ordering, and if furthermore for all a, b $\in$ S, there is always a join and a meet of a and b in S with respect to $\leq$. (They will be unique.)

Provided S is a lattice, $a \vee b$ {$a \wedge b$} is defined as the join {meet} of a and b in S with respect to $\leq$.

$\langle S, \leq \rangle$ is a *distributive lattice* iff it is a lattice, and if furthermore

$a \wedge (b \vee c) = (a \wedge b) \vee (a \wedge c)$	(distribution of $\wedge$ over $\vee$)
$a \vee (b \wedge c) = (a \vee b) \wedge (a \vee c)$	(distribution of $\vee$ over $\wedge$)

for all a, b, c $\in$ S.

$\langle S, \leq, \overline{} \rangle$ has the *De Morgan properties* iff

$a \leq b$ implies $\bar{b} \leq \bar{a}$	(inverse monotony)
$\bar{\bar{a}} = a$	(period two)

for all a, b $\in$ S.

$\langle S, \leq, \overline{} \rangle$ is a *De Morgan lattice* iff it has the De Morgan properties, and if $\langle S, \leq \rangle$ is a lattice.

T is a *truth filter for* $\langle S, \leq, \overline{} \rangle$ iff $\langle S, \leq \rangle$ is a lattice, $T \subseteq S$, and for all a, b $\in$ S,

$a \wedge b \in T$ iff $a \in T$ and $b \in T$	(filter)
$a \vee b \in T$ iff $a \in T$ or $b \in T$	(prime)
$\bar{a} \in T$ iff $a \notin T$	(consistent and exhaustive)

$\langle S, \leq, \overline{}\rangle$ is an *intensional lattice* iff it is a De Morgan lattice, and if furthermore it has a truth filter (in general not unique). Equivalently: a De Morgan lattice such that $\bar{a} \neq a$, all $a \in S$.

$\langle S, \circ\rangle$ is a *semi-group* iff for all $a, b, c \in S$,

$$a\circ(b\circ c) = (a\circ b)\circ c \qquad \text{(associativity)}$$

$\langle S, \circ\rangle$ is *commutative* iff for all $a, b \in S$,

$$a\circ b = b\circ a \qquad \text{(commutativity)}$$

$\boldsymbol{t}$ is an *identity for* $\langle S, \circ\rangle$ iff

$$a\circ \boldsymbol{t} = \boldsymbol{t}\circ a = a \qquad \text{(identity)}$$

for all $a \in S$.

$\langle S, \circ\rangle$ is a *monoid* iff it is a semi-group and has an identity. (It will be unique.)

$\langle S, \circ, \leq\rangle$ is *lattice-ordered* iff $\langle S, \leq\rangle$ is a lattice, and for all $a, b, c \in S$,

$$a\circ(b\vee c) = (a\circ b)\vee(a\circ c) \qquad \text{(distribution of } \circ \text{ over } \vee)$$

$\langle S, \circ, \leq\rangle$ is *residuated* iff for every $a, b \in S$ there are y and y' such that for all $x \in S$, $x \leq y$ iff $x\circ b \leq a$, and $x \leq y'$ iff $b\circ x \leq a$.

$\langle S, \circ, \leq\rangle$ is *upper semi-idempotent* iff for all $a \in S$,

$$a \leq a\circ a \qquad \text{(upper semi-idempotence)}$$

$\langle S, \circ, \leq, \overline{}\rangle$ is *normal* iff $\langle S, \circ\rangle$ has an identity $\boldsymbol{t}$, $\langle S, \leq, \overline{}\rangle$ has a truth filter T, and if for all $a \in S$, $\boldsymbol{t} \leq a$ iff $a \in T$.

$\langle S, \circ, \leq, \overline{}\rangle$ has *Property (A)* iff for all $a, b, c \in S$,

$$a\circ b \leq c \text{ iff } b\circ \bar{c} \leq \bar{a} \text{ iff } \bar{c}\circ a < \bar{b} \qquad \text{(Property (A))}$$

The following now give the chief kinds of structures in which we shall be interested.

$\langle S, \circ, \leq, \overline{}\rangle$ is a *De Morgan semi-group* iff

> $\langle S, \circ\rangle$ is a commutative (note well) semi-group,
> $\langle S, \leq, \overline{}\rangle$ is a De Morgan lattice,
> $\langle S, \circ, \leq\rangle$ is lattice-ordered, and upper semi-idempotent (note well),

and

> $\langle S, \circ, \leq, \overline{}\rangle$ has Property (A).

$\langle S, \circ, \leq, \overline{}\rangle$ is an *intensional semi-group* iff it is a De Morgan semi-group, and if furthermore $\langle S, \leq, \overline{}\rangle$ has a truth filter.

$\langle S, \circ, \leq, \overline{}\rangle$ is a *De Morgan monoid* iff it is a De Morgan semi-group, and if furthermore $\langle S, \circ\rangle$ has an identity.

$\langle S, \circ, \leq, \overline{}\rangle$ is an *intensional monoid* iff it is a De Morgan semi-group, if $\langle S, \leq, \overline{}\rangle$ has a truth filter, and if $\langle S, \circ\rangle$ has an identity; i.e., if it is both a De Morgan monoid and an intensional semi-group.

$\langle S, \circ, \leq, \overline{}\rangle$ is a *normal De Morgan monoid* iff it is an intensional monoid, and is normal.

[Notes by principal authors. 1. It turns out that intensional semi-groups and monoids, first proposed by us, are not known to be interesting; there is insufficient interplay between the truth filter and the semi-group operation. 2. Dunn monoids and Church monoids are defined in §28.3.2, and Ackermann groupoids in §59; idempotent De Morgan monoids are used in §29.4, and Curry lattices in §36.2.1.]

In §42 a De Morgan monoid is construed as a structure $\langle S, \circ, \vee, \overline{}\rangle$ instead of $\langle S, \circ, \leq, \overline{}\rangle$; i.e., the binary operation $\vee$ is taken as primitive. Then it is possible to give an equational characterization; the following is that of Meyer, Dunn, and Leblanc 1974.

First define

$$a \wedge b =_{df} \overline{\bar{a} \vee \bar{b}}$$
$$a{:}b =_{df} \overline{\bar{a} \circ b}$$
$$a \leq b \text{ iff } a \vee b = b$$

Then $\langle S, \circ, \vee, \overline{}\rangle$ is a De Morgan monoid iff there is a $\boldsymbol{t} \in S$ such that for all a, b, c $\in$ S,

$$\begin{aligned}
(a \circ b) \circ c &= a \circ (b \circ c) \\
a \circ b &= b \circ a \\
\boldsymbol{t} \circ a &= a \\
a \vee b &= b \vee a \\
(a \vee b) \vee c &= a \vee (b \vee c) \\
a &= a \vee (a \wedge b) \\
a \wedge (b \vee c) &= (a \wedge b) \vee (a \wedge c) \\
\bar{\bar{a}} &= a \\
a \circ (b \vee c) &= (a \circ b) \vee (a \circ c) \\
(b{:}a) \circ a &\leq b \\
a &\leq a \circ a
\end{aligned}$$

Proof that the two notions of De Morgan monoid are equivalent is left to the reader.

§28.2.2. R and De Morgan semi-groups. We may introduce a new connective into **R** (§27.1.1) by defining $A \circ B = \overline{A \rightarrow \bar{B}}$; §27.1.4. We may then define a multiplicative operation upon the Lindenbaum algebra (§18.7) of **R**, **R**/$\rightleftarrows$, by letting $[A] \circ [B] = [A \circ B]$ (note the harmless ambiguity of "$\circ$" used both syntactically and algebraically). We now show that $\mathbf{R}/\rightleftarrows$ is a De Morgan (indeed, an intensional) semi-group under $\circ$.

We first note that it was established, essentially, in §18.7 that $\mathbf{R}/\rightleftarrows$ is an intensional lattice, so it only remains to show that $\circ$ is commutative, associative, upper semi-idempotent, distributive over join, and that it satisfies (A). We sketch proofs of these, making free use of replacement and elementary properties of **R** such as double negation.

To show commutativity, it suffices to prove $A\circ B\to B\circ A$ in **R**, i.e., $\overline{A\to\overline{B}}\to\overline{B\to\overline{A}}$, which follows immediately from contraposition.

To show associativity (in the presence of commutativity), it suffices to show $A\circ(B\circ C)\to(A\circ B)\circ C$, i.e., $\overline{A\to\overline{\overline{B\to\overline{C}}}}\to\overline{\overline{A\to\overline{B}}\to\overline{C}}$, which by contraposition is equivalent to $\overline{A\to\overline{B}}\to\overline{C}\to.A\to(B\to\overline{C})$, which in turn is equivalent by another contraposition to $C\to(A\to\overline{B})\to.A\to(B\to\overline{C})$, which is equivalent by yet another contraposition to $C\to(A\to\overline{B})\to.A\to(C\to\overline{B})$, which is just an instance of permutation.

To show semi-idempotency, it suffices to prove $A\to(A\circ A)$, i.e., $A\to(\overline{A\to\overline{A}})$, which by contraposition is just $(A\to\overline{A})\to\overline{A}$, the *reductio* axiom.

To show that $\circ$ distributes over join, it suffices to prove $A\circ(B\vee C)\rightleftarrows(A\circ B)\vee(A\circ C)$, i.e., $\overline{A\to\overline{B\vee C}}\rightleftarrows\overline{A\to\overline{B}}\vee\overline{A\to\overline{C}}$, which by De Morgan and contraposition is equivalent to $(A\to.\overline{B}\&\overline{C})\rightleftarrows.(A\to\overline{B})\&(A\to\overline{C})$, which follows immediately from the conjunction introduction and elimination axioms.

Finally, we show that $\circ$ satisfies property (A). Since we have already shown commutativity, obviously it suffices to show only one part of (A). And since we have double negation, it then suffices to show $(A\circ B)\to C\to.(B\circ\overline{C})\to\overline{A}$, i.e., $\overline{A\to\overline{B}}\to C\to.\overline{B\to\overline{\overline{C}}}\to\overline{A}$, which by contraposition is $\overline{C}\to(A\to\overline{B})\to.A\to(B\to C)$, which by another contraposition is $\overline{C}\to(A\to\overline{B})\to.A\to(\overline{C}\to\overline{B})$, which is just an instance of permutation.

This completes the proof. Let us observe by the theorem of §28.2.1 that the residual of $[A]$ by $[B]$, $[A]:[B]$, is just $[\overline{B\circ\overline{A}}]$, which by definition of $\circ$ (and two double negations) is just $[B\to A]=[B]\to[A]$. Let us record these facts in

THEOREM 1. The Lindenbaum algebra of **R**, $\mathbf{R}/\rightleftarrows$, is a De Morgan (indeed, an intensional) semi-group under $\circ$ (where $[A]\circ[B]=[A\circ B]$, and where $A\circ B$ is defined in turn as $\overline{A\to\overline{B}}$). The residual of $[A]$ by $[B]$, $[A]:[B]$, is $[B]\to[A]$.

However, the Lindenbaum algebra of **R** is not a free De Morgan semi-group, not even a free intensional semi-group. This is because $A\to A\to B\to B$ is a theorem of **R**, but it is not true in intensional semi-groups in general that $b:(a:a)\leq b$. The following proof sketch of $A\to A\to B\to B$ is instructive. We start with $A \to A \to B \to .A \to A \to B$ and permute, giving us $A \to A \to .A \to A \to B \to B$. But this last, together with $A \to A$, gives us by *modus ponens* the desired theorem. Intensional semi-groups in general do not take account of such a use of *modus ponens* (which we might express algebraically as: $a \leq b$ and $b:a \leq d:c$ imply $c \leq d$).

For an argument that $b:(a:a) \leq b$ is indeed independent of the postulates for an intensional semi-group, consider the intensional semi-group consisting of the elements -1, -0, $+0$, and $+1$, where in the lattice structure $-1 \leq -0 \leq +0 \leq +1$; so it is the four element chain, where $\overline{\pm a} = \mp a$, and where $\circ$ is defined by the following matrix:

$\circ$	-1	-0	$+0$	$+1$
-1	-1	-1	-1	-1
-0	-1	$+1$	$+1$	$+1$
$+0$	-1	$+1$	$+1$	$+1$
$+1$	-1	$+1$	$+1$	$+1$

It is easily verified that this is indeed an intensional semi-group. Then, remembering $a:b = \overline{b \circ \bar{a}}$, $b:(a:a) = \overline{a \circ \overline{\bar{a} \circ \bar{b}}}$, and upon letting $a = b = +0$, we can compute $b:(a:a) = +1$, but $+1 \not\leq +0$. Observe further that $+0:+0 = -1$, and hence $+0:+0 \leq d:c$ for all c and d, even though it is not the case for all c and d that $c \leq d$; so we can see directly how this intensional semi-group violates the algebraic analogue of *modus ponens* mentioned above. It is interesting to observe that if a De Morgan semi-group is a monoid (as this one is *not*), then it must have the algebraic analogue of *modus ponens*, for by P16, $a \leq b$ implies $\boldsymbol{t} \leq b:a$ for the identity $\boldsymbol{t}$ of the monoid; but then by transitivity, $\boldsymbol{t} \leq d:c$, and hence, by P16, $c \leq d$.

This raises the question of whether $\mathbf{R}/\rightleftarrows$ has an identity. Perhaps it is free in intensional monoids. Is there then a formula A such that $[A]$ is the identity of $\mathbf{R}/\rightleftarrows$? We know by P15 and Theorem 1, that if there is such a formula A, then $[A] \leq [B] \to [B]$ for every formula B, i.e., $A \to .B \to B$ is a theorem of **R** for every formula B; and in particular, for every propositional variable p, $A \to .p \to p$ is a theorem of **R**. But it may be shown for **R** that no formula of the form $C \to D$ is a theorem unless C and D share a propositional variable. (This is shown by essentially the same argument as in §22.1.3; one only needs to observe that **R** satisfies the matrices.) But this means that A contains every propositional variable of **R**, which means that if **R** is formulated with an infinite number of propositional variables, then there is no such formula A. So in general $\mathbf{R}/\rightleftarrows$ has no identity.

However, if **R** is formulated with but a finite number of propositional variables $p_1, \ldots, p_m$ (let us call such a formulation $\mathbf{R}_m$), then it has a formula which acts as an identity, namely, $(p_1 \rightarrow p_1) \& \ldots \& (p_m \rightarrow p_m)$ (let us call this formula $\boldsymbol{t}_m$). As in §45.1, one can show that $\boldsymbol{t}_m \rightarrow (A \rightarrow A)$ is a theorem. But then by permutation, we have as a theorem $A \rightarrow (\boldsymbol{t}_m \rightarrow A)$, which means $[A] \leq [\boldsymbol{t}_m] \rightarrow [A]$, which means by P5 that $[A] \circ [\boldsymbol{t}_m] \leq [A]$. The other half of the equality follows easily from the fact that $\boldsymbol{t}_m \rightarrow A \rightarrow A$ is a theorem of **R** (it may be proven from $\boldsymbol{t}_m \rightarrow A \rightarrow . \boldsymbol{t}_m \rightarrow A$ *via* a permutation and a use of *modus ponens* with $\boldsymbol{t}_m$ as the minor premiss: see the proof of $A \rightarrow A \rightarrow B \rightarrow B$ after Theorem 1). In particular then, $\boldsymbol{t}_m \rightarrow \bar{A} \rightarrow \bar{A}$ is a theorem. But by contraposition, $A \rightarrow \overline{\boldsymbol{t}_m \rightarrow \bar{A}}$ is a theorem, i.e., by definition of "$\circ$", $A \rightarrow (\boldsymbol{t}_m \circ A)$ is a theorem, which means that $[A] \leq [\boldsymbol{t}_m] \circ [A]$, which means that $[\boldsymbol{t}_m]$ is the identity of $\mathbf{R}_m/\rightleftarrows$.

Although **R** does not in general have a formula that acts as an identity, the system $\mathbf{R}^t$ may be obtained as an extension of **R** formed by enriching the grammar with a constant proposition $\boldsymbol{t}$ and adding two axioms — $\boldsymbol{t}1'$ and $\boldsymbol{t}2$ of §27.1.2 — that ensure that $\boldsymbol{t}$ has just the properties needed to make $[\boldsymbol{t}]$ the identity.

It may be shown that $\mathbf{R}^t$ is a conservative extension of **R** in the sense that where A is a formula of $\mathbf{R}^t$ that does not contain $\boldsymbol{t}$, then A is provable in $\mathbf{R}^t$ iff **A** is provable in **R**. The argument is a simple modification of §45.1, and we omit it.

We record the following algebraic version of this result, the proof of which is trivial.

THEOREM 2. The Lindenbaum algebra of $\mathbf{R}/\rightleftarrows$ is isomorphically embeddable in the Lindenbaum algebra of $\mathbf{R}^t/\rightleftarrows$ under the mapping which sends $[A]$ in $\mathbf{R}/\rightleftarrows$ into $[A]$ in $\mathbf{R}^t/\rightleftarrows$.

Both the logical and the algebraic versions of this result establish that we can study **R** and its Lindenbaum algebra by studying $\mathbf{R}^t$ and its Lindenbaum algebra. We shall find that it is profitable to do this since $\mathbf{R}^t$ and its Lindenbaum algebra are more amenable to algebraic treatment.

§28.2.3. $\mathbf{R}^t$ and De Morgan monoids. We start this section with a theorem, proven but not recorded in the last section.

THEOREM 1. The Lindenbaum algebra of $\mathbf{R}^t$, $\mathbf{R}^t/\rightleftarrows$, is a De Morgan (indeed, an intensional) monoid, with identity $[\boldsymbol{t}]$ (where the semi-group operation is defined as in Theorem §28.2.2(1)).

Although $\mathbf{R}/\rightleftarrows$ turned out to be a De Morgan semi-group but *not* a free one, $\mathbf{R}^t/\rightleftarrows$ is a free De Morgan monoid. (Indeed, it is a free intensional monoid; so just as free De Morgan lattices and free intensional lattices turn out to be the same, free De Morgan monoids turn out to be the same as free intensional monoids.)

We shall prove this by first defining and proving that $\mathbf{R}^t$ is consistent in the class of De Morgan monoids. We accordingly say that Q is a *De Morgan monoid model* (henceforth, *model*) iff $Q = \langle\langle S, \circ, \leq, \overline{\ \ }, \boldsymbol{t}\rangle, s\rangle$, where $\langle S, \circ, \leq, \overline{\ \ }, \boldsymbol{t}\rangle$ (henceforth S) is a De Morgan monoid and where s is an *assignment function* for S, i.e., a function such that for each formula A, if A is a propositional variable, $s(A) \in S$, and if A is $\boldsymbol{t}$, $s(A) = \boldsymbol{t}$. (Note that this definition, as well as the subsequent definitions, is applicable to **R** as well as $\mathbf{R}^t$, the "if A is $\boldsymbol{t}$" clause being vacuous in application to **R**.)

Given a model $Q = \langle S, s\rangle$, we define a *valuation* determined by Q as a function v_Q defined over all formulas and having values in S as follows: for all formulas A,

if A is a propositional variable or $\boldsymbol{t}$, $v_Q(A) = s(A)$;
if A has the form $\overline{B}$, $v_Q(A) = \overline{v_Q(B)}$;
if A has the form $B \vee C$, $v_Q(A) = v_Q(B) \vee v_Q(C)$;
if A has the form $B \rightarrow C$, $v_Q(A) = v_Q(C) : v_Q(B)$.

Note that these conditions determine that

if A has the form $B \& C$, $v_Q(A) = v_Q(B) \wedge v_Q(C)$;
if A has the form $B \circ C$, $v_Q(A) = v_Q(B) \circ v_Q(C)$.

Let us then say that a formula A is *true in a model* $Q = \langle S, s\rangle$ iff $v_Q(A) \geq \boldsymbol{t}$ (this definition is motivated by the fact that A is a theorem of $\mathbf{R}^t$ iff $[\boldsymbol{t}] \leq [A]$, which is an immediate consequence of the postulates governing $\boldsymbol{t}$). Otherwise A is *false in the model.*

We then define a formula as *valid in a De Morgan monoid* S iff it is true in every model $Q = \langle S, s\rangle$, and otherwise as *falsifiable in a De Morgan monoid* S.

We finally define a formula as *valid in a class of De Morgan* monoids iff it is valid in each member of the class, and otherwise as *falsifiable in the class.* If the particular class in question is the class of all De Morgan monoids, then we shall say simply that the formula is *valid* or *falsifiable.*

We shall say that **R** or $\mathbf{R}^t$ is *consistent with respect to a class of De Morgan* monoids iff, if a formula A is provable, then it is valid in that class. We shall say that **R** or $\mathbf{R}^t$ is *complete with respect to a class of De Morgan* monoids iff each formula A valid in the class is provable. We shall call **R** or $\mathbf{R}^t$ just *consistent* or *complete* iff the particular class in question is the class of all De Morgan monoids.

We may now prove

THEOREM 2. Both **R** and $\mathbf{R}^t$ are consistent; hence both **R** and $\mathbf{R}^t$ are consistent in the class of intensional monoids.

Since **R** is contained in $\mathbf{R}^t$, and since intensional semi-groups are De Morgan semi-groups, it suffices to show $\mathbf{R}^t$ consistent. We show that the axioms §§27.1.1–2 are valid and that the rules preserve validity. Since all of the axioms have the form of implications, we may show an axiom $A \to B$ valid by showing that for every valuation v_Q, $v_Q(A) \leq v_Q(B)$, for then by P16, $\boldsymbol{t} \leq v_Q(B){:}v_Q(A)$. The validity of R1, R5, R6, R8, R9, R11, and R13 then follows immediately from De Morgan lattice properties, since (schematically) these axioms contain the arrow only as the principal connective. As an illustration, we consider R8: $A \to . A \vee B$. By inductive definition of v_Q, $v_Q(A \vee B) = v_Q(A) \vee v_Q(B)$. But $v_Q(A) \leq v_Q(A) \vee v_Q(B)$ is a lattice property. The validity of the remaining axioms follows trivially from the properties of De Morgan monoids that we developed in §28.2.1. Thus R2 is treated by P4, R3 by P8, R4 by P18, R7 by P10, R10 by P11, R12 by (A), and $\boldsymbol{t}1'$ and $\boldsymbol{t}2$ by P14 and P15, respectively. As an illustration, we shall demonstrate the validity of an algebraically important theorem: $A \to (B \to C) \to . B \to (A \to C)$. By inductive definition of v_Q, $v_Q(A \to (B \to C)) = (v_Q(C){:}v_Q(B)){:}v_Q(A)$, and $v_Q(B \to (A \to C)) = (v_Q(C){:}v_Q(A)){:}v_Q(B)$, but by P6 these are identical.

We next show that the rules preserve validity. For *modus ponens*, suppose $\boldsymbol{t} \leq v_Q(A)$ and $\boldsymbol{t} \leq v_Q(A \to B)$. But then $\boldsymbol{t} \leq v_Q(B){:}v_Q(A)$, and by P14, $v_Q(A) \leq v_Q(B)$. But then by transitivity, $\boldsymbol{t} \leq v_Q(B)$. For adjunction, we assume that $\boldsymbol{t} \leq v_Q(A)$ and $\boldsymbol{t} \leq v_Q(B)$; but then $\boldsymbol{t} \leq v_Q(A) \wedge v_Q(B) = v_Q(A \& B)$.

From Theorem 2, we obtain

THEOREM 3. The Lindenbaum algebra of $\mathbf{R}^t$ formulated with n propositional variables is the free De Morgan monoid with n free generators, and it is also the free intensional monoid with n free generators.

We prove the first part by considering any formulation of $\mathbf{R}^t$ with n propositional variables p_i. Clearly the $[p_i]$ generate $\mathbf{R}^t/\rightleftarrows$. Now consider any mapping f of these generators into an arbitrary De Morgan monoid S. Now define an assignment function s for S so that $s(p_i) = f([p_i])$ and $s(\boldsymbol{t}) = \boldsymbol{t}$. We thus have a model $Q = \langle S, s \rangle$ and its valuation v_Q. Define $h([A]) = v_Q(A)$. It is easy to see that if h is a well-defined (single-valued) mapping then it is a $\boldsymbol{t}$-homomorphism. But that h is well defined follows from Theorem 2. For suppose $[A] = [B]$, i.e., that both $A \to B$ and $B \to A$ are theorems of $\mathbf{R}^t$. But then by Theorem 2, $\boldsymbol{t} \leq v_Q(A \to B) = v_Q(B){:}v_Q(A)$ and $\boldsymbol{t} \leq v_Q(B \to A) = v_Q(A){:}v_Q(B)$. But then by P16, $v_Q(A) = v_Q(B)$, and $h([A]) = h([B])$. The second part of the theorem follows immediately from the fact that $\mathbf{R}^t/\rightleftarrows$ is an intensional monoid (Theorem 1).

We remark that Theorem 3 identifies (up to $\boldsymbol{t}$-isomorphism) the free De Morgan monoid with n free generators, and that it is an immediate consequence of this fact that every De Morgan monoid is a $\boldsymbol{t}$-homomorphic image of an intensional monoid.

We next observe that it is a consequence of the fact that $\mathbf{R}^{\boldsymbol{t}}/\rightleftarrows$ is an intensional monoid that

THEOREM 4. $\mathbf{R}^{\boldsymbol{t}}$ is complete with respect to the class of intensional monoids; hence, $\mathbf{R}^{\boldsymbol{t}}$ is complete.

The second part of the theorem follows from the first in virtue of the fact that an intensional monoid is a De Morgan monoid. We prove the first part by showing that, if a formula A is not provable in $\mathbf{R}^{\boldsymbol{t}}$, then it is falsifiable in $\mathbf{R}^{\boldsymbol{t}}/\rightleftarrows$. We define the *canonical model* $\mathrm{C} = (\mathbf{R}^{\boldsymbol{t}}/\rightleftarrows, \mathrm{c})$, where c is the *canonical assignment function*, i.e., if A is a propositional variable or $\boldsymbol{t}$, $\mathrm{c}(A) = [A]$. We next define the *canonical valuation* $\mathrm{v_C}$ as the valuation determined by C. It is easy to see that $\mathrm{v_C}(A) = [A]$. We next observe that if A is not provable in $\mathbf{R}^{\boldsymbol{t}}$, then $[\boldsymbol{t}] \nleq [A]$. This is because otherwise $\boldsymbol{t}\rightarrow A$ would be provable, and since $\boldsymbol{t}$ is also provable by R1 and $\boldsymbol{t}1$, A would be provable by *modus ponens*. Since $[\boldsymbol{t}]$ is the identity of $\mathbf{R}^{\boldsymbol{t}}/\rightleftarrows$, this completes the theorem.

Putting Theorems 2 and 4 together gives us

THEOREM 5. A formula A is provable in $\mathbf{R}^{\boldsymbol{t}}$ iff it is valid, and iff it is valid in the class of intensional monoids.

Even though **R** does not have a formula like $\boldsymbol{t}$ that acts like an identity, we get by virtue of the embedding of $\mathbf{R}/\rightleftarrows$ into $\mathbf{R}^{\boldsymbol{t}}/\rightleftarrows$ of Theorem §28.2.2(2), as an immediate consequence of Theorem 4,

THEOREM 6. **R** is complete with respect to the class of intensional monoids; hence **R** is complete.

Combining Theorem 6 and Theorem 2, we get

THEOREM 7. A formula A is provable in **R** iff it is valid, and iff it is valid in the class of intensional monoids.

§28.2.4. An algebraic analogue to the admissibility of (γ). The admissibility of (γ) to the system $\mathbf{R}^{\boldsymbol{t}}$, expressed in terms of the Lindenbaum algebra of $\mathbf{R}^{\boldsymbol{t}}$, takes the following form:

(I) $[\boldsymbol{t}] \leq [A]$ and $[\boldsymbol{t}] \leq [\bar{A}]\vee[B]$ implies $[\boldsymbol{t}] \leq [B]$.

(I) suggests a conjecture concerning De Morgan monoids in general, that is an obvious algebraic analogue of the admissibility of (γ), namely,

(I′) $t \leq a$ and $t \leq \bar{a} \vee b$ implies $t \leq b$.

We shall find at the end of this section that such a conjecture would be false, but we postpone this denouement until we develop some interesting equivalents of property (I′).

We first observe that (I′) is equivalent to a simpler property

(II) $t \leq \bar{t} \vee b$ implies $t \leq b$.

Property (II) is obviously only a special case of (I′), substituting t for a. To see that (II) implies (I′), we assume (II) and the hypotheses of (I′). The hypotheses of (I′) easily imply $t \leq a \wedge (\bar{a} \vee b)$, which in turn implies via a distribution $t \leq (a \wedge \bar{a}) \vee (a \wedge b)$. By P17, $a \wedge \bar{a} \leq a \circ \bar{a}$. But by Theorem §28.2.1(1) $\overline{a{:}a} = a \circ \bar{a}$, so $\bar{a} \wedge a \leq \bar{t}$, which means that $\bar{t} \leq t \vee (a \wedge b)$. But then by (II), $t \leq a \wedge b$, and so $t \leq b$, which completes the equivalence.

We next show that (II) is equivalent to

(III) $b \nleq \bar{t}$ implies that there exists a truth filter (see §18.2) T such that $b \in T$ but $\bar{t} \notin T$.

It is easy to see that (II) is a necessary condition for (III), for suppose that $t \leq \bar{t} \vee b$, but $t \nleq b$. Then $\bar{b} \nleq \bar{t}$, and by (III) there exists a truth filter T such that $\bar{b} \in T$ but $\bar{t} \notin T$. But since $\bar{t} \notin T$, $t \in T$, and hence $\bar{t} \vee b \in T$. But since truth filters are prime, then either $\bar{t} \in T$ or $b \in T$. But we know that $\bar{t} \notin T$, and since $\bar{b} \in T$, then $b \notin T$ as well. But this is a contradiction, so contrary to hypothesis, $t \leq b$.

The sufficiency of (II) is a bit harder. We suppose that $b \nleq \bar{t}$ and consider the filter generated by b together with t, F(b, t). It is easy to see that F(b, t) is consistent in the sense that it contains for no element a, both a and $\bar{a}$. If it did, then $t \wedge b \leq a \wedge \bar{a}$, and since $a \wedge \bar{a} \leq \bar{t}$ (see the argument that showed the equivalence of (I′) and (II)), then $t \wedge b \leq \bar{t}$. But then $t \leq \bar{t} \vee \bar{b}$, and by (II), $t \leq \bar{b}$, i.e., $b \leq \bar{t}$, contrary to the hypothesis of (III).

Now that we know there is at least one consistent filter containing both b and t, we consider the set of all such consistent filters, ordered by set inclusion. It is easy to see that the union of any chain of consistent filters containing both b and t is itself a consistent filter containing both b and t. So every chain has an upper bound, and we have the hypothesis of Zorn's lemma, so we may conclude that there is a consistent filter T that contains both b and t and is maximal with respect to these properties. We now show that T is exhaustive in the sense that it contains for every element a, at least one of a and $\bar{a}$.

Suppose that T contained for some element a, neither a nor $\bar{a}$. Then consider the filter generated by T together with a, F(T, a), and the filter generated by T together with $\bar{a}$, F(T, $\bar{a}$). Since these are filters which properly include

T, and since T was shown maximal with respect to being a consistent filter containing both b and t, then these filters must be inconsistent. But this means that for some elements b_1, $b_2 \in T$, and for some elements c and d,

$$b_1 \wedge a \leq c \wedge \bar{c}, \quad \text{and} \quad b_2 \wedge \bar{a} \leq d \wedge \bar{d}.$$

But since for any element x, $x \wedge \bar{x} \leq \bar{t}$ (see the proof of the equivalence of (I′) and (II)), then

$$b_1 \wedge a \leq \bar{t} \quad \text{and} \quad b_2 \wedge \bar{a} \leq \bar{t}.$$

Then setting $b = b_1 \wedge b_2$, we have for some element $b \in T$, and

$$b \wedge a \leq \bar{t} \quad \text{and} \quad b \wedge \bar{a} \leq \bar{t}.$$

But then we have

$$(b \wedge a) \vee (b \wedge \bar{a}) \leq \bar{t},$$

which leads by distribution to

$$b \wedge (a \vee \bar{a}) \leq \bar{t}$$

But since $t \leq a \vee \bar{a}$ (which follows from $a \wedge \bar{a} \leq \bar{t}$), then $b \wedge (a \vee \bar{a}) \in T$, and hence $\bar{t} \in T$. But this is a contradiction, since T was shown consistent. So T is both consistent and exhaustive and hence a truth filter, which completes the proof.

It is interesting to observe that (III) is a generalization of Stone's 1936 theorem concerning maximal filters in Boolean algebras. Stone proved that for any element $b \neq \boldsymbol{0}$, there exists a maximal (proper) filter M such that $b \in M$. The generalization is obvious once it is recalled that a Boolean algebra is a De Morgan monoid, and that a filter is maximal in a Boolean algebra iff it contains for every element, exactly one of a and $\bar{a}$. Indeed, our proof that (II) implies (III) was essentially based upon Stone's proof, for (II) is trivially true of Boolean algebras.

Now let us at last discuss the truth of (I′). It is an easy consequence of the equivalence of (I′) with (III) that (I′) does not hold of any De Morgan monoid whose underlying De Morgan lattice does not have a truth filter, i.e., whose underlying De Morgan lattice is not an intensional lattice. One might next conjecture that perhaps (I′) holds of intensional monoids, but this conjecture too would be false. For a counterexample, consider the intensional monoid whose underlying intensional lattice is the subintensional lattice of M_0 of §22.1.3 generated by $\{+\boldsymbol{1}, +\boldsymbol{0}, +\boldsymbol{3}\}$, and whose multiplication is defined by the following matrix:

∘	−3	−1	−0	+0	+1	+3
−3	−3	−3	−3	−3	−3	−3
−1	−3	−1	−0	−3	−1	−0
−0	−3	−0	−0	−3	−0	−0
+0	−3	−3	−3	+0	+0	+0
+1	−3	−1	−0	+0	+1	+3
+3	−3	−0	−0	+0	+3	−3

It may be routinely, though laboriously, verified that this is indeed an intensional monoid, and indeed it may be shown that on this underlying intensional lattice it is the only intensional monoid the identity of which is $+1$.

We now observe that property (II) (and at the same time property (I′)) fails in it, since $+1 \leq -1 \vee +0$, and yet $+1 \nleq +0$. It is interesting to note that the underlying intensional lattice served as an earlier counterexample to a (γ)-like property of intensional lattices. Belnap and Spencer 1966 showed that it is not true of every intensional lattice that any consistent filter is extendible to a consistent, exhaustive filter (unlike the case with Boolean algebras), and they did this by observing that the filter generated by $+1$ in this intensional lattice is consistent, but not so extendible. What we have done with our present counterexample is to introduce a De Morgan multiplication upon this intensional lattice so as to make $+1$ the identity.

§28.2.5. The algebra of E and $\mathbf{R}^{\Box t}$: closure De Morgan monoids. We may define a multiplication operation upon the Lindenbaum algebra of **E**, as we did upon **R**/⇄, but the result is not so pleasant since the multiplication then turns out not to be associative. Associativity is equivalent to permutation, but it is exactly the addition of permutation to **E** which gives **R**. These remarks do not merely reflect a prejudice against non-associative algebraic systems, but are based upon the repeated failure to find perspicuous axioms involving multiplication that capture the restricted permutation axiom $(A{\rightarrow}.B{\rightarrow}C{\rightarrow}D){\rightarrow}.B{\rightarrow}C{\rightarrow}.A{\rightarrow}D$ of **E**. (Consider as an example of an unperspicuous axiom $\overline{a_1 \circ \overline{a_2 \circ (b \circ \overline{c_1 \circ c_2})}} = (\overline{a_1 \circ a_2} \circ b) \circ \overline{c_1 \circ c_2}$.)

However, Meyer 1966 made a (false — see §28.1) conjecture which, if true, gave a graceful way out. As reported in §§27.1.3 and 28.1, he offered for consideration a logical system $\mathbf{R}^{\Box}$, obtained from **R** by adding a necessity operator □ and the axioms and rules of §27.1.3.

Meyer then conjectured that $A{\rightarrow}B$ in **E** can be defined by □(A→B) in $\mathbf{R}^{\Box}$, so that a formula is provable in **E** iff its obvious definitional transform is provable in $\mathbf{R}^{\Box}$. Meyer pointed out that this definition is analogous to Lewis's definition of strict implication in terms of necessary material implication.

All of this suggested that just as McKinsey and Tarski 1948 studied the algebra of strict implication in **S4** by considering Boolean algebras with a Kuratowski closure operation, so we may study the algebra of entailment by studying De Morgan semi-groups with an appropriate closure operation. Indeed, since it is easy to prove that $\boldsymbol{t}$ and its axioms can be added conservatively to $\mathbf{R}^{\Box}$, we may study the algebra of the resulting system $\mathbf{R}^{\Box\boldsymbol{t}}$.

Let us then define a *De Morgan closure operation* C upon a De Morgan monoid as a unary operation satisfying (where Ia — the *De Morgan interior operation* — is defined as $\overline{\mathrm{C}\bar{\mathrm{a}}}$).

C1 $\mathrm{a} \leq \mathrm{Ca}$
C2 $\mathrm{CCa} = \mathrm{a}$
C3 $\mathrm{C(a \vee b)} = \mathrm{Ca \vee Cb}$
C4 $\mathrm{C}(\bar{\boldsymbol{t}}) = \bar{\boldsymbol{t}}$
C5 $\mathrm{Ca \circ Ib} \leq \mathrm{C(a \circ b)}$,

and call the resulting structure a *closure De Morgan monoid.* We remark that axiom C5 is easily shown to be equivalent to

C5′ $\mathrm{I(a{:}b)} \leq \mathrm{Ia{:}Ib}$.

The operation C is a specialization of a very general notion of a *closure operation* on a lattice defined by C1, C2, and

C6 $\mathrm{a} \leq \mathrm{b}$ implies $\mathrm{Ca} \leq \mathrm{Cb}$,

for it is easily shown that C6 follows from C3. (See especially Ward 1942, p. 192; Certaine 1943, p. 22; and Birkhoff 1948, p. 49.) We remark further that C has marked similarities to the *Kuratowski closure operation* of McKinsey and Tarski 1944, which is defined upon a Boolean algebra by C1, C2, C3, and

C7 $\mathrm{C}(\boldsymbol{0}) = \boldsymbol{0}$.

Indeed, when a Boolean algebra is considered as a De Morgan monoid in the way described in §28.2.3, then $\wedge$ becomes $\circ$ and $\boldsymbol{1}$ becomes $\boldsymbol{t}$, C7 becomes a special case of C4, and C5 becomes derivable. So a closure De Morgan monoid may be looked at as a generalization of a closure Boolean algebra.

To make a long story short, upon defining $\Diamond A$ (A is possible) in $\mathbf{R}^{\Box\boldsymbol{t}}$ as $\overline{\Box\bar{A}}$, and then defining on the Lindenbaum algebra of $\mathrm{R}^{\Box\boldsymbol{t}}$, $\mathrm{R}^{\Box\boldsymbol{t}}/\rightleftarrows$,

$\mathrm{C}[A] = [\Diamond A]$,

it may be easily shown that the resulting structure is a closure De Morgan monoid. We may then restrict our notion of a model in §28.2.3 appropriately and add to our inductive definition of a valuation:

if A is of the form $\Box B$, $\mathrm{v_Q}(A) = \mathrm{Iv_Q}(B)$, i.e.,
if A is of the form $\Diamond B$, $\mathrm{v_Q}(A) = \mathrm{Cv_Q}(B)$.

It is then easy to prove that $\mathbf{R}^{\Box t}$ and hence $\mathbf{R}^{\Box}$, are appropriately consistent and complete in the class of closure De Morgan monoids and hence that $\mathbf{R}^{\Box t}/\rightleftarrows$ is appropriately free in the class. The only trick is to see that the effect of the rule $\Box$I of necessitation is obtained by showing that $\boldsymbol{t} \leq$ a implies a $\leq$ Ia, which follows easily from C6 and C4.

To make a short story even shorter, everything that was done in the previous sections for $\mathbf{R}^{t}$ and $\mathbf{R}$ in their relation to De Morgan monoids can be mimicked for $\mathbf{R}^{\Box t}$ and $\mathbf{R}^{\Box}$ in their relation to closure De Morgan monoids — time, space, and patience permitting.

§28.3. Conservative extensions in R (by Robert K. Meyer). In §27.1.1 it was claimed that **R** is well axiomatized in the sense that in the interesting cases the theorems determined by the particular axioms involving a group of connectives constitute all the theorems of **R** involving those connectives. We make good on most of this claim: in §28.3.1 we prove that the positive fragment of **R** with cotenability is determined by the positive axioms, while in §28.3.2 we summarize conservative extension results and prove one new one: $\mathbf{R}_{\rightarrow}$ really is the pure implicational fragment of **R**. (For these and some other results with different methods, see §60.4.)

§28.3.1. On conserving positive logics II. The theorem used in §24.4.2 to establish that **E** is a conservative extension of the system $\mathbf{E}_{+}$ cannot be used to answer the analogous question for **R**, since its proof breaks down when one attempts to verify the **R**-theorem $A \rightarrow . A \rightarrow A \rightarrow A$.

The result of this failure and of the failure of other efforts is that for now we postpone the question: is **R**, formulated with $\rightarrow$, &, $\vee$, and $\overline{}$, a conservative extension of the system $\mathbf{R}_{+}$ determined by the negation-free axioms R1–R11 of §27.1.1? Only with the semantic developments of Chapter IX do we obtain a "Yes" (§60.4).

We can, however, resolve with methods already at our disposal the question for the formulation of **R** in which the connective $\circ$ for cotenability is taken as an additional positive primitive; a modification of the theorem of §24.4.2 does the trick. This yields an additional good reason to formulate **R** in this way.

So let $\mathbf{R}_{+}$ be formulated with $\rightarrow$, &, $\vee$, and $\circ$ primitive, with rules, *modus ponens* and adjunction, and with axioms R1–R11 of §27.1.1 and $\circ$1 and $\circ$2 of §27.1.4. The negation completion **R** of $\mathbf{R}_{+}$ results when R12–R14 of §27.1.1 are added.

We truck now with the appropriate modifications in the proofs of the theorem of §24.4.2. Let $\mathrm{M}^{+} = \langle \mathrm{M}^{+}, \mathrm{O}^{+}, \mathrm{D}^{+} \rangle$ be an $\mathbf{R}_{+}$–matrix. Let M^{-} and $*$ be as in §24.4.2(I). Let $\boldsymbol{F}$ and $\boldsymbol{T}$ be distinct non-members of $\mathrm{M}^{+} \cup \mathrm{M}^{-}$. We define the *relevant enlargement* $\mathrm{M} = \langle \mathrm{M}, \mathrm{O}, \mathrm{D} \rangle$ of M^{+} as follows:

VIII. 1. $M = (M^{+} \cup M^{-} \cup \{\boldsymbol{F}, \boldsymbol{T}\})$;
2. $D = (D^{+} \cup M^{-} \cup \{\boldsymbol{T}\})$
3. $O = \{\rightarrow, \&, \vee, \circ, \overline{}\}$, where
a. M shall be rigorously compact — i.e., when at least one argument to an operation is $\boldsymbol{F}$ or $\boldsymbol{T}$, the value shall be given by §24.4.2(Va-f).
b. On M^{+}, operations of O shall agree with corresponding operations of O^{+}.
c. For all $a \in M^{+}$, $\bar{a} = a^{*}$ and $\overline{a^{*}} = a$.
d. For all $a, b \in M^{+}$, IIa–IIf and IIIa–IIId of §24.4.2 shall hold; furthermore
$(a \rightarrow b^{*}) = (a \circ b)^{*}$;
$(a^{*} \rightarrow b) = \boldsymbol{F}$;
$(a \circ b^{*}) = (b^{*} \circ a) = (a \rightarrow b)^{*}$
$(a^{*} \circ b^{*}) = \boldsymbol{T}$

(The first entry under d may be thought of as an answer to the question posed by §24.4.2(IIIc) alternative to VIIa, which doesn't work here.) The specifications VIII suffice to define M. The reader should note that the strategy of relevant enlargement is in a sense opposite to that of rigorous enlargement of §24.4.2, for what we essentially did there was to take a matrix L^{+} and to add $\boldsymbol{F}$ and $\boldsymbol{T}$ to get M^{+}; by copying M^{+} we got M. The technique of the present construction may be viewed the other way round — first we copy, and then we add $\boldsymbol{F}$ and $\boldsymbol{T}$.

We now apply the proof of the theorem of §24.4.2 to **R**, *mutatis mutandis*.

THEOREM. **R** (with co-tenability) is a conservative extension of $\mathbf{R}_{+}$ (with co-tenability).

PROOF. Strategy is as in §24.4.2, so we shall be brief. If A is a non-theorem of $\mathbf{R}_{+}$, it is not true on an interpretation α^{+} in some $\mathbf{R}_{+}$-matrix M^{+} — e.g., the canonical one. Form the relevant enlargement M of M^{+} and show, by verifying the axioms and rules of **R**, that M is an **R**-matrix; the interpretation α which agrees with α^{+} on sentential variables agrees with α^{+} on all negation-free sentences of **R**, so in particular A is not true on α; hence A is a non-theorem of **R**. So all negation-free theorems of **R** are already theorems of $\mathbf{R}_{+}$, which was to be proved.

There are two interesting corollaries to our results, which we shall draw in conclusion. First, where **L** is a logic, we mean as elsewhere by a *regular* **L**-*theory* any set T of sentences of **L** which contains all axioms of **L** and which is closed under the rules of **L**; we write $\vdash_{T} A$ if $A \in T$, and we call T complete {consistent} {normal} if for every sentence A of **L** at least one {not both} {exactly one} of A, $\bar{A}$ is in T. Then

COROLLARY 1. Let **L** be one of the rigorous logics, or **R**. Then there is a complete **L**-theory T such that, for all negation-free sentences A of **L**, $\vdash_{T} A$ iff $\vdash_{L} A$. Furthermore, where **L** is **E**, **R**, or **T**, there is a consistent and complete **L**-theory T* with this property.

PROOF. We prove the corollary for the rigorous logics, leaving the reader to handle **R** in like manner. Given $\mathbf{L}_{+}$, construct the matrix M and the interpretation α_{M} as in the proof of the theorem of §24.4.2, and consider the set T of all sentences of **L** which are true on α_{M}. Since M is an **L**-matrix, it is easy to show that T is an **L**-theory; furthermore T is complete, since by the construction of M at least one of a, $\bar{a} \in$ D for all a $\in$ M. Finally, the restriction of α_{M} to sentences of $\mathbf{L}_{+}$ is the canonical interpretation of $\mathbf{L}_{+}$; hence by §24.4.2, Lemma 1, $\vdash_{T} A$ iff $\vdash_{L} A$, for all negation-free sentences of **L**.

Though, by the construction of T that theory is complete, it is nevertheless woefully inconsistent; in fact $\vdash_{T} \bar{B}$ whenever B is negation-free. Suppose, however, that **L** is **E**, **R**, or **T**. Then application of the methods of §28.4 yields the result that T has a consistent and complete sub-**L**-theory T*. (Interdependence between this section and §28.4 involves no circularity.) Since T* must at any rate contain all theorems of **L** (since it's an **L**-theory) and since it cannot contain any non-theorems of its extension T, when A is negation-free $\vdash_{T*} A$ iff $\vdash_{L} A$. This ends the proof of Corollary 1.

Corollary 1 sheds interesting light on the relevance logics. First, the construction of T shows that the means of blocking the so-called "implicational" paradoxes really work; the philosophical point, worked out nicely by Dunn 1966, is that a sentence is not necessarily relevant to its negation. Second, the corollary shows that all of the negation-free non-theorems of one of the relevant logics **L** may be rejected together in a single consistent and complete **L**-theory T*. It would be nice to find a recursive axiomatization of such a T*, since that would imply a positive solution of the decision problem, not yet solved for any of the relevant logics, for at least the negation-free fragment $\mathbf{L}_{+}$ of **L**. (We are less hopeful of *general* methods than before, given §60.8, however.)

COROLLARY 2. Let **L** be **R** or a rigorous logic. Then all negation-free theorems of **L** are intuitionistically valid.

PROOF. It suffices to note that all negation-free axioms and rules of **L** (including those for $\circ$) are intuitionistically valid, whence the corollary follows by the conservative extension results.

§28.3.2. R is well axiomatized. We — not me, but the principal authors of this book — had long wondered whether we had axiomatized the relevant logics correctly, in the following sense: is it the case, for the principal relevant logics, that in interesting cases all theorems in a given set of connectives can be derived using *only* axioms and rules which contain those connectives? I — meaning me — intend by "interesting cases" those that have some philosophical and technical significance; since all axioms are formulated with arrows, for example, evidently one is not going to be able to get all, or any, arrow-free theorems without using axioms that have arrows in them. Indeed, we — i.e., them again — had long ago picked three cases as the most interesting philosophically:

1. For a given relevant logic **L**, is its *implicational fragment* exactly determined by its implicational axioms and *modus ponens?*
2. For a given relevant logic **L**, is its *positive* (i.e., *negation-free*) fragment exactly determined by its negation-free axioms, with *modus ponens* and adjunction?
3. For a given relevant logic **L**, is its *implication-negation fragment* exactly determined by its implication-negation axioms, and *modus ponens?*

The reason that these questions are the most interesting is the following: 1 is interesting because the distinctive characteristic of relevant logics is *supposed* to be their treatment of implication; indeed, the *pure* calculi of entailment and relevant implication $\mathbf{E}_{\to}$ and $\mathbf{R}_{\to}$ are independently motivated in §§3–4; evidently this motivation would be largely wasted if there were, say, formulas of $\mathbf{R}_{\to}$ which were theorems of **R** but not theorems of $\mathbf{R}_{\to}$.

2 is interesting because of the constructive character of the relevant logics; i.e., their negation-free axioms are intuitionistically acceptable. But that fact alone wouldn't commend them very much if there were intuitionistically unacceptable negation-free theorems, proved by taking a detour through negation.

3 is interesting because of the Gentzen system of §13. I write at a time of rapid solution of outstanding problems, but to date $\mathbf{E}_{\overset{\sim}{\to}}$ and $\mathbf{R}_{\overset{\sim}{\to}}$ are the largest fragments of the relevant logics *firmly* under control — i.e., known to be decidable and admitting normal form proof techniques, and disproof techniques. Again, this profits little if what may be disproved in, say, $\mathbf{R}_{\overset{\sim}{\to}}$ is by some crafty disjunctive argument provable in all of **R**.

It is unreasonable that any of 1–3 should be answered in the negative. Appealing again to the third person "we," we saw this long ago. My own intuitions are much worse than ours are, so that I wasted time looking for counterexamples to what we knew all along. That this was not a wholly fruitless search — and that it does require something approaching special revelation to know when to make these conjectures — may be gleaned by reflection on the system **RM** of §§27.1.1 and 29.3. For what *I* thought were unconvincing reasons, *we* never did like **RM**, as *we* say in §29.5. Dunn and I like it though, as *I* say in §29.3 — Dunn may be excused on ground of paternity, but my affection is simple and innocent. But we were right, at least on the following point: **RM** is funny, in that on a natural axiomatization, every one of 1–3 is false for it; i.e., it takes *all* of negation, conjunction, and disjunction to develop the peculiar properties of **RM**; accordingly, the natural Dunn axiomatization of §27.1.1 does not well-axiomatize **RM**. (**RM** is indeed a laboratory of counterexamples to *a priorism* in logic; every step toward its understanding lay in finding false what one took to be an obvious property of the system. "Darum keine Uberraschung," says Old Testament Wittgenstein, and his New Testament disciples agree that there are no genuine surprises in logic. That's the *wrong* attitude, at least for those of us off the special revelation pipeline.)

RM may at best be characterized as a *semi*-relevant logic, however, and for the *proper* relevant logics special revelation has paid off in every case closely examined to date. Affirmative answers to question 2 are given for **R** in §28.3.1, and for a large number of other relevant logics, including **E** and **T**, in §24.4.2. Similarly, affirmative answers can be given to question 1 for **R**, **E**, **T** and other relevant logics in Chapter IX using the semantical methods developed there. (These answers presuppose §28.3.1 and §24.4.2.) In like manner, the semantics of Chapter IX may be used to answer 3 affirmatively for **R**. This argument extends to $\mathbf{R}^{\Box}$ and to **E**, so (barring an unexpected hole) §60.4 will record a "Yes" for **E** also to question 3. See §60.4 for the most up-to-date information about conservative extension results for relevant logics in general.

It is also the case that certain less interesting questions may be settled on application of the new semantical methods. For example, the cotenability operator $\circ$ may be added conservatively to any fragment of **R** without it; i.e., to the system determined by implicational axioms alone (also proved below), by implication and conjunction axioms, or by implication, conjunction, and disjunction axioms. This is significant in the last case because it is the system with &, $\vee$, $\rightarrow$, *and* $\circ$ which is proved negation-conservative in §28.3.1. Similar remarks may be made about the constant $\boldsymbol{t}$ — any interesting fragment of **R** without it is *inessentially* enriched on its addition with appropriate axioms. Likewise, adding $\vee$ (with its axioms) to any positive fragment with & is conservative. (Framing the $\rightarrow$, $\vee$ fragment of **R** (without &) has proved oddly recalcitrant, over the lingering effects of distribution; this is surprising in that $\mathbf{R}_{\rightarrow\&}$ is straightforward. But the following, in Polish notation, is a theorem of $\mathbf{R}_{\rightarrow\vee}$ such that it's not easy to find natural axioms and rules in $\rightarrow$, $\vee$ which yield it:

CCCCACprqrrACprqCCACACprqACprqCACprqrrACACAC prqACprqprq.)

Results that employ the semantics of §§48–60 are of course as much Routley's as mine; indeed, he made the original suggestion that questions of conservative extension might readily be settled on application of these semantical methods. In the presence of conjunction, that addition of t is conservative is *our* result (third person, again). For other remarks on conservative extensions in relevant logics, see also Prawitz 1965.

I devote the remainder of this section to giving another argument, independent of §60.4, that $\mathbf{R}_{\rightarrow}$ is the implicational fragment of **R**. The argument is interesting in that it probes a little more deeply into the algebraic structure of **R** by considering algebras associated with fragments thereof. It then turns out, by a simple lattice theoretic construction, that we can get & and $\vee$ virtually for free, in the sense that an algebra which rejects a given implicational non-theorem of $\mathbf{R}_{\rightarrow}$ can be painlessly transformed into an algebra which continues to reject that non-theorem while verifying all axioms of $\mathbf{R}_{+}$. Certain uninteresting parts of the argument will be skipped over in a hurry; these serve to introduce, again for free, the intensional conjunction $\circ$ and the minimal true sentential constant t.

(Maksimova 1971 uses essentially the same argument, which she found first.)

Let $\mathbf{R}_{\rightarrow}$ be axiomatized as in §3 or §8.3.4. Let $\mathbf{R}_{\rightarrow}'$ be got by adding t and $\circ$ to the formation apparatus, together with additional axiom schemes $A \rightarrow .t \rightarrow A$, $t \rightarrow A \rightarrow A$, $A \rightarrow .B \rightarrow .A \circ B$ and $(A \rightarrow .B \rightarrow C) \rightarrow .(A \circ B) \rightarrow C$.

Lemma 1. $\mathbf{R}_{\rightarrow}'$ is a conservative extension of $\mathbf{R}_{\rightarrow}$.

Proof. Gentzen methods like those of §7 (but simpler) suffice.

De Morgan monoids, in the sense of §28.2, algebraize **R**. Correspondingly algebraizing $\mathbf{R}_{\rightarrow}'$ are *Church monoids;* a Church monoid is just like a De Morgan monoid except that its order is conferred by any old *partially ordered set*, not necessarily a De Morgan lattice. More specifically, a structure $\mathrm{D} = \langle \mathrm{D}, t, \circ, \leq, \rightarrow \rangle$ is a *Church monoid* provided that

(i) $\langle \mathrm{D}, t, \circ \rangle$ is a *commutative monoid*, with monoid operation $\circ$ and identity t. (That is, $\circ$ is commutative and associative, and $t \circ \mathrm{x} = \mathrm{x}$ for all x in D.)

(ii) As a monoid, D is *partially ordered* by $\leq$. (I.e., $\leq$ is reflexive, transitive, and antisymmetric, and whenever $\mathrm{a} \leq \mathrm{b}$, for any c in D, $\mathrm{a} \circ \mathrm{c} \leq \mathrm{b} \circ \mathrm{c}$.)

(iii) D is *residuated* with respect to $\rightarrow$; i.e., for all a, b, c in **D**, $(\mathrm{a} \circ \mathrm{b}) \leq \mathrm{c}$ iff $\mathrm{a} \leq (\mathrm{b} \rightarrow \mathrm{c})$.

(iv) D is *square-increasing;* i.e. $\mathrm{a} \leq \mathrm{a} \circ \mathrm{a}$, for all a in D.

A Church monoid is called a *Dunn monoid* provided that its order is conferred by a distributive lattice; i.e., provided

(v) For all a, b in D, least upper and greatest lower bounds $a \vee b$ and $a \wedge b$ exist, and

(vi) For all a, b, c in D, $a \wedge (b \vee c) = (a \wedge b) \vee (a \wedge c)$.

("Intensional distribution," $a \circ (b \vee c) = (a \circ b) \vee (a \circ c)$, is a consequence of (i)–(v).) In this section, we shall let $\mathbf{R}_+$ be the system one gets by adding adjunction and the &, $\vee$ axioms to $\mathbf{R}_{\rightarrow}'$; i.e., R5–R11 of §27.1.1.

As Church monoids algebraize $\mathbf{R}_{\rightarrow}'$, so Dunn monoids algebraize $\mathbf{R}_+$. We give specific content to these claims as follows: an *interpretation* of $\mathbf{R}_{\rightarrow}'$ in a Church monoid {of $\mathbf{R}_+$ in a Dunn monoid} D shall be any function h defined on all formulas of $\mathbf{R}_{\rightarrow}'$ {of $\mathbf{R}_+$} with values in D, subject to the conditions

(a) $h(\boldsymbol{t}) = \boldsymbol{t}$
(b) $h(A \circ B) = h(A) \circ h(B)$
(c) $h(A \rightarrow B) = h(A) \rightarrow h(B)$
(d) $h(A \& B) = h(A) \wedge h(B)$
(e) $h(A \vee B) = h(A) \vee h(B)$

(d) and (e), of course, apply only in the $\mathbf{R}_+$ case.

A formula A is *true* on an interpretation h in a Church monoid {Dunn monoid} iff $\boldsymbol{t} \leq h(A)$. A formula A is *Church-valid* {*Dunn-valid*} iff A is true on all interpretations in all Church {Dunn} monoids. The methods of §28.2 may then be applied *mutatis mutandis* to show almost immediately

LEMMA 2. A is a theorem of $\mathbf{R}_{\rightarrow}'$ iff A is Church-valid, and A is a theorem of $\mathbf{R}_+$ iff A is Dunn-valid, for all formulas A of the respective calculuses.

It will now suffice to prove our desired conservative extension result for $\mathbf{R}_{\rightarrow}$ if we can embed every Church monoid in a Dunn monoid. For if A is a non-theorem of $\mathbf{R}_{\rightarrow}$, it is by Lemma 1 a non-theorem of $\mathbf{R}_{\rightarrow}'$. By Lemma 2, there is a Church monoid D and an interpretation h such that $\boldsymbol{t} \nleq h(A)$ in D. Given the embedding result we seek, we can extend every Church monoid to a Dunn monoid by adding extra elements — essentially, enough elements to make the underlying partially ordered set a distributive lattice; but then the same interpretation h in the thus augmented D — call it D′ to show that it's bigger — continues to reject A, in the sense that $\boldsymbol{t} \nleq h(A)$ in D′. By Lemma 2 again, A is a non-theorem of $\mathbf{R}_+$. By the conservative extension result for negation of §28.3.1, A is a non-theorem of **R** with $\circ$, and *a fortiori* a non-theorem of plain **R**. Contraposing, all formulas of $\mathbf{R}_{\rightarrow}$ if theorems of **R** are already theorems of $\mathbf{R}_{\rightarrow}$. We accordingly finish the proof that the pure calculus of relevant implication is indeed the pure calculus of relevant implication by proving the promised embedding lemma.

LEMMA 3. Every Church monoid may be embedded in a Dunn monoid; i.e., if C is a Church monoid then there is a Dunn monoid D, such that C is isomorphic to a submonoid of D.

PROOF. Suppose the Church monoid C = $\langle C, \boldsymbol{t}, \circ, \leq, \rightarrow\rangle$ is given. A subset X of C is a C-semi-ideal iff X is *closed down* — i.e., iff, for all a, b in C, if a $\leq$ b and b is in X then a is in X. Let I(C) be the set of all C-semi-ideals. Evidently, ordered by set inclusion I(C) is a distributive lattice, with intersection as lattice meet and union as lattice join. Define X$\circ$Y as the set of all a such that, for some b $\in$ X and c $\in$ Y, a $\leq$ b$\circ$c; evidently I(C) is closed under $\circ$, which is easily seen to be commutative and associative by applying Church monoid properties. Noting that I(C) is closed under *arbitrary* unions, define X$\rightarrow$Y as the union of all the Z such that X$\circ$Z is a subset of Y. Proof that I(C) is residuated thereby — i.e., that (iii) holds — that it is square-increasing — (iv) holds — and that it is lattice-ordered — (vi) holds — is quite straightforward. The function f which assigns to each element a of C its *principal ideal* — the set f(a) of elements b such that b $\leq$ a — is evidently a 1–1 function from C to I(C); showing that f($\boldsymbol{t}$) is the identity of I(C) completes the proof that I(C) is a Dunn monoid; one finishes the proof of the lemma by showing f(a$\circ$b) = f(a)$\circ$f(b), f(a$\rightarrow$b) = f(a)$\rightarrow$f(b), and a $\leq$ b iff f(a) is a subset of f(b), which shows that C has an isomorphic copy in D.

The proof of the main theorem was given in the remarks preceding the key Lemma 3. Accordingly we merely state it. (More details are given in Meyer 1973a.)

THEOREM. **R** is a conservative extension of $\mathbf{R}_{\rightarrow}$.

§28.4. On relevantly derivable disjunctions (by Robert K. Meyer). Where *A* and *B* are negation-free formulas of one of the relevant logics **T**, **E**, or **R** (§27.1.1), we show that

1. $\vdash A \vee B$ iff $\vdash A$ or $\vdash B$.

Results from §§24.4.2. 25.2, and 28.3.1 will be presupposed.

Our strategy in proving that 1 holds for the relevant logics will be as follows. First, we shall determine a set of conditions such that the negation-free formulas of any logic which simultaneously satisfies all of these conditions have property 1. Second, using results from the sections just mentioned, we show that the relevant logics satisfy all of these conditions. We close with observations related to the intuitionist logic **H** and the Lewis system **S4**, noting now that 1 is one of the more famous properties of **H** (Gödel 1933).

For present purposes, a *logic* **L** is a triple ⟨F, O, T⟩, where $\{\rightarrow, \wedge, \vee, \overline{}\}$ = O, F is a set of formulas built up from sentential variables and the operations of O, and T is the set of theorems of **L**, which we require to be closed under *modus ponens* for →, adjunction, and substitution for sentential variables. Where **L** is ⟨F, O, T⟩, an L-*theory* is any triple ⟨F, O, T′⟩, where T⊆T′ and T′ is closed under *modus ponens* for → and adjunction. Where no ambiguity results, we identify a theory with its set T′ of theorems, and we write $\vdash_{T'} A$ if $A \in$ T′.

Let T be an **L**-theory. T is *consistent* if for no A both $\vdash_T A$ and $\vdash_T \overline{A}$; *complete* if for all A, $\vdash_T A$ or $\vdash_T \overline{A}$. If both T and T′ are **L**-theories and T⊆T′, we call T′ an *extension* of T and T a *subtheory* of T′. If T′ is an extension of **L**, i.e., every theorem of **L** is in T′, T′ is a *regular* **L**-theory. If T′ is regular, consistent, and complete, it is a *normal* **L**-theory.

We are interested in logics **L** which have the following properties:

2. Every complete regular **L**-theory has a normal subtheory.
3. $\vdash_{\mathbf{L}} \overline{A}\wedge\overline{B}\rightarrow\overline{A\vee B}$.
4. There is a complete regular **L**-theory T such that, for all negation-free formulas A of **L**, $\vdash_T A$ iff $\vdash_{\mathbf{L}} A$.

Property 2 is interesting, worth some notice on its own. Although it holds trivially in the classical case (since the inconsistent classical theory is itself trivial), in general 2 is a non-trivial converse to Lindenbaum's lemma (i.e., every regular consistent theory has a normal extension); for relevance logics, both Lindenbaum's lemma and 2 hold; in particular, since the relevance logics admit non-trivial inconsistent theories, 2 holds non-trivially for them.

LEMMA 1. Let **L** be a logic for which 2–4 hold, and let A and B be negation-free formulas of **L**. Then if $\vdash_{\mathbf{L}} A\vee B$, $\vdash_{\mathbf{L}} A$ or $\vdash_{\mathbf{L}} B$.

PROOF. Suppose A and B are negation-free non-theorems of **L**. By 4 there is a complete, and by 2 a consistent and complete, **L**-theory T such that $\vdash_T \overline{A}$ and $\vdash_T \overline{B}$. By adjunction, 3, and *modus ponens*, $\vdash_T \overline{A\vee B}$. Hence not $\vdash_T A\vee B$, since T may be assumed consistent; *a fortiori*, not $\vdash_{\mathbf{L}} A\vee B$. So if $\vdash_{\mathbf{L}} A\vee B$, $\vdash_{\mathbf{L}} A$ or $\vdash_{\mathbf{L}} B$, which was to be proved.

We turn now to the determination of sufficient conditions for 2, and to proof that the relevance logics satisfy these conditions. This determination rests on a sharpening of the normalization technique of §25.2, to which the reader is referred for motivating remarks.

The reader is presumed familiar with the notion of a matrix M for a logic **L** (henceforth, **L**-*matrix*), and with associated standard notions. A *complete* **L**-matrix is here a triple M = ⟨M, O, D⟩, where (i) M is a non-empty set, (ii) O is the set of matrix operations (corresponding to the operations of **L**), (iii) D is a non-empty subset (of designated elements of) M closed under &I and →E and containing I(A) when I is an interpretation of **L** in M and $A \in$ **L**, and (iv) for all a $\in$ M, a $\in$ D or $\bar{a} \in$ D. A *normal* **L**-*matrix* is here a complete **L**-matrix which is also consistent, satisfying in addition to (i)–(iv), (v) for all a $\in$ M, a $\notin$ D or $\bar{a} \notin$ D.

Let M = ⟨M, O, D⟩ be a complete **L**-matrix, where O = $\{\rightarrow, \wedge, \vee, \bar{\ }\}$, and let *the normalization* M* = ⟨M*, O*, D⟩ *of* M be exactly as in §25.2.2.

LEMMA 2. Let **L** = ⟨F, O, T⟩ be a logic such that, for every complete **L**-matrix M, its normalization M* is an **L**-matrix. Then every regular complete **L**-theory has a normal **L**-subtheory — i.e., 2 holds for **L**.

PROOF. Let T′ = ⟨F, O, T′⟩ be a complete **L**-theory. We may consider T′ itself a matrix (i.e., we identify T′ with its so-called *Lindenbaum matrix*), and we note that T′ is a complete **L**-matrix. Let ⟨F*, O*, T′⟩ be the normalization of ⟨F, O, T′⟩. Consider the interpretation I of **L** in ⟨F*, O*, T′⟩ such that I(p) = p for each sentential variable p. Let T_I be the set of formulas A of **L** such that I(A) $\in$ T′; it is readily verified that ⟨F, O, T_I⟩ is an **L**-theory, since on the hypothesis of the lemma the normalization of ⟨F, O, T′⟩ is an **L**-matrix; moreover by the definition of normalization exactly one of I(B), $\overline{\text{I}(B)}$ (= $\overline{\text{I}(B)}$*) is designated, so accordingly ⟨F, O, T_I⟩ is consistent and complete.

We complete the proof of Lemma 2 by proving T_I a subtheory of T′. Let h : F*→F be as in the definition of normalization in §25.2.2, and consider the composite hI : F→F. It is readily proved by induction that hI(A) = A for all $A \in$ F, using the definitions of the operations on M* given in §25.2.1. But $A \in T_I$ iff I(A) $\in$ T′ (by definition of T_I), so hI(A) $\in$ T′ (by definition of h) and accordingly $A \in$ T′ (by the observation just made); so $T_I \subseteq$ T′, which was to be shown.

We now prove the hypothesis of Lemma 2 for the relevance logics.

LEMMA 3. Let M = ⟨M, O, D⟩ be a complete **T**-matrix. (We refer to the system **T** of §27.1.1.) Then the normalization M* = ⟨M*, O*, D⟩ of M is a **T**-matrix. If in addition M is an **E**-matrix or an **R**-matrix, so is M*.

Proof. It was proved in §25.2.2 (item 11) and §25.2.3 that Lemma 3 holds for **E**, for **R**, and for **T** under the slightly stronger hypothesis that M is prime — i.e., whenever $(a \vee b) \in D$, $a \in D$ or $b \in D$. (Alas, this hypothesis is not available to us in the present case, on pain of circularity in the proof of 1.) We employ a similar strategy here, showing that the axioms of **T** hold under all interpretations I of **T** in M*, and that D is closed in M* under *modus ponens* for $\rightarrow^*$ and adjunction for $\wedge^*$, given that M is known to be a **T**-matrix. Since no difficulties of principle arise, the task of applying the technique of §25.2.2–3 in detail may be safely left to the reader; the generality of the present result requires special attention only in that, with respect to the §25.2.2 truth partition of the underlying complete **T**-matrix M, the possibilities $a \in F$, $b \in F$, $(a \vee b) \in N$, and $a \in T$, $b \in T$, $(a \wedge b) \in N$ arise. A similar proof establishes the result for **E**-matrices, and for **R**-matrices.

We can now prove 1 for the relevant logics. Half of 1 is trivial since all instances of $A \rightarrow . A \vee B$ and $B \rightarrow . A \vee B$ are provable. Furthermore,

Theorem 1. Suppose A and B are negation-free and $\vdash_{\mathbf{R}} A \vee B$; then $\vdash_{\mathbf{R}} A$ or $\vdash_{\mathbf{R}} B$. Furthermore if negation-free $A \vee B$ is a theorem of **E** or of **T**, so is at least one of A, B.

Proof. By Lemma 1, we need only show that 2–4 hold. But 2 holds by Lemmas 2 and 3. Furthermore all forms of the De Morgan laws are easily proved in the relevant logics, so the particular law 3 holds for **E**, **R**, and **T**. Moreover, it is observed as Corollary 1 in §28.3.1 that 4 holds for **E**, **R**, and **T**, which completes the proof.

It was observed at the outset that 1 is a familiar property of the intuitionist logic **H** and of the negation-free fragment $\mathbf{S4}_+$ of **S4** (formulated with strict implication primitive). Direct proof of these facts may be had from the present result, since it is shown in §36.2.1 that **H** is an exact subsystem of the positive part of **R**, and in §36.2.4 that $\mathbf{S4}_+$ is an exact subsystem of the positive part of **E**.

We return to the topic of this section in §62.4, where development of the methods of §22.3 yields a snappier and more general argument for Theorem 1.

§28.5. Consecution formulation of positive R with co-tenability and t (by J. Michael Dunn). Let $\circ$ be the binary co-tenability connective of §27.1.4, and t the constant of §27.1.2. Let $\mathbf{R}_+$, in this section, be the system whose sentences are formed using the connectives $\circ$, $\rightarrow$, &, and $\vee$, and the constant $\boldsymbol{t}$, and whose axiom schemata and rules are those of **R** that do not involve negation (i.e., $\rightarrow$E, &I, and R1–R11 of §27.1.1) together with $\circ$1 and $\circ$2 of §27.1.4, and $\boldsymbol{t}$1 and $\boldsymbol{t}$2 of §27.1.2.

Meyer (§28.3.2) has shown, incidentally, that one may add the **R** axiom schemata involving negation to $\mathbf{R}_+$ so as to obtain a conservative extension of $\mathbf{R}_+$.

We now develop a consecution calculus $\mathbf{LR}_+$ for $\mathbf{R}_+$, [Note by principal authors: this section is a substantial revision by us of some notes by Dunn. So although the result is wholly Dunn's, mistakes in the details and in the presentation should be charged to us.]

§28.5.1. The consecution calculus $\mathbf{LR}_+$. Belnap 1960 suggested that the antecedent of a consecution be a sequence of sequences. In this section we will take nesting even more seriously by not stopping at the second level. We have in effect a sequence of sequences of . . . sequences.

And there are two kinds: *intensional* sequences (or I–*sequences*) $\mathrm{I}(\alpha_1, \ldots, \alpha_n)$ corresponding to co-tenability and *extensional* sequences (or E-*sequences*) $\mathrm{E}(\alpha_1, \ldots, \alpha_n)$ corresponding to conjunction. If we count a plain formula as both, then it would suffice to consider intensional sequences of extensional sequences of . . . , or a similar alternation beginning with an extensional sequence, and what we call below an "antecedent in normal form" can in fact be so described. But it is convenient to allow either kind of sequence to occur as a member of either kind. Hence we give the following definition of $\mathbf{LR}_+$-*antecedent*, or hereafter simply *antecedent:* Each formula (in &, $\vee$, $\circ$, $\rightarrow$, and $\boldsymbol{t}$) is an antecedent; and if $\alpha_1, \ldots, \alpha_n$ are antecedents, so are (where $n \geq 1$)

$$\mathrm{I}(\alpha_1, \ldots, \alpha_n)$$

and

$$\mathrm{E}(\alpha_1, \ldots, \alpha_n).$$

Then a *consecution* in $\mathbf{LR}_+$ has the form $\alpha \vdash A$, with α an antecedent and A a formula. (Note: α cannot be empty in $\mathbf{LR}_+$; it is the role of $\boldsymbol{t}$ to allow us to so manage things.)

We use *small Greek letters* as ranging over antecedents, and *capital Greek letters* as ranging over (possibly empty) sequences of symbols drawn from the following: the formulas, the symbols I and E, the two parentheses, and the comma. We shall use "V" as standing indifferently for I or E so as to be able to state rules common to both. And we agree that displayed parentheses are always to be taken as paired. (The reader may wish to reread these conventions.)

We now state the axioms and rules for $\mathbf{LR}_+$.

The *axioms* have the usual form:

$A \vdash A$ (Id)

The structural rules are manifold. First the familiar ones.

Permutation (CV⊢)

$$\frac{\Gamma_1 \mathrm{V}(\alpha_1, \ldots, \alpha_i, \alpha_{i+1}, \ldots, \alpha_n)\Gamma_2 \vdash A}{\Gamma_1 \mathrm{V}(\alpha_1, \ldots, \alpha_{i+1}, \alpha_i, \ldots, \alpha_n)\Gamma_2 \vdash A}(\mathrm{CV}\vdash)$$

Contraction (WV⊢)

$$\frac{\Gamma_1 \mathrm{V}(\alpha_1, \ldots, \alpha, \alpha, \ldots, \alpha_n)\Gamma_2 \vdash A}{\Gamma_1 \mathrm{V}(\alpha_1, \ldots, \alpha, \ldots, \alpha_n)\Gamma_2 \vdash A}(\mathrm{WV}\vdash)$$

Weakening (KE⊢)

$$\frac{\Gamma_1 \alpha \Gamma_2 \vdash A}{\Gamma_1 \mathrm{E}(\alpha, \beta)\Gamma_2 \vdash A}(\mathrm{KE}\vdash)$$

Note that the α_i, α, and β must be antecedents (*a fortiori* non-empty), whereas Γ_1 and Γ_2 in general will not be antecedents; e.g., Γ_2, if non-empty, will begin either with a comma or with a right parenthesis. Note further that permutation and contraction are available equally for E- and I-sequences (one has (CE⊢), (CI⊢), (WE⊢), and (WI⊢)), but that weakening is tied exclusively to E-sequences (there is no rule (KI⊢)). Because of the former fact, permutation and contraction might have been given as

$$\frac{\Gamma_1 \alpha, \beta \Gamma_2 \vdash A}{\Gamma_1 \beta, \alpha \Gamma_2 \vdash A}(\mathrm{C}\vdash) \qquad \frac{\Gamma_1 \alpha, \alpha \Gamma_2 \vdash A}{\Gamma_1 \alpha \Gamma_2 \vdash A}(\mathrm{W}\vdash)$$

for these rules are indeed a summary of CE⊢, CI⊢, WE⊢, and WI⊢. But this presentation of the system would have obscured the logical differences between the E and I versions. Finally, remark that WE⊢ is reversible by means of KE⊢; but WI⊢ is irreversible.

Now for some structural rules, peculiar to nested sequences, insuring that each antecedent is in fact equivalent to an I-sequence of E-sequences of I-sequences . . . (counting formulas as either); or perhaps an alternation starting with an E-sequence. The first pair guarantees that a unit-sequence does the same work as its member, whereas the second has the effect of dissolving I within I, or E within E.

$$\frac{\Gamma_1 \mathrm{V}(\alpha)\Gamma_2 \vdash A}{\Gamma_1 \alpha \Gamma_2 \vdash A}(\mathrm{V}_1 \text{ elim})$$

$$\frac{\Gamma_1 \alpha \Gamma_2 \vdash A}{\Gamma_1 \mathrm{V}(\alpha)\Gamma_2 \vdash A}(\mathrm{V}_1 \text{ int})$$

We remark that (E_1 int) is derivable via (KE⊢) and (WE⊢) and that (I_1 int) is probably useless.

$$\frac{\Gamma_1 V(\alpha_1, \ldots, V(\Sigma), \ldots, \alpha_n)\Gamma_2 \vdash A}{\Gamma_1 V(\alpha_1, \ldots, \Sigma, \ldots, \alpha_n)\Gamma_2 \vdash A}(V_2 \text{ elim})$$

$$\frac{\Gamma_1 V(\alpha_1, \ldots, \Sigma, \ldots, \alpha_n)\Gamma_2 \vdash A}{\Gamma_1 V(\alpha_1, \ldots, V(\Sigma), \ldots, \alpha_n)\Gamma_2 \vdash A}(V_2 \text{ int})$$

We say that a consecution is in *normal form* if none of the (reversible) rules (V_1 elim), (V_2 elim), and (WE⊢) is applicable. To allow consecutions not in normal form would appear to be only a convenience; we may if we like regard a consecution not in normal form as a mere notational variant of the consecution to which it reduces by (V_1 elim), (V_2 elim), and (WE⊢), since the result of such a reduction is effective and unique.

The *logical* rules of $\mathbf{LR}_+$ fall into the usual two groups (but there is no right rule for $\boldsymbol{t}$; it would be just $\vdash \boldsymbol{t}$ as an axiom if we allowed empty antecedents).

$$\frac{\Gamma_1 I(A, B\Sigma)\Gamma_2 \vdash C}{\Gamma_1 I((A \circ B)\Sigma)\Gamma_2 \vdash C}(\circ\vdash) \qquad \frac{\alpha \vdash A \quad \beta \vdash B}{I(\alpha, \beta) \vdash (A \circ B)}(\vdash\circ)$$

$$\frac{\Gamma_1 E(A, B\Sigma)\Gamma_2 \vdash C}{\Gamma_1 E((A \& B)\Sigma)\Gamma_2 \vdash C}(\&\vdash) \qquad \frac{\alpha \vdash A \quad \beta \vdash B}{E(\alpha, \beta) \vdash (A \& B)}(\vdash\&)$$

$$\frac{\Gamma_1 A \Gamma_2 \vdash C \quad \Gamma_1 B \Gamma_2 \vdash C}{\Gamma_1 (A \vee B)\Gamma_2 \vdash C}(\vee\vdash) \qquad \frac{\alpha \vdash A}{\alpha \vdash (A \vee B)}(\vdash\vee) \qquad \frac{\alpha \vdash B}{\alpha \vdash (A \vee B)}(\vdash\vee)$$

$$\frac{\alpha \vdash A \quad \Gamma_1 B \Gamma_2 \vdash C}{\Gamma_1 I((A \rightarrow B), \alpha)\Gamma_2 \vdash C}(\rightarrow\vdash) \qquad \frac{I(\Sigma\alpha, A) \vdash B}{I(\Sigma\alpha) \vdash (A \rightarrow B)}(\vdash\rightarrow)$$

$$\frac{\Gamma_1 \alpha \Gamma_2 \vdash C}{\Gamma_1 I(\boldsymbol{t}, \alpha)\Gamma_2 \vdash C}(\boldsymbol{t}\vdash)$$

REMARKS. 1. Note the exact correspondence between the cotenability and conjunction rules; we chose a formulation emphasizing this. 2. The α exhibited in (⊢→) insures that the rules preserve nonemptiness of antecedents. In practice one can always use ($\boldsymbol{t}$⊢) to do the job. 3. The I-and-E notation, though easy enough to talk about, is awkward to use. In actually constructing proofs we recommend

$(\alpha_1; \ldots; \alpha_n)$ for $I(\alpha_1, \ldots, \alpha_n)$,

and

$(\alpha_1, \ldots, \alpha_n)$ for $E(\alpha_1, \ldots, \alpha_n)$, allowing semicolons to dominate over commas, and being sloppy about all of the structural rules except (WI⊢) and (KE⊢), which are irreversible. And — by and large — writing down only steps in normal form.

§**28.5.2. Translation.** I-sequences are to be translated into $\mathbf{R}_+$ via cotenability and E-sequences via conjunction, as the definition of the following translation-function, T, exhibits.

$$\mathrm{T}(A) = A$$
$$\mathrm{T}(\mathrm{V}(\alpha)) = \mathrm{T}(\alpha)$$
$$\mathrm{T}(\mathrm{E}(\alpha, \Delta)) = (\mathrm{T}(\alpha)\&\mathrm{T}(\mathrm{E}(\Delta)))$$
$$\mathrm{T}(\mathrm{I}(\alpha, \Delta)) = (\mathrm{T}(\alpha)\circ\mathrm{T}(\mathrm{I}(\Delta)))$$
$$\mathrm{T}(\alpha \vdash A) = (\mathrm{T}(\alpha)\rightarrow A)$$

THEOREM 1. Part 1. If $\alpha \vdash A$ is provable in $\mathbf{LR}_+$, then $\mathrm{T}(\alpha \vdash A)$ is provable in $\mathbf{R}_+$. Part 2. If A is provable in $\mathbf{R}_+$, $\boldsymbol{t} \vdash A$ is provable in $\mathbf{LR}_+$. Accordingly, since $\vdash\boldsymbol{t}$ in $\mathbf{R}_+$, A is provable in $\mathbf{R}_+$ iff $\boldsymbol{t} \vdash A$ is provable in $\mathbf{LR}_+$.

For Part 1 we must show that we can get the effect of the rules of $\mathbf{LR}_+$ in $\mathbf{R}_+$. This relies chiefly on the following derived rules of $\mathbf{R}_+$:

$$\frac{A\rightarrow C \quad B\rightarrow D}{A\&B \rightarrow C\&D} \qquad \frac{A\rightarrow C \quad B\rightarrow D}{A\circ B \rightarrow C\circ D}$$

The theorem $(A\circ(A\rightarrow B))\rightarrow B$ of $\mathbf{R}_+$ is useful in dealing with (→⊢). Further details are left to the reader.

For Part 2 we must tediously prove $\boldsymbol{t} \vdash A$ in $\mathbf{LR}_+$ for every axiom A of $\mathbf{R}_+$, a procedure we omit; and we must show the admissibility in $\mathbf{LR}_+$ of the rules

$$\frac{\boldsymbol{t} \vdash A \quad \boldsymbol{t} \vdash B}{\boldsymbol{t} \vdash A\&B} \qquad \frac{\boldsymbol{t} \vdash A \quad \boldsymbol{t} \vdash A\rightarrow B}{\boldsymbol{t} \vdash B}$$

answering to &I and →E. The former is trivial; for the latter we must, as usual, prove an Elimination theorem, which in this case has the form of claiming the following rule as admissible:

$$\frac{\gamma \vdash M \quad \Gamma_0 M \Gamma_1 \ldots \Gamma_{n-1} M \Gamma_n \vdash D}{\Gamma_0 \gamma \Gamma_1 \ldots \Gamma_{n-1} \gamma \Gamma_n \vdash D}$$

for $n \geq 0$. To give a briefer statement of the rule we show admissible, and for other uses, we introduce some notation.

Let α and β be antecedents, and let X be a (possibly empty) set of formula-occurrences in α. Then we define

$$\alpha(\beta/\mathrm{X})$$

as the result of replacing, in α, every formula-occurrence in X by β. (Note: formalizing talk about "occurrences" is a tedious business, especially in regard to how best to reify them. The job is done with maximum elegance in Quine 1940, §56.)

Let $\alpha \vdash A$ be a consecution. Then

$$\alpha \vdash_{\mathbf{LR}_+} A$$

means that $\alpha \vdash A$ is provable in $\mathbf{LR}_+$.

Elimination theorem. Let X be a set of formula-occurrences of M in δ; then $\gamma \vdash_{\mathbf{LR}^+} M$ and $\delta \vdash_{\mathbf{LR}_+} D$ imply $\delta(\gamma/\mathrm{X}) \vdash_{\mathbf{LR}_+} D$.

§28.5.3. Regularity. To prove the theorem we subject the rules of $\mathbf{LR}_+$ to a somewhat closer analysis. We define a *rule* as a set of inferences, and an *inference* as an ordered pair consisting of a sequence (or if you like, a set) of consecutions — the premisses — as left entry and a consecution — the *conclusion* — as right entry. An inference is said to be an *instance* of each rule of which it is a member.

Let *Inf* be an inference which is an instance of some rule *Ru* of $\mathbf{LR}_+$. We define a *conclusion-parameter* to be a formula occurrence (note both words) in the conclusion of *Inf* which is *not* the newly introduced "principal formula" for one of the logical rules, nor newly introduced in β by (K$\mathrm{E}\vdash$) (see statement of the rule). We further define a *premiss-parameter* to be a formula occurrence in a premiss of *Inf* which "matches" in an obvious (but awkward to define) way a conclusion-parameter in *Inf*. (What we mean is that inspection of our statement of the primitive rules makes it perfectly clear which premiss formula occurrences match which conclusion formula occurrences, but that making a formal definition out of this would be too tedious to be quite worthwhile for our purposes.) We shall use *match* in this technical sense, allowing also that a conclusion-parameter matches itself. For every conclusion-parameter there is at least one premiss containing at least one matching premiss-parameter.

We use these ideas in order to state a property of $\mathbf{LR}_+$ useful in proving the Elimination theorem. Let *Inf* be an inference, say

$$\frac{\alpha_1 \vdash C_1,\ \ldots,\ \alpha_n \vdash C_n}{\alpha_{n+1} \vdash C_{n+1}}$$

and let X be a set of conclusion-parameters all of which occur in α_{n+1}. For $1 \leq i \leq n+1$, let X_i be the set of premiss or conclusion parameters in α_i matching at least one of those in X. For β an antecedent, define *Inf*(β/X) as

$$\frac{\alpha_1(\beta/X_1) \vdash C_1, \ \ldots, \ \alpha_n(\beta/X_n) \vdash C_n}{\alpha_{n+1}(\beta/X_{n+1}) \vdash C_{n+1}}$$

That is, *Inf*(β/X) is obtained by replacing all parameters matching those in X by β. Now say that a rule *Ru* is *left regular* if for every inference *Inf* and set X of conclusion-parameters occurring in the antecedent of the conclusion of *Inf*, if *Inf* is an instance of *Ru* then *Inf*(β/X) is also an instance of *Ru* for every antecedent β.

We need a little more pizzaz for the mate concept of "right regularity." Let *Inf* be as displayed above. Let D be a formula, let β be an antecedent, and let X be the unit (note well) set of some formula occurrence in β. Define S_i as either $\beta(\alpha_i/X) \vdash D$ or $\alpha_i \vdash C_i$ according as C_i does or does not match the conclusion parameter C_{r+1}. Then *Inf*(D, β, X) is defined as just

$$\frac{S_1, \ \ldots, \ S_n}{S_{n+1}}$$

That is, *Inf*(D, β, X) is obtained by systematically replacing parameters matching those in the consequent of the conclusion by D, and simultaneously embedding antecedents (of consecutions containing matches of that parameter) in a larger context represented by an occurrence of an (arbitrary) formula occurrence in β.

Then say a rule *Ru* is *right regular* if whenever it has *Inf* as an instance, it has *Inf*(D, β, X) as well as for all D, β, and X a unit set of a formula occurrence in β.

Now call a rule *regular* if it is both left and right regular. Let us nevertheless state the regularity of the rules of **LR**$_+$ as a consequence of two lemmas.

LEFT REGULARITY LEMMA. Every rule of **LR**$_+$ is left regular.

RIGHT REGULARITY LEMMA. Every rule of **LR**$_+$ is right regular.

Proof is by inspection of the rules.

§28.5.4. Elimination theorem. First the concept of rank. Let *Der* be a derivation (in tree form); let S be the final consecution in *Der;* unless S is an axiom, let *Inf* be the inference in *Der* of which S is conclusion; and let X be a set of formula occurrences in S. Now define *the rank of* X *in Der* as follows. If X is empty, its rank is 0. If X is non-empty and contains only formula-occurrences either newly introduced by (KE$\vdash$) or by a logical rule, or if S is an axiom, so that X contains no conclusion-parameters, the rank of X is 1. Otherwise, let $S_1, \ldots, S_n$ be the premisses of *Inf*, and for each i let *Der*$_i$ be the subderivation of *Der* terminating in S_i, and let X_i be the (possibly empty) set of premiss-parameters in S_i which match at least one member of X. (At least one X_i will be non-empty). Let i be such that the rank of X_i in Der_i is at least as great as the rank of any X_j in Der_j. Then the rank of X in *Der* is defined as one greater than the rank of X_i in Der_i.

We write

$$\gamma \vdash_{\mathbf{LR}_+} M \text{ with rank}_M = k$$

to mean that $\gamma \vdash M$ is provable in a derivation in $\mathbf{LR}_+$ such that the rank of the set consisting of the displayed occurrence of M in that derivation is k. And we write

$$\delta \vdash_{\mathbf{LR}_+} D \text{ with rank}_{\mathrm{X}} = j$$

to mean that X is a set of occurrences in δ, and $\delta \vdash D$ is provable in a derivation such that the rank of X in that derivation is j.

To prove the Elimination theorem as stated at the end of §28.5.2, it suffices for us to show that for all M, k, j, γ, δ, X, and D, if $\gamma \vdash_{\mathbf{LR}_+} M$ with $\text{rank}_M = k$, and $\delta \vdash_{\mathbf{LR}_+} D$ with $\text{rank}_{\mathrm{X}} = j$, where X is a set of occurrences of M in δ, then $\delta(\gamma/\mathrm{X}) \vdash_{\mathbf{LR}_+} D$. The argument is by a nested induction.

First we choose an arbitrary M and assume as *Outer hypothesis* that the theorem true for all M' shorter than M; i.e., we assume that for all M' shorter than M, for all k, j, γ, δ, X, and D, if $\gamma \vdash_{\mathbf{LR}_+} M'$ with $\text{rank}_{M'} = k$ and $\delta \vdash_{\mathbf{LR}_+} D$ with $\text{rank}_{\mathrm{X}} = j$, where X is a set of occurrences of M' in δ, then $\delta(\gamma/\mathrm{X}) \vdash_{\mathbf{LR}_+} D$.

Next we choose arbitrary k and j and assume as *Inner hypothesis* that the theorem is true for M and for all k', j' whose sum is less than that of k and j; i.e., we assume that for all k', j' with sum less than that of k and j, and for all γ, δ, X and D, if $\gamma \vdash_{\mathbf{LR}_+} M$ with $\text{rank}_M = k'$ and $\delta \vdash_{\mathbf{LR}_+} D$ with $\text{rank}_{\mathrm{X}} = j'$, where X is a set of occurrences of M in δ, then $\delta(\gamma/\mathrm{X}) \vdash_{\mathbf{LR}_+} D$.

Finally, we choose arbitrary γ, δ, X, and D, and suppose as *Step hypothesis* that $\gamma \vdash_{\mathbf{LR}_+} M$ with $\text{rank}_M = k$ and $\delta \vdash_{\mathbf{LR}_+} D$ with $\text{rank}_{\mathrm{X}} = j$, where X is a set of occurrences of M in δ.

If under these hypotheses we can succeed in showing that $\delta(\gamma/\mathrm{X}) \vdash_{\mathbf{LR}_+} D$, then we shall be done. For convenience we define L, R, and C:

$$L\colon \gamma \vdash M \qquad R\colon \delta \vdash D$$
$$C\colon \delta(\gamma/\mathrm{X}) \vdash D$$

CASE 1. $k = 1$ and L is an axiom, or $j = 0$, or $j = 1$ and R is an axiom. Then either $C = R$ or $C = L$, so that the Step hypothesis suffices.

CASE 2. $j = 1$ and the derivation leading to R terminates in

$$R\colon \frac{\Gamma_1\alpha\Gamma_2 \vdash D}{\Gamma_1 \mathrm{E}(\alpha, \beta)\Gamma_2 \vdash D}(\mathrm{KE}\vdash)$$

Then all occurrences of M in X must lie in β. Then $C = \Gamma_1(\alpha, \beta(\gamma/X))\Gamma_2 \vdash D$ comes from the same premiss by (KE$\vdash$).

CASE 3. $j \geq 2$. Let the derivation terminate in R and in an inference *Inf* which is an instance of a rule *Ru*. Let X′ be the set of occurrences of M in X which are conclusion-parameters in *Inf* (so that X–X′ contains those occurrences just introduced by *Inf*). By the Inner hypothesis, every premiss of *Inf*(γ/X') is provable. But by Left regularity, *Inf*(γ/X') is an instance of *Ru*, so that $\delta(\gamma/X') \vdash_{LR_+} D$. If X–X′ is empty then $\delta(\gamma/X) \vdash_{LR_+} D$ as desired. Otherwise, let X_0 be the set of occurrences of M in $\delta(\gamma/X')$ which correspond to the occurrences of M in X–X′. Evidently all these occurrences were introduced by *Inf*(γ/X'), so that $\delta(\gamma/X') \vdash_{LR_+} D$ with $\text{rank}_{X_0} = 1$. So we may again use the Inner hypothesis (1 being less than $2 \leq j$) to obtain $(\delta(\gamma/X'))(\gamma/X_0) \vdash D$, i.e., $\delta(\gamma/X) \vdash D$.

CASE 4. $k \geq 2$ and $j = 1$, but R is neither an axiom nor by (KE$\vdash$). So R is by a logical rule, and X is the unit (note well) set containing the formula occurrence newly untroduced into δ. Let the derivation terminating in L end in an inference *Inf* which is an instance of a rule *Ru*. By the Inner hypothesis applied just to those premisses of *Inf* containing a parameter matching the M displayed in L, one obtains the provability of every premiss of *Inf*(D, δ, X). By Right regularity *Inf*(D, δ, X) is an instance of *Ru*; so C, as required.

CASE 5. $k = 1$ and $j = 1$, but R is neither an axiom nor by (KE$\vdash$), nor is L an axiom. So R is by a logical rule and X is the unit set containing the formula occurrence newly introduced into δ, and the displayed M in L is also newly introduced by a logical rule. The rules yielding L and R will evidently match.

5.1. $M = (A \circ B)$ or $(A\&B)$, with the ends of the derivations leading to L and R being as follows, and with $\cdot$ being $\circ$ or & according as V is I or E.

$$L\colon \frac{\alpha \vdash A \quad \beta \vdash B}{V(\alpha, \beta) \vdash (A \cdot B)} \qquad R\colon \frac{\Gamma_1 V(A, B\Sigma)\Gamma_2 \vdash D}{\Gamma_1 V((A \cdot B)\Sigma)\Gamma_2 \vdash D}$$

By two uses of the Outer hypothesis we obtain

$$\Gamma_1 V(\alpha, \beta\Sigma)\Gamma_2 \vdash D$$

and then

$$C\colon \Gamma_1 V(V(\alpha, \beta)\Sigma)\Gamma_2 \vdash D$$

by (V_2 int).

5.2. M $= (A \vee B)$, with ends of the L and R derivations being

$$L\colon \frac{\alpha \vdash A}{\alpha \vdash (A \vee B)} \qquad R\colon \frac{\Gamma_1 A \Gamma_2 \vdash D \qquad \Gamma_1 B \Gamma_2 \vdash D}{\Gamma_1 (A \vee B) \Gamma_2 \vdash D}$$

Or the premiss for L might be $\alpha \vdash B$. In either case we use the Outer hypothesis to obtain C.

5.3. $M = (A \rightarrow B)$, with ends of the L and R derivations being

$$L\colon \frac{\mathrm{I}(\Sigma\beta, A) \vdash B}{\mathrm{I}(\Sigma\beta) \vdash (A \rightarrow B)} \qquad R\colon \frac{\alpha \vdash A \qquad \Gamma_1 B \Gamma_2 \vdash D}{\Gamma_1 \mathrm{I}((A \rightarrow B), \alpha)\Gamma_2 \vdash D}$$

One use of the Outer hypothesis yields

$$\mathrm{I}(\Sigma\beta, \alpha) \vdash B$$

and another gives

$$\Gamma_1 \mathrm{I}(\Sigma\beta, \alpha)\Gamma_2 \vdash D$$

Then (I_2 int) yields

$$C\colon \Gamma_1 \mathrm{I}(\mathrm{I}(\Sigma\beta), \alpha)\Gamma_2 \vdash D.$$

We remark that $M = \boldsymbol{t}$ does not arise as a special case; for given $\gamma \vdash \boldsymbol{t}$, either we have an axiom (Case 1) or $\boldsymbol{t}$ is parametric (Case 3).

This completes the proof of the Elimination theorem, and of the equivalence of $\mathbf{LR}_+$ with $\mathbf{R}_+$ under the stated translation. Alas, we do not know how to convert this equivalence into a decision procedure.

Here is an example of a derivation suggesting the power of $\mathbf{LR}_+$. We use the notational convention mentioned at the end of §28.5.1 (semicolons intensional, commas extensional).

$$\cfrac{A \vdash A \qquad \cfrac{A \vdash A \qquad \cfrac{\cfrac{\cfrac{B \vdash B \qquad C \vdash C}{B, C \vdash B\&C}(\vdash\&)}{B, C \vdash (B\&C)\vee(B\&D)}(\vdash\vee) \qquad \cfrac{\cfrac{B \vdash B \qquad D \vdash D}{B, D \vdash B\&D}(\vdash\&)}{B, D \vdash (B\&C)\vee(B\&D)}(\vdash\vee)}{B, C\vee D \vdash (B\&C)\vee(B\&D)}(\vee\vdash)}{B, (A\rightarrow C\vee D; A) \vdash (B\&C)\vee(B\&D)}(\rightarrow\vdash)}{(A\rightarrow B; A), (A\rightarrow C\vee D; A) \vdash (B\&C)\vee(B\&D)}(\rightarrow\vdash)$$

$$\cfrac{\cfrac{(A\rightarrow B; A), (A\rightarrow C\vee D; A) \vdash (B\&C)\vee(B\&D)}{((A\rightarrow B, A\rightarrow C\vee D; A), ((A\rightarrow B, A\rightarrow C\vee D); A) \vdash (B\&C)\vee(B\&D)}\text{(KE}\vdash\text{) twice and (CE}\vdash\text{)}}{(A\rightarrow B, A\rightarrow C\vee D); A \vdash (B\&C)\vee(B\&D)}(\text{WE}\vdash)$$

$$\cfrac{\cfrac{\cfrac{\cfrac{(A\rightarrow B, A\rightarrow C\vee D); A \vdash (B\&C)\vee(B\&D)}{(A\rightarrow B)\&(A\rightarrow C\vee D); A \vdash (B\&C)\vee(B\&D)}(\&\vdash)}{(A\rightarrow B)\&(A\rightarrow C\vee D) \vdash A\rightarrow.(B\&C)\vee(B\&D)}(\vdash\rightarrow)}{\boldsymbol{t}; (A\rightarrow B)\&(A\rightarrow C\vee D) \vdash A\rightarrow.(B\&C)\vee(B\&D)}(\boldsymbol{t}\vdash)}{\boldsymbol{t} \vdash ((A\rightarrow B)\&(A\rightarrow C\vee D))\rightarrow.A\rightarrow.(B\&C)\vee(B\&D)}(\vdash\rightarrow)$$

Note the critical use of (WE⊢); although the conclusion antecedent is an I-sequence of E-sequences, its proof requires an additional level: an E-sequence of I-sequences of E-sequences. The question of decidability seems to reduce to finding a bound on the number of levels we must go up (relative to a given candidate consecution).

§29. Miscellany. This section is analogous to §8; see the Analytical table of contents for a list of topics treated.

§29.1. Goble's modal extension of R. Goble 1971 considers the following question (which we put not in his words, but in a way which we hope appreciates his intent): "What would happen if the earth suddenly stopped revolving about the sun?" He anticipates an answer of T. C. Mits (Lieber and Lieber 1944) as being something like, "God only knows; *anything* might happen." Erected into a principle, the reply might reflect a feeling that "If A is true, then if A were not true, then anything B would be possible," which Goble puts as the principle (P):

$$A \rightarrow . \sim A \rightarrow \Diamond B,$$

for a suitable sense of the arrow.

After canvassing several senses of "if . . . then — " which might make the principle plausible (the "material sense" makes it trivial, and the Lewis senses destroy modality), he observes that what is required is a sense of the arrow which is (a) paradox-free, and (b) non-apodictic: to wit, that of **R**.

§29.1.1. The system G. He therefore considers an extension **G** of **R** which is characterized by the axioms of **R**$^{\Box}$ as in §27.1.3 (with or without □4 $\Box A \rightarrow \Box\Box A$; this has to do with modality reduction, and is inessential to Goble's enterprise), together with the axiom

$$A \rightarrow . \Box B \rightarrow A,$$

which, in **R**$^{\Box}$, sneaks subtly between modal fallacies and fallacies of relevance. The resulting system is shown to be reasonably stable (in the sense that the fragment lacking conjunction and disjunction has consecution and natural deduction formulations), and a number of interesting observations are made (e.g. that addition of the **S5** axiom $\sim\Box A \rightarrow \Box\sim\Box A$ reduces the arrow to absurdity, i.e., material "implication").

§29.1.2. Dunn's translation of G into R$^{\Box}$. As was pointed out to us by Dunn in April, 1969, Goble's extension **G** is already available in **R**$^{\Box}$, in the sense that a definable connective will do the same work. We consider the result $\mathbf{R}^{\Box t}$ of adding t to $\mathbf{R}^{\Box}$ as in §27.1.2, with the feature (there discussed) that

$$A\rightleftarrows.t\rightarrow A.$$

Then define

$$\mathrm{L}A =_{\mathrm{df}} \Box A\&t.$$

An inductive argument then shows that L then behaves exactly as does $\Box$ in Goble's extension of $\mathbf{R}^{\Box}$. If L replaces $\Box$ in the axioms and rules of **G**, the results are all theorems and rules of $\mathbf{R}^{\Box}$, so **G** is a subsystem of $\mathbf{R}^{\Box t}$ on this interpretation. And if A is a formula of **G**, and A' is its translation in $\mathbf{R}^{\Box t}$, then if A' is a theorem of $\mathbf{R}^{\Box t}$, then A is a theorem of **G**, since

$$\mathrm{L}A\rightleftarrows.\mathrm{L}A\&t$$

is a theorem of **G** (using Goble's additional axiom of the previous subsection).

§29.2. The bounds of finitude. Meyer 1970b shows that among the pure arrow formulas of **R** containing only one variable, say p, exactly six are non-equivalent. Temporarily define $M = p\rightarrow.p\rightarrow p$; then the six, with the strongest at the bottom, may be pictured as follows:

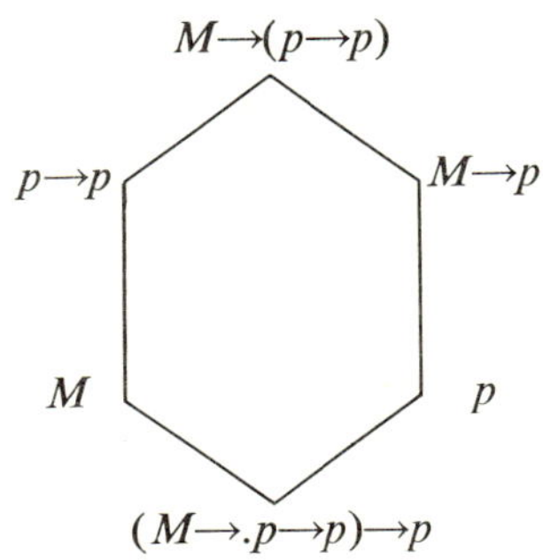

But that is the end of finitude. For Meyer 1970b also shows that non-equivalence in **R** becomes infinite if we go either to pure arrow formulas with two variables, or to arrow negation formulas with only one variable. The following sequences are given as examples, where in the latter case $A\circ B = \overline{A\rightarrow\overline{B}}$ in the arrow negation part of **R**: p, $p\rightarrow q$, $p\rightarrow.p\rightarrow q$, . . . ; p, $p\circ p$, $p\circ(p\circ p)$, ... Meyer's proof of non-equivalence involves a truly Leibnizian idea (see Leibniz 1679, or Łukasiewicz 1951) in which irrelevance is represented by relative primeness of positive integers. He remarks "that those who have argued that it is impossible to be formally precise about relevance seem to have overlooked this simple and natural realization of a logic which was framed for just that purpose." So *calculemus!*

PROBLEM. Is there in **E** an infinitude of pairwise non-equivalent formulas in a single variable?

§29.3. Sugihara is a characteristic matrix for RM (by Robert K. Meyer). The object of the present section will be to present the system **RM** of §27.1.1 as a system intermediate between the paradox-free and the classical logics, which nevertheless lacks the more noisome paradoxes. (See also §§8.15, 14.7 and 27.2.) We then show that **RM** has a simple, and in a certain sense a natural, interpretation. (The casual reader, after getting clear on the interpretation early in §29.3.2, may skip to §29.3.3 for some easy but slightly wacky facts.) The interpretation in fact antedates the axiomatic system, having been supplied by Sugihara 1955 for other but related purposes, namely the rejection of certain non-theorems from his own system of strict implication, as explained in §26.9.

On the whole, I agree with the principal authors on matters logical. But there are points of difference, and the worth of **RM** is one of them; they'll have their say in §29.5; here, I have mine. Still, the paper from which this section is taken is several years old, and I should put some things differently today. (Rewriting it is out of the question, however, since I've been threatened with maximum mutilation by the principal authors if I further delay publication of this book.) So some prefatory comments are in order.

I like Mingle, as the reader will discover, for two reasons. (There is tension between them.) First, its theory of deduction, due to Dunn and McCall both in motivation and formulation, simplifies the Church theory, set out in §3, in a reasonable way. Second, it is useful for many of the things that **R** might be good for, while being much more easily visualized. On the other hand, as noted in §8, this involves the breakdown of the relevance principle, undermining the *raison d'etre* of the enterprise.

As I argue, sometimes one doesn't need the whole relevance principle, and, on these occasions, **RM** is good enough, when some relevance is desirable. Indeed, it has proved very useful by illuminating in a context that is technically simple some of the relevance features shared among **RM**, **R**, and **E**. Nevertheless, one must confess that the system happened by accident; nobody thought that a mild modification of **R** would produce a system this strong.

Accordingly, it now seems to me that the two poles of motivation must be viewed as conflicting. Despite my affection for **RM** (I like its initials, especially), I agree in the end with the principal authors; all things considered, **R** is the superior system. But the breakdowns that afflicted **RM** rested on **R**-style negation, which on the semantical viewpoint of §48 ff. is not as transparent as the other truth-functional connectives. Accordingly, further pursuit of the original Dunn-McCall insights, dropping the **R**-style negation and perhaps proceeding as in §56, appears an interesting present alternative.

§29.3.1. Development and comparison of RM and R. The system **R** of relevant implication may be axiomatized on the basis of the ten axioms R1–R7 and R11–R13 of §27.1.1, if we allow ourselves to introduce disjunction by definition. For convenience we renumber these as R1–R10 *for this section only*, and continue by giving the rules as

R11 From A and $A\rightarrow B$, infer B ($\rightarrow$E)
R12 From A and B, infer AB (&I)

(We often use AB for $(A\&B)$.) Introduced by familiar definition are:

R13 $A\vee B =_{\text{df}} \overline{\overline{A}\&\overline{B}}$
R14 $A\rightleftarrows B =_{\text{df}} (A\rightarrow B)(B\rightarrow A)$
R15 $A\supset B =_{\text{df}} \overline{A}\vee B$
R16 $A\equiv B =_{\text{df}} (A\supset B)(B\supset A)$

Also useful are intensional disjunction and cotenability as in §27.1.4.

R17 $A+B =_{\text{df}} \overline{A}\rightarrow B$
R18 $A\circ B =_{\text{df}} \overline{\overline{A}+\overline{B}}$

(We warn the reader that in what follows "+" is sometimes used in the sense defined by R17, and sometimes in its ordinary arithmetical sense; the context should suffice to make the usage clear.)

RM is obtained from **R** by adding the single axiom scheme

RM0 $A\rightarrow .A\rightarrow A$ (expansion)

to axioms and rules R1–R12 and accompanying definitions.

The heart of **R** is, as we have noted, the weak theory of implication $\mathbf{R}_{\rightarrow}$ determined by axioms and rules R1–R4, R11. In motivating $\mathbf{R}_{\rightarrow}$, the essential consideration of Church 1951 was that

$A\rightarrow .B\rightarrow C$

is to count as a theorem of that system if there is a derivation of C from the sequence A, B in which both A and B are *used* (see §3).

On this criterion RM0 would be motivated if there were a derivation from

(i) A, A

of

(ii) A

in which both members of (i) are *used.*

The question, it is easy to see, boils down to the following: if we simply write down (ii), have we used each member of the sequence (i)? If so, RM0 is Church-motivated; if not, RM0 is not Church-motivated.

Since RM0 is demonstrably not a theorem of the Church system $\mathbf{R}_{\rightarrow}$, or of its extension **R**, it is clear that on Church and §3 criteria, we have not used both members of (i) in deriving (ii). The reasoning seems to be that, inasmuch as a sequence is characterized by its order, to use its first member is not to use its second, even if for other purposes we want to consider members of the sequence identical.

But though the reasoning is clear and in fact produces a very interesting system, it is easy to view it also as somewhat exasperating. For in the first place, it appears rather more natural to think of premisses as coming in sets rather than in sequences, at least where the esoteric considerations sometimes invoked by modal logicians do not intervene. And indeed, the permutation theorem $(A\rightarrow.B\rightarrow C)\rightarrow.B\rightarrow.A\rightarrow C$ guarantees that changing the order of a sequence of premisses is itself of no effect; what counts is the number of times each individual formula turns up in the sequence. But it seems strange to suppose that an otherwise valid argument should become invalid because we assumed its premiss too many times, though this seems precisely the sort of conclusion toward which a serious attempt to base logic on $\mathbf{R}_{\rightarrow}$ would lead.

In fact, there are other systematic reasons, as we shall soon see, which might lead one to prefer the Church criterion to the slightly more liberal criterion that we shall now introduce to motivate RM0. Let $\mathbf{RM0}_{\rightarrow}$ be the smallest set of sentences (in which no connective occurs but the arrow) which is closed under R11 and which contains $A_1\rightarrow(A_2\rightarrow \ldots (A_n\rightarrow B) \ldots)$, $n \geq 1$, whenever there is a derivation of B from $A_1, \ldots, A_n$ in which every member of the *set* $\{A_1, \ldots, A_n\}$ is used in an application of →E. It is then easy to show, following Church's arguments in principle, that $\mathbf{RM0}_{\rightarrow}$ is precisely the system determined by RM0, R1–R4, and R11. (But note that it is now known that $\mathbf{RM0}_{\rightarrow}$ is *not* the pure implicational fragment of **RM**; see §8.15.)

It might be good to linger here, for $\mathbf{RM0}_{\rightarrow}$ is on its own hook an interesting system. Let us, however, hasten on to consider the full system **RM**. We begin by noting some conspicuous theorems, some of which are surprising. (Numerous persons besides the author, in particular Ackermann, Anderson,

Belnap, Church, and Dunn, are responsible for the proof of one or another of these theorems.)

R19. Suppose A contains none but the truth functional connectives &, $\vee$, $\supset$, and $\equiv$. Then A is a theorem of **RM** iff A is a two-valued tautology.

R20	$A \rightleftarrows \bar{\bar{A}}$	Double negation
R21	$A \rightarrow B \rightleftarrows . \bar{B} \rightarrow \bar{A}$	Contraposition
R22	$A \rightarrow \bar{A} \rightarrow \bar{A}$	*Reductio*
R23	$(A \rightarrow . B \rightarrow C) \rightarrow . B \rightarrow . A \rightarrow C$	Permutation
R24	$A \rightarrow A \rightarrow B \rightarrow B$	Specialized assertion
R25	$B \rightarrow C \rightarrow . A \rightarrow B \rightarrow . A \rightarrow C$	Prefixing
R26	$A(A \vee B) \rightleftarrows A$	Absorption
R27	$(A \vee AB) \rightleftarrows A$	Absorption
R28	$A(B \vee C) \rightleftarrows . AB \vee . AC$	Distribution
R29	$(A \vee . BC) \rightleftarrows . (A \vee B)(A \vee C)$	Distribution
R30	$A \rightarrow (B \rightarrow C) \rightleftarrows . A \circ B \rightarrow C$	Portation
R31	$(A \circ B \rightarrow C) \rightarrow . AB \rightarrow C$	& Importation
R32	$A(BC) \rightleftarrows . (AB)C$	Assoc &
R33	$A \vee (B \vee C) \rightleftarrows . (A \vee B) \vee C$	Assoc $\vee$
R34	$A \circ (B \circ C) \rightleftarrows . (A \circ B) \circ C$	Assoc $\circ$
R35	$A + (B + C) = (A + B) + C$	Assoc $+$
R36	$AB \rightleftarrows . BA$	Comm &
R37	$A \vee B \rightleftarrows . B \vee A$	Comm $\vee$
R38	$A \circ B \rightleftarrows . B \circ A$	Comm $\circ$
R39	$A + B \rightleftarrows . B + A$	Comm $+$
R40	$(A + . BC) \rightleftarrows . (A + B)(A + C)$	$+$& Distribution
R41	$A \circ (B \vee C) \rightleftarrows . (A \circ B) \vee (A \circ C)$	$\circ\vee$ Distribution
R42	$\bar{A} \& \bar{B} \rightleftarrows . \overline{A \vee B}$	De Morgan
R43	$\bar{A} \vee \bar{B} \rightleftarrows . \overline{A \& B}$	De Morgan
R44	$\bar{A} \circ \bar{B} \rightleftarrows . \overline{A + B}$	De Morgan
R45	$\bar{A} + \bar{B} \rightleftarrows . \overline{A \circ B}$	De Morgan
R46	$A \rightarrow . A \vee B$	$\vee$I
R47	$B \rightarrow . A \vee B$	$\vee$I
R48	$(A \rightarrow C)(B \rightarrow C) \rightleftarrows . A \vee B \rightarrow C$	$\vee$E
R49	$(A \rightarrow B)(A \rightarrow C) \rightleftarrows . A \rightarrow . BC$	&I
R50	$A \rightarrow B \rightleftarrows . \bar{A} + B$	
R51	$A \rightarrow B \rightleftarrows . \overline{A \circ \bar{B}}$	
R52	$(A \rightleftarrows B) \rightleftarrows . (A \rightarrow B) \circ (B \rightarrow A)$	
R53	$(A \rightarrow B) \rightarrow \bar{A} \vee B$	
R54	$(A + B) \rightarrow . A \vee B$	

R55	$AB \rightarrow (A \circ B)$	
R56	$A \rightleftarrows . AA$	Id &
R57	$A \rightleftarrows . A \vee A$	Id $\vee$
R58	$A \rightarrow . A \circ A$	
R59	$A + A \rightarrow A$	

All of the above are theorems of **R** as well as **RM**. The following are theorems of **RM** but not of **R**.

RM60	$A + A \rightleftarrows A$	Id +
RM61	$A \circ A \rightleftarrows A$	Id $\circ$
RM62	$A + \bar{A} + B + \bar{B}$	
RM63	$(A \rightarrow B) + (B \rightarrow A)$	Simple order
RM64	$(A \rightarrow B) \vee (B \rightarrow A)$	Simple order
RM65	$A + \bar{A} \rightleftarrows . A \vee \bar{A}$	
RM66	$A \circ \bar{A} \rightleftarrows . A\bar{A}$	
RM67	$A \rightarrow A \rightleftarrows . A \supset A$	
RM68	$(A \rightarrow . B \vee C) \rightleftarrows . (A \rightarrow B) \vee (A \rightarrow C)$	
RM69	$A + (B \vee C) \rightleftarrows . (A + B) \vee (A + C)$	$+\vee$ Distribution
RM70	$A \circ (BC) \rightleftarrows . (A \circ B)(A \circ C)$	$\circ$& Distribution
RM71	$A \circ B \rightarrow . A + B$	

A few comments are in order with respect to the effect of our "minor" liberalization of the Church relevance criterion. RM60 expresses a welcome idempotence for our suggested "intensional" disjunction. (On the principle that Quine calls minimum mutilation, we suggest that if a non-truth-functional disjunction is wanted, that defined in **RM** has a great deal to offer.) RM61 does the same for $\circ$. RM62 is the ultimate source of most of the surprises in **RM**, so that it is fitting here to give rather specific hints toward its proof.

1	$(A + \bar{A})(B + \bar{B})$	Identity R17 Adjunction
2	$(A + \bar{A})(B + \bar{B}) + (A + \bar{A})(B + \bar{B})$	1 Id +
3	$(A + \bar{A})(B + \bar{B}) + (B + \bar{B})$	2 +& Distribution &E
4	$(A + \bar{A}) + (B + \bar{B})$	3 Comm+ +& Dist. &E
5	RM62	4 Assoc+

RM62 is perhaps a little surprising, but it really becomes shocking when, by applying the commutativity and associativity of + and the pseudo-definition R50 of $\rightarrow$, we transform it into RM63, and then go on to trade in RM63 for RM64 using R54.

RM63 and RM64, in fact, decide that **RM** represents a much longer step in the direction of classical logic (and, for that matter, in the direction of an extensional approach to sentential logic) than one would have thought from the heuristic considerations by which we motivated its axioms and rules. For it suggests that under normal semantical conditions the sentences of a *complete* theory T may be *simply ordered* by the strong implication relation formalized by **RM**. This point will figure heavily in future developments.

We call this an extensional step for the following reasons: 1. There is no reason to suppose a connection in *meaning* between two arbitrary sentences of a given complete theory. 2. Those logics which one thinks of as motivated by an insistence on meaning connections (in some sense) — e.g., the Lewis systems, intuitionist logic, the paradox-free logics — uniformly reject RM64. 3. On the other hand, well-known attempts to generalize classical logic by extending the truth functional point of view — e.g., the many-valued logics of Łukasiewicz defined in Łukasiewicz and Tarski 1930 — hold RM64 to be valid; *a fortiori*, it is classically valid.

RM64 also draws us in the classical direction in another respect, namely it shows that **RM** is not a conservative extension of the system determined by its positive axioms (following §27.1.1 by introducing R46–R48 as axioms for "$\vee$," revoking R13). For all the positive axioms and the rules are intuitionistically valid, while RM64 is not. Similarly, when a reasonable deduction-theoretic approach is taken to classical logic (as e.g. in Kleene 1952, p. 82), one must appeal to negation to prove theorems like RM64.

RM64 leaves shattered in the dust much of the motivation to which previous opponents of the paradoxes have appealed. But this just goes to show that one can have many reasons for disliking the paradoxes; one very plausible ground for disliking them is that they turn every minor inconsistency into a catastrophe. From this charge, **RM** is yet free. If in other respects it moves in the direction of classical logic, there is as yet no reason to rue that fact.

Let us spell this out a little more carefully. So far as either *usage* or *intuition* is concerned, it seems to us unreasonable to hold, either with the classicist or with **RM**, that for two *normally* interpreted sentences A and B either A implies B or B implies A, even if we follow **Principia** in denying the synonymy of implication and entailment. But it is not unreasonable to abstract, as the classical logician does, from the muddles of language and thought in search of a well-behaved characterization of implication; it becomes unreasonable only when the characterization becomes itself so well-behaved that it interferes with certain uses of logic. Such is the case, we hold, when falsehoods with which our quarrels are minor are held to imply falsehoods with which we cannot live under any circumstances.

A simple remedy, in this case, is to *grade* our sentences, specifying that a sentence of grade $|k|$ shall be, if true, of grade $+|k|$; if false, of grade

$-|k|$. A simple criterion of (material) consequence is then the following: all sentences of grade $\geq k$ are consequences of a sentence of grade k.

This in fact is the classical situation, except that the only grades considered are $+1$ and -1. On the old claim that a man (or a theory) asserting a sentence is committed to its consequences, *logical or non-logical*, this makes most of us (and most of our theories) even more stupid than we are. On the other hand, the grading device commits us to nothing more outrageous than the most outrageous falsehood we hold to be the case. Accordingly, the man who "makes one mistake when the moon is bright" is not committed, in a theory based on **RM**, to "make it every night" unless it was a particularly outrageous mistake. For some, this offers hope.

No such grading device appears to be *explicit* in language or thought. Neither, we add, are temperatures of 38.73°C explicit in our intuitive notions of heat and cold. But the possibility of the sharper distinction is there in the one case as in the other, for the qualitative distinctions we make between a slight chill and biting cold are paralleled in the qualitative distinctions we make between trivial and serious error. Logic, no less than physics, has the task of making explicit what is already implicit in the areas where it is applied.

RM64 goes far toward spelling out an intuitive semantics for **RM**; the semantics in fact antedate the system, as has been noted. But the Sugihara matrix, which was merely plausible for Sugihara's own system, turns out to be characteristic for **RM**. We are jumping the gun, however, because the semantics turns out to be a pleasant by-product of the choice of natural axioms and rules to govern the strong conditional, just as Post's completeness proof for the classical sentential calculus was a pleasant by-product of the natural (from their point of view) axioms and rules chosen by Russell, Whitehead, and their predecessors. Let us therefore spend a little more time on the deductive consequences of the axioms.

RM65–RM67 show that, although the truth functional connectives must in general be distinguished from their non-truth-functional analogues, in the particular case of excluded middles, contradictions, and identities we may equivalently make either choice. RM68–RM70 state additional laws of distribution not provable in **R**; in conjunction with R28–R29 and R40–R41, these principles make possible conjunctive and disjunctive normal forms in which (i) negation is confined to sentential variables, (ii) no truth functional connective occurs within the scope of a non-truth-functional connective, and (iii) no connectives occur but $\sim$, $\circ$, $+$, $\vee$, and &. Finally, RM71 shows, together with R54 and R55, that the connectives can be ranked; an item on the left on the following list implies all items to its right, but not necessarily conversely: $A\&B$, $A\circ B$, $A+B$, $A\vee B$.

§29.3.2. Syntactic and semantic completeness of RM. In this section, we provide both syntactic and semantic completeness proofs for **RM**. We begin on the semantic side. Let I be any set of integers which includes $-j$ whenever it includes j. The Sugihara matrix S_I shall be the septuple

$$\langle I, \wedge, \oplus, -, \rightarrow, \circ, \vee \rangle$$

where

i. I is as above;

ii. The *extensional sequence* of integers I shall be the integers arranged in natural order, and where for arbitrary members m, n of I,

a. $m \wedge n$ is the minimum of m, n in the extensional sequence;
b. $m \vee n$ is the maximum of m, n in the extensional sequence;
c. $-m$ is the negative of m.

iii. The *intensional sequence* of integers I shall be the integers arranged in order of increasing absolute value, with negative integers preceding positive integers of the same absolute value, and where for arbitrary members m, n of I,

a. $m \oplus n$ is (not the sum but) the maximum of m, n in the intensional sequence;
b. $m \circ n$ is $-(-m \oplus -n)$;
c. $m \rightarrow n$ is $-m \oplus n$.

($m \rightarrow n$ has the following alternative definition due to Sugihara 1955, who did not introduce $\oplus$; if $m \leq n$, $m \rightarrow n$ is $-m \vee n$; if $m > n$, $m \rightarrow n$ is $-m \wedge n$. We note also that $m \circ n$ is *not* the minimum of m, n in the intensional sequence.)

A special part in the ensuing developments will be played by certain finite Sugihara matrices S_I, where I is the set of all *non-zero* integers whose absolute value is less than or equal to a given positive integer i; such matrices shall be called *normal* (i.e. $0 \notin$ I) *finite Sugihara matrices* and shall be noted by S_i, where i is as above.

Intuitively, the points of a normal Sugihara matrix correspond to the graded truth values discussed above. $\vee$, $\wedge$, and $-$ will be linked, respectively, to the *extensional* connectives $\vee$, &, and $\sim$, and they yield about what one would expect. The truth value of the (ordinary) disjunction of two sentences will be that of the "truer" disjunct; of the (ordinary) conjunction, the "falser" conjunct; of the negate, the negative. For $\rightarrow$, the natural strong connective, the idea is that if B is at least as true as A, $A \rightarrow B$ is to have as value the value of the corresponding material conditional, and hence is to be true. If A, on the other hand, is "truer" than B, $B \rightarrow A$ will be true and will have as value the value of $\bar{B} \vee A$. In the light of RM63, $A \rightarrow B$ has accordingly the value of the denial of $B \rightarrow A$; i.e., the value of $\bar{A}B$.

In presenting these considerations, which are of course purely heuristic, we have to some extent anticipated the semantics which we have led the reader to expect. An S_I-*interpretation* f shall be a function f defined on the sentential variables of **RM**, and then recursively on all sentences, with values according to the following specifications:

If B is a sentential variable, $f(B) \in I$.
If B is CD, $f(B)$ is $f(C) \wedge f(D)$.
If B is $C+D$, $f(B)$ is $f(C) \oplus f(D)$.
If B is $\sim C$, $f(B)$ is $-f(C)$.
If B is $C \vee D$, $f(B)$ is $f(C) \vee f(D)$.
If B is $C \rightarrow D$, $f(B)$ is $f(C) \rightarrow f(D)$.
If B is $C \circ D$, $f(B)$ is $f(C) \circ f(D)$.

Since some of these connectives are defined in **RM**, we leave it to the reader to check that the values of f are well-defined. We pass on to appropriate semantic concepts.

A sentence B is *true on interpretation* f if $f(B)$ is *non-negative;* S_I-*consistent*, if true on *some* S_I-interpretation f; S_I-*valid*, if true on *all* S_I-interpretations f. Finally, a sentence is *logically valid* if S_I-valid for all S_I.

The reader is right in supposing that all and only **RM** theorems are logically valid. In order to show this, we turn to the promised syntactic notions. It will prove convenient to deal with **RM**, **R**, and the classical two-valued logic **TV** (see §33.1) together for purposes of comparison and contrast; we shall suppose **TV** formulated not as in §33.1, but by the addition of the *axiom of paradox* $B \rightarrow (A \rightarrow A)$ as a new scheme to the axioms of **RM**. (We shall extend the previous conventions by prefixing the number of items by "TV" if they hold in **TV** but not **RM**.)

Let **L** be one of the systems **TV**, **RM**, **R**. Let S be any set of sentences of **L**. Let T be the smallest set which contains all members of S, all axioms of **L**, and all consequences of members by R11 (*modus ponens*) and R12 (adjunction): we call T *the* (*regular*) **L**-*theory determined by S*. Members of an **L**-theory T determined by S, we call *theorems* of T; of S, *proper axioms* of T; the axioms of **L** are the *logical axioms* of T. T is an **L**-*theory* if it is the **L**-theory determined by some set S; the **L**-theory determined by an empty S is of course just **L**. We write, as usual,

$$S \vdash_{\mathbf{L}} A$$

when A is a theorem of the **L**-theory determined by S; in addition, we write

$$S \vdash_{\mathrm{T}} A$$

when A is a theorem of the **L**-theory determined by S∪T. If S is empty, we write simply $\vdash_{\mathbf{L}} A$ or $\vdash_{\mathrm{T}} A$. Finally, if T′ is the **L**-theory determined by S∪T, we call T′ the *axiomatic extension of* T *by* S, or sometimes simply *an axiomatic extension of* T. Subscripts and distinguishing pieces of nomenclature will be dropped when clear from context.

All of this hopefully sounds very Official (it was written to sound very Official), but the familiar underlying idea is that a regular theory based on a given system of logic shall contain (1) all the theorems of the logic, (2) all the non-logical assumptions of the theory, and (3) whatever follows in accordance with the logic from (1) and (2). (This point of view gets significantly modified in §55 by dropping (1). "Regular **L**-theory" accordingly becomes standard when (1) holds, too.)

Among theories, logicians have long considered two kinds worthy of special consideration: (i) consistent theories, which contain both A and $\bar{A}$ for no sentence A, and (ii) complete theories, which contain at least one of A, $\bar{A}$ for each sentence A. Of particular interest from the classical point of view are those theories which are both consistent and complete. These have both semantic and algebraic significance — semantic significance, since a consistent and complete theory corresponds to a classical *interpretation* which makes all theorems true and all non-theorems false; algebraic significance, because consistent and complete theories correspond to maximal filters when **TV** is turned into a Boolean algebra in the style of Lindenbaum (§18.7).

Consistent and complete theories are similarly of particular importance in the investigation of the semantics and algebra of **RM** (for the algebra see §29.4). From the semantical point of view, it is presumably only a consistent and complete theory which can correspond precisely to how the world is. One should not, however, jump to the conclusion that only consistent and complete theories are of interest from the deductive point of view; indeed, it has long been recognized that incomplete theories have conspicuous uses — e.g., we may wish to study theories which admit a variety of interpretations, or we may be prevented from being able to specify recursively the complete theory which we believe to characterize some given domain (e.g., arithmetic), and so forth. Leaving for elsewhere the case for taking inconsistent systems seriously as well, let it simply be suggested here that there is an essential *symmetry* between incompleteness and inconsistency, in that a sentence is equally undecidable according to a given theory when both it and its negation are theorems as when neither are. **TV** obscures, indeed obliterates, this symmetry.

RM, like **R**, is a logic designed for foul weather as well as fair. But, happy as we shall be to see that inconsistency does not signal total collapse in **RM**, it remains to be established that this system is also semantically stable under normal conditions.

In order to provide a measure of such stability, we introduce the following notions. A theory is *syntactically consistent* provided that it has a consistent and complete axiomatic extension. A theory is *syntactically complete* provided that it is the intersection of all of its consistent and complete axiomatic extensions.

It is readily noted that all **TV**-theories are syntactically consistent and complete, provided that they are simply consistent. Indeed, that **TV** itself is semantically *weakly complete* may be taken as a reformulation, given the identification of consistent and complete theories with classical interpretations, of the fact that it is syntactically complete. Similarly, that **TV** is strongly complete may be reduced to the assertion that an arbitrary consistent **TV**-theory is syntactically consistent. (The importance of syntactical consistency and completeness for **TV** was noted by Polish workers, in particular Tarski and Lindenbaum, in the thirties. And it is to be noted that our syntactical analogues of the completeness theorems can be used as lemmas in their proof; see Mendelson 1964, Chapter 2.)

To understand the importance of syntactical completeness, an importance first noted by Dunn for **E** and **R**, we wish first to note that there are **RM**-theories, and hence *a fortiori* **R**-theories, which lack this property. In particular, the **RM**-theory T whose single proper axiom is $p\bar{p}\vee q$ is syntactically incomplete; for it is easy to see that every consistent and complete extension of T will contain q, since the addition of $\bar{q}$ to T renders provable $p\bar{p}\vee q\bar{q}$ by adjunction and distribution, while the negation of the latter is already a theorem of T and indeed of **R**. On the other hand, q is not a theorem of T; for proof, note that the following S_I-interpretation f makes all axioms of T true and its rules truth-preserving while making q false: let I be $\{-\mathit{1}, \mathit{0}, +\mathit{1}\}$, and let f$(p)$ = $\mathit{0}$, f(q) = $-\mathit{1}$.

That syntactical completeness does not hold in general for **RM**-theories is neither surprising nor unwanted. For these logics are intended to make room for wild theories along with the tame; in the theory just presented, for example, our axiom presents $p\bar{p}$ as an *alternative* to q; an effort to complete the formal semantics would presumably investigate extensions containing $p\bar{p}$ as well as those containing q. (It does; see §§49 and 54.1.) If we are to look contradictions squarely in the face and not shunt them off the logical stage, the requirement of syntactical completeness for arbitrary **RM**-theories is simply too restrictive.

As Dunn has noted, syntactical completeness for a consistent **R**-theory (and hence a consistent **RM**-theory) T is secured iff the following condition is fulfilled:

(γ) If A and $A\supset B$ are theorems of T, so is B.

(See §25, especially §§25.2–3, where it is shown that (γ) holds for **T**, **E** or

R. We remark that (γ) and syntactic completeness was proved for **T**, **R** and E only after the upcoming positive solution for **RM**.)

To sharpen the issues, let us call an **RM**-theory for which (γ) holds a *normal* theory. Such a theory is, by the previous remarks, syntactically complete if consistent. To take inconsistent theories seriously is to eschew normality for *arbitrary* theories. Nevertheless, one expects well-behaved theories to be normal; it is only when inconsistencies occur (or at least threaten, as in the example) that normality ceases to be desirable. Though we want our logic to be useful in pathological cases, we do not wish the theory determined by the logical axioms and rules to be itself pathological; i.e., we wish **RM** itself to be a *normal* theory.

After noting a few facts about the behavior of **R**-theories under axiomatic extension, we proceed to syntactical consistency and completeness proofs.

$\rightarrow$E, &I, and R55 guarantee the following, for all **R**-theories T and sets S of sentences. (Again, since all **RM**-theories are **R**-theories, in which all instances of $A\rightarrow(A\rightarrow A)$ are theorems, what is true of all **R**-theories is automatically true of all **RM**-theories.)

R72. If $\vdash_T A\rightarrow B$, then $A \vdash_T B$.
R73. If $S \vdash_T A_1 \ldots A_n\rightarrow B$, then $S, A_1, \ldots, A_n \vdash_T B$.
R74. If $S \vdash_T A_1\circ \ldots \circ A_n\rightarrow B$, then $S, A_1, \ldots, A_n \vdash_T B$.

Constantly used hereafter will be the replacement rule:

R75. Let T be an **R**-theory, and let C' be the result of replacing one or more occurrences of A by B in C. Then for all sets S of sentences

(a) $S, A\rightleftarrows B \vdash_T C\rightleftarrows C'$, and
(b) $S, A\rightleftarrows B, C \vdash_T C'$.

Proof of R75. (b) follows trivially from (a); the classical proof of (a), on the other hand, is unavailable, since $(A\rightleftarrows B)\rightarrow(C\rightleftarrows C')$ is not a theorem scheme of **R** or even of **RM**. Let $t_{C,\, C'}$, however, be the conjunction $(p_1\rightarrow p_1)\& \ldots \&(p_n\rightarrow p_n)$, where $p_1, \ldots, p_n$ are all the sentential variables occurring in either of C, C'. Then a straightforward inductive argument like that of §45.1 establishes

$$\vdash_R (A\rightleftarrows B)\& t_{C,\, C'}\rightarrow(C\rightleftarrows C')$$

for *zero* or more replacements of A by B, whence (a) follows easily.

The replacement rule may be strengthened by distinguishing consequent and antecedent parts of a sentence (§§5.1.2 and 22.1.1). Proof is similar.

R76. Let T be an **R**-theory, and let C' be the result of replacing antecedent occurrences of B by A or consequent occurrences of A by B in C, or both. Then

S, $A \to B \vdash_T C \to C'$.

Finally, for all **R**-theories T and sets S,

R77 S, $A \to B \vdash_T AB \rightleftarrows A$
R78 S, $A \to B \vdash_T A \vee B \rightleftarrows A$
R79 If S, $A \vdash_T C$ and S, $B \vdash_T C$, then S, $A \vee B \vdash_T C$.

Proof of R79 is given in §25.2.1.

We can now prove syntactic consistency for **R**-theories, making no assumptions about the cardinality of the set of sentential variables of a given theory.

THEOREM 1. Each consistent **R**-theory is syntactically consistent; i.e., it has a consistent and complete extension.

PROOF. We employ the method of Henkin 1949. Let T be an arbitrary consistent **R**-theory, and use the well-ordering theorem (if necessary) to index the sentences of **R** by ordinals 1, 2, etc. We define correlated theories $T_0, \ldots, T_i, \ldots, T^*$ for a given T: Let T_0 be T, and for each limit ordinal i (for each successor ordinal i) let T_i be the axiomatic extension of the union of its predecessors (of T_{i-1}) by A_i or by $\overline{A}_i$ according as the extension by A_i is or is not consistent. Let T^* be the union of all the T_i thus defined. If T^* is inconsistent, clearly there is a least i such that T_i is inconsistent; invoking, if need be, R79 and R19 (excluded middle), we show that if $i \neq 0$, a predecessor of T_i is inconsistent, contradicting the leastness of i. On the other hand, T_0 is assumed consistent; hence so is T^*. But T^* is complete, since on our swing through the ordinals we ensured that A or $\overline{A}$ would be added somewhere whatever its index. This proves Theorem 1.

Syntactic completeness does not hold for arbitrary **R**-theories, as noted. Nevertheless, there is an analogue of syntactic completeness (and hence of classical weak completeness) which holds for all **R**-theories. Let us call a theory *very weakly complete* if it is the intersection of *all* its complete extensions, consistent or not. Then

THEOREM 2. Every **R**-theory, hence every **RM**-theory, is very weakly complete.

PROOF. Obviously a theory is included in the intersection of all its extensions. On the other hand, suppose for a given **R**-theory T that there is a sentence A which is in every complete extension of T but not in T. Using the method of proof of Theorem 1, well-order and index the sentences of **R**. We now define a progression of theories $T_0, \ldots, T_i, \ldots T^*$ in like manner, except that each time we extend the theory we shall pick that one of B, $\bar{B}$ which leaves A unprovable. As before, it turns out that if A is provable in T* it must have been provable way back in T_0, which it wasn't; furthermore, T* is a complete theory without A, contradicting the assumption that A can be in all complete extensions of T without being in T itself. Note that this time, however, we have no guarantee that T* is consistent, since at some point in the game we may have had to make a choice which rendered a contradiction provable to avoid getting A. For instance, to avoid getting q in every complete extension of the **RM**-theory displayed above, with single axiom $p\bar{p}\vee q$, we must, for example put in $\bar{q}$ when its turn comes; this makes our complete extension inconsistent, but it doesn't require q to be in it. Of course, classically this option is not open to us, since to buy any contradiction is automatically to buy A.

Very weak completeness is itself an indication of stability amid the storms we are imagining our logic subject to. But we promised the reader normality for **RM** itself, and in Theorem 3 we deliver.

THEOREM 3. **RM** is syntactically complete; i.e., **RM** is the intersection of all its *consistent* and *complete* extensions.

PROOF. The method of proof of Theorem 2 will do in broad outline; we dispose of half the proof at once by noting that **RM** is consistent (since truth functionally consistent) and is hence included in the intersection of its consistent and complete extensions.

The other half of the proof also follows that of Theorem 2. The object is to show a sentence A is in **RM** if it is in every consistent and complete extension of **RM**. It will suffice to show that every sentence A unprovable in **RM** is unprovable in some consistent and complete extension of **RM**. For the duration of the proof, A shall be a *particular* unprovable sentence of **RM**; we give a recipe for concocting a consistent and complete **RM**-theory T which doesn't have A either.

Since we already have by Theorem 2 a complete theory without A, all our subsequent labor is devoted to showing that we can find a T which is also consistent. On seeing the labor which it has cost us, the reader who has grasped the spirit of this section will find self-evident our corollary to Theorem 3: Consistency is too darned much trouble!

Leaving such deeper insights until later, we proceed again to build progressions of theories. These progressions, happily, will be finite. Let the sentential variables in our unprovable sentence A be in alphabetical order $p_1, \ldots, p_n$, and let T_0 be formed from **RM** by adding as a proper axiom

1 $q \rightleftarrows p_1$

for every sentential variable q which does not occur in A. Clearly T_0 is a consistent extension of **RM** without A, since any proof of A from these axioms would be a proof from **RM** theorems on the substitution of p_1 for each q that does not occur in A. The point of these proper axioms is trivial; they enable us to go about our theory-building in complete indifference to sentences with sentential variables not in A, since these are immediately equivalent to sentences containing only variables that occur in A.

We now form theories $T_1, \ldots, T_n$ by applying the recipe of Theorem 2. If T_{i-1} may be extended by p_i without getting A, so extend it to form T_i; otherwise form T_i by adding $\overline{p_i}$ to T_{i-1} as an additional proper axiom. A is not in T_n, for if it were there would be a least i such that $\vdash_{T_i} A$. By the rules for constructing T_i, $\overline{p_i}$ is the extra axiom, so both

2 $p_i \vdash_{T_{i-1}} A$, and
3 $\overline{p_i} \vdash_{T_{i-1}} A$,

whence by 2, 3, and R79,

4 $p_i \vee \overline{p_i} \vdash_{T_{i-1}} A$,

i.e., A is in T_{i-1}, contrary to the leastness of i. This drives A all the way back to T_0, where we already know it isn't, hence it's not in T_n either.

The proper axioms of T_n, besides the equivalences 1, are the literals

5 $a_1, \ldots, a_n$

where for each i, a_i is either p_i or $\overline{p_i}$, and where each sentential variable of A turns up, possibly negated, on the list 5. Let us call the items 5 the *positive literals determined by A*, where each positive literal is a sentential variable occurring in A, or its negate. (Thus, positive literals here are atoms in the sense of §15.1; later we shall have "negative literals," which will be negates of atoms in the sense of §15.1, and "literals," which will be either.)

Enumerate pairs of positive literals in some definite order — e.g., let a_i, a_j precede a_k, a_m if $i < k$ or, if $i = k$, when $j < m$. We shall now build theories $T_{n+1}, \ldots, T_{n+n^2}$ using this enumeration. Let T_{i+1} be formed from T_i here by adding, when its turn comes up, as a new proper axiom,

6 $a_k \rightleftarrows a_m$

unless A is a theorem of the resulting theory. (If the addition would result in the proof of A, let T_{i+1} be T_i.)

The resulting theory T_{n+n^2} still lacks A, since, if the addition of an axiom candidate threatens a proof of A, we don't add the candidate. In fact, T_{n+n^2} is like T_n except that some of the positive literals have been made equivalent; furthermore, making any additional ones equivalent would render A provable in T_{n+n^2}. Let us use these equivalences to make up a (in general) shorter list of positive literals,

7 $b_1, \ldots, b_k$

to be called the *distinct* positive literals. (The plan is the following: let b_1 be a_1, and let b_{i+1} be the first positive literal in the list 5 not equivalent to any of its predecessors in T_{n+n^2}. By the way we constructed this theory, this can only be the case if there is no axiom stating such equivalence.)

Enumerate pairs of distinct positive literals in some definite order — e.g., on the same plan as before. We shall now build theories $T_{n+n^2+1}, \ldots, T_{n+n^2+k^2}$ using this enumeration. Where i is the number attached to the pair b_j, b_m, let T_{n+n^2+i} be formed from its predecessor by adding

8 $b_j \rightarrow b_m$

as an additional proper axiom, *unless* A is a theorem of the resulting theory. In this case, add

9 $b_m \rightarrow b_j$

as a proper axiom. (This latter specification is redundant, since if the addition of $b_j \rightarrow b_m$ yields A we should be forced anyway to add $b_m \rightarrow b_j$ in its turn. Putting the matter thus, however, simplifies the statement of the proof.)

The result of adding axioms 1, 5, 6, and 8 to **RM** is still a consistent theory, since it is truth functionally consistent on making all of the positive literals true and giving the sentential variables not in A the value of p_1. A little work is now required to show that the theory doesn't contain A. If it does, then there is a least i greater than $n+n^2$ such that $\vdash_{T_i} A$. By our rules for forming T_i, however, in T_{i-1}

10 $b_j \rightarrow b_m \vdash A$, and
11 $b_m \rightarrow b_j \vdash A$,

whence by R79,

12 $(b_m \rightarrow b_j) \vee (b_j \rightarrow b_m) \vdash A$ in T_{i-1}.

But in view of RM64, 12 makes A already a theorem of T_{i-1}, contradicting the leastness of i.

Hence our most recent theory, which we henceforth call T*, is still consistent and without A. Let us define

B T*-*implies* C

by

$\vdash_{T^*} B \rightarrow C$.

Note that T*-implication is a simple order relation on distinct positive literals 7, since the choice of axioms for T* has assured, for $j \neq m$, $\vdash b_j \rightarrow b_m$ or $\vdash b_m \rightarrow b_j$, but not both. We take this opportunity, accordingly, to re-name again the distinct positive literals in the induced order: c_1 shall be the b_j which T*-implies all the others, and c_{i+1} shall be the b_j which T*-implies all but $c_1, \ldots, c_i$. We leave it to the reader to verify that the sequence

13 $c_1, \ldots, c_k$

of *ordered* distinct positive literals has been well-defined.

We have been building theories patiently, but we get our final theory T in a rush. Form T from T* by adding, for distinct positive literals c_j, c_m,

14 $\overline{c_m \rightarrow c_j}$

whenever c_j precedes c_m in the sequence 13 — i.e., whenever $c_m \rightarrow c_j$ is not a theorem of T*.

This completes the construction of the **RM**-theory T. To complete the proof, we must show

I. A is not a theorem of T.
II. T is complete.
III. T is consistent.

We tackle the problems in that order. Suppose, for *reductio*, that A is a theorem of T. Since A isn't a theorem of T*, we may resume our old habit of adding items of the form 14 one at a time in some definite order. Let T′ be the last theory in the resulting sequence which lacks A. Then for some c_j, c_m such that c_j precedes c_m in the sequence 13,

15 $\overline{c_m \rightarrow c_j} \vdash A$ in T′.

But since $m > j$, we already have as an axiom

16 $\vdash_{T'} c_j \rightarrow c_m$.

Accordingly, by 16 and adjunction

17 $c_m \rightarrow c_j \vdash_{T'} c_j \rightleftarrows c_m$.

But c_j and c_m are *distinct* positive literals; i.e., by our rule for selecting axioms 6,

18 $c_j \rightleftarrows c_m \vdash_{T'} A$.

So by 17 and 18,

19 $c_m \rightarrow c_j \vdash_{T'} A$.

Putting 15 and 19 together by the familiar appeal to R79 and excluded middle,

20 $\vdash_{T'} A$,

which contradicts the assumption that T′ lacks A. So the hypothesis that A is a theorem of the final theory T is refuted, and I is proved.

To prove that T is complete and consistent, we return to the sequence 13 of ordered distinct positive $c_1, \ldots, c_k$, where k is a positive integer whose value is fixed by the construction of T. We shall do so by showing that T has a *characteristic interpretation* in the Sugihara matrix S_k; i.e., that there is an S_k-interpretation f such that, for all sentences B of **RM**, B is true on the interpretation f if and only if B is a theorem of T. (What we shall show essentially is that the Lindenbaum algebra (§18.7, *mutatis mutandis*) of T is isomorphic to S_k.) For $1 \leq i \leq k$, introduce *distinct negative literals* by definition as follows:

21 $c_{-i} =_{\text{df}} {\sim}c_i$.

Leaving i, j to range over positive integers $\leq k$, we draft m, n to range over non-zero integers, positive and negative, of absolute value less than or equal to k. c_m shall be a *distinct literal* just in case it is a distinct positive literal or a distinct negative literal.

(i) If q is a sentential variable which does not occur in A and p_1 is as in axioms 1 of T, let $\mathrm{f}(q) = \mathrm{f}(p_1)$.

(ii) Where p is one of the positive literals 5, and where $p \rightleftarrows c_i$ is one of the axioms 6, let $\mathrm{f}(p) = +i$.

(iii) Where $\bar{p}$ is one of the positive literals 5, and where $\bar{p} \rightleftarrows c_i$ is one of the axioms 6, let $\mathrm{f}(p) = -i$.

Since each positive literal is equivalent by one of the axioms 6 to exactly one distinct positive literal, the reader may easily verify that f is well-defined.

We now show, for every sentence B of **RM**,

22 If $\mathrm{f}(B) = m$, $\vdash_{\mathrm{T}} B \rightleftarrows c_m$.

Proof of 22, by a structural induction on B: if p is a positive literal, $p \rightleftarrows c_i$ is an axiom where, by (ii), $\mathrm{f}(p) = +i$. If $\bar{p}$ is a positive literal, $\bar{p} \rightleftarrows c_i$ is an axiom where, by (iii), $\mathrm{f}(p) = -i$. By transposition and 21, $\vdash_{\mathrm{T}} p \rightleftarrows c_{-i}$. If p does not occur in A, $p \rightleftarrows p_1$ is an axiom 1 where $\mathrm{f}(p) = \mathrm{f}(p_1) = m$ by (i); since $\vdash_{\mathrm{T}} p_1 \rightleftarrows c_m$ by the remarks just made, $\vdash_{\mathrm{T}} p \rightleftarrows c_m$ by the replacement theorem R75.

The rule R75 of replacement will suffice for the induction step provided that the following equivalences can be proved in T.

(a) ${\sim}c_m \rightleftarrows c_{-m}$
(b) $c_m \& c_n \rightleftarrows c_{m \wedge n}$
(c) $c_m \rightarrow c_n \rightleftarrows c_{(-m) \oplus n}$

Ad (a). Use 21 and perhaps double negation.

Ad (b) and (c). By R50 and (a), $\vdash_T c_m \rightarrow c_n \rightleftarrows . c_{-m} + c_n$. In place of (c), it accordingly suffices to establish

(d) $c_m + c_n = c_{m \oplus n}$.

We treat (b) and (d) together, noting three subcases.

(i) m and n are both positive, $m \leq n$.
(ii) m and n are both negative, $m \leq n$.
(iii) m is negative and n is positive.

In subcase (i) $c_m \rightarrow c_n$ is an axiom 8. By R77, $\vdash_T c_m c_n \rightleftarrows c_m$, which proves (b) for this subcase. By R78, $\vdash_T c_m \vee c_n \rightleftarrows c_n$. Hence by R54 and transitivity, $\vdash_T c_m + c_n \rightarrow c_n$. On the other hand, by RM71 and R30, $\vdash_T c_m \rightarrow (c_n \rightarrow . c_m + c_n)$; but c_m is an axiom of T; detaching, $\vdash_T c_n \rightarrow . c_m + c_n$. Adjoining, $\vdash_T c_m + c_n \rightleftarrows c_n$; but $m \oplus n$ is n on the assumption (i), establishing (d) for this subcase.

In subcase (ii), $c_m \rightarrow c_n$ follows by contraposition from an axiom of T. By R77, $\vdash_T c_m c_n \rightleftarrows c_m$, disposing of (b). By this, R55, RM71, and transitivity, $\vdash_T c_m \rightarrow . c_m + c_n$. On the other hand, by R23 $\overline{c_n} \rightarrow . (c_n + c_m) \rightarrow c_m$. But $\overline{c_n}$ is the double negate of an axiom 5; detaching and permuting, $\vdash_T (c_m + c_n) \rightarrow c_m$. Adjoining, $\vdash_T c_m + c_n \rightleftarrows c_m$; but $m \oplus n$ is m on the assumption (ii), establishing (d) for this subcase.

In subcase (iii), $c_m \rightarrow \overline{c_1}$ by contraposing an axiom of T. Similarly, $c_1 \rightarrow c_n$ is an axiom. But c_1 is also an axiom of T; hence by idempotence $\vdash_T c_1 + c_1$; i.e., by R50, $\vdash_T \overline{c_1} \rightarrow c_1$. Then by two steps of transitivity, $\vdash_T c_m \rightarrow c_n$. The argument of previous subcases then yields $\vdash_T c_m c_n \rightleftarrows c_m$, completing the argument for (b). The proof of (d) here has two further subcases, according as

(iii)′ $|m| \leq n$, or
(iii)″ $|m| > n$

For (iii)′, we have by R23 $\vdash \overline{c_m} \rightarrow . c_m + c_n \rightarrow c_n$. But $\overline{c_m}$ is the double negate of an axiom 5; detaching, $\vdash_T c_m + c_n \rightarrow c_n$. On the other hand, since $|m| \leq n$ we have $\vdash_T \overline{c_m} \rightarrow c_n$ by replacement in an axiom 8. But by RM0, $c_n \rightarrow . c_n \rightarrow c_n$. Commuting antecedents by R4 and introducing + by definition, we get from the result $\overline{c_m} \rightarrow . c_n \rightarrow c_n$ of applying transitivity to the last two sentences, $\vdash_T c_n \rightarrow . c_m + c_n$. But $m \oplus n$ is n on the assumption (iii)′, so that the usual adjunction disposes of (d) here as in previous subcases.

For (iii)″, a slight modification of the argument of subcase (i) produces $\vdash_T c_m \rightarrow . c_m + c_n$. On the other hand, since $|m| > n$, $\overline{\overline{c_m} \rightarrow c_n}$ is trivially equivalent to an axiom 14. (This is the only point in the argument into which axioms 14 enter explicitly.) Hence $\vdash_T c_m + c_n$. But by excluded middle and idempotence, $\vdash_{\mathbf{RM}} c_m + c_n + \overline{c_m + c_n} + c_m$; i.e. $\vdash_{\mathbf{RM}} \overline{c_m + c_n} \rightarrow . c_m + c_n \rightarrow c_m$. Detaching, $\vdash_{\mathrm{T}} c_m + c_n \rightarrow c_m$, whence the usual adjunction establishes (d).

This exhausts the cases for $m \leq n$; in view of the commutative laws R36 and R39, we may argue similarly when $n \leq m$. The completion of the inductive argument for 22 is straightforward and is left to the reader.

II now follows easily — i.e., T is complete. For by 22, for each sentence B there is a distinct positive literal c_i such that

23 $\vdash_{\mathrm{T}} B \rightleftarrows c_i$ or $\vdash_{\mathrm{T}} \overline{B} \rightleftarrows c_i$.

But c_i is an axiom 5 of T. Hence for each B,

24 $\vdash_{\mathrm{T}} B$ or $\vdash_{\mathrm{T}} \overline{B}$.

T is truth functionally consistent until the addition of axioms 14, but these axioms make the proof of consistency not quite trivial. The following lemma is semantically important.

RM80. Let S_I be an arbitrary Sugihara matrix and f any S_I-interpretation. Then the logical axioms and definitions of **RM** are true on f; furthermore, if the premisses of a rule R11 or R12 are true on f, so is the conclusion. Hence all theorems of **RM** are logically valid in the sense defined above.

Proof of RM80. Proof is by tedious but straightforward verification. The following lemma now suffices for III.

25. Let A and T be as above, and let f be the S_k-interpretation for which 22 holds. Then if B is a theorem of T, B is true on f; in particular, $\overline{A}$ is true on f.

Proof of 25. It suffices to show the axioms of T true on f and its rules truth-preserving. The axioms are of six kinds:

Logical axioms. These are true on f by RM80.

Axioms $q \rightleftarrows p_1$ (1). These are true on f by (iii) of its definition.

Axioms p_i or $\overline{p_i}$ (5). True on f by (i) or (ii) of its definition.

Axioms $a_i \rightleftarrows a_j$ (6). Clearly there is one and only one distinct positive literal c_h such that $c_h \rightleftarrows a_i$ and $c_h \rightleftarrows a_j$ are axioms 6. Under specifications (i)–(ii), $f(a_i) = f(a_j) = h$. It is then easily verified that $f(a_i \rightleftarrows a_j)$ is the non-negative integer h, so that $a_i \rightleftarrows a_j$ is true on f.

Axioms $c_i \rightarrow c_j$ (8, 9), where c_i, c_j are distinct positive literals (13), $i \leq j$. But then $f(c_i) = i \leq j = f(c_j)$; consulting Sugihara's alternative characterization of $\rightarrow$, we note immediately that $c_i \rightarrow c_j$ is true on f.

Axioms $\overline{c_j \rightarrow c_i}$ (14), where c_i, c_j are items 13, $i \leq j$. But then $f(c_j) = j > i = f(c_i)$, which makes $c_j \rightarrow c_i$ negative and its negation positive (i.e. true) on f. That the rules of →E and &I preserve truth on f is also part of the content of RM80; hence all theorems of T are true on f. Since T is complete and without A by II and I, respectively, $\bar{A}$ in particular is true on f.

III now follows easily — i.e., T is consistent. For, every sentence B of **RM**, it follows from 25 that

26. It is not the case both that $\vdash_T B$ and $\vdash_T \bar{B}$.

Proof of 26. Suppose for some B, $\vdash_T B$ and $\vdash_T \bar{B}$. Then by 25, both B and $\bar{B}$ are true on the S_k-interpretation f defined just above 22, which is impossible.

This completes the proof of Theorem 3; for an arbitrary non-theorem A of **RM**, we have constructed a consistent and complete theory T which lacks A. By contraposition, what is in all consistent and complete **RM**-theories is in **RM**; hence **RM** is syntactically complete.

We now gather in the corollaries to Theorem 3.

Corollary 3.1. Suppose a sentence A contains n sentential variables. Then

RM81. $\vdash_{\mathbf{RM}} A$ iff A is S_n valid.

Proof. For necessity, we observe that **RM**-theorems are S_I-valid for all I by RM80; hence in particular valid for the normal matrix S_n (where I is $\{-n, \ldots, -1, +1, \ldots, +n\}$). For sufficiency, suppose A is not a theorem of **RM**. By 25 in the proof of Theorem 3, for some positive integer $k \leq n$ there is an S_k-interpretation f such that $\bar{A}$ is true on f; hence $f(A) < 0$; since f is automatically an S_n-interpretation (which happens, accidentally as it were, to have no integers m such that $m > k$ in its range), A is S_n-invalid. This completes the proof of Corollary 3.1.

Corollary 3.2. **RM** is decidable.

Proof. Every sentence A has a fixed, finite number n of sentential variables. By 3.1, for given A, $\vdash_{\mathbf{RM}} A$ iff A is S_n-valid — i.e., true on every S_n-interpretation. There are $(2n)^n$ such interpretations; checking the truth of A on each of them terminates. Such a finite check serves to refute every refutable A; **RM** is decidable.

Corollary 3.3. There is an efficient method of proving an arbitrary **RM**-theorem. ("Efficient" is meant here relatively, since we can always enumerate proofs. And the reader will quickly grasp that the procedure outlined in the text is none too efficient.)

PROOF. Note first that if A is an **RM**-theorem with n sentential variables, it will be a theorem of the version of **RM** formulated with just n sentential variables. Furthermore, there is only a finite number of ways to pick proper axioms for consistent and complete **RM**-theories (leaving out axioms 1) according to the recipe of Theorem 3. Since the sets of these proper axioms will be finite, each of them may be expressed as a single axiom B_i. Following the proof of Theorem 3, we can establish $\vdash_{\mathbf{RM}} B_i \rightarrow A$ for each such choice. Letting $\vee B_i$ stand for the disjunction of the B_i, repeated use of R48 yields $\vdash_{\mathbf{RM}} \vee B_i \rightarrow A$. Copious use of distribution, excluded middle, and RM64 yields a proof of $\vee B_i$. By $\rightarrow$E we get A in this way if it's an **RM**-theorem.

COROLLARY 3.4. Let **RM** be formulated with finitely many sentential variables $p_1, \ldots, p_n$. Then S_n is characteristic for **RM** — i.e., all and only the theorems are S_n-valid.

PROOF by corollary 3.1.

COROLLARY 3.5. Let **RM** be formulated with infinitely many sentential variables of unspecified cardinality. Let I be any infinite set of integers which contains $-m$ iff it contains m. Then S_I is characteristic for **RM** — i.e., all and only the theorems of **RM** are S_I-valid.

PROOF. By RM80, all theorems of **RM** are S_I-valid for all I. Conversely, suppose A is not a theorem of **RM**. We show first that A is S_{Z*}-invalid, where Z^* is the set of non-zero integers.

In fact, where A has n sentential variables there is an S_n-interpretation f by Corollary 3.1 such that A is not true on f. Clearly f is also an S_{Z*}-interpretation. Hence A is S_{Z*} invalid.

Let I now be an arbitrary infinite set of integers that contains $-m$ whenever it contains m, and let h be a mapping from Z^* into the non-zero integers of I such that (i) if $m < n$, $\mathrm{h}(m) < \mathrm{h}(n)$ and (ii) $\mathrm{h}(-m) = -\mathrm{h}(m)$. Clearly such an h exists. Define now an S_I-interpretation g by specifying, for all sentential variables p, $\mathrm{g}(p) = \mathrm{h}(\mathrm{f}(p))$, where f is the S_{Z*}-interpretation of the preceding paragraph. It is easily verified that A is not true on g, completing the proof of 3.5.

Corollary 3.5 gives us our characteristic matrix for **RM**. In fact, it gives us a raft of them. Isomorphisms aside, two stand out: S_{Z*}, just defined, and S_Z, where Z is the set of all integers, including zero. S_{Z*} can be pressed into immediate service; S_Z will occupy us in the next section.

Corollary 3.6. **RM** is a *normal* theory; i.e., whenever $\vdash_{\mathbf{RM}} A$ and $\vdash_{\mathbf{RM}} A \supset B$, then also $\vdash_{\mathbf{RM}} B$.

Proof. By the remarks above, this follows immediately on the observation that **RM** is consistent and syntactically complete. Alternatively, we may use S_{Z*} directly to establish normality. Suppose $\vdash_{\mathbf{RM}} A$ and $\vdash_{\mathbf{RM}} \overline{A} \vee B$. Since for all S_{Z*} interpretations f, $f(A) > 0$, $f(\overline{A}) < 0$. By 3.5, $\vdash_{\mathbf{RM}} B$.

We summarize the various observations of this section in the following theorem. (Again, **RM**-*theory* here means *regular* **RM**-*theory* in the sense of §28.4 and elsewhere.)

Theorem 4. The following conditions are equivalent:

(i) $\vdash_{T} A$ for all **RM**-theories T.
(ii) $\vdash_{T} A$ for all complete **RM**-theories T.
(iii) $\vdash_{T} A$ for all consistent and complete **RM**-theories T.
(iv) $\vdash_{\mathbf{RM}} A$.
(v) A is S_I-valid for all Sugihara matrices S_I.
(vi) A is S_{Z*}-valid, where Z^* is the set of integers $\neq 0$.
(vii) A is S_Z-valid, where Z is the set of integers.
(viii) A is S_i-valid, where the number of sentential variables occurring in A is not greater than i.

Proof by the various theorems and corollaries.

§29.3.3. Glimpses about. We were tempted to follow in the footsteps of a great logician by entitling this section, "Glimpses beyond." But we do not plan to go very far beyond the results of Theorem 4, so perhaps "Glimpses about" is more accurate. In the relatively unexplored domain of paradox-free logics, **RM** has a number of interesting features, and it is worth calling attention to some of them.

First, despite the fact that the Sugihara matrices are *characteristic* for **RM**, they are not as described at the outset of §29.3.2 wholly adequate for the semantics of arbitrary **RM**-*theories*, even those formulated with only a denumerable number of sentential variables.

Consider, e.g., the following theory T. Let the sentential variables of T be indexed by the positive rationals, and let p_x be a proper axiom of T for all positive rationals x; furthermore, let $p_x \rightarrow p_y$ and $\overline{p_y \rightarrow p_x}$ be proper axioms for all positive rationals such that $x < y$. If a contradiction were derivable from these axioms, it would have to follow from a finite subset thereof; clearly, however, all finite subsets are S_i-consistent for some finite i, and hence consistent. On the other hand, the theory T itself is S_{Z*}-inconsistent.

Suppose for *reductio* that it is rather S_{Z*}-consistent; then there is an S_{Z*}-interpretation f such that $f(p_1) = i$ and $f(p_2) = i+j$ for some positive integers i and j. But then to make the axioms of T true the denumerably many sentential variables p_x, $1 < x < 2$, must take distinct values under f among the finitely many integers $i+1, \ldots, i+j-1$; obviously this is impossible; so T is S_{Z*}-inconsistent. Adaptation of the last part of the argument for corollary 3.5 then shows T to be S_I-inconsistent for all normal Sugihara matrices S_I.

One way of putting this observation is that, although the semantics suggested in the first part of §29.3.2 suffice for a proof of semantical weak completeness for **RM**, this semantics does not suffice for a semantical strong completeness proof, in the sense that every consistent theory comes out true on some interpretation. In view of the syntactical consistency result of Theorem 1, however, it appears clear that an appropriate extension of the semantics of §29.3.2 should produce a semantical strong completeness result as well.

Such an extension has been supplied by Dunn. In constructing the matrices S_I, let us, instead of limiting ourselves to sets I of integers, allow I to be *any* simply ordered set with an operation "$-$" which is properly antitone and of period two — i.e., such that $--a = a$ and $a < b$ iff $-b < -a$. In particular, Dunn has proved that all consistent denumerable **RM**-theories T come out true on an S_{Q*}-interpretation f where Q* is the set of non-zero rationals and where otherwise operations are defined as before. The reader can perhaps see how the method of proof of Theorem 3 may be generalized to establish this result. (The key step lies in the realization that the consistent and complete extension T* guaranteed by Theorem 1 for an arbitrary consistent T may have denumerably many ordered distinct positive literals, in the sense of Theorem 3.)

RM has many surprising features; one of the most astonishing is that the Craig 1957 interpolation lemma fails for it. (One of the principal authors has suggested that therefore **RM** is "unreasonable in the sense of Craig.")

RM81. There are sentences A and B of **RM** such that $\vdash_{\mathbf{RM}} A \rightarrow B$, neither $\vdash_{\mathbf{RM}} \bar{A}$ nor $\vdash_{\mathbf{RM}} B$, and such that there is no sentence C whose variables are just those common to A and B which satisfies the conditions $\vdash_{\mathbf{RM}} A \rightarrow C$ and $\vdash_{\mathbf{RM}} C \rightarrow B$.

Proof of RM81. Let A be $s \vee .pq\bar{q}$ and let B be $(s \vee p)(s \vee r \vee \bar{r})$. It is easily verified that $\vdash_{\mathbf{RM}} A \rightarrow B$ but neither $\vdash_{\mathbf{RM}} \bar{A}$ nor $\vdash_{\mathbf{RM}} B$. Suppose then that there is a C whose only variables are among p, s such that $\vdash_{\mathbf{RM}} A \rightarrow C$ and $\vdash_{\mathbf{RM}} C \rightarrow B$. Define an S_2-interpretation f by setting $f(p) = +2$, $f(s) = -2$, and $f(q) = f(r) = +1$. Then $f(A) = -1$ and $f(B) = +1$. But since the values of the variables of C are confined to ± 2 under f, an induction on the complexity of C shows $f(C) = +2$ or $f(C) = -2$. Since $A \rightarrow C$ is true on f it cannot be the latter; since $C \rightarrow B$ is true on f it cannot be the former. Conclusion: there is no C satisfying the supposition, which proves RM81.

For another curious fact, compare the following statements of the *relevance principle* for **R** and **TV**, respectively:

R82. Suppose $\vdash_{\mathbf{R}} A \rightarrow B$. Then A and B share a sentential variable (see §22.1.3).

TV83. Suppose $\vdash_{\mathbf{TV}} A \supset B$. Then either (i) A and B share a sentential variable or (ii) *either* $\vdash_{\mathbf{TV}} \bar{A}$ *or* $\vdash_{\mathbf{TV}} B$.

Since one gets **RM** from **R** by adding a very relevant *looking* axiom scheme, $A \rightarrow (A \rightarrow A)$, I (and Dunn too) supposed initially that the analogue of R82 would hold for **RM** also; in fact, $\vdash_{\mathbf{RM}} p\bar{p} \rightarrow q \vee \bar{q}$ is a counterexample. The correct form of the relevance principle for the intermediate logic **RM** is indeed intermediate between R82 and TV83.

RM84. Suppose $\vdash_{\mathbf{RM}} A \rightarrow B$. Then either (i) A and B share a sentential variable or (ii) *both* $\vdash_{\mathbf{RM}} \bar{A}$ *and* $\vdash_{\mathbf{RM}} B$.

PROOF of RM84. Suppose $\vdash_{\mathbf{RM}} A \rightarrow B$. Suppose further that A and B do not share a sentential variable. Consider an arbitrary S_{Z*}-interpretation f, and suppose first that A is true on f. Define a new S_{Z*}-interpretation g as follows: if p occurs in A, $g(p)$ shall be the sum of $f(p)$ and $f(p)$; otherwise, $g(p)$ shall be $+1$. It is easy to verify that $g(A) > +1$ while $g(B) = +1$. But then $A \rightarrow B$ is not true on g, contradicting by Theorem 4 the hypothesis that it is **RM**-provable. Accordingly A cannot be true on f; but f was arbitrary, hence $\bar{A}$ is true on all S_{Z*} interpretations and is by Theorem 4 provblea in **RM**. By parity of reasoning, $\vdash_{\mathbf{RM}} B$, proving RM84.

We make two further remarks about relevance. First, in the light of RM84 it is even more surprising that the Craig interpolation lemma fails for **RM**, for the parity between TV83 and RM84 would have led one to suspect a stronger version of that theorem to have held in **RM**. Accordingly we conjecture that there is an appropriate version of that theorem, perhaps involving sentential constants, which does hold for **RM**.

Having disgusted some of our friends by proving **RM** unreasonable in the sense of Craig, and others by noting that **RM** is unreasonable also in the sense of Anderson and Belnap, we shall finish a job well begun by showing that **RM** permits *everyone* to be unreasonable, without suffering the psychotic break which follows classically on a single mistake. Let us call a sentence B of a theory *lamentable* if both B and $\bar{B}$ are provable in that theory. Then in the first place,

RM85. Let B and C be lamentable sentences of an **RM**-theory T. Then $\vdash_T B \rightleftarrows C$.

RM85 shows that an inconsistent theory identifies all lamentable sentences, in the sense that all such sentences T-imply each other; in particular, each lamentable sentence is thus identified with its own negation. (For proof of RM85, note the S_2-validity, and hence by Theorem 4 the **RM**-theoremhood, of $A\bar{A}\rightarrow(A\rightleftarrows\bar{A})$ and $A\bar{A}B\bar{B}\rightarrow(A\rightleftarrows B)$).

All lamentable sentences of T are of course theorems. But let us call B a *proper theorem* of T in case $\vdash_T B$ but *not* $\vdash_T \bar{B}$, and let us call B *properly refutable* in T in case $\vdash_T \bar{B}$ but not $\vdash_T B$. We now show that inconsistencies do not multiply the supply of proper theorems beyond necessity.

RM86. Let T be an inconsistent **RM**-theory, and let A be a sentence of T none of whose sentential variables occur among the sentential variables of the proper axioms of T. Let A be a non-theorem of **RM**. Then A is a non-theorem of T.

PROOF of RM86. A, if a non-theorem of **RM**, is falsifiable in S_Z by Theorem 4. Let f be an S_Z-interpretation so defined on sentential variables of A that $f(A) < \boldsymbol{0}$, and let $f(p) = \boldsymbol{0}$ for all sentential variables not in A. It is clear that all theorems of T are true on f, but A is not. This proves RM86. An immediate corollary is

RM87. Let T be an inconsistent **RM**-theory, and let A be a sentence none of whose sentential variables occur among the sentential variables of the proper axioms of T and let S be a set, possibly empty, of such sentences. Then

$$S \vdash_T A \text{ iff } S \vdash_{\mathbf{RM}} A$$

PROOF of RM87. Sufficiency is trivial. For necessity, assume $S \vdash_T A$. It's quickly seen in this case that either $\vdash_T A$ or else there are in S formulas $B_1, \ldots, B_n$ such that n is finite and that, where B is the conjunction of the B_i, $\vdash_T B\rightarrow A$, on an appropriate form of the deduction theorem for **RM**. But by hypothesis neither A nor a possible $B\rightarrow A$ contains any sentential variables occurring in proper axioms of T, whence by RM86, $S \vdash_{\mathbf{RM}} A$, ending the proof.

RM86 and its generalization RM87 are stronger results than the relevance principle RM84 or even its stronger version R82. (Though the theorems are stated in particular for inconsistent systems, it is clear by the method of proof that they hold for arbitrary T, since the hypothesis of inconsistency is nowhere invoked.) For the relevance principles merely claim that strong conditionals of the form $A \rightarrow B$ are unprovable in the respective systems unless certain relevance requirements are met, whereas RM86 and RM87 show that if A is added as a proper axiom, or even if an infinite set of proper axioms is added, an irrelevant B remains unprovable. For so far as RM84 and R82 are concerned, there might be some devious trick (using, say, adjunction) by which B becomes newly provable if an irrelevant A is added as an axiom; RM86 and RM87 show that no such trick will work for **RM**-theories. (The argument may be readily adapted to show, *a fortiori*, that no such trick will work for **R** either.)

As our use of S_Z and S_Q above suggests, the extension of the Sugihara semantics to include *non-normal* matrices permits us to form the notion of a *model* of an inconsistent theory. Paralleling our syntactic talk, we may speak of a sentence as *properly* true on an interpretation f if its value is *greater than* $\mathit{0}$, and as *true* if, as before, its value is greater than or equal to zero. Falsehood and proper falsehood may be characterized conversely; A is properly false if $f(A) < \mathit{0}$, and false if $f(A) \leq \mathit{0}$ on a given interpretation f.

If we now characterize a model of a theory as an interpretation which makes all of its sentences true, every theory, no matter how bad, automatically has a model — just interpret all variables as $\mathit{0}$. Of more interest, accordingly, is the problem of making as many sentences as possible *properly* true for a given theory. RM87 is one result in this direction; its sharpening would be of considerable interest.

And we note finally, recalling RM85, that the deduction-theoretic structure of **RM** automatically identifies lamentable sentences and in particular identifies each lamentable sentence with its negation. For classical logic, there is in a certain sense but one truth and one falsehood, with nothing in between. **RM**, by using other points on the number line besides $+\mathit{1}$ and $-\mathit{1}$, permits us to allow in the same sense that there are as many truths and falsehoods as we please; if, however, there is an item in between, it is the only such item. From **RM**'s point of view, truths and falsehoods we have aplenty, but genuine neutrality is rare indeed.

RM87 is in a sense the chief result of this section; it shows that we can have quite a strong logic, well-motivated on the deduction-theoretic side, in which contradictions are *isolable* — stronger, in fact, than systems having this property which are much better known. On the other hand, RM87, and **RM** itself, represents but a *starting point* and not an *ending* point; by extending **RM** to a theory of quantification and by proving stronger versions of RM87, it is to be hoped that we can establish what at present cannot be shown for strong theories — the absolute unprovability of certain sentences in these theories — i.e., we might hope to show that, whatever contradictions might afflict such high-powered mathematical enterprises as a formalized analysis or a formalized set theory, these contradictions cannot render provable such repulsive candidates as $0 = 1$, and perhaps other falsehoods less repulsive, thus saving at least part of an enterprise which logicians have pursued since the days of Hilbert.

Dreams aside, what are we to make of **RM**? Neither too much nor too little, I hope. **RM** is, as I have stressed throughout, an *intermediate* logic, satisfying fully neither the classical yearning for simplicity nor the demands of entailment theorists that a formal logic should "tell it like it is." Yet as our completeness proof shows, **RM** is an exceptionally simple system which nevertheless meets many of the motivating conditions which have been put forth in favor of logics apparently much more complex. Łukasiewicz and Tarski 1930 suggested that sentential calculuses in general might serve as a kind of "laboratory" in which ideas about the general methodology of the deductive sciences could be tested; we suggest that **RM** in particular might well serve this function for the purpose of testing what insights, if any, are to be garnered from the attempt to apply paradox-free logics to concrete mathematical and scientific questions. For by a curious twist **RM** comes to possess virtues claimed for the logics between which it stands; it is more intelligible than **R**, which is supposed to be intuitive; safer than **TV**, which is supposed to be reliable. But it is not as intelligible as **TV** or as safe as **R**; such, it must be presumed, is the price of moderation.

(Added March 1973: Since this was written, **R** has in fact become rather more intelligible, see §47 on. But the "laboratory" argument was a good one, work in the simpler context of **RM** having produced ideas, especially with respect to (γ) — see §29.4 particularly — that would have been more difficult to come by in **R** or **E**.)

§29.4. Extensions of RM (by J. Michael Dunn). By an *extension* of **RM**, we mean one that is closed under substitution and the rules of **RM** ($\rightarrow$E and &I). Scroggs 1951 showed that every similarly closed consistent proper extension of **S5** has a finite characteristic matrix (even though **S5** does not), that every such may be obtained by adding an appropriate "Dugundji formula," and that the classical propositional calculus is the only Post consistent, Post complete extension of **S5**, the last having been originally proven by McKinsey from a general result of Tarski. We show the same things about **RM** (the last having been originally proven by Ulrich 1967 from the same general result of Tarski). In the following proof of these results we mimic the techniques of Scroggs, and then rely upon Meyer's completeness result for **RM** of §29.3 in much the same way that Scroggs used a completeness result of McKinsey and Tarski for **S5**. There is an alternative proof in Dunn 1970 which is less imitative of Scroggs's proof and which does not use Meyer's result as a lemma. Indeed, there Meyer's result (generalized to strong completeness) follows as a special case from a completeness theorem for all extensions of **RM** (including **RM** itself). Of course we would never have thought of any of our completeness results for **RM** and its extensions if it had not been for Meyer's pioneering discovery of the connection between **RM** and the Sugihara matrix. We point out to any reader interested in comparing the proof that follows to the alternative proof in Dunn 1970 that an important distinction between the two lies in the way that the alternative proof manages to avoid the apparatus of "subdirect products." Scroggs's own results could be obtained in an analogous alternative way, as are Scroggs-type results for yet another system in Dunn and Meyer 1971.

In addition to §29.3, we presuppose familiarity with the algebraic notions of §§18 and 28.2 — especially the latter.

We begin by introducing two key algebraic structures related to **RM**. $S = \langle S, \rightarrow, \wedge, \vee, \overline{} \rangle$ is an *idempotent De Morgan monoid* iff the structure $S' = \langle S, \circ, \leq, \overline{} \rangle$ obtained by defining $a \circ b = \overline{a \rightarrow \overline{b}}$ and $a \leq b$ iff $a \vee b = b$ is a De Morgan monoid, with $\wedge$ its meet, in the sense of §28.2.1, which further satisfies the principle

$$a \circ a = a \qquad \text{(Idempotence)}.$$

(We retain the definitions of $\circ$ and $\leq$ throughout.) It is easy to see that just as the Lindenbaum algebra of **R** is a De Morgan monoid (§28.2.3), so that of **RM** is an idempotent De Morgan monoid. Of course $[t]$ is the monoid identity, $[A\&B]$ and $[A \vee B]$ give us lattice bounds, and $[A] \leq [B]$ iff $\vdash A \rightarrow B$. The class of idempotent De Morgan monoids can be defined equationally (see the end of §28.2.1), so that it is closed under subalgebras, homomorphic images, and direct products.

Adapting and generalizing definitions of Sugihara 1955 and §29.3 to our purposes, we shall say that a *Sugihara chain* is a structure $S = \langle S, \rightarrow, \wedge, \vee, \overline{} \rangle$ satisfying the following conditions. 1. It is a De Morgan lattice. 2. It is a *chain* with respect to $\leq$; i.e., $a \leq b$ or $b \leq a$, all $a, b \in S$. 3. It contains an element t such that $t \leq a$ iff $\bar{a} \leq a$. 4. If $a \leq b$, then $a \rightarrow b = \bar{a} \vee b$, and if $a \nleq b$ (i.e., if $b < a$), then $a \rightarrow b = \bar{a} \wedge b$ — this last being the Sugihara 1955 definition. We observe that every Sugihara chain is an idempotent De Morgan monoid. Particular examples are the Sugihara matrices of §29.3 (but see §26.9):

$$S_{Z^*} = \{\ldots, -i, \ldots, -2, -1, +1, +2, \ldots, +i, \ldots\}$$

and

$$S_i = \{-i, \ldots, -2, -1, +1, +2, \ldots, +i\}$$

treated as lattices with $\leq$ the natural ordering, with $\bar{a}$ the negative of a, and with $\rightarrow$ defined as in 4 above. $+1$ is the monoid identity.

Now an idea of universal algebra. Let S be an algebra. S is a *subdirect product* of $S_1, \ldots, S_n$ iff (1) S is a subset of the direct product $S_1 \times \ldots \times S_n$, with operations defined "pointwise"; and (2) for each i, and for each $a \in S_i$, at least one element $\langle s_1, \ldots, s_i, \ldots, s_n \rangle$ in S is such that $s_i = a$. In plain language, every a in every S_i gets "used" at least once. In other words, the homomorphism $h: S \rightarrow S_i$ defined by $h(\langle s_1, \ldots, s_i, \ldots, s_n \rangle) = s_i$ is *onto* S_i (Birkhoff 1948; but he says "union" for "product"). It is a theorem of universal algebra that every algebra is isomorphic to a subdirect product of algebras which are in a reasonable sense not further decomposable (Birkhoff 1948). Our first result is along these lines.

Embedding theorem. If S is an idempotent De Morgan monoid, then S is isomorphic to a subdirect product S′ of Sugihara chains.

This result includes the Stone 1936 embedding theorem for Boolean algebras as a special case once it is realized that idempotent De Morgan monoids are generalizations of Boolean algebras, and that the only Boolean algebra which is a chain (excluding the trivial one-element algebra) is the two element algebra.

Proof. We show that for every S, if there are $a, b \in S$ such that $a \nleq b$, then there exists a homomorphism h of S onto a Sugihara chain so that $h(a) \nleq h(b)$. This follows because if $a \nleq b$, then $t \nleq a \rightarrow b$. Then by the Stone 1937 prime filter theorem for distributive lattices, there is a prime filter P such that $t \in P$ and $(a \rightarrow b) \notin P$. Define S/P as the algebra obtained by identifying into equivalence classes [x] the set of all y such that $(x \rightleftarrows y) \in P$, and defining operations upon the classes via operations upon their representatives. It is easy to see that S/P is the desired homomorphic image of S under the natural homomorphism, $h(a) = [a]$, once we observe the following three theorems about idempotent De Morgan monoids:

1 $t \leq (a \rightarrow b) \vee (b \rightarrow a)$
2 $(a \rightarrow b) \leq ((a \rightarrow b) \rightleftarrows (\bar{a} \vee b))$
3 $\overline{a \rightarrow b} \leq ((a \rightarrow b) \rightleftarrows (\bar{a} \wedge b))$.

1 ensures that S/P is a chain, and 2 and 3 ensure that $\rightarrow$ is defined as on the Sugihara matrix. Of course $[t]$ is the identity for S/P, so that S/P is a Sugihara chain.

So for any two distinct elements $a, b \in S$ we can find a homomorphic image of S under a homomorphism h so that $h(a) \neq h(b)$. By a familiar construction of Stone's we show that S is embeddable in the direct product of all the S/P's. Let Q be the set of prime filters P, and let h_P be the natural homomorphism onto S/P. Define $f(a) = \{h_P(a)\}_{P \in Q}$; f is the desired isomorphism. And its image S′ is a subdirect product of the various S/P's, because all of the h_P's are onto.

It is easy to see that, if S is finite then S is isomorphic to a subdirect product of a finite number of finite chains. And the only finite chains available (besides the degenerate one element chains) are isomorphic to some S_i with perhaps *0* thrown in (*0* made its own negation, *0* hence becoming the identity). Since we have used S_i for the chain $\{-i, \ldots, -1, +1, \ldots, +i\}$ *without 0*, let us use S_i0 for the chain $\{-i, \ldots, -1, 0, +1, \ldots, +i\}$ *with 0*; and let us use $S_i(0)$ as ranging over both S_i and S_i0. We thus have

FINITE EMBEDDING THEOREM. Every finite idempotent De Morgan monoid S with more than one element is isomorphic to a subdirect product S′ of a finite number of $S_i(0)$'s.

Remark that we can give a constructive proof of the Finite embedding theorem by observing that every prime filter containing the identity will be the principal filter determined by some join-irreducible element $\leq t$.

The algebras we have considered can be made to relate to logics in the usual way (see §28.2.3). In the first place we turn an algebra into a *matrix* by supplying designated elements. This is easy for monoids; given any monoid, we construe it as a matrix by choosing the set $\{a : t \leq a\}$ as the designated elements. Then a formula A is *valid* in a monoid iff every valuation v carries it into a designated element. For the next theorem we need one new idea: say two matrices are equivalent if they render valid exactly the same set of formulas.

EQUIVALENCE THEOREM. Every finite idempotent De Morgan monoid with more than one element is equivalent (as a matrix) to some $S_i(0)$.

We can now tell you what the equivalent $S_i(0)$ in question will be for a given idempotent De Morgan monoid S. By the Finite embedding theorem, we know that S is isomorphic to some subdirect product S′ of $S_{i_1}(0), \ldots, S_{i_n}(0)$. Without loss of generality, assume that $S_{i_n}(0)$ is maximal in size among the algebras $S_{i_1}(0), \ldots, S_{i_n}(0)$. We will show that S′ is equivalent to $S_{i_n}(0)$. This will complete the proof because we know that S is isomorphic, and hence equivalent, to S′.

We show first that if a formula *A* is valid in S′, then it is valid in $S_{i_n}(0)$, by contraposition. This will be just a bit of universal algebraic logic. Suppose then that a formula *A* is falsifiable in $S_{i_n}(0)$ by a valuation v. Then define a valuation v′ on the propositional variables p, so that $v'(p) = \langle x_{i_1}, \ldots, v(p)\rangle$, for some elements $x_{i_1} \in S_{i_1}(\mathbf{0}), \ldots$. There must be such an element with last component $v(p)$, since we are dealing with a subdirect product. And clearly, since operations in the subdirect product are defined component-wise, $v'(A) = \langle y_{i_1}, \ldots, v(A)\rangle$, for some elements $y_{i_1}, \ldots$. And so since $v(A)$ is undesignated, so must $v'(A)$ be undesignated.

So we next show the other half: If A is valid in $S_{i_n}(\mathbf{0})$, then A is valid in the subdirect product S′ of $S_{i_1}(\mathbf{0}), \ldots, S_{i_n}(\mathbf{0})$. It is convenient to prove the following lemma, which is crucial not only here, but later on.

SEQUENCE LEMMA. Recall that $S_i\mathbf{0}$ is S_i with $\mathbf{0}$. Consider the sequence of matrices $S_1, S_1\mathbf{0}, S_2, S_2\mathbf{0}, \ldots, S_i, S_i\mathbf{0}, \ldots$. Let us relabel these matrices in order as $M_2, M_3, M_4, M_5, \ldots$. (The new subscripts indicate the number of elements.) Then if A is valid in M_i, then for all $j \leq i$, A is valid in M_j.

PROOF. It obviously suffices to show that, if A is valid in M_i, then A is valid in its predecessor (if it exists) M_{i-1}. We distinguish two cases.

CASE 1. M_i has the form $S_n\mathbf{0}$. This is easy, because M_{i-1} is then just S_n, which is a submatrix of $S_n\mathbf{0}$.

CASE 2. M_i has the form S_n. The algebraically minded reader may also properly regard this case as easy since $M_{i-1} = S_{n-1}\mathbf{0}$, and he undoubtedly sees that $S_{n-1}\mathbf{0}$ is a homomorphic image of S_n and that validity is preserved under homomorphic images. For others of us the proof could go as follows.

The proof is by contraposition. We suppose that A is not valid in $M_{i-1} = S_{n-1}\mathbf{0}$. Then there is a valuation v_{n-1} into $S_{n-1}\mathbf{0}$, so that $v_{n-1}(A) < \mathbf{0}$. We define a new valuation v_n into S_n on the propositional variables by cases: if $v_{n-1}(p) \geq \mathbf{0}$, $v_n(p) = v_{n-1}(p)+1$ (this is straight addition, *not* the $\oplus$ of §29.3); if $v_{n-1}(p) < \mathbf{0}$, $v_n(p) = v_{n-1}(p)-1$. It is then possible to show by induction, that for any formula A: if $v_{n-1}(A) > \mathbf{0}$, then $v_n(A) = v_{n-1}(A)+1$; if $v_{n-1}(A) < \mathbf{0}$, then $v_n(A) = v_{n-1}(A)-1$; and if $v_{n-1}(A) = \mathbf{0}$, then $v_{n-1}(A) \in \{+\mathbf{1}, -\mathbf{1}\}$. (The idea of the "disjunctive" values in the last clause, which seems to be the most economical way of getting the induction to go through, is due to Woodruff.) The inductive argument is straightforward, as we say in this racket; i.e., it would take several pages to do it. But let us convince you of its plausibility. A value $n > \mathbf{0}$ is really indistinguishable from $n+1$, and a value $n < \mathbf{0}$ is really indistinguishable from $n-1$, so the raising up and lowering down ought to be O.K. The only other relevant remark to make is that we are in effect "splitting" $\mathbf{0}$ into two "halves," $+\mathbf{1}$, and $-\mathbf{1}$, both of which have (in a matrix without $\mathbf{0}$) certain aspects of $\mathbf{0}$ (in a matrix with $\boldsymbol{0}$). And the splitting is harmless since $+\boldsymbol{1}$ and $-\boldsymbol{1}$ are closed among themselves. (See §25.2.2 and — especially — §42.3.4 for more splitting. The present case was the modest genesis of splitting. The reader may wish to compare Dunn 1970 for its use in showing Ackermann's rule (γ) admissible for **RM**. This splitting of $S_n\boldsymbol{0}$'s for **RM** suggested to Meyer the more general splitting techniques for other relevant logics.)

The upshot of all this is that it is possible to define v_n from the given v_{n-1} so that if $v_{n-1}(A) < \boldsymbol{0}$, then $v_n(A) \leq -\boldsymbol{1}$. So if A is valid in S_n, it is valid in $S_{n-1}\boldsymbol{0}$.

Having established the Sequence lemma, we return to showing the other half of the Equivalence theorem. We know that if A is not valid in the subdirect product S′ of $S_{i_1}(\boldsymbol{0}), \ldots, S_{i_n}(\boldsymbol{0})$, there is a valuation v such that $v(A) = \langle x_{i_1}, \ldots, x_{i_n}\rangle$ and such that $\langle x_{i_1}, \ldots, x_{i_n}\rangle \not\geq \langle \boldsymbol{t}_{i_1}, \ldots, \boldsymbol{t}_{i_n}\rangle$, where the $\boldsymbol{t}_i$'s are the identities again. There must exist x_{i_j} such that $x_{i_j} \not\geq \boldsymbol{t}_{i_j}$ (since order is defined component-wise). We can then falsify A in $S_{i_j}(\boldsymbol{0})$ as follows. Define the i_j-th projection h_{i_j} so that for an arbitrary element of the subdirect product, $h_{i_j}(\langle y_{i_1}, \ldots, y_{i_j}, \ldots, y_{i_n}\rangle) = y_{i_j}$. h_{i_j} is a homomorphism of the subdirect product onto $S_{i_j}(\boldsymbol{0})$. So we now define a valuation v_{i_j} into $S_{i_j}(\boldsymbol{0})$ such that $v_{i_j}(B) = h_{i_j}(v(B))$. $v_{i_j}(A) = x_{i_j} \not\geq \boldsymbol{t}_{i_j}$, falsifying A in $S_{i_j}(\boldsymbol{0})$. Now notice: from the way we chose $S_{i_n}(\boldsymbol{0})$ as maximal in size, it follows that $S_{i_j}(\boldsymbol{0})$ precedes $S_{i_n}(\boldsymbol{0})$ in the sequence M_2, M_3, . . . , of the Sequence lemma (unless they are identical). So by the Sequence lemma it follows that A is falsifiable in $S_{i_n}(\boldsymbol{0})$. So the Equivalence theorem is finished.

REJECTABILITY THEOREM. If X is an extension of **RM**, and if A is not a theorem of X, then there exists an $S_n(\boldsymbol{0})$ such that A is rejected in $S_n(\boldsymbol{0})$ and such that $S_n(\boldsymbol{0})$ is adequate for X (i.e., all of X's theorems are valid in $S_n(\boldsymbol{0})$).

This theorem is essentially just the "finite model property" investigated by Harrop 1958, 1959, and others, but with the models in question importantly restricted to the $S_n(\boldsymbol{0})$'s. We begin by simply proving that X has the finite model property. We consider the sub-Lindenbaum algebra of X generated by the (finitely many) propositional variables in the given non-theorem A. This sub-Lindenbaum algebra is finite, as may be seen by observing that it follows from Meyer's completeness results for **RM** (§29.3) that there are but finitely many non-equivalent formulas in **RM** in a finite number of variables. *A fortiori* the same holds true of any extension X of **RM**. This sub-Lindenbaum algebra L is adequate for X and rejects S. Since L is a finite idempotent De Morgan monoid, by the Equivalence theorem it follows that L is equivalent to some $S_n(\boldsymbol{0})$, which completes the proof.

We now show the

EXTENSION THEOREM. If X is a consistent proper extension of **RM**, X has a finite characteristic matrix, namely, some $S_n(\boldsymbol{0})$ (= some M_i).

PROOF. Let G be the set of indices i of the matrices M_i in the Sequence lemma such that M_i is adequate for X. G is either infinite or finite. If G is infinite, then it follows by the Sequence lemma that G contains every index. But by Meyer's completeness result, X is identical with **RM** (since the theorems of X are then valid in each of S_1, S_2, . . .), contradicting the assumption that X is a *proper* extension.

So suppose that G is finite. Now by the Rejectability theorem, any non-provable formula is rejectable in some M_i such that M_i is adequate for X. Since X is consistent it has at least one non-provable formula, so G is non-empty. Let n be the largest index in G. By the Sequence lemma, if a formula is rejectable in M_i ($i \leq n$), it is rejectable in M_n. So any non-provable formula is rejectable in M_n. On the other hand, by construction, any provable formula is valid in M_n, so M_n is characteristic for X, and the proof is done.

Now for some corollaries.

DUGUNDJI FORMULA THEOREM. Every consistent proper extension X of **RM** may be obtained by adding to **RM** one the following so-called "Dugundji formulas," each giving rise to a distinct extension:

$$P_3 \quad (p_1 \rightleftarrows p_2) \vee (p_1 \rightleftarrows p_3) \vee (p_2 \rightleftarrows p_3)$$
$$P_4 \quad (p_1 \rightleftarrows p_2) \vee (p_1 \rightleftarrows p_3) \vee (p_1 \rightleftarrows p_4) \vee (p_2 \rightleftarrows p_3) \vee (p_2 \rightleftarrows p_4) \vee (p_3 \rightleftarrows p_4)$$
$$\vdots \qquad \vdots$$

PROOF. Let P_n denote the Dugundji formula which contains exactly n distinct variables. Clearly P_n is rejectable in a matrix M_i iff $i \geq n$, i.e., iff M_i has n or more elements. Now by the Extension theorem, any consistent proper extension X has a characteristic matrix M_i. So any Dugundji formula P_n is valid in M_i iff $i < n$. So if we add P_{i+1} to **RM**, it is valid in M_i, but not valid in M_{i+1}, M_{i+2}, M_{i+3}, ... So the extension with P_{i+1} must have M_i as its characteristic matrix (a smaller matrix won't do, because then the formula P_i would become valid; but it is rejectable in M_i, which we have already seen to satisfy **RM**+P_{i+1}).

NORMAL EXTENSIONS THEOREM. Ackermann's rule (γ) is admissible for all those consistent proper extensions which have a characteristic matrix of the form S_n, and is inadmissible for all those with a characteristic matrix of the form $S_n\boldsymbol{0}$.

PROOF. Ackermann's rule (γ) is easily seen to be admissible for all of the extensions with characteristic matrices of the form S_n because these are all normal in the sense of §29.2. The second part of the theorem may be shown by consideration of an example that we owe to Meyer. Consider the matrix $S_1\boldsymbol{0}$ in relation to the formula

$$(p_2\&\sim p_2)\vee(p_1\vee\sim p_1\rightarrow.p_2\vee\sim p_2).$$

It is easy to see that this formula is valid in $S_1\boldsymbol{0}$, because although the right disjunct is rejectable in $S_1\boldsymbol{0}$, the only way this can be done is to ensure that $p_2\vee\sim p_2$ gets the value $\boldsymbol{0}$. But then the left disjunct gets the value $\boldsymbol{0}$ as well, and is designated. We can generalize this example so that for the matrix $S_n\boldsymbol{0}$ we get the same effect with the formula

$$(p_{n+1}\&\sim p_{n+1})\vee(p_1\vee\sim p_1\rightarrow.p_2\vee\sim p_2)\vee\ \ldots\ \vee(p_n\vee\sim p_n\rightarrow.p_{n+1}\vee\sim p_{n+1}).$$

To motivate the next theorem, we recall that it was shown by Meyer (see §29.3.3) that **RM** does not satisfy the strong relevance principle satisfied by **R**, namely, that $A\rightarrow B$ is a theorem only if A and B share a sentential variable. Rather it was shown that **RM** does satisfy the weaker relevance principle that $A\rightarrow B$ is a theorem only if either A and B share a sentential variable or *both* $\sim A$ and B are theorems (RM84 of §29.3.3).

WEAK RELEVANCE THEOREM. Every proper extension of **RM** with a characteristic matrix of the form S_n fails to satisfy the weak relevance principle mentioned above, although all those with a characteristic matrix of the form $S_n\boldsymbol{0}$ do.

PROOF. The first part of the theorem is obvious for the extension with characteristic matrix S_1, since this is just the classical sentential calculus. So the first interesting case arises in considering that extension which has S_2 as its characteristic matrix. But the following formula is easily seen to be valid in this matrix:

$$p\&\sim p\rightarrow.q_1\vee(q_1\rightarrow q_2).$$

And yet the consequent can be falsified by itself in S_2. In general, for the matrix S_n the following formula has the same effect:

$$p\&\sim p\rightarrow.q_1\vee(q_1\rightarrow q_2)\vee\ \ldots\ \vee(q_{n-1}\rightarrow q_n).$$

We leave the "although" part of the theorem to the reader.

REMARK. The last two theorems in combination show that **RM** sits in a precarious position, being maximal among those systems which both have (γ) admissible and which satisfy the weak relevance principle. The "odd-numbered" extensions of **RM** fail the (γ) test, and the "even-numbered" extensions fail the weak relevance test.

COMPLETE, CONSISTENT EXTENSION THEOREM. The only Post consistent, Post complete extension of **RM** is the classical propositional calculus. (X is Post consistent if not all formulas are theorems, and Post complete if every proper extension is Post inconsistent.)

As we remarked at the beginning, Ulrich 1967 originally obtained this theorem from a result of Tarski. But it is an obvious consequence of the last theorem. For it is clear that any consistent extension of **RM** that is not obtainable by adding the Dugundji formula P_3, i.e., that is not already the classical propositional calculus, can be further extended by adding P_3.

In closing, let us mention that if we were to write the section over again, we would have included the matrix S_0 (0 its only element) in the sequence setting $M_1 = S_0$, for the sake of generality. By doing this we can remove the restriction in the Finite embedding theorem that the idempotent De Morgan monoids have more than one element, and similarly for the Equivalence theorem. The Extension theorem would run just as before, except now we could remove the restriction that the extension be consistent (we only used that fact to show G non-empty, but now we always have $M_1 \in G$, since every formula is valid in S_0.

All of this allows for a treatment of all extensions of **RM** — not just the consistent ones. It is obvious that every *Post inconsistent* extension of **RM** has S_0 as its characteristic matrix. But this is also true for *inconsistent* extensions in the sense that we have used throughout (often called *negation inconsistency*) — some formula and its negation both provable. Let us state this as the

INCONSISTENT EXTENSIONS THEOREM. Every (negation) inconsistent extension of **RM** has S_0 as its characteristic matrix.

PROOF. Before we proceed, note that the syntactic correlate of what we are arguing is that every negation inconsistent extension of **RM** is Post inconsistent. Meyer has a prior syntactic proof of this for the case not only of **RM** but also for **R** (see §29.11). However, our semantic proof for **RM** is easy, so we proceed.

Suppose X is a negation inconsistent extension of **RM**. Then by the Extension theorem (with the consistency restriction removed as indicated above) it has some M_i as its characteristic matrix. If $i = 1$, then $M_i = S_0$, and we are through. Suppose $M_i = S_n (n > 0)$. Then since S_n is normal in the sense of §29.2, no formula and its negation can both be valid in S_n and so X must be negation consistent, contrary to hypothesis. The only chance for a negation inconsistent extension which is not a Post inconsistent extension is if $M_i = S_n\boldsymbol{0}$ $(n > 0)$. But then the extension X can be further extended to an extension with characteristic matrix S_n, and we have just seen that any such extension must be negation consistent.

§29.5. Why we don't like mingle. As is clear from the previous two subsections, **RM** (which has been investigated principally by McCall, Dunn, and Meyer) is stable, and has some interesting properties. The mingle systems have also been of importance in suggesting results for other systems. But there is one respect in which mingle systems fail completely to be "of a kind" with the systems **T**, **E**, and **R** of this volume.

Meyer has shown that, in view of the unhappy theorem $A\rightarrow.{\sim}A\rightarrow A$ of **RM** (it comes from the mingle axiom in the form ${\sim}A\rightarrow.{\sim}A\rightarrow{\sim}A$ by contraposing the consequent, and then permuting), we have

$$(A\rightarrow A)\&(B\rightarrow B)\rightarrow.\overline{(A\rightarrow A)\&(B\rightarrow B)}\rightarrow.(A\rightarrow A)\&(B\rightarrow B).$$

Whence by identity, &I, →E, and De Morgan we come to

$$\overline{A\rightarrow A}\vee\overline{B\rightarrow B}\rightarrow.(A\rightarrow A)\&(B\rightarrow B),$$

and finally by properties of disjunction and conjunction to

$$\overline{A\rightarrow A}\rightarrow.B\rightarrow B,$$

which loses all semblance of either relevance or necessity. And such theorems as

$$(A\rightarrow B)\vee(B\rightarrow A)$$

bring us so close to the dread material "implication," that, in spite of the formal interest **RM** has, we are inclined to echo, from Alighieri 1307, *Lasciate ogni sperenza, voi ch'entrate* (*Inferno*, III, 9), at least if *voi* are intent on entering into a theory of "if ... then —," and not some other interesting topic.

§29.6. "... the connection of the predicate with the subject is thought through identity. ..." This celebrated phrase, drawn from Kant 1781, was intended as an explanation of analyticity: "S is P" is to be *analytic* just in case the subject is in some sense partially identical with the predicate. His attitude was of course influenced by his ignorance of how meager logic was at the time of his writing. "It is remarkable also that to the present day this [Aristotelian] logic has not been able to advance a single step, and is thus to all appearance a closed and completed body of doctrine" (1781, preface to second edition). We contend that though logic *has* advanced several steps since 1781, Kant's insight can still be construed as correct.

§29.6.1. Perry's analytic implication. This view might also be attributed to Parry, at least in the sense that we may look at Parry's 1933 system of *analytische Implikation* (mentioned in §15.1) as one way of understanding Kant's dictum that the "predicate is contained in the subject." Parry's axioms are as follows (our notation and numbering), where we take $\rightarrow$, &, and $\sim$ as primitive, with $\vee$ and $\rightleftarrows$ defined as usual. (The axioms given here are not exactly those of Parry's original paper. They form an equivalent and somewhat simpler set which Professor Parry presented at a 1963 Colloquium in Buffalo; he was kind enough to communicate them to us in correspondence. We label the system **PAI** for reference.)

POSTULATES FOR **PAI**

&:

1 $(A\&B) \rightarrow (B\&A)$
2 $(A\rightarrow(B\&C))\rightarrow.A\rightarrow B$
3 $(A\rightarrow B)\&(A\rightarrow C)\rightarrow.A\rightarrow(B\&C)$

$\sim$:

4 $A \rightarrow \sim\sim A$

$\vee$ and &:

5 $(A\rightarrow C)\&(B\rightarrow C)\rightarrow.(A\vee B)\rightarrow C$
6 $(A\&(B\vee C))\rightarrow.(A\&B)\vee(A\&C)$

$\vee$ and $\sim$:

7 $A\rightarrow B\rightarrow.\sim A\vee B$

$\vee$, &, and $\sim$:

8 $(A\vee(B\&\sim B)) \rightarrow A$

Arbitrary formulas $(\ldots A \ldots)$ containing A:

9 $(\ldots A \ldots)\rightarrow.A\rightarrow A$
10 $((A\rightleftarrows B)\&(\ldots A \ldots))\rightarrow(\ldots B \ldots)$

Rules: $\rightarrow$E and &I.

Many questions concerning this system are open. Consistency is of course assured, but no one has yet investigated a number of other natural problems. Can the pure implicational theory (or any other interesting fragments) of the system be isolated? Is the system or any of its interesting fragments decidable? What semantical theory corresponds to the syntax? Gödel is quoted (in Parry 1933) as saying

> "*p* impliziert *q* analytisch," kann man vielleicht so interpretieren: "*q* ist aus *p* und den logischen Axiomen ableitbar und enthält keine anderen Begriffe als *p*" und es wäre, nachdem man diese Definition genauer präzisiert hat, ein Vollständigkeitsbeweis für die *Parry*schen Axiome zu erstreben, in dem Sinn, dass alle Sätze, welche für die obige Interpretation von → gelten, ableitbar sind.

What does this mean?

We leave these and other questions for future investigation by ourselves or others (but see §29.6.2), proving only Parry's theorem, which suggests the Kantian interpretation mentioned above.

The system **PAI** is motivated by the idea that the consequent of an *analytische Implikation* should simply "unpack" the antecedent, and that in consequence such formulas as

$$A\rightarrow .A\vee B$$

and

$$A\rightarrow B\rightarrow .C\rightarrow A\rightarrow .C\rightarrow B$$

should fail, since the consequents of these might refer to information not contained in the antecedents. Parry in fact states this property formally, and shows that his system has it:

THEOREM. If $A\rightarrow B$ is provable in **PAI**, then all variables in B also occur in A.

The proof uses the following matrices, which satisfy the axioms and rules:

~	
−1	+0
−0	+1
*+0	−1
*+1	−0

&	−1	−0	+0	+1
−1	−1	−1	−1	−1
−0	−1	−0	−1	−0
+0	−1	−1	+0	+0
+1	−1	−0	+0	+1

→	−1	−0	+0	+1
−1	+0	+0	+0	+0
−0	−1	+1	−1	+1
+0	−1	−1	+0	+0
+1	−1	−0	−1	+1

Then if B contains variables not in A, we give such variables the value -1, and all other variables -0. Then A takes the values -0 or $+1$, and B takes $+0$ or -1; by the → matrix, $A\rightarrow B$ then takes the value -1, and is therefore unprovable. Hence if $A\rightarrow B$ is provable, B contains no variables not present in A.

This way of looking at analyticity in Kant's sense has strong initial plausibility, but it seems to force on us some choices which we may not want to make. Certainly Kant would like to have "All brothers are siblings" be analytic, and one standard way of showing that it is so, strongly recommended by tradition, is to explain that with the help of appropriate definitions we can reduce the statement to a logical truth. If we take "brother" to mean "male sibling" we get a logical truth of the form

$$\forall x[(\mathrm{Male}(x)\&\mathrm{Sib}(x))\rightarrow\mathrm{Sib}(x)],$$

which should be provable in an appropriate quantificational extension of Parry's system.

But surely it is equally open to us to explain the analyticity of the statement in another way, i.e., by defining "sibling" as "brother or sister," in which case we get

$$\forall x[\mathrm{Bro}(x)\rightarrow(\mathrm{Bro}(x)\vee\mathrm{Sis}(x))],$$

which seems equally acceptable, but is not available in Parry's system.

Just how Kant's "thinking through identity" is to be understood is a topic we do not know how to discuss — but either of these examples seems as good a case of it as the other. We claim, at any rate, that the two stand or fall together, which is entailed by the fact that they stand together, as was shown in §15.1.

Another system with Parry's property is that of Hintikka 1959. He considers a system in which equivalence is defined metalogically in such a way that $A\rightleftarrows B$ holds only when A and B have exactly the same variables: "Formulae which are tautologically equivalent by the propositional calculus are equivalent provided that they contain occurrences of exactly the same free variables, and so are expressions obtained from them by replacing one or more free individual variables by bound ones." He then writes $A\rightleftarrows B$ when this metalogical equivalence holds, and he lets $A\rightarrow B$ abbreviate $A\rightleftarrows(A\&B)$. Hintikka nowhere suggests that his $\rightarrow$ is to be understood as entailment, but we take the opportunity of pointing out that his condition is neither necessary nor sufficient for tautological entailmenthood: $B\&A\&{\sim}A\rightarrow B$ satisfies his condition, and $A\rightarrow.A\vee B$ fails.

Still another study based on related ideas came to our attention too late to permit more than citation: Zinov'ev 1967, especially chapters six and seven.

§29.6.2. Dunn's analytic deduction and completeness theorems. Although exactly what Parry's system **PAI** (§29.6.1) comes to as it stands remains unclear, Dunn 1972 has demonstrated that one obtains a system with nice properties answering to the underlying intuitions of analytic implication by adding a couple of axioms.

In the first place, Dunn observes that Parry's own system is "modal"; in particular, **PAI** is a subsystem of **S3**. So in order to disentangle considerations of analytic relevance from considerations of modality, Dunn proposes demodalizing the system by adding to the axioms of §29.6.1

11 $A \to . A \to A \to A$ (or, equivalently, $A \to . \overline{A} \to A$).

It is to be noted that 11 satisfies the matrix of §29.6.1, so that the Parry property is not disturbed. The move from Parry's own system to that of Dunn is then partly analogous to the move from the system **E** of entailment to the system **R** of relevant implication as in §27.1.1

Secondly, Dunn adds

12 $A \& \bar{B} \to . \overline{A \to B}$, or equivalently, $(A \to A \& . B \to B) \to . A \to B \to . A \to B)$,

which is known to be unprovable in Parry's original system, even with 11 added (the fact, due to Meyer, is reported in Dunn 1972). 12 also satisfies the Parry matrix of §29.6.1. The reason for this addition is simple: it is needed in order to sustain the revealing theorems Dunn proves about the modified system. Letting **DAI** be Dunn's modification (i.e., **PAI** + 11 + 12), we mention one proof-theoretical and one semantic result.

First, what Dunn nicely calls the

ANALYTIC DEDUCTION THEOREM. If there is in **DAI** a proof of B in the Official sense from hypotheses $C_1, \ldots, C_n, A$, and if every variable which occurs in B also occurs in A, then there is in **DAI** a proof of $A \to B$ from $C_1, \ldots, C_n$.

This gives a sharp generalization of a property queried by Gödel for the original Parry system. Gödel's question: given arrow-free A and B such that all variables in B are also in A, is $A \to B$ provable whenever B is deducible from A in the two valued calculus?

Second, Dunn gives alternative semantic characterizations of his modified system **DAI**, of which we choose one. We define a *Parry matrix* to be any structure isomorphic to the Cartesian product $S \times \{\boldsymbol{F}, \boldsymbol{T}\}$ of a family of sets S closed under set union ($\cup$) with the two element Boolean algebra $\{\boldsymbol{F}, \boldsymbol{T}\}$, choosing as designated all and only the elements $\langle a, \boldsymbol{T}\rangle$ with $a \in S$ and with the following operations, where $a, b \in S$ and $x, y \in \{\boldsymbol{F}, \boldsymbol{T}\}$

$$\langle a, x\rangle \vee \langle b, y\rangle = \langle a \cup b, x \vee y\rangle;$$
$$\overline{\langle a, x\rangle} = \langle a, \bar{x}\rangle;$$
$$\langle a, x\rangle \wedge \langle b, y\rangle = \langle a \cup b, x \wedge y\rangle;$$
$$\langle a, x\rangle \to \langle b, y\rangle = \langle a \cup b, \boldsymbol{T}\rangle \text{ or } \langle a \cup b, \boldsymbol{F}\rangle$$

according as both $b \subseteq a$ and ($x = \boldsymbol{F}$ or $y = \boldsymbol{T}$), or not.

Dunn observes that the Parry matrix of §29.6.1 is a Parry matrix with S = {∅, U}, and with ∅ the empty set and U some non-empty set, and he proves that a formula A is a theorem of his modified calculus **DAI** just in case it is valid in every Parry matrix.

Last, Dunn 1972 reports that Meyer has given an intuitively appealing account of the import of this algebraic structure, which we present in our own way as follows.

1. Let a proposition be reified as an ordered pair ⟨a, x⟩ consisting of a "content" in the sense of a *set of objects*, a, "mentioned" by the proposition, and a *truth value*, x.
2. Let the set of objects "mentioned" by a complex proposition be always the union of the sets of objects mentioned by its parts: it mentions all and only those objects mentioned by at least one of its parts.
3. Let the truth value of truth functional compounds depend upon that of their parts in the usual way.
4. Let the truth value of an analytic implication ⟨a, x⟩→⟨b, y⟩ be counted true just in case not only does the first materially "imply" the second but also the "content" of the second (the "predicate") is contained (literally, in the set theoretical sense) in the first (the "subject").

Then we obtain exactly the semantics indicated above.

§29.7. Co-entailment again. Following up §8.7, we mention Meyer's observation that in **R**, but not in **E**, we can define $A \rightleftarrows B$ as $(A \to B) \circ (B \to A)$. We know from properties of ∘ in **R** (§27.1.4) that

$$A \to B \to . B \to A \to . A \rightleftarrows B$$

under this definition, and it remains only to secure the elimination axioms for $A \rightleftarrows B$; we prove one of them:

1	$A \to B \to . \overline{B \to A} \to . \overline{A \to B}$	See §5.1.1
2	$A \to B \to . \overline{A \to B} \to . \overline{B \to A}$	1 contraposition in consequent
3	$\overline{A \to B} \to . A \to B \to \overline{B \to A}$	2 perm

Then contraposition and the definition of ∘ in **R** yield $E_{\rightleftarrows}1$ of §8.7. The other axiom of §8.7 comes similarly from $B \to A \to . A \to B \to . A \to B$.

Note that this proof is blocked in **E** at step 3.

§29.8. Connexive implication (by Storrs McCall). Thanks to the generosity of its principal authors (which they may already regret), this book has been expanded to include a number of neighbors of **E** in addition to **E** itself. This is not to say that the authors are really very fond of **E**'s neighbors. As they say, they growl a bit about them from time to time. Nor is it to say that the neighbors always resemble **E** very closely. In the belief that good fences make good neighbors, this section will emphasize the difference between **E** and connexive logics as much as the similarity. But the latter still fall within the embrace of **E**'s Good Neighbor Policy (or so we hope).

§29.8.1. Connexive logic and connexive models. The search for a satisfactory connexive logic is motivated by different considerations from those which led to **E**. Connexive logic represents an attempt to formalize the species of implication recommended by Chrysippus:

> And those who introduce the notion of connexion say that a conditional is sound when the contradictory of its consequent is incompatible with its antecedent. (Sextus Empiricus, translated in Kneale and Kneale 1962, p. 129.)

Using "$A \circ B$" to signify that A is compatible with B, we have Chrysippus's definition:

1 $A \rightarrow B =_{\text{df}} \overline{A \circ \overline{B}}$.

Combining 1 with the plausible thesis that if A implies B, A is compatible with B:

2 $A \rightarrow B \rightarrow . A \circ B$,

we obtain, using classical negation theses:

3 $A \rightarrow B \rightarrow . \overline{A \rightarrow \overline{B}}$.

This latter expression, named *Boethius's Thesis*, may be regarded along with *Aristotle's Thesis* $\overline{\overline{A} \rightarrow A}$ as one of the characteristic marks of a system of connexive logic. (For Boethius and Aristotle see McCall 1963, 1964 and 1966. Chrysippus's definition 1 and theses 2 and 3 are to be found in Nelson 1930.)

Three points are quickly made. (i) Aristotle's and Boethius's theses are non-classical in the sense of not forming part of two valued logic. (ii) They cannot be consistently added to any system of logic in which formulas imply or are implied by their own negations. (iii) Systems of the latter type include all the well-known alternatives to classical logic, such as intuitionistic logic, the Lewis systems, Lukasiewicz's many-valued logics, **E**, **R**, **T**, etc. It follows that the search for a connexive logic must extend more widely and that the sought-for system (assuming it to comprise only the connectives $\rightarrow$, &, $\vee$, $\overline{}$ and $\rightleftarrows$) will not, unlike the family of systems **H**, **S4**, **E**, **R**, **T**, be a fragment of classical logic in the sense of containing only two-valued theses. Instead it will stand to classical logic rather as Riemannian and Lobatchevskian geometries stand to Euclidean.

It is easy to see how **H**, **S4**, **E**, **R**, **T**, etc., come to be non-connexive. The most obvious difficulty lies with the theses $A\&B \to A$ and $A \to A \vee B$:

1	$A\&\bar{A} \to A$	
2	$A \to \bar{A} \vee A$	
3	$A\&\bar{A} \to \bar{A} \vee A$	transitivity
4	$A\&\bar{A} \to \overline{A\&\bar{A}}$	De Morgan and double negation

However, although connexive logic must refrain from asserting $A\&B \to A$ or $A \to A \vee B$, even this degree of self-denial is not enough. From a connexive point of view, the rot has already set in with $\mathbf{E}_{\to}$, which allows the formula $\overline{A\to A}\to(A\to A)\to(A\to A)\to\overline{A\to A}$ to entail its own negation (see McCall 1966). Even the feeble system $\mathbf{T}_{\to}$ is strong enough to yield a contradiction with Boethius's thesis is added to it:

1	$(\bar{A}\to A\to.\bar{A}\to A)\to.\bar{A}\to A\to\bar{A}\to.\bar{A}\to A\to A$	$\mathbf{T}_{\to 1}4$ (§8.3.2)
2	$\bar{A}\to A\to\bar{A}\to.\bar{A}\to A\to A$	1 $\mathbf{T}_{\to 1}1$ →E
3	$\bar{A}\to A\to A\to\overline{\bar{A}\to A\to\bar{A}}$	Boethius
4	$\bar{A}\to A\to\bar{A}\to\overline{\bar{A}\to A\to\bar{A}}$	2 3 transitivity
5	$\overline{\bar{A}\to A\to\bar{A}\to\overline{\bar{A}\to A\to\bar{A}}}$	Boethius, Ax. 1 →E

If this be so, what pure implicational system *can* be used as a foundation for connexive logic? Answer: a weakening $\mathbf{C}_{\to}$ of $\mathbf{E}_{\to}$ which differs from $\mathbf{T}_{\to}$ in that the axiom $(A\to.A\to B)\to.A\to B$ is dropped instead of the axiom $A\to A\to B\to B$. The following matrices, found in Angell 1962, show that various connexive systems can be consistently based upon $\mathbf{C}_{\to}$:

→	1	2	3	4	~
*1	1	4	3	4	4
*2	4	1	4	3	3
3	1	4	1	4	2
4	4	1	4	1	1

&	1	2	3	4
1	1	2	3	4
2	2	1	4	3
3	3	4	3	4
4	4	3	4	3

We say *various* connexive systems because no one system satisfying the matrices has yet emerged which appears to be wholly satisfactory. The reason will appear below. For the moment let us see how systems based on Angell's matrices stand with regard to **E**'s criteria of necessity and relevance.

Angell's matrices for implication and negation satisfy the criterion of necessity of §5.2.1 and §12: no theorem has the form $A\to.B\to C$ unless A contains an arrow. When conjunction is introduced, however, the criterion is no longer satisfied, as is shown by the fact that the formula $A\&A\to.B\to B$ satisfies the matrices. This is one respect in which connexive logics based on Angell's matrices are deficient. In another respect, however, they cleave more faithfully than **E** to the principle that contingent propositions should not entail necessary propositions. (See §22.1.2 and Routley and Routley 1969. As the Routleys point out, the principle itself may well be erroneous — try putting "2+2 = 4" for p in the true entailment "$\exists x(x$ knows that $p) \rightarrow p$"). A prime example is $A \rightarrow A \vee \bar{A}$, which connexive logics avoid. Turning to relevance, connexive systems based on the Angell matrices violate reasonable relevance criteria on the one hand, while adhering to an even stricter criterion on the other. Exemplifying the former is the doubly damned $A \& A \rightarrow . B \rightarrow B$, while the absence of $A \& B \rightarrow A$ illustrates the latter. The remarks in Nelson 1930 about $A \& B \rightarrow A$ (with minor alphabetic changes) reflect an intuitive criterion of relevance stricter than that applied to **E**:

> It cannot be asserted that the conjunction of *A* and *B* *entails* *A*, for *B* may be totally irrelevant to and independent of *A*, in which case *A and B* do not entail *A*, but it is only *A* that entails *A*. I can see no reason for saying that *A* and *B* entail *A*, when *A* alone does and *B* is irrelevant, and hence does not function as a premiss in the entailing.

A final point of comparison between **E** and systems based on Angell's matrices concerns the theses corresponding to disjunctive syllogism $A \& (\bar{A} \vee B) \rightarrow B$ (see §25.1), and antilogism $A \& B \rightarrow C \rightarrow . A \& \bar{C} \rightarrow \bar{B}$ (§22.2.3). Unlike **E**, Angell-type systems retain these not too implausible theses, which lead to fallacies of relevance only in conjunction with $A \& B \rightarrow A$ and its congeners.

In McCall 1966 Angell's matrices are axiomatized, and the resulting axiomatic system **CC1** is shown to be (a) functionally incomplete, and (b) Post complete. These results are mainly of formal interest, however, as **CC1** leaves much to be desired with regard to our intuitions of what a connexive logic ought to be. Besides containing such unsatisfactory theses as $A \& A \rightarrow . B \rightarrow B$ and $A \& A \rightarrow . A \rightarrow A \rightarrow . A \& A$, whose presence in **CC1** is doubtless due to the phenomenon of "matrix size" (matrix size, the plague of many propositional logics, derives from the fact that any system with a characteristic n-valued matrix contains theses which assert in effect that the system has n and only n values), **CC1** also lacks the extremely plausible theses $A \rightarrow A \& A$ and $A \& A \rightarrow A$. Worse still, Routley and Montgomery 1968 show that adding the latter to even relatively weak subsystems of **CC1** leads to inconsistency. What is needed is a new approach to the problem of capturing within a formal system the *idée maîtresse* of connexive logic — that no proposition can be incompatible with itself, and hence cannot entail, or be entailed by, its own negation.

As a first step toward constructing a more satisfactory connexive system, note that most if not all of the interesting features of connexivity occur within the sphere of first-degree formulas. To recall §19, a zero-degree formula contains only the connectives &, $\vee$, and $\sim$, while first degree formulas add ingredients of the form $A \rightarrow B$, where A and B are of zero degree. The remainder of this section will be devoted to formulating a system of connexive implication containing only zero and first degree formulas, which avoids some of the objectionable features of **CC1** and other systems based on a finite matrix.

The *idée maîtresse* of connexive logic, as was mentioned before, is that no proposition should imply or be implied by its own negation. In the algebraic model for relevant implication of §22.1.3, the structure of the distributive lattice is such that $-3 \leq +3$, and hence any formula which uniformly takes the value -3 *will* imply its own negation. What is perhaps the most natural way of avoiding this model-feature is to have two distinct isomorphic lattices D and U, unconnected by the ordering relation $\leq$, which have the property that each element x of D has a unique complement $\bar{x}$ in U and vice versa. (An alternative and in the writer's present opinion less satisfactory way of avoiding $x \leq \bar{x}$ is exhibited in the algebraic model depicted on p. 85 of McCall 1967.) Such a situation might be represented using two Hasse diagrams for four-element Boolean lattices as follows:

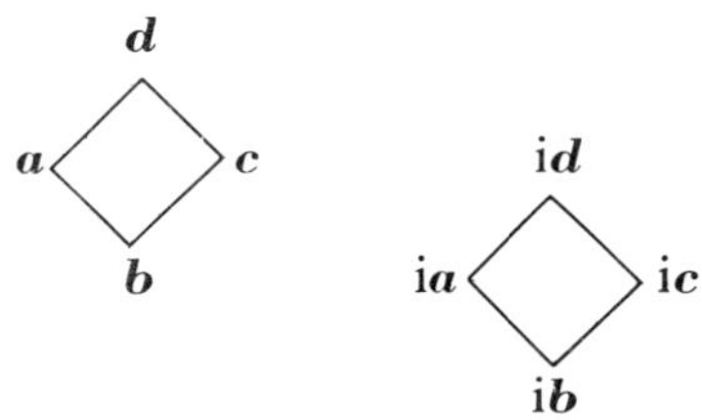

D U

Since D and U are isomorphic, each element x will have a unique image ix such that iix = x. Before defining the type of complementation that is needed, the operations of meet and join will be extended to apply to pairs of elements from different lattices. At present the meet $x \wedge y$ and join $x \vee y$ of two elements from the same lattice are Boolean. If x and y are in different lattices we assign $x \wedge y$ to the "lower" lattice U, and $x \vee y$ to the "upper" lattice D, as follows:

If $x \in D, y \in U; x \wedge y = ix \wedge y$ and $x \vee y = x \vee iy$
$x \in U, y \in D; x \wedge y = x \wedge iy$ and $x \vee y = ix \vee y$

(Since we shall eventually wish to take D as the class of "designated" values of our model, and to associate "$\wedge$" with "and" and "$\vee$" with "or," the definition of meet and join satisfies the requirement that $v(A\&B) \in D$

("is true") iff v(A) ∈ D and v(B) ∈ D; and that v(A∨B) ∈ D iff v(A) ∈ D or v(B) ∈ D.) Our Hasse diagrams may be modified to represent the operations of meet and join as follows:

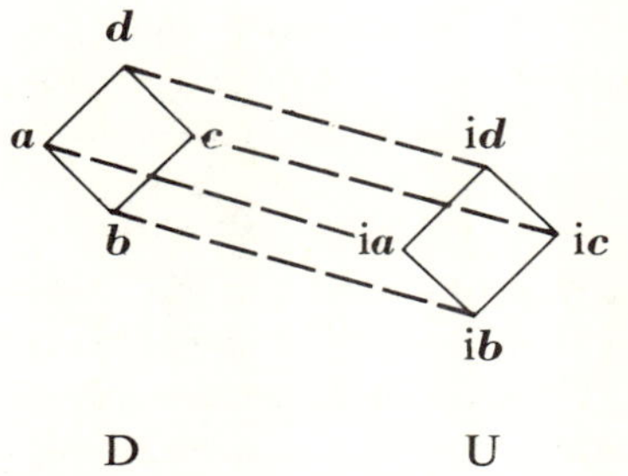

Thus ***a*∧i*d*** = i***a***, i***a***∨***c*** = ***d***, etc. Note that the dotted lines do *not* denote the ordering relation: we do not have for example i***c*** ≤ ***d***.

Turning now to complementation, remark that the structure of the diagram just above is such that the most obvious form of complementation is Boolean. (This represents one of the principal differences between the algebra we are constructing and the algebraic model for relevant implication of §22.1.3, the complementation of which is not Boolean.) This requires that we introduce the elements ***1*** and ***0***, which emerge naturally as the uppermost of the "designated" elements D and the least of the "undesignated" elements U, respectively. We then associate with each x an $\overline{x}$ such that $x \wedge \overline{x} = \mathbf{0}$ and $x \vee \overline{x} = \mathbf{1}$. This results in the following:

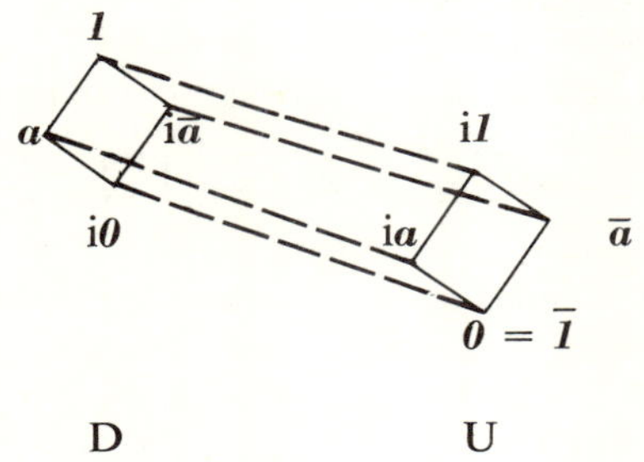

It remains to show that $\overline{x}$ is unique, and a proof of this is indicated by the fact that the last diagram is nothing more than a Boolean algebra with certain ordering relations missing. A structure like this may be formed from any Boolean algebra by bisecting it into (i) a prime filter (possibly but not necessarily based on one of the atoms of the algebra, e.g. i***0*** in the diagram just above), and (ii) the residue of the algebra, consisting of a prime ideal. All relations of partial ordering extending between elements of (i) and (ii) are canceled, although the elements of the algebra remain closed under exactly the same operations of meet and join as before. It is not difficult to show that the distributive lattices (i) and (ii), now renamed D and U, each contain exactly one of any pair of elements $\{x, \bar{x}\}$ of the original algebra. (Suppose for example that D contained both of such a pair. Then since $x \wedge \bar{x} = \mathit{0}$ and D is a filter, D would contain $\mathit{0}$, contrary to hypothesis.) Hence every element in D possesses a unique complement in U and vice versa, as was to be proved.

What we have done is to construct a "connexive algebra" which is connexive in the sense that $x \leq \bar{x}$ never holds, while at the same time a number of other plausible laws of algebraic logic continue to hold.

Codifying this construction in a definition, we define a *connexive algebra* as any septuple $\langle B, \wedge, \vee, \overline{}, D, U, \leq \rangle$, where B is a Boolean algebra with the usual operations $\wedge$, $\vee$ and $\overline{}$ defined on it, D is a prime filter in B, U is the complement of D in B, and $\leq$ is a binary relation on B defined as follows:

$$x \leq y \text{ iff (i) } x \vee y = y \text{ and (ii) either } x, y \in D \text{ or } x, y \in U.$$

(Note that condition (ii) can be simplified to read "either $x \in D$ or $y \in U$," since if $x \in D$ and $x \vee y = y$ then $y \in D$, and if $y \in U$ and $x \vee y = y$ then $x \in U$.) In what follows we shall make use of a certain subclass of the class C of all connexive algebras, namely a certain set of finite connexive algebras $\{C_n\}$. This set derives from the corresponding set of Boolean algebras $\{B_n\}$, in which each B_k has 2^k elements and is constructed by taking the Cartesian product of the two-element algebra with itself k times. (Each element is a string of k zeros and ones, with operations defined on the strings in normal Boolean fashion.) Each B_k is converted into a connexive algebra C_k by selecting a prime filter D_k based on one of B_k's atoms, namely, the atom whose string has a *1* in the last position and *0*'s elsewhere.

We shall now associate with these connexive algebras a system of propositional logic, interpreting $\leq$ as $\rightarrow$ and focusing for the time being on the relational properties of the arrow rather than on its properties as a connective or an operator. No nesting of $\rightarrow$ will be allowed, and the logic will contain zero and first degree formulas only.

The first step in associating a propositional logic with our family of algebras will be to define a new operation $\Rightarrow$ on the elements of each algebra as follows:

$$x \Rightarrow y = \mathit{1} \quad \text{if } x \leq y$$
$$x \Rightarrow y = \mathit{0} \quad \text{otherwise}$$

(The choice of *1* {or *0*} is relatively unimportant so long as *some* element of D {or its complement U} is chosen.) We then convert the algebra into a model by singling out D as the class of designated elements, and we assign values to propositional variables arbitrarily and to complex propositional formulas as follows:

$$\mathrm{v}(A \rightarrow B) = \mathrm{v}(A) \Rightarrow \mathrm{v}(B)$$
$$\mathrm{v}(A \& B) = \mathrm{v}(A) \wedge \mathrm{v}(B)$$
$$\mathrm{v}(\bar{A}) = \overline{\mathrm{v}(A)}$$

Where X is any connexive algebra, D is the prime filter of X which constitutes the class of designated elements, and v is a valuation as defined above, ⟨X, D, v⟩ is a *connexive model*. Let C* be the class of all connexive models corresponding to the class of all connexive algebras C, and let $\{C_n\}^*$ be the class of models corresponding to $\{C_n\}$. Then the set of propositional formulas receiving a designated value for all valuations in all models C* will be denoted by ΣC*; similarly for the set of formulas $\Sigma\{C_n\}^*$.

It is easy to see that the set ΣC* is consistent: if $\mathrm{v}(A) \in \mathrm{D}$, $\mathrm{v}(\bar{A}) \in \mathrm{U}$. §29.8.2 will be devoted to giving an axiomatic formulation **CFL** of ΣC*, and §29.8.3 to showing that **CFL** is complete in a fairly strong sense, namely that any consistent proper extension of **CFL** is characterized by a finite model.

§29.8.2. Axiomatization of the family of connexive models. The first thing to be noted about ΣC* and $\Sigma\{C_n\}^*$ is that they contain the full two-valued calculus in & and ¯. That this is so may be seen from the following tables for conjunction and negation, which spring in turn from the definitions of the meet and complementation operators: d and u are any elements such that d ∈ D and u ∈ U.

&	d	u	¯
*d	d	u	u
u	u	u	d

Defining ⊃ in terms of conjunction and negation, we see that the rule of ⊃-detachment holds, and this rule together with the rule of substitution will comprise the only rules of inference for the system **CFL**, whose axiomatic basis is as follows. (We recall the limitation that there shall be no arrow in the scope of another arrow, and we use the following definitions: 1. the standard definitions of ⊃, ∨, and ≡. 2. $(A \rightleftarrows B) =_{\mathrm{df}} (A \rightarrow B) \& (B \rightarrow A)$. 3. $\boldsymbol{T} =_{\mathrm{df}} p \supset p$. 4. $\boldsymbol{F} =_{\mathrm{df}} p \& \bar{p}$.)

POSTULATES FOR CFL

Axioms

1 $(q \rightarrow r) \supset [(p \rightarrow q) \supset (p \rightarrow r)]$
2 $(p \rightarrow q) \supset [(p \& r) \rightarrow (r \& q)]$
3 $[p \& (q \vee r)] \rightarrow [(p \& q) \vee (p \& r)]$
4 $[p \& (q \& r)] \rightarrow [(p \& q) \& r]$
5 $(p \rightarrow \bar{q}) \supset (q \rightarrow \bar{p})$
6 $(p \& \bar{p}) \rightarrow (q \& \bar{q})$
7 $p \rightarrow (p \& p)$
8 $(p \& p) \rightarrow p$
9 $\bar{\bar{p}} \rightarrow p$
10 $(p \rightarrow q) \supset \overline{p \rightarrow \bar{q}}$
11 $\{[(p \& \bar{q}) \rightarrow (p \& \bar{p})] \& (p \vee \bar{q})\} \supset (p \rightarrow q)$
12 $p \supset [p \rightarrow (p \supset p)]$
13 $[p \rightarrow (p \supset p)] \supset p$
14 $(p \supset q) \supset [(q \supset r) \supset (p \supset r)]$

Rules of inference

R1. Substitution for propositional variables, with the restriction that no nesting of arrows may result.
R2. From A, A⊃B to infer B.

Axioms 1–9 and 14 require no special comment. 10 is one of the characteristic theses of connexive implication. Axiom 11 reflects the fact that, in the system of connexive algebra that has been constructed, $\mathrm{x} \wedge \bar{\mathrm{y}} = \boldsymbol{0}$ is a necessary but not a sufficient condition of $\mathrm{x} \leq \mathrm{y}$. For example, in the last diagram in §29.8.1 we have $\mathbf{i}\boldsymbol{a} \wedge \bar{\boldsymbol{a}} = \boldsymbol{0}$ but not $\mathbf{i}\boldsymbol{a} \leq \boldsymbol{a}$. Similarly in **CFL** $(p \& \bar{q}) \rightarrow \boldsymbol{F}$ is a necessary but not a sufficient condition of $p \rightarrow q$. We need in addition to exclude those cases in which $\mathrm{v}(p) \in \mathrm{U}$ and $\mathrm{v}(q) \in \mathrm{D}$, and this is accomplished by the requirement that $p \vee \bar{q}$. Since $p \vee \bar{q}$ holds for all values of p and q except when $\mathrm{v}(p) \in \mathrm{U}$ and $\mathrm{v}(q) \in \mathrm{D}$, axiom 11 says in effect that if $\mathrm{v}(p \& \bar{q}) = \boldsymbol{0}$ and either $\mathrm{v}(p), \mathrm{v}(q) \in \mathrm{D}$ or $\mathrm{v}(p), \mathrm{v}(q) \in \mathrm{U}$, then $\mathrm{v}(p \rightarrow q) \in \mathrm{D}$. Finally, axioms 12 and 13 state that just as in connexive algebra the designated elements x are those such that $\mathrm{x} \leq \boldsymbol{1}$, in **CFL** the true formulas are those that entail $p \supset p$ (*pace* the principal authors).

We now show that **CFL** is an axiomatic formulation of ΣC*. It is not difficult to verify that the axioms and rules of inference of **CFL** are satisfied in all connexive models, and we have as a consequence

THEOREM 1. If $\vdash_{\mathbf{CFL}} A$, then $A \in \Sigma\mathrm{C}^*$.

We now note that $\{C_n\}^* \subseteq C^*$, hence $\Sigma C^* \subseteq \Sigma\{C_n\}^*$. Therefore, instead of proving the converse of Theorem 1, it suffices to show

THEOREM 2. If $A \in \Sigma\{C_n\}^*$, then $\vdash_{\mathbf{CFL}} A$.

The proof requires the following lemmas.

LEMMA 1. Given any expression U of **CFL** we can find an expression V in *normal form* such that $\vdash_{\mathbf{CFL}} U \equiv V$. V is a conjunction

$$W_1 \& W_2 \& \ldots \& W_l$$

$l \geq 1$, where each conjunct W_i is of the form

$$(X_1 \vee \ldots \vee X_m) \vee (\overline{Y_1 \rightarrow \boldsymbol{F}} \vee \ldots \vee \overline{Y_n \rightarrow \boldsymbol{F}}) \vee [(Z_1 \rightarrow \boldsymbol{F}) \vee \ldots \vee (Z_r \rightarrow \boldsymbol{F})],$$

$m, n, r \geq 0$. Each of the X's, Y's and Z's in turn is a conjunction of propositional variables and negations of variables in which each variable of V occurs at least once and at most twice.

PROOF. To put U, which we shall assume to be in primitive notation, into normal form, perform the following steps.

(1) Replace every subformula $A \rightarrow B$ of U by the equivalent formula $[(A \& \bar{B}) \rightarrow \boldsymbol{F}] \& (A \vee \bar{B})$ according to thesis 61 of **CFL** (theses above 14 — the last axiom — and rules above R2 — the last rule — bear numbers according to their position in an as yet unpublished deductive elaboration of **CFL**):

61 $(A \rightarrow B) \equiv \{[(A \& \bar{B}) \rightarrow \boldsymbol{F}] \& (A \vee \bar{B})\}$

(2) Consider the antecedent A of each subformula $A \rightarrow \boldsymbol{F}$ of U. If A does not contain at least one occurrence of every variable of U then add any missing ones B to A by means of

114 $A \rightleftarrows [A \& (B \vee \bar{B})]$

and the rule for the replacement of connexive equivalents (derived rule R3 of **CFL**).

(3) Expand the antecedent A of each subformula $A \rightarrow \boldsymbol{F}$ of U into "perfect" disjunctive normal form in such a way that each disjunct is an iterated conjunction in which each variable of U occurs at least once. The theses required (together with rule R3) are:

64 $(A\&B) \rightleftarrows (B\&A)$
125 $(A\vee B) \rightleftarrows (B\vee A)$
73 $[A\&(B\&C)] \rightleftarrows [(A\&B)\&C]$
127 $[A\vee(B\vee C)] \rightleftarrows [(A\vee B)\vee C]$
106 $\overline{A\&B} \rightleftarrows (\bar{A}\vee\bar{B})$
79 $\overline{A\vee B} \rightleftarrows (\bar{A}\&\bar{B})$
78 $A \rightleftarrows \bar{\bar{A}}$
90 $[A\&(B\vee C)] \rightleftarrows [(A\&B)\vee(A\&C)]$
92 $[A\vee(B\&C]] \rightleftarrows [(A\vee B)\&(A\vee C)]$.

(4) Replace each subformula $(A\vee B)\rightarrow \boldsymbol{F}$ by $(A\rightarrow \boldsymbol{F})\&(B\rightarrow \boldsymbol{F})$ according to

103 $[(A\vee B)\rightarrow \boldsymbol{F}] \equiv [(A\rightarrow \boldsymbol{F})\&(B\rightarrow \boldsymbol{F})]$.

(5) Put the whole expression U into conjunctive normal form (see Hilbert and Ackermann 1959, p. 12) using reduction theses of two valued logic. The elements which are the disjuncts of this conjunctive normal form will be propositional variables A_i, negations of same $\overline{A_j}$, and expressions of the form $\overline{Y\rightarrow \boldsymbol{F}}$ and $Z\rightarrow \boldsymbol{F}$.

(6) Consider the variables A_i and negations $\overline{A_j}$ under (5) above. Add each of the variables of U by successive applications of $A \equiv [A\&(B\vee\bar{B})]$. Expand the result into a disjunction of conjunctions as in (3) above. Use $A \equiv (A\&A)$ and 62 $A \rightleftarrows (A\&A)$ to eliminate duplicate pairs of variables and negated variables. When step (6) is completed U will be in normal form. We rename it V and continue.

LEMMA 2. Every expression V in normal form is such that either $\vdash_{\mathbf{CFL}} V$ (we shall in future write $\vdash A$ for $\vdash_{\mathbf{CFL}} A$) or $V \notin \Sigma\{C_n\}^*$.

PROOF. We have that $V = W_1\&W_2\& \ldots \&W_l$. Since

$$\vdash V \text{ if } \vdash W_1 \text{ and } \vdash W_2 \text{ and } \ldots \text{ and } \vdash W_l,$$

and

$$V \in \Sigma\{C_n\}^* \text{ if } W_1 \notin \Sigma\{C_n\}^* \text{ or } W_2 \notin \{\Sigma C_n\}^* \text{ or } \ldots \text{ or } W_l \notin \Sigma\{C_n\}^*$$

it suffices to consider each of the conjuncts W_i separately. We recall that W_i is of the form:

$$(X_1\vee \ldots \vee X_m)\vee(\overline{Y_1\rightarrow \boldsymbol{F}}\vee \ldots \vee \overline{Y_n\rightarrow \boldsymbol{F}})\vee[(Z_1\rightarrow \boldsymbol{F})\vee \ldots \vee(Z_r\rightarrow \boldsymbol{F})]$$

and conduct the argument by cases.

CASE 1. W_i contains a disjunct A, or a disjunct $\overline{A \rightarrow F}$, $A = B_1 \& B_2 \& \ldots \& B_j$, where the set $\{B_1, B_2, \ldots, B_j\}$ contains a pair $\{B_k, \overline{B_k}\}$. We replace the disjunct A, or $\overline{A \rightarrow F}$, by F in virtue of the following equivalences:

68 $(B \& \overline{B}) \rightleftarrows F$
76 $(B \& F) \rightleftarrows F$
120 $\overline{F \rightarrow F} \equiv F$

following which they are omitted from W_i since

121 $(B \vee F) \equiv B$.

CASE 2. W_i contains a disjunct $A \rightarrow F$ where A reduces to F as in Case 1. In this case $\vdash W_i$.

CASE 3. W_i contains a mutually contradictory pair of disjuncts A and $\overline{A}$. In this case also $\vdash W_i$.

In the remaining cases we assume that cases 1–3 do not obtain. Cases 4 and 5 require the definition of a *Venn set* of zdfs, as follows:
The set $A = \{A_1, A_2, \ldots, A_n\}$ is a *Venn set* iff

(1) Each zdf $A_i = B_1 \& B_2 \& \ldots \& B_m$, where each B_j is a propositional variable or the negation of a variable.
(ii) Every variable represented in A occurs once and only once in each A_i.
(iii) No members A_i and A_j of A are identical unless $i = j$.
(iv) $n = 2^m$.

For example, the set $\{p \& q, \bar{p} \& q, p \& \bar{q}, \bar{p} \& \bar{q}\}$ is a Venn set. The name derives from the fact that when the corresponding set of elements {ab, āb, a$\bar{\text{b}}$, ā$\bar{\text{b}}$} of a Boolean algebra with two free generators is simultaneously represented on a Venn diagram, the elements cover the whole diagram.

Cases 1–3 having been dealt with, we note that each of the subformulas X_h, Y_k, Z_j of W_i contains exactly one occurrence of every propositional variable of W_i.

CASE 4. The set $\{X_1, \ldots, X_m, Y_1, \ldots, Y_n\}$ of subformulas of W_i comprises a Venn set. In this case we have $\vdash W_i$. To show this replace each disjunct $\overline{Y_k \rightarrow F}$, $0 \leq k \leq n$, by Y_k, obtaining a new expression W_i^*. W_i^* will include among its disjuncts a Venn set. By repeatedly applying the equivalence

122 $[(A \& B) \vee (A \& \overline{B})] \equiv A$

to selected pairs of disjuncts of W_i*, we reduce W_i*'s Venn set to the pair $\{A, \bar{A}\}$, where A is a propositional variable. But $\vdash A \vee \bar{A} \vee B$, hence $\vdash W_i^*$. Furthermore, because of the conditionals

124 $Y_k \supset \overline{Y_k \rightarrow \boldsymbol{F}}$

we see that if $\vdash W_i^*$ then $\vdash W_i$. Hence $\vdash W_i$.

CASE 5. The set $\{X_1, \ldots, X_m, Y_1, \ldots Y_n\}$ does not comprise a Venn set. In this case $W_i \notin \Sigma\{\mathrm{C}_n\}^*$. To falsify W_i, we require a connexive model $\langle \mathrm{C}_{r+1}, \mathrm{D}, \mathrm{v}\rangle$, where r is the number of subformulas $Z_j \rightarrow \boldsymbol{F}$ of W_i. Let B_{r+1} be defined as in §29.8.1 (i.e., let its elements consist of strings of ones and zeros, each string of length $r+1$, with Boolean operations defined on the strings). Let D be defined as the set of all elements having a *1* in the last column; this is obviously a prime filter. U and $\leq$ are defined in the connexive way in terms of D.

Next we must define the valuation v. We define it on all propositional variables by saying what its value is for each column for each variable.

(i) *Last column.* We devise a formula U which is not a member of the set $\{X_1, \ldots, X_m, Y_1, \ldots Y_n\}$, but which would have to be added if it were to constitute a Venn set. Assign the value *1* to each unnegated, and *0* to each negated variable of U, so that $\mathrm{v}(U) = 1$. This makes every X_h and every Y_k zero in the last column. (Thus in a Venn diagram, if we blow up one region to the size of the universe, the other regions automatically shrink to zero.)

(ii) *j-th column* $(1 \leq j \leq r)$. We relate the jth column to the subformula $Z_j \rightarrow \boldsymbol{F}$ of W_i as follows: in the jth column, assign *1* to a variable if it is unnegated in Z_j and *0* to a variable if it is negated in Z_j. This results in every formula Z_j being assigned a *1* in the jth column, so that no Z_j is the zero of the Boolean algebra. This falsifies all the disjuncts $Z_j \rightarrow \boldsymbol{F}$ of W_i.

Our previous treatment of the last column makes the value of every disjunct X_h of W_i undesignated. Finally, noting that every Y_k of W_i is distinct from every Z_j because of Case 3 above, it will be seen that the valuation v assigns every Y_k zero in the jth column, $1 \leq j \leq r$. Since it also has zero in the last column, every Y_k takes zero throughout, and is hence the zero of the algebra. It follows that all disjuncts $\overline{Y_k \rightarrow \boldsymbol{F}}$ of W_i are falsified.

Since all the disjuncts of W_i receive undesignated values by v, W_i itself is falsified. That is to say, $W_i \notin \Sigma\{\mathrm{C}_n\}^*$. This completes the proof of Lemma 2.

We are now in a position to prove Theorem 2. Let A be any formula, and consider the normal form B of A. By Lemma 2 either $\vdash B$ or $B \notin \{\mathrm{C}_n\}^*$. If $\vdash B$, then since $\vdash A \equiv B$, $\vdash A$. If $B \notin \{\mathrm{C}_n\}^*$, then for some connexive algebra C_k and some valuation v, $\mathrm{v}(B) \in \mathrm{U}$, and we cannot have $\mathrm{v}(A) \in \mathrm{D}$, since if so $A \equiv B$. Hence $\mathrm{v}(A) \in \mathrm{U}$, and $A \notin \{\mathrm{C}_n\}^*$. Therefore if $A \in \{\mathrm{C}_n\}^*$, then $B \in \{\mathrm{C}_n\}^*$, whence $\vdash B$, whence $\vdash A$. Q.E.D.

§29.8.3. Scroggs property. We now proceed to obtain a somewhat stronger completeness result, namely that every consistent proper normal extension of **CFL** has a finite characteristic matrix. (A normal extension is defined as being closed under substitution and ⊃-detachment: we note that the rule for →-detachment is derivable in **CFL**.)

Since as far as is known no name is in current use for systems (i) which have no finite characteristic model or matrix, (ii) all of whose consistent proper normal extensions have finite characteristic matrices, let us speak of such systems either as possessing the *Scroggs property*, or as being *saturated*. (Note that the latter term admits of the following generalization: a system **S** is *saturated with respect to a property P* if (i) **S** has P, (ii) no (consistent, normal) proper extension of **S** has P.) In Scroggs 1951 it is shown that **S5** is saturated, and the same result is obtained for **RM** in §29.4 and for Dummett's **LC** in Dunn and Meyer 1971. Systems which are saturated are complete in a fairly strong sense. No proper subsystem of a saturated system is saturated; hence **E**, **R**, **S4**, etc. are not. In what follows we sketch a proof that **CFL** is saturated.

Corresponding to the well-known fact that all finite Boolean algebras can be arranged in a chain in the sense of being isomorphic to a member of the sequence of algebras $\{B_n\}$, where B_n has 2^n elements, we demonstrate the following three theorems.

THEOREM 3. Let $\{C_n\}$ be the sequence of the connexive algebras corresponding to $\{B_n\}$. Then $\{C_n\}$ can be arranged in a chain, each member C_j of which is isomorphically embeddable in C_{j+1}.

PROOF. Embed B_j in B_{j+1} by mapping $\langle a_1, \ldots, a_j\rangle$ in B_j into $\langle a_1, \ldots, a_j, a_j\rangle$ in B_{j+1} (each a_i is a *0* or a *1*). This is not only a Boolean isomorphism, but preserves D and U (hence $\leq$), which were determined (§29.8.1) by the last column.

THEOREM 4. Any formula falsifiable in C_j is falsifiable in C_{j+1}, and any formula valid in C_{j+1} is valid in C_j. This follows immediately from the fact that C_j is a subalgebra of C_{j+1}.

THEOREM 5. Every finite connexive algebra is connexively isomorphic to some member of the sequence $\{C_n\}$. This is a consequence of the fact that every connexive algebra is based upon a Boolean algebra, and that every finite Boolean algebra is isomorphic to some member of the sequence $\{B_n\}$ of §29.8.1. Furthermore, all prime filters of a finite Boolean algebra are isomorphic.

Define an X-algebra as an algebra in which all theses of a system X are valid. The heart of the proof that **CFL** possesses the Scroggs property is then enshrined in the following

THEOREM 6. Let T be a consistent, proper, normal extension of **CFL**. Then if a formula E is not a thesis of T, there is a finite connexive algebra C_n such that C_n is a T-algebra and E is not satisfied by C_n.

PROOF. T is a **CFL**-theory in a sense analogous to §25.2.1. We construct the *Lindenbaum algebra* L(T) of T as follows. Let T′ be an extension of T which (i) is closed under $\supset$-detachment (though not under substitution), (ii) does not contain E, (iii) is *maximal consistent* in the sense that $(A\&B) \in$ T′ iff $A, B \in$ T′; $(A \vee B) \in$ T′ iff $A \in$ T′ or $B \in$ T′; $\sim A \in$ T′ iff $A \in$ T′. (The existence of at least one such maximal consistent extension of T is guaranteed by Zorn's lemma.) The elements of L(T) will be equivalence classes of formulas $[A]$, where A is a zdf, the required congruence between A and B defined in terms of theoremhood in T′: $\vdash_{T'} A \rightleftarrows B$, and with $[A] =_{df} \{B : \vdash_{T'} B \rightleftarrows A\}$. It is not difficult to show that $\vdash_{T'} A \rightleftarrows B$ is a congruence with respect to &, $\vee$, and $\overline{}$, so that $[A] \wedge [B] = [A\&B]$, $[A] \vee [B] = [A \vee B]$, $\overline{[A]} = [\bar{A}]$. Finally we define D as $\{[A] : A \in T'\}$, U as the complement of D, and stipulate that $[A] \leq [B]$ iff (i) $[A] \vee [B] = [B]$ and (ii) either $[A] \in$ D or $[B] \in$ U.

Since **CFL** (and hence T′) contains all Boolean identities in the form of connexive equivalences (for example, $a \vee (b \wedge c) = (a \vee b) \wedge (a \vee c)$ corresponds to $A\&(B \vee C) \rightleftarrows (A \vee B)\&(A \vee C)$), it is easy to show that the elements of L(T) form a Boolean algebra. Furthermore, because of the maximal properties of T′, D is seen to be a prime filter of L(T), so that $\langle$L(T), $\wedge$, $\vee$, $\overline{}$, D, U, $\leq\rangle$ is a connexive algebra. We now show that every theorem of T is valid in L(T). For this we require the definition of a *canonical valuation* on a Lindenbaum algebra as one which assigns to every propositional variable A the value $[A]$, and the following two lemmas:

LEMMA 1. If v is any canonical valuation on L(T) and A any zdf, then $v(A) = [A]$. Proof is by induction on the number of logical connectives in A.

LEMMA 2. Where v is a canonical valuation on L(T) and A is any formula,

$\vdash_{T'} A$ iff $v(A) \in$ D

The proof is again by induction on the number of connectives in A, the only case liable to cause any trouble being that in which $A = (B\to C)$:

$\vdash_{T'} B\to C$	iff $(B\to C) \in T'$	
	iff $[(B\&\bar{C}\to \boldsymbol{F})\&(B\vee \bar{C})] \in T'$	Thesis 61 of ***CFL***
	iff $(B\&\bar{C}\to \boldsymbol{F}) \in T'$ and $(B\vee \bar{C}) \in T'$	Max. cons. of T′
	iff $(B\&\bar{C} \rightleftarrows \boldsymbol{F}) \in T'$ and $(B\vee \bar{C}) \in T'$	Thesis 134 of **CFL**
	iff $(B\&\bar{C} \rightleftarrows \boldsymbol{F}) \in T'$ and $(B \in T'$ or $\bar{C} \in T')$	Max. cons. of T′
	iff $[B\&\bar{C}] = [\boldsymbol{F}]$ and $(v(B) \in D$ or $v(\bar{C}) \in D)$	Df[ ], and induction hyp.
	iff $[B]\wedge\overline{[C]} = [\boldsymbol{F}]$ and $([B] \in D$ or $\overline{[C]} \in D)$	Df$\wedge$, Df^{-} and Lemma 1
	iff $[B]\vee[C] = [C]$ and $([B] \in D$ or $[C] \in U)$	L(T) Boolean
	iff $[B] \leq [C]$	Df$\leq$
	iff $([B] \Rightarrow [C]) \in D$	Df$\Rightarrow$
	iff $(v(B) \Rightarrow v(C)) \in D$	Lemma 1
	iff $v(B\to C) \in D$	Df of valuation of $\to$-formulas

The proof that every theorem of T is valid in L(T) now proceeds as follows. Let A be a theorem of T, containing propositional variables $p_1, \ldots, p_n$. Let v be an arbitrary valuation of A over L(T) — not necessarily canonical — such that $v(p_1) = [B_1], \ldots, v(p_n) = [B_n]$. Where $A(B_1/p_1, \ldots, B_n/p_n)$ is the result of substituting B_1 for $p_1, \ldots, B_n$ for p_n, in A, we note that $\vdash_T A(B_1/p_1, \ldots, B_n/p_n)$, since T is closed under substitution. It follows that $\vdash_{T'} A(B_1/p_1, \ldots, B_n/p_n)$, and hence $v'(A(B_1/p_1, \ldots, B_n/p_n)) \in D$ by Lemma 2, where v′ is canonical. But this implies that $v(A) \in D$, since a canonical valuation of a substitution-instance of a formula is tantamount to a valuation of that formula. Since v was arbitrary, it follows that A is valid in L(T).

LEMMA 3. E is not valid in L(T). The proof is immediate when we recall that $E \notin T'$. By Lemma 2 above $v(E) \notin D$, where v is canonical. Hence E is not valid.

What we have shown up to this point is that L(T) is a T-algebra which falsifies E. Consider now the subalgebra $L(T)_0$ of L(T) generated by $[A_1], \ldots, [A_n]$, where $A_1, \ldots, A_n$ are all the propositional variables of E. Close $L(T)_0$ under the operations $\wedge$, $\vee$, $^{-}$, and define $D_0 = D \cap L(T)_0$ and $\leq_0$ in terms of D_0. We now demonstrate

LEMMA 4. $L(T)_0$ as defined above (i) is a connexive algebra, (ii) is finite, (iii) is a T-algebra, and (iv) falsifies E.

PROOF. (i) follows from the fact that subalgebras of Boolean algebras are Boolean, and that the intersections of prime filters with such subalgebras are themselves prime filters of the subalgebras. (ii) holds because $L(T)_0$ is finitely generated. (iii) follows from the fact that L(T) is a T-algebra, and that $L(T)_0$ is a subalgebra of L(T). Finally, we deduce (iv) by observing that the canonical valuation which falsified E in L(T) in lemma 3 made use only of the elements generated by $[A_1], \ldots, [A_n]$, all of which are in $L(T)_0$.

To complete the proof of Theorem 6, all we need is to note that by Theorem 5, $L(T)_0$ is isomorphic to some finite connexive algebra C_n. C_n is then the sought-for T-algebra which falsifies E.

THEOREM 7. (Scroggs property). Every consistent, normal, proper extension of **CFL** has a finite characteristic matrix, namely some C_n.

PROOF. Let K be the set of indices of those connexive matrices C_n that are X-matrices, where X is a consistent, normal, proper extension of **CFL**. Since X is consistent, K is non-empty by Theorem 6.

If K contains infinitely many indices (hence all indices), then X is identical with **CFL** by Theorem 2 of §29.8.2, contrary to the assumption that X is a proper extension of **CFL**.

If K contains finitely many indices, consider the greatest such index k. C_k is an X-matrix, and by Theorem 4 the set of X-matrices is $\{C_j : j \leq k\}$. Now suppose that a formula A is not a thesis of X. Then, by Theorem 6, A fails to be satisfied by some X-matrix C_i, and since C_i is an X-matrix, $i \leq k$. But since by Theorem 4, if A fails in C_i it fails in all C_n such that $n \geq i$, it follows that A fails in C_k. Hence C_k is a characteristic matrix for X.

§29.8.4. Whither connexive implication? So far in our discussion of connexive logic, we have confined ourselves to the consideration of first degree formulas: formulas in which no nesting of arrows is permitted. Although this approach has its advantages as far as simplicity is concerned, first degree systems being generally more manageable than systems of higher degree, it also has its drawbacks. One of the disadvantages of the system **CFL** has been pointed out by Meyer 197+c, who demonstrates that **CFL** is derivable, using appropriate definitions, in the first degree fragment of any of the systems of strict implication **S1–5**. Confining himself to formulas in which no nesting of the $\prec$ of strict implication is permitted, Meyer is able to show that connexive implication is definable as follows:

$$A \rightarrow B =_{\text{df}} (A \prec B) \& (A \equiv B).$$

Making use of this definition, the set of provable first degree formulas containing $\rightarrow$ in **S1–5** coincides exactly with the set of theorems of **CFL**. (Incidentally, Meyer is also able to axiomatize the former set with only five axioms, which as he points out leaves him nine to the good as compared with **CFL**.)

[Note by principal authors. Though reluctant to take sides with one rather than another of our co-authors, in this case we must express at least a certain mystic fellow-feeling with the present writer; when axiomatizing, subtractions from fourteen are as much to be avoided as additions. See the end of §11.]

We referred to the fact that first degree connexive implications are definable *à la* Meyer in **S1–5** as a "disadvantage" of **CFL** because of the presence in the definition of $A \rightarrow B$ above of the conjunct $A \equiv B$. What the definition shows is that in **CFL** a necessary (though not a sufficient) condition of A's connexively implying B is that A and B be materially equivalent. Is this a disadvantage? Well, it means that what we are confronted with in **CFL** is a subclass of the class of valid material equivalences, and although this is perhaps no real cause for alarm (the systems **E** and **R**, after all, do no more than confront us with a subclass of the class of valid material implications) the present author does confess to a slight uneasiness concerning this fact.

What can be done? For those who feel that the study of connexive implication should extend further than the study of a certain interesting class of material equivalences, two alternatives are open. Either the field of valid first degree connexive formulas must be extended to include some that are not equivalential, or we must pass from first degree to higher degree logic. Let us briefly consider each of these alternatives.

(1) One of the most natural interpretations of connexive implication is as a species of physical or "causal" implication. Angell 1962 speaks of the incompatibility of $A \rightarrow B$ and $A \rightarrow \overline{B}$ as the *principle of subjunctive contrariety*, and McCall 1969 proposed the connexive formula $\mathrm{Oxt_1} \rightarrow \mathrm{Oyt_2}$ as a symbolization of "the occurrence of an event of type x at time t_1 is a sufficient condition for the occurrence of an event of type y at time t_2." Subjunctive contrariety is illustrated by the incompatibility of the following pair of conditionals:

> If Hitler had invaded England in 1940, he would have conquered her.
>
> If Hitler had invaded England in 1940, he would not have conquered her.

The connexive character of $\mathrm{Oxt_1} \rightarrow \mathrm{Oyt_2}$ hinges upon the fact that if x at t_1 is sufficient for y at t_2, then it is *not* the case that x at t_1 is also sufficient for the non-occurrence of y at t_2. Note that it makes little if any sense to embed arrows of causal implication. In the formula $A \rightarrow B$, what A and B denote are events or states of affairs. The logic of causal implication is inherently first degree. Now, are there formally valid first degree causal conditionals that are not equivalential? A possible candidate is $[A\&(\bar{A} \vee B)] \rightarrow B$ and, oddly enough, $A\&B \rightarrow A$, upon which aspersions were cast (perhaps prematurely) in §29.8.1. The reason for the disparagement was that $A\&B \rightarrow A$ yielded $A\&\bar{A} \rightarrow A$ by substitution, and eventually $A\&\bar{A} \rightarrow \overline{A\&\bar{A}}$ by contraposition, De Morgan's laws, and transitivity. But if, in $A\&B \rightarrow A$, A and B denote events or states of affairs occurring at the same moment in time, then A and B will have to be consistent with one another, and the substitution which yields $A\&\bar{A} \rightarrow A$ cannot be performed. If we made an attempt to construct a formal system of first degree causal conditionals, the required restriction on the rule of substitution would be something like the following: uniform substitution of zdf's for propositional variables, provided that no variable in the substituted zdf occurs elsewhere in the formula to which the rule of substitution is applied.

(2) The first degree system **CFL** is not as it stands very well suited to extension to higher degrees, since although the definition of the operator $\Rightarrow$ in connexive algebra permits nesting of arrows, the actual values assigned to $\mathrm{x} \Rightarrow \mathrm{y}$ in **CFL** result in certain well-known laws of propositional logic such as suffixing ($A \rightarrow B \rightarrow .B \rightarrow C \rightarrow .A \rightarrow C$) receiving undesignated values. The operation $\Rightarrow$ is defined as follows:

$$\begin{aligned} \mathrm{x} \Rightarrow \mathrm{y} &= \mathit{1} \text{ if } \mathrm{x} \leq \mathrm{y} \\ &= \mathit{0} \text{ otherwise,} \end{aligned}$$

and suffixing takes the value $\mathit{0}$ when $\mathrm{v}(A) = \mathrm{v}(C) = \mathit{1}$, $\mathrm{v}(B) = \mathrm{d} \in \mathrm{D}$ where $\mathrm{d} < \mathit{1}$. However, this can be changed by altering the definition of $\Rightarrow$. The following alteration, for example, will result in the validation of such **E**-like laws as suffixing, weak permutation and weak assertion:

$$\begin{aligned} \mathrm{x} \Rightarrow \mathrm{y} &= \mathit{1} \text{ if } \mathrm{x} \leq \mathrm{y} \\ &= \boldsymbol{t} \text{ if } \mathrm{x} \nleq \mathrm{y} \text{ and } (\mathrm{x} \in \mathrm{D} \text{ iff } \mathrm{y} \in \mathrm{D}) \\ &= \mathit{0} \text{ otherwise,} \end{aligned}$$

where $\boldsymbol{t}$ is the least element of D. However, enough has been said about connexive implication at this point, and the detailed investigation of connexive systems with nested arrows must await another occasion.

§29.9. Independence (by John R. Chidgey). We know from §8.4 that the axioms in a formulation of a system may or may not be independent, and that there may be many sets of independent axioms for the same system.

In this section we present one set of independent axioms for **R**, and one set for **T**; further results can be found in Chidgey 1974. The style of presentation is explained in §26.2.

INDEPENDENCE OF THE AXIOMS AND RULES OF **R** (formulation of §27.1.1).

R1	$A\to A$	Not independent (§14.1.1).	
R2	$A\to B\to.B\to C\to.A\to C$	M VIII.1	$A = 2$; $B = 1$; $C = 0$.
R3	$A\to.A\to B\to B$	M XIII.1	$B = 1$; $A = 1$.
R4	$(A\to.A\to B)\to.A\to B$	M XXIV	$B = 0$; $A = 1$.
R5	$A\&B \to A$	M X.1	$A = 0$; $B = 1$.
R6	$A\&B \to B$	M X.2	$A = 1$; $B = 0$.
R7	$(A\to B)\&(A\to C)\to.A\to B\&C$	M X.3	$A = 1$ or $A = 2$; $B = 1$; $C = 2$.
R8	$A \to A\vee B$	M XI.1	$A = 2$; $B = 1$.
R9	$B \to A\vee B$	M XI.2	$A = 1$; $B = 2$.
R10	$(A\to C)\&(B\to C)\to.(A\vee B)\to C$	M XI.3	$A = 0$; $B = 0$; $C = 0$ or $C = 1$.
R11	$A\&(B\vee C) \to (A\&B)\vee C$	M XII	$A = 2$; $B = 3$ or $B = 4$; $C = 1$; or $A = 4$; $B = 1$ or $B = 2$; $C = 3$.
R12	$A\to\bar{B}\to.B\to\bar{A}$	M X.4	$A = 2$; $B = 1$.
R13	$\bar{\bar{A}}\to A$	M X.5	$A = 0$.
$\to$E	From A and $A\to B$ to infer B	M XIV	$A = 2$; $B = 0$.
&I	From A and B to infer $A\&B$	M XI.4	$A = 2$; $B = 2$.

INDEPENDENCE OF THE AXIOMS AND RULES OF **T** (formulation of §27.1.1).

A1	$A\to A$	Not independent (§14.1.1).	
A2	$A\to B\to.B\to C\to.A\to C$	Not independent (§14.1.1).	
A3	$A\to B\to.C\to A\to.C\to B$	M VIII.1	$A = 1$; $B = 0$; $C = 2$.
A4	$(A\to.A\to B)\to.A\to B$	M IX	$A = 1$; $B = 0$.
A5	$A\&B \to A$	M X.1	$A = 0$; $B = 1$.
A6	$A\&B \to B$	M X.2	$A = 1$; $B = 0$.
A7	$(A\to B)\&(A\to C)\to.A\to B\&C$	M X.3	$A = 1$ or $A = 2$; $B = 1$; $C = 2$.
A8	$A \to A\vee B$	M XI.1	$A = 2$; $B = 1$.
A9	$B \to A\vee B$	M XI.2	$A = 1$; $B = 2$.
A10	$(A\to C)\&(B\to C)\to.(A\vee B)\to C$	M XI.3	$A = 0$; $B = 0$; $C = 0$ or $C = 1$.
A11	$A\&(B\vee C) \to (A\&C)\vee C$	M XII	$A = 2$; $B = 3$ or $B = 4$; $C = 1$; or $A = 4$; $B = 1$ or $B = 2$; $C = 3$.

A12	$A \to \bar{A} \to \bar{A}$	M XIII.1	$A = 1.$
A13	$A \to \bar{B} \to .B \to \bar{A}$	M X.4	$A = 2; B = 1.$
A14	$\bar{\bar{A}} \to A$	M X.5	$A = 0.$
$\to$E	From A and $A \to B$ to infer B	M XIV	$A = 2; B = 0.$
&I	From A and B to infer $A \& B$	M XV	$A = 1; B = 2$ or $A = 2; B = 1.$

The nineteen matrix sets referred to here and in §26.2 are as follows. (The non-consecutiveness of the matrix numbers is to be explained by the fact that they are drawn from Chidgey 1974 without renumbering.)

MATRIX SET M I.5

$\to$	0	1	2	3
0	2	2	2	2
1	0	2	2	2
*2	0	0	2	2
*3	0	0	0	2

$\sim$
3
2
1
0

&	0	1	2	3
0	0	0	0	0
1	0	0	1	1
*2	0	1	2	2
*3	0	1	2	3

$\vee$	0	1	2	3
0	0	1	2	3
1	1	1	2	3
*2	2	2	2	3
*3	3	3	3	3

MATRIX SET M VII

$\to$	0	1	2	3	4	5
0	5	5	5	5	5	5
*1	0	1	3	3	3	5
*2	0	0	3	3	3	5
*3	0	0	0	3	3	5
*4	0	0	0	0	1	5
*5	0	0	0	0	0	5

$\sim$
5
4
3
2
1
0

&	0	1	2	3	4	5
0	0	0	0	0	0	0
*1	0	1	1	1	1	1
*2	0	1	2	2	2	2
*3	0	1	2	3	3	3
*4	0	1	2	3	4	4
*5	0	1	2	3	4	5

$\vee$	0	1	2	3	4	5
0	0	1	2	3	4	5
*1	1	1	2	3	4	5
*2	2	2	2	3	4	5
*3	3	3	3	3	4	5
*4	4	4	4	4	4	5
*5	5	5	5	5	5	5

Matrix Set M VIII.1

→	0	1	2
0	2	2	2
1	2	2	2
*2	0	2	2

	~
0	2
1	2
*2	0

&	0	1	2
0	0	0	0
1	0	2	2
*2	0	2	2

∨	0	1	2
0	0	0	2
1	0	0	2
*2	2	2	2

Matrix Set M IX

→	0	1	2	3
0	2	2	2	2
1	1	2	2	2
*2	0	1	2	2
*3	0	0	1	2

	~
0	3
1	2
*2	1
*3	0

&	0	1	2	3
0	0	0	0	0
1	0	1	1	1
*2	0	1	2	2
*3	0	1	2	3

∨	0	1	2	3
0	0	1	2	3
1	1	1	2	3
*2	2	2	2	3
*3	3	3	3	3

Matrix Set M X.1

→	0	1	2
0	2	2	2
*1	0	2	2
*2	0	2	2

	~
0	2
*1	0
*2	0

&	0	1	2
0	0	2	0
*1	0	2	2
*2	0	2	2

∨	0	1	2
0	0	2	2
*1	2	2	2
*2	2	2	2

MATRIX SET M X.2

→	0	1	2
0	2	2	2
*1	0	2	2
*2	0	2	2

	~
0	2
*1	0
*2	0

&	0	1	2
0	0	0	0
*1	2	2	2
*2	0	2	2

∨	0	1	2
0	0	2	2
*1	2	2	2
*2	2	2	2

MATRIX SET M X.3

→	0	1	2
0	2	2	2
*1	0	2	2
*2	0	2	2

	~
0	2
*1	0
*2	0

&	0	1	2
0	0	0	0
*1	0	2	0
*2	0	2	2

∨	0	1	2
0	0	2	2
*1	2	2	2
*2	2	2	2

MATRIX SET M X.4

→	0	1	2
0	2	2	2
*1	0	2	2
*2	0	2	2

	~
0	2
*1	2
*2	0

&	0	1	2
0	0	0	0
*1	0	2	2
*2	0	2	2

∨	0	1	2
0	0	2	2
*1	2	2	2
*2	2	2	2

Matrix Set M X.5

→	0	1	2
0	2	2	2
*1	0	2	2
*2	0	2	2

	~
0	2
*1	2
*2	2

&	0	1	2
0	0	0	0
*1	0	2	2
*2	0	2	2

∨	0	1	2
0	0	2	2
*1	2	2	2
*2	2	2	2

Matrix Set XI.1

→	0	1	2
0	2	2	2
1	2	2	2
*2	0	0	2

	~
0	2
1	2
*2	0

&	0	1	2
0	0	0	0
1	0	0	0
*2	0	0	2

∨	0	1	2
0	0	0	2
1	0	0	2
*2	2	0	2

Matrix Set M XI.2

→	0	1	2
0	2	2	2
1	2	2	2
*2	0	0	2

	~
0	2
1	2
*2	0

&	0	1	2
0	0	0	0
1	0	0	0
*2	0	0	2

∨	0	1	2
0	0	0	2
1	0	0	0
*2	2	2	2

Matrix Set M XI.3

→	0	1	2
0	2	2	2
1	2	2	2
*2	0	0	2

	~
0	2
1	2
*2	0

&	0	1	2
0	0	0	0
1	0	0	0
*2	0	0	2

∨	0	1	2
0	2	0	2
1	0	0	2
*2	2	2	2

Matrix Set M XI.4

→	0	1	2
0	2	2	2
1	2	2	2
*2	0	0	2

	~
0	2
1	2
*2	0

&	0	1	2
0	0	0	0
1	0	0	0
*2	0	0	0

∨	0	1	2
0	0	0	2
1	0	0	2
*2	2	2	2

Matrix Set M XII

→	0	1	2	3	4	5
0	5	5	5	5	5	5
1	0	3	4	0	0	5
2	0	0	3	0	0	5
*3	0	1	2	3	4	5
*4	0	0	1	0	3	5
*5	0	0	0	0	0	5

	~
0	5
1	4
2	3
*3	2
*4	1
*5	0

&	0	1	2	3	4	5
0	0	0	0	0	0	0
1	0	1	1	0	0	1
2	0	1	2	0	0	2
*3	0	0	0	3	3	3
*4	0	0	0	3	4	4
*5	0	1	2	3	4	5

∨	0	1	2	3	4	5
0	0	1	2	3	4	5
1	1	1	2	5	5	5
2	2	2	2	5	5	5
*3	3	5	5	3	4	5
*4	4	5	5	4	4	5
*5	5	5	5	5	5	5

Matrix Set M XIII.1

→	0	1	2
0	2	2	2
1	0	2	2
*2	0	0	2

	~
0	2
1	1
*2	0

&	0	1	2
0	0	0	0
1	0	1	1
*2	0	1	2

∨	0	1	2
0	0	1	2
1	1	1	2
*2	2	2	2

Matrix Set M XIV

→	0	1	2
0	2	2	2
1	0	2	2
*2	2	2	2

	~
0	2
1	1
*2	0

&	0	1	2
0	0	0	0
1	0	0	0
*2	0	0	2

∨	0	1	2
0	0	2	2
1	2	2	2
*2	2	2	2

Matrix Set XV

→	0	1	2	3
0	2	2	2	2
*1	0	2	0	2
*2	0	0	2	2
*3	0	0	0	2

	~
0	3
*1	2
*2	1
*3	0

&	0	1	2	3
0	0	0	0	0
*1	0	1	0	1
*2	0	0	2	2
*3	0	1	2	3

∨	0	1	2	3
0	0	1	2	3
*1	1	1	3	3
*2	2	3	2	3
*3	3	3	3	3

Matrix Set XXI

→	0	1	2	3	4	5
0	5	5	5	5	5	5
*1	0	4	4	4	4	5
*2	0	0	2	0	4	5
*3	0	0	0	3	4	5
*4	0	0	0	0	4	5
*5	0	0	0	0	0	5

	~
0	5
*1	4
*2	2
*3	3
*4	1
*5	0

&	0	1	2	3	4	5
0	0	0	0	0	0	0
*1	0	1	1	1	1	1
*2	0	1	2	1	2	2
*3	0	1	1	3	3	3
*4	0	1	2	3	4	4
*5	0	1	2	3	4	5

∨	0	1	2	3	4	5
0	0	1	2	3	4	5
*1	1	1	2	3	4	5
*2	2	2	2	4	4	5
*3	3	3	4	3	4	5
*4	4	4	4	4	4	5
*5	5	5	5	5	5	5

Matrix Set XXIV

→	0	1	2
0	2	2	2
1	1	2	2
*2	0	1	2

	~
0	2
1	1
*2	0

&	0	1	2
0	0	0	0
1	0	1	1
*2	0	1	2

∨	0	1	2
0	0	1	2
1	1	1	2
*2	2	2	2

§29.10 Consecution formulation of $\mathbf{R}^{\Box}_{+}$. The system $\mathbf{R}^{\Box}_{+}$ is just $\mathbf{R}^{\Box}$ of §27.1.2–4, with t and co-tenability, but without negation; i.e., we have →, &, ∨, ∘, and t. Define $\mathrm{LR}^{\Box}_{+}$ by adding to the rules of §28.5 the following two:

$$\frac{\Gamma_1 A \Gamma_2 \vdash C}{\Gamma_1 \Box A \Gamma_2 \vdash C}(\Box\vdash) \qquad \frac{\alpha \vdash A}{\alpha \vdash \Box A}(\vdash\Box)$$

where for ($\vdash\Box$) every formula-occurrence in α must lie within the scope of a $\Box$, except that occurrences of t need not so lie. For example, α might be I($\Box A$, E(t, $\Box C$)). The appropriate equivalence between $\mathbf{R}^{\Box}_{+}$ and $\mathbf{LR}^{\Box}_{+}$ is stated by a theorem exactly like that of §28.5.2, and as there, an Elimination theorem is needed in order to establish the admissibility of an analogue of →E in the consecution calculus. The exact form of the §28 Elimination theorem, however, will not work for $\mathbf{LR}^{\Box}_{+}$, since *that* argument requires reducing the right rank (rank_X) whenever possible, only treating the case of a large left rank (rank_M) when the right rank is 1; otherwise there would have been trouble about (∨⊢). But the argument for $\mathbf{LR}^{\Box}_{+}$ requires just the opposite treatment in order to avoid trouble about (⊢□); one must first reduce left rank, treating large right rank only when left rank is 1.

This problem of the competing claims of (∨⊢) and (⊢□) can be solved by proving a more general

ELIMINATION THEOREM. If $\gamma_1 \vdash M, \ldots, \gamma_n \vdash M$, and $\delta \vdash D$ are all provable in $\mathbf{LR}^{\Box}_{+}$, and if $X_1, \ldots, X_n$ are pairwise disjoint sets of occurrences of M in δ, then $\delta(\gamma_1/X_1, \ldots, \gamma_n/X_n) \vdash D$ is provable in $\mathbf{LR}^{\Box}_{+}$.

The replacement notation is meant to be a self-explanatory generalization of that of §28.5.2.

As for the structure of the argument, we would have to keep track of the rank of all premisses. In working on the left premisses (note plural), one would count the rank of each, and guarantee to reduce *either* the number of premisses at maximum rank *or* the maximum rank. So one would choose to "back up" with a premiss at maximum rank in which *M* is parametric. In case of (∨⊢) one would have to increase the number of left premisses, but all the newly added premisses would have a lower rank. Hence either the maximum rank or the number of premisses at maximum rank will be reduced, and therefore one will be able to apply an appropriate induction hypothesis. In any event, the procedure would reduce the left to the case when *every M* has just been introduced. Only then would one turn to the right premiss, reducing its rank. So the case (⊢□) would be all right. Details are given in Belnap, Gupta, and Dunn 197+.

§29.11. Inconsistent extensions of R. Everyone who has come this far knows that contradictions do not in general imply everything, and it can further be extracted from an argument in §25.1 that not everything can be derived from an arbitrary contradiction, even when we liberalize the notion of derivation in Official ways. One way of putting the second fact, relative to **R**, is this: that an extension of **R** is negation inconsistent does not imply that it is Post inconsistent in the sense of permitting the proof of everything.

But we may still ask whether or not every negation inconsistent "logic" which is an extension of **R** allows the proof of arbitrary formulas; and Meyer has shown that this is indeed the case, where we follow him in meaning by a *logic* a family of formulas which is (a) an extension of **R**, (b) closed under the rules $\to$E and &I of **R**, and (c) furthermore — this is what makes it a logic — closed under substitution for propositional variables. In short, we have the following

THEOREM. Let **L** satisfy (a)–(c) above, and have as theorems both A and $\overline{A}$, for some formula A. Then every formula B is a theorem of **L**.

PROOF (Meyer). We depend on the following facts. First, where A contains only the variable p, $\vdash_{\mathbf{R}} p\to p\to.A\to A$. Hence $\overline{A\to A}\to.\overline{p\to p}$ by contraposition, and so

$$1 \quad \vdash_{\mathbf{R}} A\&\overline{A}\to\overline{p\to p}$$

by transitivity from $A\&\overline{A}\to\overline{A\to A}$. These moves are all available in **T**. Also available in **T** is a certain converse of contraction (§8.13), $(q\to.q\to q)\to .q\to(q\to.q\to q)$. Hence we may use the permutation of **R** to obtain $q\to .(q\to.q\to q)\to(q\to.q\to q)$ and thus

$$2 \quad \vdash_{\mathbf{R}} \overline{(q\to.q\to q)\to(q\to.q\to q)}\to\overline{q}$$

by contraposition.

Now let A, $\sim A$ be theorems of an **L** satisfying (a)–(c), and let A' result from A by substituting p for every variable in A. By (c) and (b) we get $\vdash_{\mathbf{L}} A'\&\sim A'$, whence by 1, (a), and the $\to$E part of (b),

$$3 \quad \vdash_{\mathbf{L}} \overline{p\to p}.$$

Now substitution in 3 yields $\vdash_{\mathbf{L}} \overline{(q\to.q\to q)\to(q\to.q\to q)}$, which from 2 with (a) and $\to$E produces

$$4 \quad \vdash_{\mathbf{L}} \overline{q}$$

One more substitution, now of $\overline{B}$, together with double negation, gets us to the conclusion that for arbitrary B, $\vdash_{\mathbf{L}} B$.

§29.12. Relevance is not reducible to modality (by Robert K. Meyer). C. I. Lewis, in proposing his systems of strict implication, did so because he thought that material implication was too weak, as we all know. In the current vocabulary, he objected to the reading "A entails B" for "$A\supset B$," where $\supset$ is the material conditional. His solution, as we would presently put it, was to add a necessity operator $\Box$ to his formation apparatus and to subject $\Box$ to further axioms and rules; on picking the *right* axioms and rules (and Lewis himself offered a choice, which subsequent research has considerably broadened), the idea was that "$\Box(A\supset B)$" would be a reasonable formal facsimile of "A entails B."

The value of the contribution that Lewis made to logic is not diminished by the fact that his favorite project was a flop; for no sooner had banishment overtaken his so-called paradoxes of material implication than he discovered a host of others — the paradoxes of strict implication, like $A\&\bar{A}\rightarrow B$ — that seem equally unpalatable. There is, however, no ultimate reason to knock Lewis on this point — the study of the logical modalities is interesting on its own hook, whether or not modalized truth functional expressions yield satisfying formal correlates to the informal notion of *following from*. Furthermore, it isn't true that one can't use the ideas of Lewis to build up entailment; the leading idea of $\mathbf{R}^{\Box}$ (§28.1) involves defining "A entails B" as "$\Box(A\rightarrow B)$," where the logic of $\rightarrow$ is given by the (non-Lewis-modal) system **R** of relevant implication and that of $\Box$ by, essentially, **S4.** But, from the point of view of Lewis, that system $\mathbf{R}^{\Box}$ is still cheating, since the ground logic **R** which supports the intuitions about relevance is not itself built up by modalizing truth functions.

The question accordingly arises whether or not we can *really* bring Lewis up-to-date by defining $\rightarrow$ in a relevant logic as a modalized truth functional expression. For the purpose of concreteness, let us look at the system **E** of entailment in particular, and let us assume it formally acceptable as an attempt to formalize entailment. Is it possible to *reformulate* **E** with just the truth functional connectives &, $\vee$, $\bar{\ }$ and an additional unary operation $\Box$ so that the $\rightarrow$ of **E** is definable by some scheme

1 $A\rightarrow B =_{\text{df}} \Box\varphi(A, B)$,

where $\varphi(A, B)$ is some truth functional combination of A and B? We shall call the question just put the *Lewis problem* for **E**. The principal result of this section is that, when understood in a sensible way, the answer to the Lewis problem is "No" for **E** and for related relevant logics.

Before putting the Lewis problem more sharply, let us say something about the interest and utility of both affirmative and negative solutions. An affirmative solution would in the first place show in some sense that the Lewis intuitions about strict implication were correct; if one understands logical necessity plus the truth functions, one understands entailment; we might go on to say that Lewis himself just didn't understand logical necessity too well, since he gave us an unacceptable theory of entailment (indeed, several such), but his intuitions were after all in the right place; correcting him is just a matter of cranking up the formal machinery and improving his axioms for $\Box$. We add moreover that an affirmative solution to the Lewis problem would have been welcome from a purely technical point of view; unary operations like $\Box$ are on almost all accounts easier to work with than binary ones like $\rightarrow$, and the resulting expected simplification of the algebra and semantics of **E** would presumably have clarified a number of problems.

The negative solution, however, is also extremely interesting. In the first place it shows, as one might on general grounds have thought anyway, that entailment is an *essentially relational* notion, depending on a genuine connection between antecedent and consequent that cannot be simulated by compounding these truth functionally and then extrinsically and externally modalizing the whole. (Indeed, it might well be thought otiose that modern logic, which has made a good part of its fortune by taking seriously real relations among *terms*, has been truly reluctant to take with equal seriousness relations among *sentences*, preferring to keep sentence-building machinery as simple and trivial as possible; that course, to be sure, disposes quickly of the logical problems that might have arisen at the sentential level, even as Aristotelian syllogistic disposes quickly of logical problems at the level of terms; the price in both instances is that the problems disposed of crop up just as quickly somewhere else — e.g., in trying to find the logical form of the laws of science.) And a negative solution suggests, too, that attempts to analyze entailment semantically will rest on features perhaps suggested by, but not naturally reducible to, the kinds of analyses of modal logics given by Kripke 1959, Hintikka 1963, and others. (The latter conclusion is at least partly *ex post facto*, given the semantical analyses of Chapter IX.)

We return now to a sharper characterization of the Lewis problem for **E**. First, we wish to distinguish it from a couple of other problems, whose solutions would also be interesting, that have been suggested by Belnap and by Massey. By requiring that $\varphi(A, B)$ in the suggested definition 1 be *truth functional*, we explicitly exclude that $\Box$ itself should occur in $\varphi(A, B)$, requiring rather that $\Box$ be extrinsically added to a truth functional expression. This restriction presumably conforms to what Lewis was trying to do and is consistent with our thesis that entailment isn't any kind of strict classical implication; it leaves open, however, the possibility that there exists a unary operation, or perhaps a number of unary operations, that will serve in combination with truth functions, no doubt in an intuitively artificial way, to define entailment. An affirmative solution to this more general problem would perhaps have part of the effect of an affirmative solution to the Lewis problem proper, in providing technical simplification. Clarification of intuitions, on the other hand, would seem unlikely given an affirmative solution of the wider problem; though one never knows. Massey suggests a further condition on the wider problem — that $\Box$ itself be definable within **E** — not being here imposed on the Lewis problem.

Second, even the Lewis problem has presumptively a trivial affirmative solution unless we place some conditions on $\Box$. We think about the problem, temporarily, in Polish notation, where $\supset$ is the material conditional defined as usual. Then we might think of $\Box\supset$ as simply a funny way of writing $\rightarrow$, replacing $\rightarrow AB$ wherever it occurs in the axioms of **E** with $\Box\supset AB$, and in the rule of $\rightarrow$E, making no other assumptions about $\Box$. But since $\Box\supset$ is, on this approach, being treated as an indissoluble unit, any other truth functional connective would presumably do just as well; if $\Box\supset$ works, so presumably will $\Box\&$, for example, and a view of entailment which makes it turn out indifferently as strict implication or strict conjunction doesn't do much for the view that entailment is really strict implication.

Accordingly, we shall view the Lewis problem as carrying a condition on $\Box$. We have only one condition in mind, which is both a minimal one and of a sort to be assumed acceptable to Lewis. It is that $\Box$ respect, essentially, substitutivity of equivalents, which we may put as follows:

2 If $\vdash A\rightleftarrows B$ then $\vdash \Box A\rightleftarrows\Box B$.

$\rightleftarrows$ in 2 is to be understood as defined by co-entailment (*not*, in particular, mere material equivalence) in whatever system is under consideration — in the present illustrative case, **E**. The idea of imposing condition 2 is that if one has truly formalized entailment, it is to be presumed that two formulas which provably co-entail each other have the same logical content, as §40.3 puts it, and accordingly that sameness of logical content will be preserved under the connectives and operations of the system, including $\Box$. (Another way of putting 2, and analogous conditions on the other connectives, is that it requires provable equivalence in the sense appropriate to a logical system — e.g., for **E**, co-entailment — to be a congruence relation with respect to the connectives and operations admitted in the system. The principal authors moreover suggest that systems with this property be called *Fregean*, which allows our principal result to be stated succinctly in the form, "There is no Fregean reformulation of **E**, or of any relevant logic, in &, $\vee$, $\bar{\ }$, $\Box$, which permits an adequate definition of $\rightarrow$ by the scheme 1 above.")

Just as the condition 2 is inserted to prevent a trivial but unilluminating affirmative solution of the Lewis problem, we must ward off an equally unilluminating negative solution. Accordingly, we don't require, as noted, that $\Box$ already be definable in **E**; in particular, it's trivial that when $\Box A$ is naturally defined in **E** as $(A\rightarrow A)\rightarrow A$, as in §4.3, that no strict implication defined according to the scheme 1 comes to the same thing as entailment in the sense of **E**. In other words, the conditions of the problem are that we are allowed to start fresh with an analysis of logical necessity and the truth functions, which *turns out* on application of the scheme 1 to characterize entailment.

Problems of definition, in their clearest form, usually turn out to be problems of conservative extension (§14.4). We have now the machinery to put the Lewis problem for **E**, as we have interpreted it, in this form. The question is, "Are there sentential logics **E*** and **E**** with the following properties?"

(i) **E*** is formulated with &, $\vee$, $\overline{\ \ }$, $\Box$, and **E**** is formulated with &, $\vee$, $\overline{\ \ }$, $\Box$, $\rightarrow$;

(ii) There is a truth functional scheme $\varphi(A, B)$ as in 1 above such that, abbreviating $\Box\varphi(A, B)$ by $A\Rightarrow B$, and $(A\Rightarrow B)\&(B\Rightarrow A)$ by $A\Leftrightarrow B$, **E**** is got from **E*** by adding the defining axiom scheme

3 $(A\rightarrow B)\Leftrightarrow(A\Rightarrow B)$;

(iii) The condition 2 holds for **E*** and for **E**** in the form, "If $\vdash A\Leftrightarrow B$ then $\vdash \Box A\Leftrightarrow\Box B$"; and similarly for other connectives.

(iv) E** is a conservative extension of **E** (as formulated, e.g., in §21, taking $\rightarrow$, &, $\vee$, $\overline{\ \ }$ as primitive) in the sense that a formula of **E**** in the connectives of **E** is a theorem of **E**** iff it is already a theorem of **E**. And E** is a conservative extension of **E***.

The content of (i)–(iv) implies, of course, that we can already view E** as but a notational variant of E* and that, accordingly, if they hold, **E** is in a reasonable sense exactly contained in the Lewis-style modal system **E***.

Let us accordingly assume, for *reductio*, that there do exist systems **E***, **E**** that satisfy (i)–(iv). (We take it for granted, also, that *modus ponens* works for $\Rightarrow$ and $\Leftrightarrow$ in **E****.) To get a contradiction from this assumption, yielding the promised negative solution to the Lewis problem, let us examine first the obvious candidate for $\varphi(A, B)$ presupposed in (ii) — namely, $\bar{A}\vee B$. The intuitive condition 1 then becomes

1′ $A\rightarrow B =_{\text{df}} \Box(\bar{A}\vee B)$.

We reason now as follows: In **E****, by 1′ (or its more official form 3 in (ii) above) the formulas

(a) $\bar{A}\rightarrow A$, and
(b) $\Box(\bar{\bar{A}}\vee A)$

entail each other. But we have as theorems of **E**

(c) $\bar{\bar{A}}\rightleftarrows A$, and
(d) $(A\vee A)\rightleftarrows A$,

whence by the definability of $\rightleftarrows$ in **E**** as $\Leftrightarrow$ and the replacement principle 2 (made official in (iii)), (b) and hence (a) co-entails in E**

(e) $\Box A$.

In some sense the logical equivalence of (a) and (e) in E** was to be expected; Lewis-style logics rather naturally define $\Box A$ as the strict implication of A by its denial, and we see that this natural definition rests on very minimal principles: characterization of strict implication via 1′, the replacement principle 2, double negation, and idempotence of $\vee$. But under these assumptions it has turned out, though we explicitly refrained from imposing it as a condition, that $\Box A$ is definable in **E** after all for arbitrary A, by (a). This imposes more conditions on

(f) $A \rightarrow B$

than it can handle. For on the one hand, (f) has been *defined* in **E**** by

(g) $\Box(\bar{A} \vee B)$

On the other hand, since (a) and (e) amount to the same thing, (g) is equivalent to, tacitly but inessentially using De Morgan laws,

(h) $(A \& \bar{B}) \rightarrow (\bar{A} \vee B)$.

But (f) and (h) are not logically equivalent, in the sense of provable co-entailment, in **E**, though evidently they are so in general in **E****. So **E**** is not a conservative extension of **E** when $\rightarrow$ is defined by 1′; hence there is no system **E*** in which entailment is naturally definable as strict material implication.

Having rejected the obvious candidate 1′ as a definition of $\rightarrow$, it might be thought that our arguing is now over. On the contrary, it has only begun. Evidently, the only reasonable candidates for $\varphi(A, B)$ in the condition 1 are $\bar{A} \vee B$ and its truth functional equivalents. Now in the Lewis systems proper, all truth functional equivalents of $\bar{A} \vee B$ are also strictly equivalent to $\bar{A} \vee B$, in which case the argument just gone through disposes of all reasonable alternatives. In the relevant logics, however, truth functional equivalence is strictly weaker than co-entailment, even between wholly truth functional formulas, and so the possibility is left open that there is a candidate for $\varphi(A, B)$ in 1 that leads to the satisfaction of (i)–(iv); even though the obvious candidate $\bar{A} \vee B$ won't do, perhaps some candidate truth functionally but not relevantly equivalent thereto will do. In fact, the general problem, where $\varphi(A, B)$ is *any* truth functional formula, whether truth functionally equivalent to $\bar{A} \vee B$ or not, is no more difficult to solve; this accounts for 1 having been put in the form it took and shows moreover, if anyone wants to know, that not only is entailment not strict implication but also that it is not strict conjunction, strict contradiction, or strict anything-else truth functional.

We now repeat our *reductio* argument in the general framework. Suppose again that there are **E***, **E**** such that (i)–(iv) hold, where, intuitively,

1 $A \rightarrow B =_{\text{df}} \Box\varphi(A, B)$

for $\varphi(A, B)$ some truth functional combination of A, B. For reasons of symmetry, it is easier to introduce, and to think about, a relevant disjunction $+$ in place of $\rightarrow$, defining

4 $A+B =_{\text{df}} \bar{A} \rightarrow B.$

Thus the formula (a) above is in the present notation $A+A$. (Note that 4 makes sense for both **E** and **E****; we treat it, as we have treated $\rightleftarrows$, $\Rightarrow$, $\Leftrightarrow$, as purely abbreviatory, rather than as a new connective in an appropriate extension of **E**.)

Since we get $+$ rather than $\rightarrow$ out of 1 simply by plugging $\bar{A}$ into φ in place of A, the plugging in is still truth functional, so, trivially, 1 yields

5 $A+B =_{\text{df}} \Box\psi(A, B),$

for $\psi(A, B)$ some truth functional combination of A, B. The reasoning analogous to that of the special case considered above now goes like this. In **E****, the formulas

(j) $A+A$, and
(k) $\Box\psi(A, A)$

entail each other, by 5. Now $\psi(A, A)$ is some truth functional expression built up from A, by &, $\vee$, $\bar{\ }$. But, since considerations of relevance do not enter where only one variable is at issue, there are, just as in the classical case, only four non-equivalent formulas in general A and the truth functions; i.e., *every formula* built up truth functionally in **E** from A alone co-entails one of the following:

(m) A
(n) $\bar{A}$
(o) $A \& \bar{A}$
(p) $A \vee \bar{A}$

Since by (iii) if B and C co-entail each other in the appropriate sense in **E**** then so do $\Box B$ and $\Box C$, evidently (k) and hence (j) co-entails

(q) $\Box B$, where B is one of (m)–(p).

We show first that we may disregard the cases (o)–(p); since they are symmetrical between A and $\bar{A}$, we examine only case (o), leaving (p) to the reader. Suppose in fact that $A+A$ is equivalent in general to $\Box(A \& \bar{A})$. Then, for p a sentential variable, $p+p$ is equivalent to $\Box(p \& \bar{p})$ and $\bar{p}+\bar{p}$ is equivalent to $(\bar{p} \& \bar{\bar{p}})$. But by double negation and commutativity of &, $p \& \bar{p}$ and $\bar{p} \& \bar{\bar{p}}$ of course co-entail each other in **E** and hence in **E****, whence by the Fregean principle (iii), $p+p$ and $\bar{p}+\bar{p}$ — i.e., $\bar{p}\rightarrow p$ and $\bar{\bar{p}}\rightarrow\bar{p}$ — co-entail each other in **E****. Evidently they do not do so in **E**, since all theorems of **E** are truth table valid while the co-entailment in question is a truth table contradiction.

(o), and similarly (p) having been disposed of, we now have by the equivalence of (j) and (q) that in **E****, (j) co-entails one of

(r) $\Box A$, or
(s) $\Box\bar{A}$.

Evidently the case (s) is unreasonable, but let us put up with it, noting that if (j) is equivalent to (s), (r) by Frege co-entails

(t) $\bar{A}+\bar{A}$.

So in any case (r) co-entails in **E**** either (j) or (t).

We wish now to show again that we have overloaded, switching $\rightarrow$ to $+$,

(u) $A+B$.

On the one hand, (u) co-entails in **E****, by 5,

(v) $\Box\psi(A, B)$.

But on the other hand, depending on which of the co-entailments (r) with (j), or (r) with (t), hold, (v) itself co-entails one of

(w) $\psi(A, B)+\psi(A, B)$, or
(x) $\overline{\psi(A, B)}+\overline{\psi(A, B)}$.

Since (x) still reflects a truth functional compounding of A, B, in either case there is in **E**** a scheme $\pi(A, B)$, built up truth functionally from A and B, such that (u) schematically co-entails in **E****

(y) $\pi(A, B)+\pi(A, B)$.

We wish now to finish the *reductio* argument, as before, by showing that there is no such truth functional scheme $\pi(A, B)$ such that the schematic co-entailment (u) with (y) holds in **E**, whence, since both (u) and (y) can be chosen to be formulas of **E**, **E**** is once again not a conservative extension of **E** and the conditions for an affirmative solution of the Lewis problem have not been met.

Indeed, the schematic co-entailment (u) with (y) does not even hold, for any choice of $\pi(A, B)$, in the system got from **E** by taking as additional axiom schemes

A1 $A \rightleftarrows (A+A)$
A2 $A\rightarrow.(A\rightarrow B)\rightarrow B$
A3 $A\vee(A\rightarrow B)$

Let **RM3** be the extension of **E** got by adding A1–A3. **RM3** is also an extension of **RM**, and so it has by the Extension theorem of §29.4 a characteristic Sugihara matrix, which on a simple application of Dunn's argument is easily seen to be the 3-point Sugihara matrix $M_3 = \langle M_3, O, D\rangle$, where M_3 consists of the numbers $\boldsymbol{-1}, \boldsymbol{0}, \boldsymbol{+1}$;

O is a set of operations corresponding to the connectives of **RM3**, including $\oplus$ corresponding to the defined connective $+$, with values given as follows for all a, b in M_3:

> $\bar{a}$ is the arithmetic inverse of a;
> a&b is the arithmetic minimum of a, b;
> a$\vee$b is the arithmetic maximum of a, b;
> a$\oplus$b is the maximum of a, b in the *absolute* ordering $\boldsymbol{0}, \boldsymbol{-1}, \boldsymbol{+1}$;
> a$\rightarrow$b is evaluated as $\bar{a}\oplus$b.

D is the set of *designated* elements of M_3, and its members are $\boldsymbol{0}, \boldsymbol{1}$. An *interpretation* of **RM3** in M_3 is, as usual, a function from formulas of **RM3** to M_3 that respects the connectives: i.e., for an interpretation I, $I(A\&B) = I(A)\&I(B)$, etc. And that M_3 is *characteristic* for **RM3** means, of course, that all theorems of **RM3** take values in D on all interpretations, and that every non-theorem takes the sole undesignated value $\boldsymbol{-1}$ on some interpretation.

We return to the demonstration that (u) and (y) don't co-entail each other, even in **RM3** and *a fortiori* not in **E**, for any truth functional choice of π. Because of the idempotence axiom A1 of **RM3**, and choosing A, B as sentential variables p, q, this reduces to a demonstration that

(z) $p+q \rightleftarrows \pi(p, q)$

is invalid in M_3 for any truth functional choice of π. (Put otherwise, intensional disjunction, and hence application, is not truth functionally definable in **RM3**.)

Checking the matrix specifications, it is clear that (z) will be valid in M_3, and hence a theorem of **RM3**, iff for all interpretations I of **RM3** in M_3, $I(p+q) = I(\pi(p, q))$. Since the appropriate equivalences hold in **RM3**, we may assume $\pi(p, q)$ in disjunctive normal form: i.e., that it is of the form $A_1\vee \ldots \vee A_n$, where each of the A_i is a conjunction of sentential variables and negates thereof; evidently, we may assume that no variables occur as conjuncts of an A_i but p, q and no negates but $\bar{p}$, $\bar{q}$.

We note first that $\pi(p, q)$ must be *truth functionally* equivalent to $p\vee q$; for, on all interpretations that don't assign $\boldsymbol{0}$ to either of p, q, $p+q$ behaves like $p\vee q$ (identifying $\boldsymbol{+1}$ with $\boldsymbol{T}$ and $\boldsymbol{-1}$ with $\boldsymbol{F}$), and so, if (z) is to hold, $\pi(p, q)$ must behave like $p\vee q$ also. We now consider the interpretation which assigns $\boldsymbol{0}$ to p and $\boldsymbol{-1}$ to q, calling it I. Because $+$ is evaluated in the absolute order, $I(p+q) = -1$. We now inspect what I does to $\pi(p, q)$, normal-formed as $A_1 \vee \ldots \vee A_n$. Disjunctions go to the maximum in the natural order, whence if (z) is to hold on I, we must have $I(A_i) = -1$ for each disjunct A_i. But each of the A_i is a conjunction of sentential variables and negates; but since $I(p) = I(\bar{p}) = 0$ and $I(\bar{q}) = +1$, while only $I(q) = -1$, to make each of the A_i turn out -1 under I it is required that q be a conjunct of each of the A_i, since & on interpretation goes to natural arithmetical minimum. That is, if (z) is to hold on I, $\pi(p, q)$ must admit a normal form $q\&B_1 \vee \ldots \vee q\&B_n$. But evidently this is inconsistent with the other requirement that $\pi(p, q)$ must meet, namely that it be truth functionally equivalent to $p \vee q$. (For now assign p, $+1$; and q, -1.) Too much is demanded of $\pi(p, q)$, and it must give up; there is no choice of π, under the imposed constraints, that makes (z) valid in RM3. *A fortiori*, there is no choice of π that makes (u) co-entail (y), in **E**. Accordingly, there is no system **E*** in $\Box$, &, $\vee$, $\bar{\ }$ in which entailment in the sense of **E** is definable, as a strict truth function. Accordingly, strict implication is irrelevant in the task of defining entailment.

A last word. We have pretended to carry through our arguments for the system **E** of entailment, but in fact we used very few particular properties of **E**. All of the above arguments continue to go through in the very weak relevant logic **B** of §48.5.3 and in all of the relevant extensions thereof considered there, including the systems **T** of ticket entailment of **R** of relevant implication. On the other hand, there are some very strong systems that admit our arguments; **RM3**, which we used for the hardest part of the proof (or at least the most intricate), is just a shade off being classical. So for every logic intermediate between **B** and **RM3**, and this includes every logic thus far proposed as relevant, or even wrongly conjectured to be relevant, like **RM**, the answer to the Lewis problem, at least in the Fregean form we have given it, is "No." No relevant implication, in brief, can be conceived as a Lewis-style strict implication. While it would still be interesting to know what is the weakest form of the Fregean condition 2 under which the answer is still, "No," and what is the next weakest form (under which it becomes, "Yes"), from the point of view of natural motivation the present result must be taken as definitive: relevant logics are not modal logics in the sense of Lewis even by courtesy, but incorporate on the other hand genuinely new features not intuitively reducible to a Lewis-style analysis of entailment. That was what was claimed from the first (§3); relevance is another ingredient in the analysis of logical consequence, unnoticed by Lewis and irreducible to those vital and fertile modal intuitions of Lewis for which we are all in his debt. (Thanks are due Massey and Segerberg for helpful suggestions.)

APPENDIX

Grammatical Propaedeutic. The principal aim of this piece is to convince the reader that it is philosophically respectable to "confuse" implication or entailment with the conditional, and indeed philosophically suspect to harp on the dangers of such a "confusion." (The suspicion is that such harpists are plucking a metaphysical tune on merely grammatical strings.)

Our strategy is as follows. First we lay out as clearly as we can the grammatical background — call it logical grammar — against which the distinction between conditional and implication should be made, for here as elsewhere confusion is only admissible when clarity is possible. This effort terminates in a table laying out a variety of grammatical possibilities for certain key philosophical concepts: truth, negation or falsity, conjunction, conditional, and entailment. Second, we discuss the logical grammar of truth, negation or falsity, and conjunction, in order to elicit easy-to-overlook features of the application of logical grammar to English, summarizing our views in eight theses. Last, having prepared the way (and after a brief word about quantifiers), we discuss the grammar of the conditional and entailment.

A1. Logical grammar. By a "logical" grammar we mean one of the sort that guides the thinking of most logicians; such a grammar is a more or less language independent family of grammatical categories and rules clearly deriving from preoccupation with formal systems but with at least prospective applications to natural languages. A logical grammar is accordingly one which is particularly simple, rigorous, and tidy; one which suppresses irregular or non-uniform or hard to handle details (hence differentiating itself from a "linguistic" grammar whether of the MIT type or another); one which idealizes its subject matter, one which by ignoring as much as possible leads us (perhaps wrongly) to the feeling of "Aha; that's what's *really* going on!" Among such logical grammars, we think a certain one lies unexpressed at the back of the head of most logicians. Our task is to make it explicit, not in order to criticize it — indeed, though it would be dangerous to suppose it eternally ordained, we think very well of it — but because only when it is brought forth in all clarity can we sensibly discuss how it ought to be applied. The version here presented derives from Curry and Feys 1958 and Curry 1963.

Of course here and hereafter application of any *logical* grammar will be far more straightforward and indisputable in the case of formal languages than in the case of English; for in the case of the former, possession of a tidy grammar is one of the design criteria. But English is like Topsy, and we should expect a fit only with a "properly understood" or "preprocessed" English — the preprocessing to remove, of course, the hard cases. So read on with charity.

"Logical grammar," as we understand it, begins with three fundamental grammatical categories: the "sentence," the "term," and the "functor." The first two are taken as primitive notions, hence undefined; but we can say enough to make it clear enough how we plan to apply these categories to English and to the usual formal languages.

By a *sentence* is meant a declarative sentence in pretty much the sense of traditional grammar. Some part of logical opinion would probably exclude traditional sentences like "It is a dog," which are fit only to receive a truth-value-in-context instead of an outright truth value; but in order to cast our net as widely as possible, we generously allow such candidates full status as sentences in our logical grammar. On the formal side, we intend for the the same reason to include "*Fa*" even when the denotation of the individual constant "*a*" is possible-world- or context-dependent (in a modal or tense logic), and even "*Fx*," where "*x*" is squarely a free variable.

Traditional grammar gives us less help in articulating the concept of a *term*, although the paradigm cases of both traditional nouns and logical terms are proper names such as "Wilhelm Ackermann."

Some more examples of terms from English and near-English:

Terms	What some linguists call them
Wilhelm Ackermann	proper noun
the present king of France	noun phrase
your father's moustache	noun phrase
triangularity	(abstract) noun
Tom's tallness	(abstract) noun phrase
that snow is white	that clause; factive nominal; nominalized sentence
what John said	factive nominal; nominalized sentence
his going	gerund; nominalized sentence
he	pronoun

We add some further examples from formal languages.

$3+4$	closed term
x	variable
$3+x$	open term
$\{x\colon x \text{ is odd}\}$	set abstract
ιxFx	definite description

What the logical grammarians contrast with these are so-called "common nouns" such as "horse," as well as plural noun phrases such as "Mary and Tom," mass nouns such as "water," and indefinite noun phrases such as "every man" and "a present king of France." And although we take the category of terms to be grammatical, it is helpful to heighten the contrast by a semantic remark: the terms on our list purport — at least in context or when fully interpreted ("assigned a value") — to denote some single entity, while "horse" and "Mary and Tom" do not (*pace* those who want to calculate with individuals). Perhaps the common noun is the most important English grammatical category not represented anywhere in logical grammar. Contemporary logicians (including us) uniformly torture sentences containing common nouns, such as "A horse is a mammal," into either "The-set-of-horses is-included-in the-set-of-mammals" or "For-anything-you-name, if it is-a-horse then it is-a-mammal," where the role of the common noun "horse" is played by either the term (our sense) "the-set-of-horses" or the predicate (see below) "______ is-a-horse."

So much for sentence and term The third fundamental grammatical category is constituted by the "functors." By a *functor* is meant a way of transforming a given ordered list of grammatical entities (i.e., a list the members of which are terms, sentences, or functors) into a grammatical entity (i.e., into either a term, a sentence, or a functor). That is to say, a functor is a function — a grammatical function — taking as inputs (arguments) lists of items from one or more grammatical categories and yielding uniquely as output (value) an item of some grammatical category. For each functor, as for any function, there is defined its *domain*, that is, the set of its input lists, and its *range*, which is the set of its outputs.

We are really only interested in what we will eventually call "pure elementary functors," but in order to appreciate the limitations we lay on ourselves, and so as not to fossilize our grammatical (philosophical?) imaginations, it will be useful to survey the functorial landscape by means of some definitions and examples. Those uninterested in landscapes should skip four paragraphs to the displayed table of pure elementary functors.

Given the domain of a functor, it can happen that the length of every input list is the same, or it can happen that there are input lists of different length. In the first case we say that the functor is *fixed.* For example, the "if-then" functor is fixed because it always takes as input a list of length two. Fixity will be part of the definition of "elementary," but there are important cases of non-fixed (variable) functors. For example, consider the functor in set theory which from a list of entity-names produces a name of the set of those entities, i.e. the functor represented by the notation "$\{x_1, \ldots, x_n\}$," where n is allowed to vary. This functor is not fixed. Of course one *could* define "$\{x_1, \ldots, x_n\}$" as representing a different functor for each n; each such functor would then be fixed. But this is a trivial point, and does not remove the possibility of someone else presenting the grammar of the language of set theory using the non-fixed functor we described. So much for fixity.

As we have defined them, functors can take other functors as arguments. The importance of such functors is enormous; e.g., a logical grammarian would probably construe "very" as taking a predicate (see below), e.g., the function represented by "______ is heavy," into a predicate, e.g., the function represented by "______ is very heavy." But we shall not be talking about such higher level functors, a fact we emphasize by calling a functor *first level* if its input lists are wholly confined to sentences and terms. The functors represented by "if ______ then ______" and "______+______" are first level in this sense, as is also "______ believes that ______". In order to avoid considering examples like the last, where an input list can be a mixture of terms and sentences, as well as examples like quotation marks which in English can take, among entities in logical grammar, either a term, as in " 'Tom' ", or a sentence, as in " 'Tom is tall' ", we call *pure* those among the first level functors which have input lists consisting always either wholly of terms or else wholly of sentences, and when furthermore the output is either always a term or always a sentence. When we use examples from English, they are charitably to be construed as pure first level functors.

We turn now to the last subdivision of functors. By a *substitution* functor we mean a functor for which there is a non-empty pattern (or possibly a family of patterns) of words with blanks, the blanks being ordered such that, for each input list in the domain of the functor, the output can be obtained by substituting the elements of the input list, in order, into the blanks. "If ______ then ______" and "______+______" represent substitution functors. Ordinary negation in English, if treated as a connective, is *not* a substitution functor: the passage from "Snow is white" to its ordinary negation, "Snow is not white," cannot be accomplished by substituting the former into a blank. In contrast, the logician's negation *is* a substitution functor:

one can pass from "Snow is white" to "It is not the case that snow is white" by substitution into "It is not the case that ______". Substitutionality will be part of our definition of "elementary."

A substitution functor is defined once one knows (1) the pattern of words with (ordered) blanks defining the pattern of substitution, and (2) the domain of inputs. It is therefore often convenient to speak as if the pattern of words with blanks ordered from left to right were itself the functor, especially in giving examples; but in such cases one must keep in mind that the domain of the functor then has to be either separately specified or gathered from context. For example, we will allow ourselves to say

the functor "______+______",

with the understanding that only number-terms are to be allowed as inputs.

We can now define the elementary functors: an *elementary* functor is a fixed first level substitution functor. Elementary functors do not have to be pure, but we shall confine our attention to those that are; and evidently there are four kinds of pure elementary functors, since the inputs can be either terms or sentences while the outputs may vary in the same way. The four kinds of pure elementary functors are exhibited, named, and exemplified in the table below:

Inputs	Output	Name	Examples
Terms	Term	Operator	______+______; ______'s father
Terms	Sentence	Predicate	______<______; ______ is nice
Sentences	Sentence	Connective	______ and ______; John surmises that ______
Sentences	Term	Subnector	'______'; that ______

(We note in passing that individual quantification is elementary, but not pure: "(∃______) ______" is a fixed first level (but not pure) substitution functor, taking variables (a species of term) in the first blank and sentences in the second, and producing a sentence.)

Subnectors — the last entry in the table — are seldom mentioned and seldom occur in formal languages, but they are of particular interest for the present discussion. A subnector is a way of converting a sentence into a term; the examples given in the table above are the two subnectors in English of most importance for us. We shall call them, respectively, the *quote-subnector*, and the *that-subnector*. Of course quotation in English is mixed; we make it a subnector by restricting its domain, treating " '______' " as "the sentence '______' ". (We shall use single quotes here, reserving double quotation to be used informally in the language we are using. Note that we always suppress double quotes (used quotes) from displays, and that single quotes, in displays or not, are always mentioned and never used.) Observe that putting a sentence into the blank in either

‘______’

or

that ______

invariably results in a term, e.g.

‘Tom is tall’

or

that Tom is tall.

So they satisfy the definition of “subnector.” The first display above may be taken as the name of the sentence, the second, perhaps, as the name of a proposition.

Terms in logical grammar may usefully be subdivided into finer grammatical categories. For example, number-terms, physical-object terms, set-terms, city-terms, etc., would all very likely prove useful categories. In giving a category of terms a name, we find it most convenient to use something suggesting the domain of entities which will be taken as the denotations of these terms; but the idea is that in logical grammar the category shall be defined independently of semantics. (This may well be impossible for English.)

Of special interest in what follows will be the categories of “sentence-term” and “proposition-term.” One sort of *sentence-term* is obtained as the result of applying the quote-subnector, “ ‘______’ ”, to a sentence, and one sort of *proposition-term* as the result of applying the that-subnector, “that ______”, to a sentence. Examples:

Sentence-term: ‘Tom is tall’
Proposition term: that Tom is tall

But we would naturally suppose that “the last sentence in *Paradise Lost*” would also be counted as a sentence-term, and that “what he said” or “Euclid’s first proposition” or perhaps even “Tom’s tallness” would count as proposition-terms.

Whenever there is a subdivision of terms, it is possible and usually useful to subdivide the various functors involving terms according to what sort of terms they take as input and what sort they produce as output. For example, “______ + ______” is a number-into-number-operator: its inputs must be number-terms, and its output is always a number-term. And

“______<______” is a number-predicate: its inputs must be number-terms. Also, the subnector “the probability that ______” is a number-subnector, since its output is a number-term.

Of special interest in logical theory are the functors whose inputs or outputs are restricted to either sentence-terms or proposition-terms. A *sentence-predicate* is one in whose blanks you can put only sentence-terms, and a *proposition-predicate* one in which only proposition-terms can be put. *Sentence-operator*, *proposition-operator*, *sentence-subnector*, and *proposition-subnector* are similarly defined. Examples:

______ is true

is a sentence-predicate or a proposition-predicate according as the separately specified domain is restricted to sentence-terms or proposition-terms. As a sentence-predicate, “______ is true” yields “ ‘Tom is tall’ is true,” and as a proposition-predicate it yields “That Tom is tall is true.” It has often been pointed out that “______ is true” cannot be a connective, since “Tom is tall is true” has too many main verbs.

‘______ is true’

is a sentence-operator when the domain is restricted to sentence-terms, yielding e.g., “ ‘ ‘Tom is tall’ is true.’ ” If its domain is restricted to proposition-terms, it would be a proposition-to-sentence operator, an eventuality we ignore.

that ______ is true

is a proposition-operator when the domain is restricted to proposition-terms, giving e.g. the term “that that Tom is tall is true.”

Again there is a sentence-to-proposition-operator in the vicinity which we shall ignore; but note that “That ______ is true” is also a connective, with a sentential domain, since one can put a sentence in the blank to get e.g. the sentence “That Tom is tall is true.” This possibility arises because the scope of “that” in “that ______ is true” is ambiguous, whereas the scope of the pair of quotes in “ ‘______ is true’ ” is not. The blank in the latter can take only terms, while the scope ambiguity in the former would seem to permit either terms or sentences. Consequently, in order to reduce side remarks indicating domains, we shall sometimes use parentheses to indicate the scope we have in mind for “that”; thus writing “that (______ is true)” for the operator and “(that ______) is true” for the connective.

Lastly, the quote-subnector and the that-subnector are prime examples, respectively, of sentence-subnectors and proposition-subnectors. Two further examples:

Sentence-subnector: ‘if ______ then ______’
Proposition-subnector: that if ______ then ______

If one puts sentences in the blanks, one produces a term of the appropriate sort, thus satisfying the defining conditions.

Before applying this analysis, let us summarize our logical grammar. (1) A grammatical entity is either a sentence, a term, or a functor. (2) A functor is a mapping from lists of grammatical entities into grammatical entities. (3) There are many kinds of functors, among which we isolate the pure elementary functors. (4) There are four kinds of pure elementary functors: predicates, operators, connectives, and subnectors. (5) The isolation of the categories sentence-term and proposition-term induces a classification of predicates, operators, and subnectors into sentence-predicates vs. proposition-predicates, etc.

A2. The table. Now for the application. Whenever we approach a net of concepts with the idea of trying to get clear on them, a good question to ask is, How should we parse the fundamental terminology? Our goal is the net involving the concepts entailment or implication, and the conditional; but let us make haste slowly by simultaneously raising the question for other key logical concepts: truth, negation-falsity, and conjunction.

What the foregoing gives us is an account of the *range of possibilities* for parsing entailment (we'll stick to "entailment" even though our remarks apply equally to implication) and the conditional, truth, negation-falsity, and conjunction. We should expect to find in the vicinity of each of these concepts each of the following: a predicate, a connective, an operator, and a subnector; and usually both a proposition-variety and a sentence-variety for each. A selection of possibilities is laid out in the following table.

Truth

Operator:	that (______ is true)	'______ is true'
Predicate:	______ is true	
Connective:	(that ______) is true	'______' is true
Subnector:	that ((that ______) is true)	''______' is true'

Negation or falsity

Operator:	that (______ is false)
Predicate:	______ is false
Connective:	(that ______) is false
Subnector:	that ((that ______) is false)

Conjunction

Operator:	that (______ is true and ______ is true)	
	______⁀'and'⁀______	
Predicate:	______ is true and ______ is true	
Connective:	______ and ______	
Subnector:	that (______ and ______)	'______ and ______'

Conditional-entailment (with "if", and with "entails")

Operator:	that (if ______ is true then ______ is true)	
Predicate:	if ______ is true then ______ is true	
Connective:	if ______ then ______	
Subnector:	that if ______ then ______	'if ______ then ______'
Operator:	that (______ entails ______)	'______ entails ______'
Predicate:	______ entails ______	
Connective:	(that ______) entails that ______	'______' entails '______'
Subnector:	that ((that ______) entails that ______)	

In every place in which there is a blank for terms, the domain could be restricted to either proposition-terms or to sentence-terms. And in every place in which a "that ______" is displayed, one could have a variation with quotes instead; but we have presented the quote-versions in only a selection of the cases above.

A3. Eight theses. Using this table as a basis, we arrive at eight theses: Thesis 0, . . . , Thesis 7.

A3.1. Logical grammar and logical concepts. Our first, the zero thesis, is perhaps the most important of all; namely, that logical grammar is in fact applicable to English in the sense that we can identify various constructions in English as satisfying the conditions defining a certain kind of functor. For example, we note that "if ______ then ______" is a connective, and that "______ entails ______" (as used by logicians — we always intend this qualification) is a predicate, the former requiring sentences, and the latter requiring terms, an observation of importance to anyone who wishes to be grammatical. Indeed, we think that logical grammar should be a part of the curriculum of every introductory logic course. So:

THESIS 0. Logical grammar is applicable to English.

We next remark that the table both presupposes and leads to the claim that it makes sense to talk about a concept prior to specifying its grammatical form. For example, by "negation-falsity" we mean to refer to that generalized concept of nay-saying (we stretch for a generic term) which has not yet decided what grammatical clothes to wear. Of course logicians customarily use "negation" for a connective, and "falsity" for a predicate — although using the latter ambiguously for both a propositional and a sentential predicate — but this should not blind us to the existence of a topic of which these two are specializations. Similarly, although a certain school of logicians systematically uses "conditional" for the connective and "implication" for the predicate, it is no confusion to suggest that there is a single topic here. Analogously, there is a concept of "mass" in physics before one has decided upon its grammatical role in a tidied up reconstruction: whether to construe it as carried in the language of physics by an equivalence predicate, a physical object-into-number operator, four-place predicates representing interval or ratio similarities, or what. The same remark holds for "time" and other topics to which we are accustomed to refer by means of a kind of noun; for unless there is something antecedently there, it makes no sense to ask the extremely important question as to the most fruitful grammatical analysis of the concept (or family of concepts) in question. We are accordingly led to the following

THESIS 1. A concept (or family) can be isolated without specifying its grammar; and accordingly, one can ask concerning the advantages and disadvantages of carrying it in various grammatical forms.

A3.2. A question of fit. One can also ask about the fit between English and logical grammar with respect to a given set of concepts.

A3.2.1. Simplest functors. We consider the various logical concepts in turn, pointing out in each case that functor in the vicinity of a given concept which is most easily expressed in English. *Truth:* if we disregard the null functor, the grammatically simplest pure elementary functor in the vicinity of truth is a predicate. We also note that English is indifferent as to whether the truth predicate is construed as a sentential predicate or as a propositional predicate; both, in English, are equally grammatical. *Negation-falsity:* exactly the same comment applies here. *Conjunction:* the simplest functor in the vicinity is a connective. *Conditional and entailment:* the simplest functor involving "if" is a connective, while the simplest involving "entails" is a predicate.

We may summarize by means of the banal

THESIS 2. English permits a given logical concept to be expressed by functors which differ in simplicity.

A3.2.2. More complex functors. We continue our investigation of English grammar from the point of view of logical grammar by looking at a selection of more complex functors attaching to the various logical concepts. First as to truth and negation-falsity. Every substitution-connective for either truth or negation-falsity seems to involve in its construction some kind of pre-nominalization, via either the quote-subnector or the that-subnector. So if one wants a systematic way of negating or of truifying in English, one seems required to use one of these functors, as in "that ______ is false" or "that ______ is true." Or perhaps in the case of truth, we might use the connective "Truly, ______" as in "Truly, snow is white", or even the "null" connective "______", as a truth-connective without pre-nominalization via a subnector; but we cannot happen to think of anything analogous in the case of falsity. Of course English does have at least a partial (i.e., perhaps not everywhere defined) connective for negation-falsity that does not involve pre-nominalization: we mean the connective that takes "Snow is puce" into "Snow is not puce," "Snow is not puce" into "Snow is puce", etc. But this connective is not elementary; in particular, it is not a substitution-functor. It gives us no *uniform* way of negating A, especially when A is a complex sentence.

Now for conjunction. If we wish to have a conjunction-predicate we seem to need to insert "______ is true." This is perhaps not surprising; but it is remarkable that we need to involve "______ is true" in finding an elementary propositional conjunction-operator. But let us see first how to find a substitutional sentential conjunction-operator. We are given as inputs the sentence-terms " 'Snow is white' " and " 'Tom is tall' ", and wish to produce as output a sentence-term to be taken as naming what amounts to a conjunction of what the two input-terms name; i.e., a single sentence which implies exactly those sentences jointly implied by what the input-terms name. So we want " 'Snow is white and Tom is tall' " (" 'S and T' "). Obviously we have just done what is required; but notice that we have not employed a substitution-functor in so doing. The passage from " 'S' " and " 'T' " to " 'S and T' " is not and cannot be accomplished by substitution. There are ways out, perhaps the simplest being to use concatenation: "______⁀'and'⁀______" applied to " 'S' " and " 'T' " yields " 'S'⁀'and'⁀'T' ", which is indeed a name of the conjunction "S and T" of "S" and "T". And note — this is relevant to our next point — that this works even if one puts in the blanks a sentence-term in the wider sense; e.g. "the first sentence in *Paradise Lost*." (But is it English?)

Now for the *propositional* conjunction-operator. We are given as inputs the proposition-terms "that S" and "that T," and wish to produce a proposition-term to be taken as naming what amounts to a conjunction of what the two input-terms name; so we want "that S and T." But again we have *not* employed a substitution-functor. And in this case the only way out which occurs to us employs pre-sententialization via truth: "that ______ is true and ______ is true," yielding "that that S is true and that T is true" by substitution, and naming what amounts to a (propositional) conjunction of what "that S" and "that T" name. Furthermore, consider the problem of forming a name of the proposition amounting to the conjunction of what Peter said and what Paul said; we seem, in the presence of sentence-terms in the wider sense, to have to settle for "that what Peter said is true and what Paul said is true," thus involving "______ is true" to keep the grammar straight by pre-sententializing the proposition terms. Note that "what Peter said and what Paul said" is grammatical, but that it doesn't name the conjunction required, being instead a *plural* term and thus falling outside the scope of logical grammar. Of course in *technical* English "the conjunction of ______ and ______" would do; but few would understand "I deny the conjunction of what Peter said and what Paul said" without an explanation. So if we want to remain with "ordinary" English, "true" seems the only way.

Similar remarks apply to conditional and entailment: if we wish to have a predicate involving "if," we are going to have to pre-sententialize by sprinkling in some occurrences of "true," and if we want a connective in the vicinity of entailment, we shall have to pre-nominalize by inserting either some occurrences of "that" or some quotes.

We may summarize these considerations in the following two part

THESIS 3. (a) Given a logical concept, we can always find a way to carry it in English as any functor we like, provided we are willing to pre-sententialize and pre-nominalize by means of a sufficient number of instances of "is true" and "that."

(b) Often the only or at least the easiest way to carry a concept as a given functor is to pre-sententialize or pre-nominalize by "is true" or "that."

A3.3. Parsing logical concepts. The foregoing constitutes some more or less factual observations concerning the relations between English and logical grammar. We now want to turn more directly to the question, how *should* we parse the terminology governing the key concepts of logic?

And first let us ask Pilate's question: What is Truth? In English "truth" is sometimes a common noun, as in "We take these truths to be self-evident," and sometimes a term, as in "Truth is beauty." But we systematically abjure common nouns in logical grammar, and although we do sometimes want a term "Truth" or "the True" as a name of one of the truth-values, and although Frege (and ourselves in §§27.1.2 and 33) seems to have deployed "Truth" or "the True" as a sentence, the usual problem is as to whether we want the concept of truth carried by a predicate or a connective — and if a predicate, whether sentential or propositional. Tarski's answer is well known: the important locution in the vicinity of "truth" is the truth-predicate,

______ is true.

Furthermore, Tarski takes this predicate to be a sentence-predicate, taking in its blank only sentence-terms. Others have argued that the truth-predicate is a proposition-predicate, taking proposition-terms in its blank. English certainly accepts both:

'Snow is white' is true

and

(that snow is white) is true

are equally grammatical (though the former seldom or never occurs outside of logic). The philosophical question of which (if either) truth-predicate to take as the foundation for a theory of truth must therefore be answered on other grounds. Which is our next (obvious) point:

THESIS 4. Sometimes English is indifferent as to grammatical form, in which case the choice of functor with which to carry a concept must be made on extra-grammatical grounds.

Our further and more important point: the fact that "______ is true" is the simplest functor in the vicinity of truth (Thesis 2) does *not* entail that one should so parse truth in laying the foundations of an adequate theory; the structure of English may turn out to be a bad source for intuitions guiding the construction of a fruitful theory of truth, as it certainly is in guiding us to a good theory of nobody (Lewis Carroll). It is *open* to the logical theorist, given "that Tom is tall is true," to take as immediate constituents (1) "Tom is tall" together with the connective "that ______ is true," rather than (2) "that Tom is tall" together with the predicate "______ is true," and thus to opt for some other functor more difficult to express in English (Thesis 3). Of course some *English* grammars will proceed according to (2); but that should not dictate how we apply logical grammar.

For consider negation-falsity in parallel with truth. Both have associated with them an operator, a predicate, a connective, and a subnector, and in each case the predicate is the simplest substitutional functor. But when it comes to founding a theory on a choice of fundamental grammar, the astonishing fact is that logicians almost uniformly choose differently in these two cases: they carry truth by a sentence-predicate, and negation-falsity by a connective. The most interesting and fruitful theory of truth is taken to be Tarski's semantic one, not the theory of the null connective or of "Truly, ______", while the most interesting and fruitful theory of nay-saying would by most be taken to be the theory of the negation connective, read in English as "that ______ is false" or "it is not the case that ______" or "that ______ is not the case."

We are not challenging the decision. Our only point is that in English grammar, truth and falsity are on all fours, (or all twos), so that other considerations must be invoked in support of the logician's decision to treat them differently.

Of crucial importance, both here and later, is the fact that the presence of "that" in the falsehood connective, "that ____ is false," or its cousins mentioned above, does *not* make the typical logician feel that he must construe outputs from this connective, such as "that Tom is tall is false," as somehow "about" a proposition, that Tom is tall, or as somehow implicitly metalinguistic.

In fact given

It is not the case that Tom is tall,

logicians uniformly take the "immediate constituents" to be

Connective: "It is not the case that ____"

and

Sentence: "Tom is tall."

They do not take the immediate constituents as

Predicate: "It is not the case ____"

and

Term: "that Tom is tall,"

because, presumably, they do not wish to "commit" themselves to the existence of some entity named by the term "that Tom is tall." Again it is to be pointed out that it is theoretical considerations and not English grammar which dictate the choice, for certainly some *English* grammars will take "that Tom is tall" to be an immediate constituent of "It is not the case that Tom is tall" or of its underlying form "That Tom is tall is not the case." But that should not dictate how we apply logical grammar. We are thus led to the following

THESIS 5. In contrast to the situation envisaged in Thesis 4, sometimes English has fixed ideas about grammatical form: some English constructions definitely and uniformly require terms, and some sentences. But in formulating the theory of a given logical concept, although we should of course be grammatical, it is not required that we slavishly take the details of English grammar as a sure guide to the most fruitful way to proceed.

A3.4. Reading formal constructions into English: the roles of "True" and "That". We now wish to apply these considerations to the problem of finding readings from a formal language into English, starting with formal negation — the curl — as our example. We know from Thesis 3 that every substitution connective for either truth or negation-falsity (with the mentioned odd exceptions for truth) seems to involve pre-nominalization via either the quote-subnector or the that-subnector. So if one wants a systematic way of reading "~______" in English, one seems *required* to use one of these functors, as in "It is not the case that ______". And the same holds for a truth-connective, "T(______)"; we have to say something like "that ______ is the case."

What follows? We think: nothing. We think this is a *sheer* grammatical fact about English, having no philosophical significance. We think there is nothing "conceptual" about the fact that "Not Tom is tall" or "It is not the case Tom is tall" or "Tom is tall not" are bad English grammar, while "Tom is tall and Mary is bald" is as good a sentence as "That Tom is tall and that Mary is bald" is a bad sentence. English has a convenient substitution-connective for conjunction which does not take a detour through subnectors, but it does not appear to have a convenient substitution-connective for either truth or falsity except those involving subnectors. It is accordingly easier to find a reading into English for formal conjunction than it is to find one for formal negation. There's a fact for you, and a mere fact.

We think the matter sufficiently important to illustrate with respect to another example, this time involving conjunction as a propositional operator. To give point to the illustration, we first remark that on non-grammatical grounds we think the theory of the conjunction operator an interesting one: it increases our understanding of conjunction to have it explained that given two propositions a and b, their conjunction is (the proposition which is) the greatest lower bound of a and b with respect to the ordering of propositions by the relation of propositional implication. In other words, the conjunction of propositions a and b is the weakest proposition implying both a and b. So we introduce notation "$a \wedge b$," say, for the conjunction of a and b. But how now are we to give a simple and uniform reading of "$a \wedge b$" in English? If we mind our grammar, the only way seems to be something like "that both a is true and b is true"; in §40.3 we use "that a obtains and b obtains." But this reading *appears* to involve us with the concept of truth, although it nowhere plays a role in the theory of conjunction sketched above. And if we have been brought up by Tarski, it *appears* to be metalinguistic, although our theory was not.

We may, however, conclude from the foregoing that these appearances are illusory; we are in this case using "______ is true" or "______ obtains" only because English "happens" not to have a simple propositional conjunction operator that does not involve (something like) these words. It might have been that English was possessed of a conjunction operator, say "______ et

______", so that what Peter said et what Paul said would be the weakest proposition implying both what Peter said and what Paul said. And — we think — this is a might-have-been which would not have altered the deep conceptual structure of English, whatever that is. We are therefore at liberty to treat our forced reliance on "is true" as a philosophically uninteresting feature of English.

Someone who abhors propositions is of course also at equal liberty to abhor our entire theory of propositional conjunction; we do not mean to be arguing for it just here. Our only point is that such a person is not entitled to buttress his arguments with grammatical facts unless he is also prepared to argue non-circularly that these facts are of philosophical interest. Thus:

THESIS 6. There are important grammatical constraints on the reading of functors in formal languages into English; but one should be wary of drawing exaggerated metaphysical or conceptual implications from the rules of English grammar.

(We intend this thesis to be interpreted in such a way that it is consistent with the view that indeed *some* features of English grammar — e.g. its love for the subject-predicate form — *can* be argued to be metaphysically significant.)

It is a crucial part of the argument that, on our view, the propositional predicate "______ is true" and the propositional subnector "that ______" can be viewed as sometimes (we by no means say always) playing in English low grade and philosophically uninteresting roles: that of converting a term into a sentence, and that of converting a sentence into a term, respectively, *without semantic increment.* (Of course the thesis that such and such is of no philosophical interest can itself be interesting and debatable.) The idea is that English has certain constructions requiring terms and others requiring sentences; so if one has somehow started with one but English requires the other, one simply goes to the shelf and pulls out a "that ______" or a "______ is true." Examples go both ways, as we have illustrated above: to negate a complex sentence appears to require prior nominalization with "that ______," while to conjoin propositionally seems to require prior sententialization with "______ is true." And, we argue, that a construction in English needs in one case a term and in another a sentence and therefore calls for pre-nominalization or pre-sententialization is to be viewed as a mere, sheer accident — unless independent philosophical argument can show otherwise.

Our use of the word "accident" needs clarification. Linguists divide linguistic features into those that are present in only some languages and those that are true linguistic universals (if there are any). The non-universal features might be called "accidental," since not lawlike, since not universal; but the question as to whether or not a feature is of philosophic importance is wholly independent of the question of universality. For example, the subject-predicate form may well be of profound conceptual interest even if nonuniversal; and contrariwise, even if biological constraints should lead to a law that certain types of self-embedding constructions can never, in any language, nest more deeply than (say) depth 7, that fact, though not of course devoid of interest for certain parts of philosophy, would not have the kind of conceptual implications which would render it of interest to the logical grammarian. It would be a kind of *conceptual* accident even though an *empirical* law; and it is this sense of "accident" we have in mind. The upshot is that one can never take a feature, whether universal or not, as a philosophical import without some kind of further and distinctively philosophical argument.

Although Hiż, as a linguist, may not wish to be associated with a view which reads certain linguistic features as "accidents," we quote Hiż 1961 in support of the innocuousness of subnectors: "Nominalized forms of sentences are not names of the sentences. Rather they are different forms of the very same sentences."

A corollary is that one should not feel "committed' to propositions because he is dealing in constructions like negation requiring pre-nominalization with "that ______", nor should one feel he must be "committed" to a notion of propositional truth or to some kind of primacy of sentences because constructions like propositional conjunction require pre-sententialization with "______ is true." Such "commitments" would arise only from further and less innocuous uses of "that" and "true." So:

THESIS 7. One of the roles of each of "that" and "is true" in English is to serve as a grammatical dodge to evade the constraints of English grammar without semantic penalty.

So much for theses.

A4. A word about quantifiers. It might be thought that against our downplaying of the term-sentence distinction one should count the quantificational structure of English, which leads us to ontological commitment and all that, by means of its permission to generalize with respect to term positions, while forbidding generalization with respect to sentence positions. For example, there are the well-known difficulties concerning reading propositional quantifiers into English. We think, however, that Grover 1972 shows conclusively that these difficulties are due to "accidental" features of English grammar and, accordingly, both illustrates and supports our general thesis.

A5. Conditional and entailment. Let us, finally, proceed to a consideration of conditionals and entailments from the point of view of logical grammar. An analogue to Pilate's question about truth is: What is Entailment? We follow awhile the numbering of our theses.

0. "______ entails ______" is a two place predicate, taking sentence- or proposition-terms as input and producing a sentence as output; and of course "if ______ then ______" is a connective.

1. Without disputing the usefulness of using "conditional" to refer to English constructions based on "if," we recognize the existence of a generic conditional-entailment-implication topic.

2. The table shows that the simplest functors in the vicinity of this generic concept are the "if ______ then ______" connective and the "______ entails ______" predicate.

3a. From this it by no means follows that "Entailment is a Relation"; to say so would constitute a naive and slavish adherence to the idiosyncracies of English grammar, for also entailment could be represented by any of an operator, a connective, or a subnector, provided we allow enough pre-sententializing and pre-nominalizing.

3b. The only or easiest way to carry entailment as a connective involves pre-nominalizing by "that"; which is to say, the entailment connective in English is to be taken as "that ______ entails that ______."

We remark further that the English entailment connective, like any connective, takes sentences as both input and output, and therefore nests in English without limit. Can we responsibly avoid investigating the complexities deriving therefrom?

4. English admits "entails" as both a proposition-predicate and a sentence-predicate; grammar certainly gives us no reason to doubt that the theories of these two should be equally interesting. In fact, our work in Chapter III could be construed as providing a theory of entailment in any one of its three guises as a connective, sentence-predicate, or proposition-predicate.

5. Although "entails" in English is unmistakably a predicate, this grammatical fact should not lead us to forego investigation of entailment qua connective. Indeed, although it does not represent one of our principal interests, it is symmetrically in order to remark that the if-then predicate, "if ______ is true then ______ is true" provides an object of legitimate study. (If the underlying if-then were truth functional, we should be in the presence of the *relation* of material "implication.")

6. In our study of entailment, we use the arrow, so that, using "A" and "B" as sentences, "$A \rightarrow B$" is a sentence. Accordingly, when careful we must read a formula "$A \rightarrow B$" into English as "that A entails that B." And when we raise the question of the applicability to English of the logic of

entailment, we intend the application to be made via the "that ______ entails that ______" connective. (But see below concerning the interchangeability of "if-then" and "entails.")

7. We do not think the instances of "that" in "that ______ entails that ______" commit us to propositions. We think it an accident that, in English, entailment is carried by a functor requiring terms instead of sentences and thus requires pre-nominalization. We take the immediate constituents of "that Tom runs entails that Tom moves" to be "Tom runs," "Tom moves," and the above exhibited connective, with the occurrences of "that" playing an only grammatical role. (We like propositions all right; it is just that we don't want to be "committed" to them before we are ready; we want to be committed by our theory, and not by our grammar. We take ourselves in this respect to be in parallel with the fellow using "it is not the case that ______".)

In accordance with 7, while in partial opposition to 6, it must be said that we do not much worry about dropping an occasional "that" from time to time. We feel "snow is white does not entail that snow is white entails itself" is more readable than the more correct "that snow is white does not entail that that snow is white entails itself," and is wholly unambiguous. If we thought of "that" in this sort of example as of philosophical interest, we'd worry; but we don't. Similarly, even though our letters take the places of sentences, we may sometimes be found reading "$A \to B$" as "A entails B," instead of "that A entails that B." We hope this Grammatical Propaedeutic will enable any reader of our work to sprinkle "that" around enough to save us from solecism.

We have now run out of theses, but we find ourselves with two more important points to make. The first has to do with our endemic tendency to read "$A \to B$" as "if A then B" as often as we read it with "entails." In justification of this procedure, we argue that the most significant difference between "if-then" and "entails" (or "implies") is that the first functor is a connective — calls for sentences — whereas the latter is a predicate — calls for terms. Repeat: the *most* significant difference. And if we are right, it is not much of a difference at all except to a grammarian. We think *nothing* philosophical hangs from this famous distinction. It is a mere, sheer, clear fact of grammar.

We further think there are all kinds of conditionals (uses of "if-then") in English, and also all kinds of uses of the entailment or implication connective; and that many of them correspond in a one-to-one fashion. For example, "that A logically entails that B" has "if A then, as a matter of logic, B," with ellipsis in both cases producing, respectively, "that A entails that B" and "if A then B." In particular, we think every use of "implies" or "entails" as a connective can be replaced by a suitable "if-then"; however, the converse may not be true. But with reference to the uses in which we are primarily interested, we feel free to move back and forth between "if-then" and "entails" in a free-wheeling manner.

Our very last point is a little phony, since it presupposes having to make a decision as to whether to treat entailment as a sentence-predicate on the one hand or as a connective on the other; and after all, Peirce's warning not to block the road to inquiry should lead us to embrace both alternatives as legitimate. But suppose we have to make a decision; then the final point we want to make is that there are costs on both sides. The cost of taking entailment as a connective is often remarked and well known: we must pay attention to the nesting of entailments within entailments, and so on without end. Perhaps "must" is a bit strong here, but in any event the grammatical fact that connectives use sentences as *both* input and output strongly suggests attention to the problem of nesting. In contrast, if we take entailment as a predicate, the grammar forbids nesting; for the output of a predicate is a sentence, whereas its inputs must be terms. Again this is somewhat of an oversimplification, since we might wish to deal with entailment sentences whose terms name other entailment sentences; but, ignoring that, the theory of the entailment predicate will be simpler than the theory of the entailment connective by avoiding nesting. But the cost is by no means all on one side, a point which is seldom noticed. For to take entailment as a sentence-predicate means that one has to involve himself with all the bag and baggage of The Metalanguage, which as we know very well is not only dreadfully complicated in its syntactic department, as anyone who has set out to completely formalize these matters can testify, but also, with respect to its semantic ingredients, skirts the very edge of paradox, so that one taking this road must exercise both care and caution. So let no one suppose that departure from simplicity is a peculiar characteristic of the theory of the entailment connective.

We summarize the spirit of this Propaedeutic with a bit of advice meant to parallel some most of us received in the earlier stages of our education: learn logical grammar thoroughly, and then — but only then — be relaxed about it.

BIBLIOGRAPHY FOR VOLUME I

BOOKS AND ARTICLES are uniformly referred to in the text as, e.g., "Lewis 1914a," or "Ackermann 1956," for which the reader is invited to consult the alphabetized list below. (See also Addendum.) To provide an accurate sense of history, we have used the earliest date of publication known to us, even when we have consulted later editions. In such cases we mention later editions in this bibliography, or translations we have consulted (see e.g. under Łukasiewicz 1939.) Use of "197+" indicates a piece forthcoming, or possibly not to be published; in such cases we give whatever information we have, including journal or publisher if already accepted for publication.

We plan to include in Volume II a reasonably complete bibliography of entailment and relevance logics now being compiled by Wolf. Here we have included a work only if it is either mentioned in the text of Volume I, or is a relevant piece written by one of the co-authors of this book.

With regard to our mentioned plan for the bibliography of Volume II, we would like to ask for assistance from readers of this volume. Please write Robert Wolf, Department of Philosophy, Southern Illinois University at Edwardsville, Edwardsville, Illinois, 62025.

ACKERMANN, WILHELM

1956. *Begründung einer strengen Implikation.* **The journal of symbolic logic,** vol. 21, pp. 113–128.

1958. *Über die Beziehung zwischen strikter und strenger Implikation.* **Dialectica,** vol. 12, pp. 213–222. Also published in **Logica, studia Paul Bernays dedicata,** Neuchâtel (Éditions du Griffon) 1959, pp. 9–18.

See Hilbert and Ackermann.

ALIGHIERI, DANTE

1307. **La divina commedia. I. Inferno.** Text, and translation by Charles S. Singleton, Princeton (Princeton University Press) 1970.

ANDERSON, ALAN ROSS

1954. *On alternative formulations of a modal system of Feys-von Wright.* **The journal of computing systems,** vol. 1, pp. 211–212.

1957. Review of Ackermann 1956. **The journal of symbolic logic,** vol. 22, pp. 327–328.

1958. *Mathematics and the 'language game.'* **The review of metaphysics,** vol. 11, pp. 446–458.

1959. **Completeness theorems for the systems E of entailment and EQ of entailment with quantification.** Technical Report No. 6, Contract No. SAR/Nonr-609(16), Office of Naval Research, New Haven. Reprinted in the **Zeitschrift für mathematische Logik und Grundlagen der Mathematik,** vol. 6 (1960), pp. 201–216.

1963. *Some open problems concerning the system E of entailment.* **Acta philosophica fennica,** fasc. 16, pp. 7–18.

1972 abstract. *Negative implication formulas.* **The journal of symbolic logic,** vol. 37, p. 442.

1972a. *An intensional interpretation of truth-values.* **Mind,** n.s., vol. 81, pp. 348–371. Reprinted in **Logic, language and probability,** ed. Radu J. Bogdan and Ilkka Niiniluoto, Dordrecht (D. Reidel Publishing Company) 1973, pp. 3–28.

See Anderson and Belnap, Anderson and Belnap and Wallace, Anderson and Moore.

ANDERSON, ALAN ROSS and BELNAP, NUEL D. JR.

1958 abstract. *A modification of Ackermann's 'rigorous implication.'* **The journal of symbolic logic,** vol. 23, pp. 457–458.

1959 abstract. *A simple proof of Gödel's completeness theorem.* **The journal of symbolic logic,** vol. 24, pp. 320–321.

1959a. *Modalities in Ackermann's 'rigorous implication.'* **The journal of symbolic logic,** vol. 24, pp. 107–111.

1959b abstract. *A proof of the Löwenheim-Skolem theorem.* **The journal of symbolic logic,** vol. 24, pp. 285–286.

1959c. *A simple treatment of truth functions.* **The journal of symbolic logic,** vol. 24, pp. 301–302.

1961. *Enthymemes.* **The journal of philosophy,** vol. 58, pp. 713–723.

1962. *Tautological entailments.* **Philosophical studies,** vol. 13, pp. 9–24.

1962a. *The pure calculus of entailment.* **The journal of symbolic logic,** vol. 27, pp. 19–52.

1963. **First degree entailments.** Technical Report No. 10, Contract No. SAR/Nonr-609(16), Office of Naval Research, New Haven. Reprinted in **Mathematische Annalen,** vol. 149 (1963), pp. 302–319.

1968. *Entailment.* A composite of parts of Anderson and Belnap 1962 and 1962a, in **Logic and philosophy,** ed. G. Iseminger, New York (Appleton-Century-Crofts), pp. 76–110.

See Anderson and Belnap and Wallace.

ANDERSON, ALAN ROSS, BELNAP, NUEL D., JR. and WALLACE, JOHN R.

1960. *Independent axiom schemata for the pure theory of entailment.* **Zeitschrift für mathematische Logik und Grundlagen der Mathematik,** vol. 6, pp. 93–95.

ANDERSON, ALAN ROSS and MOORE, OMAR K.

1960. *Autotelic folk-models.* **The sociological quarterly,** vol. 1, pp. 203–216.

ANGELL, RICHARD B.

1962. *A propositional logic with subjunctive conditionals.* **The journal of symbolic logic,** vol. 27, pp. 327–343.

BACON, JOHN

1966. **Being and existence: two ways of formal ontology.** Yale University doctoral dissertation.

1971. *The subjunctive conditional as relevant implication,* **Philosophia,** vol. 1, pp. 61–80.

BARCAN, RUTH C.

See Barcan Marcus.

BARCAN MARCUS, RUTH C.

1946a. *The deduction theorem in a functional calculus of first order based on strict implication,* **The journal of symbolic logic,** vol. 11, pp. 115–118.

1953. *Strict implication, deducibility, and the deduction theorem.* **The journal of symbolic logic,** vol. 18, pp. 234–236

1963. *Modalities and intensional languages.* **Boston studies in the philosophy of science,** ed. Marx W. Wartofsky, Dordrecht (D. Reidel Publishing Company), pp. 77–96.

BARKER, JOHN A.

1969. **A formal analysis of conditionals.** Southern Illinois University Monographs, Humanities Series, No. 3. Carbondale.

BAYLIS, CHARLES A.

1931. *Implication and subsumption.* **The monist,** vol. 41, pp. 392–399.

BELNAP, NUEL D., JR.

1959 abstract. *Pure rigorous implication as a sequenzenkalkül.* **The journal of symbolic logic,** vol. 24, pp. 282–283

1959a abstract. *Tautological entailments.* **The journal of symbolic logic,** vol. 24, p. 316.

1959b. **The formalization of entailment.** Yale University doctoral dissertation.

1960. *EQ and the first order functional calculus.* **Zeitschrift für mathematische Logik und Grundlagen der Mathmatik,** vol. 6, pp. 217–218.

1960a. **A formal analysis of entailment.** Technical Report No. 7, Contract No. SAR/Nonr-609(16), Office of Naval Research, New Haven.

1960b. *Entailment and relevance.* **The journal of symbolic logic,** vol. 25, pp. 144–146.

1960c abstract. *First degree formulas.* **The journal of symbolic logic,** vol. 25, pp. 388–389.

1962. *Tonk, plonk and plink.* **Analysis,** vol. 22, pp. 130–134.

1967. *Intensional models for first degree formulas.* **The journal of symbolic logic,** vol. 32, pp. 1–22.

See Anderson and Belnap, Anderson and Belnap and Wallace, Belnap and Gupta and Dunn, Belnap and Spencer, Belnap and Wallace, Dunn and Belnap.

BELNAP, NUEL D., JR., GUPTA, ANIL and DUNN, J. MICHAEL

197+. *A consecution calculus for positive implication with necessity.*

BELNAP, NUEL D., JR. and SPENCER, JOEL H.

1966. *Intensionally complemented distributive lattices.* **Portugaliae mathematica,** vol. 25, pp. 99–104.

BELNAP, NUEL D., JR. and WALLACE, JOHN R.

1961. **A decision procedure for the system $E_{\bar{I}}$ of entailment with negation.** Technical Report No. 11, Contract No. SAR/609(16), Office of Naval Research, New Haven. Also published in **Zeitschrift für mathematische Logik und Grundlagen der Mathematik,** vol. 11 (1965), pp. 277–289.

BENNETT, JONATHAN F.

1954. *Meaning and implication.* **Mind,** n.s., vol. 63, pp. 451–463.

1959. *On a recent account of entailment.* **Mind,** n.s., vol. 68, pp. 393–395.

1965. Review of Anderson and Belnap 1962a and Smiley 1959. **The journal of symbolic logic,** vol. 30, pp. 240–241.

1969. *Entailment.* **The philosophical review,** vol. 68, pp. 197–236.

BIAŁNICKI-BIRULA, A. and RASIOWA, H.

1957. *On the representation of quasi-Boolean algebras.* **Bulletin de l'académie polonaise des sciences,** vol. 5, pp. 259–261.

BIRKHOFF, GARRETT

1942. *Lattice-ordered groups.* **Annals of mathematics,** vol. 43, pp. 298–331.

1948. **Lattice theory.** Providence (American Mathematical Society).

BLANSHARD, BRAND

1939. **The nature of thought.** London (Allen and Unwin).

BOEHNER, PHILOTHEUS

1952. **Medieval logic.** Chicago (University of Chicago Press).

BOOLE, GEORGE

1847. **The mathematical analysis of logic. Being an essay towards a calculus of deductive reasoning.** Cambridge, England (Macmillan, Barclay and Macmillan) and London (George Bell). Reprinted in **George Boole's collected works,** vol. 1 (Studies in logic and probability), ed. R. Rhees, London (Watts and Co.) 1952, pp. 45–140.

CERTAINE, J.

1943. **Lattice-ordered groupoids and some related problems.** Harvard University doctoral dissertation.

CHIDGEY, JOHN R.

1973. *A note on transitivity.* **Notre Dame journal of formal logic,** vol. 14, pp. 273–275.

1974. **On entailment.** University of Manchester doctoral dissertation.

See Parks and Chidgey.

CHURCH, ALONZO

1948. Review of Jaśkowski 1948. **The journal of symbolic logic,** vol. 13, pp. 164–165.

1951. *The weak theory of implication.* **Kontrolliertes Denken, Untersuchungen zum Logikkalkül und der Logik der Einzelwissenschaften,** ed. Menne-Wilhelmy-Angsil, Munich (Kommissions-verlag Karl Alber), pp. 22–37.

1951a abstract. *Minimal logic.* **The journal of symbolic logic,** vol. 16, p. 239.

1956. **Introduction to mathematical logic,** vol. 1. Princeton (Princeton University Press).

COLERIDGE, SAMUEL TAYLOR

1863. **The friend: a series of essays.** In two volumes, London (Edward Maxon & Company).

CRAIG, WILLIAM

1957. *Linear reasoning. A new form of the Herbrand-Gentzen theorem.* **The journal of symbolic logic,** vol. 22, pp. 250–268.

CURLEY, E. M.

1972. *Lewis and entailment.* **Philosophical studies,** vol. 23, pp. 198–204.

CURRY, HASKELL B.

1950. **A theory of formal deducibility.** Notre Dame mathematical lectures, no. 6. Notre Dame (Notre Dame University Press). Second edition, 1955.

1954. *Generalizations of the deduction theorem.* **Proceedings of the international congress of mathematicians,** vol. 2, Groningen (E. P. Noordhoff), pp. 399–400.

1959. *The interpretation of formalized implication.* **Theoria,** vol. 25, pp. 1–26.

1963. **Foundations of mathematical logic.** New York (McGraw-Hill).

See Curry and Feys.

CURRY, HASKELL B., and FEYS, ROBERT

1958. **Combinatory logic,** vol. 1. Amsterdam (North-Holland Publishing Company).

DEDEKIND, RICHARD

1877. *Über die Anzahl der Ideal-Klassen in den verscheidenen Ordnungen eines endlichen Körpers.* **Festschrift der Technischen Hochschule in Braunschweig zur Sakülarfeir des Guburtsdag von C. F. Gauss,** pp. 1–55.

DIAMOND, A. H. and MCKINSEY, J. C. C.

1947. *Algebras and their subalgebras.* **Bulletin of the American Mathematical Society,** vol. 53, pp. 959–962.

DONČENKO, V. V.

1963. *Nékotoryé voprosy, svázannye s problémoj razréšéniá diá iščisléniá strogoj implikacii Akkérmana* (Some questions connected with the decision problem for Ackermann's system of rigorous implication). **Problemy logiki** (Izdatél'stvo Akadémii SSSR), pp. 18–24.

DRAKE, F. R.

1962. *On McKinsey's syntactical characterization of systems of modal logic.* **The journal of symbolic logic,** vol. 27, pp. 400–406.

DUBISCH, ROY

1964. **Lattices to logic.** New York (Blaisdell Publishing Company).

DUNCAN-JONES, AUSTIN E.

1935. *Is strict implication the same as entailment?* **Analysis,** vol. 2, pp. 70–78.

DUNN, J. MICHAEL

1966. **The algebra of intensional logics.** University of Pittsburgh doctoral dissertation.

1967 abstract. *An algebraic completeness proof for the first degree fragment of entailment.* **Abstracts of papers: third international congress for logic, methodology and philosophy of science,** Amsterdam, p. 9.

1967a abstract. *The effective equivalence of certain propositions about De Morgan lattices.* **The journal of symbolic logic,** vol. 32, pp. 433–434.

1968 abstract. *Representation theory of normal R-mingle matrices.* **The journal of symbolic logic,** vol. 33, p. 637.

1970. *Algebraic completeness results for R-mingle and its extensions.* **The journal of symbolic logic,** vol. 35, pp. 1–13.

1970a abstract. *Extensions of RM and LC.* **The journal of symbolic logic,** vol. 35, p. 360.

1972. *A modification of Parry's analytic implication.* **Notre Dame journal of formal logic,** vol. 13, pp. 195–205.

1973 abstract. *A 'Gentzen system' for positive relevant implication.* **The journal of symbolic logic,** vol. 38, pp. 356–357.

See Belnap and Gupta and Dunn, Dunn and Belnap, Dunn and Meyer, Meyer and Dunn, Meyer and Dunn and Leblanc.

DUNN, J. MICHAEL and BELNAP, NUEL D., JR.

1967 abstract. *Homomorphisms of intensionally complemented distributive lattices.* **The journal of symbolic logic,** vol. 32, p. 446.

1968. *The substitution interpretation of the quantifiers.* **Noûs,** vol. 2, pp. 177–185.

1968a. *Homomorphisms of intensionally complemented distributive lattices.* **Mathematische Annalen,** vol. 176, pp. 28–38.

DUNN, J. MICHAEL and MEYER, ROBERT K.

1971. *Algebraic completeness results for Dummett's LC and its extensions.* **Zeitschrift für mathematische Logik und Grundlagen der Mathematik,** vol. 17, pp. 225–230.

See Meyer and Dunn.

EATON, RALPH M.

1931. **General logic, an introductory survey.** New York (Charles Scribner's Sons). See pp. 226–234.

FALKENER, EDWARD

1892. **Games ancient and oriental and how to play them.** London (Longmans, Green & Co.); reprinted New York (Dover) 1961.

FEYS, ROBERT

1937. *Les logiques nouvelles des modalites.* **Revue néoscholastique de philosophie,** vol. 40, pp. 517–553, and vol. 41 (1938) pp. 217–252.

1950. *Les systèmes formalisés des modalités aristotéliennes.* **Revue philosophique de Louvain,** vol. 48, pp. 478–509.

1956. *Un système de l'implication rigoureuse.* **Logique et analyse,** no. 7, pp. 3–18.

See Curry and Feys, Feys and Ladrière.

FEYS, ROBERT and LADRIÉRE, JEAN

1955. Notes to **Recherches sur la déduction logique** (translation of Gentzen 1934). Paris (Presses Universitares de France).

FINE, KIT

1974. *Models for entailment.* **Journal of philosophical logic,** vol. 3, pp. 347–372.

FISK, MILTON

1964. **A modern formal logic.** Englewood Cliffs (Prentice Hall).

FREGE, F. L. GOTTLOB

1879. **Begriffsschrift, eine der arithmetischen nachgebildete Formalsprache des reinen Denkens.** Halle (Nebert). English translation (*Begriffsschrift, a formula language, modeled upon that of arithmetic, for pure thought*) by Stefan Bauer-Mengelberg in van Heijenoort 1967, pp. 1–82.

FUCHS, L.

1963. **Partially ordered algebraic systems.** Oxford (Pergamon Press).

GENTZEN, GERHARD

1934. *Untersuchungen über das Logische Schliessen.* **Mathematische Zeitschrift,** vol. 39, pp. 176–210, 405–431. English translation (*Investigations into logical deduction*) in **The collected papers of Gerhard Gentzen,** ed. M. E. Szabo, Amsterdam (North-Holland Publishing Company) 1969, pp. 68–131.

GOBLE, LOUIS F.

1971. *A system of modality.* **Notre Dame journal of formal logic,** vol. 12, pp. 225–237.

GÖDEL, KURT

1931. *Über formal unentscheidbare Sätze der Principia Mathematica und verwandter Systeme I.* **Monatshefte für Mathematik und Physik,** vol. 38, pp. 173–198. English translation (*On formally undecidable propositions of Principia mathematica and related systems I*) by Jean van Heijenoort in van Heijenoort 1967, pp. 596–616.

1933. *Zur intuitionistischen Arithmetik und Zahlentheorie.* **Ergebnisse eines mathematischen Kolloquiums,** Heft 4, pp. 34–38.

GOODMAN, NELSON

1947. *The problem of counterfactual conditionals.* **The journal of philosophy,** vol. 44, pp. 113–128.

1955. **Fact, fiction, and forecast.** Cambridge, Massachusetts (Harvard University Press). Third edition, Indianapolis (The Bobbs-Merrill Company) 1973.

GROVER, DOROTHY L.

1966. **Entailment.** A critical study of system E and a critical bibliography of the subject in general. M. A. dissertation, Victoria University of Wellington.

1970. **Topics in propositional quantification.** University of Pittsburgh doctoral dissertation.

1972. *Propositional quantifiers* **Journal of philosophical logic,** vol. 1, pp. 111–136.

1973. *Propositional quantification and quotation contexts.* **Truth, syntax and modality,** ed. H. Leblanc, Amsterdam (North-Holland Publishing Company) pp. 101–110.

HACKING, IAN

1963. *What is strict implication?* **The journal of symbolic logic,** vol. 28, pp. 51–71.

HALMOS, PAUL

1962. **Algebraic logic.** New York (Chelsea Publishing Co.).

HARROP, RONALD

1958. *On the existence of finite models and decision procedures for propositional calculi.* **Proceedings of the Cambridge philosophical society**, vol. 54, pp. 1–13.

1959. *The finite model property and subsystems of classical propositional calculus.* **Zeitschrift für mathematische Logik und Grundlagen der Mathematik**, vol. 5, pp. 29–32.

HENKIN, LEON

1949. *The completeness of the first order functional calculus.* **The journal of symbolic logic**, vol. 14, pp. 159–166. Reprinted in **The philosophy of mathematics**, ed. Jaakko Hintikka, London (Oxford University Press) 1969, pp. 42–50.

1954. *Boolean representation through propositional calculus.* **Fundamenta mathematicae**, vol 41, pp. 89–96.

See Henkin and Tarski, Montague and Henkin.

HENKIN, LEON and TARSKI, ALFRED

1961. *Cylindric algebras.* **Lattice theory**, Proceedings of symposia in pure mathematics, vol. 2, Providence (American Mathematical Society), pp. 83–113.

HEYTING, AREND

1930. *Die formalen Regeln der intuitionistischen Logik.* **Sitzungsberichte der Preussischen Akademie der Wissenschaften** (Physicalischmathematische Klasse), pp. 42–56.

HILBERT, DAVID and ACKERMANN, WILHELM

1959. **Grundzüge der theoretischen Logik.** 4th ed. Berlin (Springer Verlag). (Pp. 36–40 discuss *strenge Implikation.*)

HINTIKKA, JAAKKO

1959. *Existential presuppositions and existential commitments.* **The journal of philosophy**, vol. 56, pp. 125–137.

1963. *The modes of modality.* **Acta philosophica fennica, Modal and many-valued logics**, pp. 65–81. Reprinted in Jaakko Hintikka, **Models for modalities. Selected essays**, Dordrecht (D. Reidel Publishing Company) 1969, pp. 71–86.

HIŻ, HENRY

1961. *Modalities and extended systems.* **The journal of philosophy**, vol. 68, pp. 723–731.

HOCKNEY, DONALD and WILSON, KENT

1965. *In defense of a relevance condition.* **Logique et analyse**, n.s., vol. 8, pp. 211–220.

HUNTINGTON, E. V.

1904. *Sets of independent postulates for the algebra of logic.* **Transactions of the American Mathematical Society**, vol. 5, pp. 288–309.

JAŚKOWSKI, STANISŁAW

1934. *On the rules of suppositions in formal logic.* **Studia logica,** no. 1. Warsaw, pp. 5–32. Reprinted in McCall 1967a, pp. 232–258.

1948. *Trois contributions au calcul des propositions bivalentes.* **Studia societatis scientiarum torunensis,** section A, vol. 1, no. 1, pp. 1–15.

JEVONS, WILLIAM STANLEY

1870. **Elementary lessons in logic.** London (Macmillan Company). Reprinted 1965.

1871. *On a general system of numerically definite reasoning.* **Memoirs of the Literary and Philosophical Society of Manchester,** 3 s, vol. 4, pp. 330–352.

KALMAN, J. A.

1958. *Lattices with involution.* **Transactions of the American Mathematical Society,** vol. 87, pp. 485–491.

KANT, IMMANUEL

1781. **Kritik der reinen Vernunft.** Riga (Johann Friedrich Hartknoch).

KEYNES, JOHN N.

1906. **Studies and exercises in formal logic including a generalization of logical processes in their application to complex inferences.** Fourth edition, London (Macmillan Company).

KISS, STEPHEN A.

1961. **An introduction to algebraic logic.** Westport, Connecticut (privately printed; available at 3 Laurel Lane).

KLEENE, STEPHEN C.

1952. **Introduction to metamathematics.** New York (D. Van Nostrand Company, Inc.).

KNEALE, WILLIAM

1945–46. *Truths of logic.* **Proceedings of the Aristotelian Society,** n.s., vol. 46, pp. 207–234.

See Kneale and Baillie and others, Kneale and Kneale.

KNEALE, WILLIAM, BAILLIE, J. B. and others

1956. William Kneale in **Contemporary British philosophy, personal statements,** vol. III, ed. by H. D. Lewis, London (Allen and Unwin), New York (Macmillan Company).

KNEALE, WILLIAM and KNEALE, MARTHA

1962. **The development of logic.** Oxford (Clarendon Press).

König, Dénes

1927. *Über eine Schlussweise aus dem Endlichen ins Unendliche (Punktmengen-Kartenfarben-Verwandtschaftsbeziehungen-Schachspiel).* **Acta Litterarum ac Scientiarum (Sectio Scientiarum Mathematicarum),** vol. 3, pp. 121–130. (More easily available in **Theorie der endlichen und unendlichen Graphen.** Akademische Verlagsgesellschaft, Leipzig, 1936. Republished, New York (Chelsea) 1950.)

Kripke, Saul A.

1959. *A completeness theorem in modal logic.* **The journal of symbolic logic,** vol. 24, pp. 1–15.

1959a abstract. *The problem of entailment.* **The journal of symbolic logic,** vol. 24, p. 324

1965. *Semantical analysis of modal logic II.* **The theory of models,** ed. Addison, Henkin, Tarski, Amsterdam (North-Holland Publishing Company), pp. 206–220.

1965a. *Semantical analysis of intuitionistic logic I.* **Formal systems and recursive functions,** ed. J. N. Crossley and M. A. E. Dummett, Amsterdam (North-Holland Publishing Company), pp. 92–129.

Kuratowski, C.

1922. *Sur l'opération $\overline{A}$ de l'analysis situs.* **Fundamenta mathematicae,** vol. 3, pp. 182–199.

Ladd-Franklin, Christine

1901. *The reduction to absurdity of the ordinary treatment of the syllogism.* **Science,** n.s. vol 13, pp. 574–576.

1913. *The antilogism, an emendation.* **The journal of philosophy, psychology, and scientific methods,** vol. 10, pp. 49–50.

1928. *The antilogism.* **Mind,** n.s. vol. 37, pp. 532–534.

Leibniz, G. W.

1679. Untitled fragment published in translation by Louis Couturat, **Opuscules et fragments inédits de Leibniz extraits des manuscrits de la Bibliothèque royale de Hanovre,** Paris, 1903. Discussed in Łukasiewicz 1951, §34. English translation (*Rules from which a decision can be made, by means of numbers, about the validity of inferences and about the forms and moods of categorical syllogisms*) by G. H. R. Parkinson in Leibniz, **Logical papers,** ed. G. H. R. Parkinson, London (Oxford University Press) 1966, pp. 25–32.

Lemmon, E. J.

1957. *New foundations for Lewis modal systems.* **The journal of symbolic logic,** vol. 22, pp. 176–186.

See Lemmon and Meredith and Meredith and Prior and Thomas.

Lemmon, E. J., Meredith, C. A., Meredith, D., Prior, A. N. and Thomas, I.

1956. *Calculi of pure strict implication* (mimeographed 1956). Reprinted in **Philosophical logic,** ed. J. W. Davis, D. J. Hockney and W. K. Wilson, Dordrecht (D. Reidel Publishing Company) 1969, pp. 215–250.

LEŚNIEWSKI, STANISŁAW

1929. *Gründzüge eines neuen Systems der Grundlagen der Mathematik.* **Fundamenta mathematicae,** vol. 14, pp. 1–81.

LEWIS, CLARENCE IRVING

1918. **A survey of symbolic logic.** Berkeley (University of California Press) Second edition, abridged, 1960, New York (Dover).

1943. *The modes of meaning.* **Philosophy and phenomenological research,** vol. 4, pp. 236–291.

See Lewis and Langford.

LEWIS, CLARENCE IRVING and LANGFORD, COOPER HAROLD

1932. **Symbolic logic.** New York and London (The Century Co.). Second edition, New York (Dover) 1959.

LEWIS, DAVID

1971. *Completeness and decidability of three logics of counterfactual conditionals.* **Theoria,** vol. 37, pp. 74–85.

LEWY, CASIMIR

1958. *Entailment.* **Aristotelian Society supplementary volume,** vol. 32, pp. 123–142.

LIEBER, LILLIAN R. and LIEBER, HUGH GRAY

1944. **The education of T. C. Mits,** New York (Norton).

LORENZEN, PAUL

1955. **Einführung in die operative Logik und Mathematik.** Berlin-Göttingen-Heidelberg (Springer Verlag).

1962. **Formale Logik.** Berlin (Walter de Gruyter and Company). English translation (**Formal logic**) by Frederick J. Crosson, Dordrecht (D. Reidel Publishing Company) 1965.

ŁOŚ, J.

1957. *Remarks on Henkin's paper: Boolean representation through propositional calculus.* **Fundamenta mathematicae,** vol. 44, pp. 82–93.

ŁUKASIEWICZ, JAN

1929. **Elementary logiki matematycznej.** Mimeographed, Warsaw 1929; printed, Warsaw 1958; English translation by Olgierd Wojtasiewicz, New York (Macmillan) 1963. Reprinted 1966.

1939. *Der Äquivalenzenkalkül.* **Collectanea logica,** vol. 1, pp. 145–169 (printed Sept. 1939, but all copies, except for one offprint, destroyed by the Nazis during the siege of Warsaw). English translation under the title *The equivalential calculus*, by Peter Woodruff, in McCall 1967a, pp. 88–115. Also published in Łukasiewicz 1970, pp. 256–277.

1948. *The shortest axiom of the implicational calculus of propositions.* **Proceedings of the Royal Irish Academy,** vol. 52, sec. A, no. 3, pp. 25–33. Also published in Łukasiewicz 1970, pp. 295–305.

1951. **Aristotle's syllogistic, from the standpoint of modern formal logic.** Oxford (Clarendon Press) second edition, enlarged, 1957.

1952. *A system of modal logic.* **The journal of computing systems,** vol. 1, pp. 111–149. Also published in Łukasiewicz 1970, pp. 352–390.

1970. **Selected works.** ed. L. Borkowski. Amsterdam (North-Holland Publishing Company).

See Łukasiewicz and Tarski.

Łukasiewicz, Jan and Tarski, Alfred

1930. *Untersuchungen über den Aussagenkalkül.* **Comptes rendus des séances de la Société des Sciences et des Lettres de Varsovie,** vol. 23, Classe III, pp. 30–50. Translated under the title *Investigations into the sentential calculus* by J. H. Woodger in Tarski 1956, pp. 32–39. Also published in Łukasiewicz 1970, pp. 131–152.

MacColl, Hugh

1906. **Symbolic logic and its applications.** London (Longmans, Green & Co.).

1908. *'If' and 'imply'.* **Mind,** n.s., vol. 17, pp. 151–152, 453–455.

Maksimova, L. L.

1967. *O modéláh isčisléniá E* (On models of the system E). **Algébra i logika, Séminar,** vol. 6, pp. 5–20.

1971. *Intérprétaciá i téorémy otdépeniá dlá isčisléniá E i R.* **Algébra i logika,** vol. 10, pp. 376–392. English translation (*An interpretation and separation theorems for the logical systems E and R*) in **Algebra and logic,** vol. 10, pp. 232–241.

1973. *A semantics for the calculus E of entailment.* **Bulletin of the section of logic,** Polish academy of sciences, Institute of philosophy and sociology, vol. 2, pp. 18–21.

Marcus, Ruth Barcan

See Barcan Marcus.

Matthew

80 *The gospel according to St. Matthew.* **The holy bible, King James version** (translated 1611).

Maxwell, E. A.

1959. **Fallacies in Mathematics.** Cambridge, England (University Press).

McCall, Storrs

1963. **Non-classical propositional calculi.** Oxford University doctoral dissertation.

1964 abstract. *A new variety of implication.* **The journal of symbolic logic,** vol. 29, pp. 151–152.

1966. *Connexive implication.* **The journal of symbolic logic,** vol. 31, pp. 415–433.

1967. *Connexive class logic.* **The journal of symbolic logic,** vol. 32, pp. 83–90.

1967a. **Polish logic, 1920–1939.** Oxford (Clarendon Press).

1967b. *Connexive implication and the syllogism.* **Mind,** n.s., vol. 76, pp. 346–356.

1969. *Time and the physical modalities.* **The monist,** vol. 55, pp. 426–446.

McKinsey, J. C. C.

1941. *A solution of the decision problem for the Lewis systems S2 and S4, with an application to topology.* **The journal of symbolic logic,** vol. 6, pp. 117–134.

1945. *On the syntactical construction of systems of modern logic.* **The journal of symbolic logic,** vol. 10, pp. 83–94.

1948–49. *A new definition of truth.* **Synthèse,** vol. 7, pp. 428–433.

See Diamond and McKinsey, McKinsey and Tarski.

McKinsey, J. C. C. and Tarski, Alfred

1944. *The algebra of topology.* **Annals of mathematics,** vol. 45, pp. 141–191.

1946. *On closed elements in closure algebras.* **Annals of mathematics,** vol. 47, pp. 122–162.

1948. *Some theorems about the sentential calculi of Lewis and Heyting.* **The journal of symbolic logic,** vol. 13, pp. 1–15.

Mendelson, Elliott

1964. **Introduction to mathematical logic.** Princeton (D. Van Nostrand Company, Inc.).

Meredith, C. A., and Prior, A. N.

1963. *Notes on the axiomatics of the propositional calculus.* **Notre Dame journal of formal logic,** vol. 4, pp. 171–187.

See Lemmon and Meredith and Meredith and Prior and Thomas.

Meyer, Robert K.

1966. **Topics in modal and many-valued logic.** University of Pittsburgh doctoral dissertation.

1968a. *Entailment and relevant implication.* **Logique et analyse,** n.s., vol. 11, pp. 472–479.

1968b. *An undecidability result in the theory of relevant implication.* **Zeitschrift für mathematische Logik und Grundlagen der Mathematik,** vol. 14, pp. 255–262.

1970a. *E and S4.* **Notre Dame journal of formal logic,** vol. 11, pp. 181–199.

1970b. *R_I — the bounds of finitude.* **Zeitschrift für mathematische Logik und Grundlagen der Mathematik,** vol. 16, pp. 385–387.

1970c abstract. *Some problems no longer open for E and related logics.* **The journal of symbolic logic,** vol. 35, p. 353.

1971. *On coherence in modal logics.* **Logique et analyse,** n.s., vol. 14, pp. 658–668.

1971a. *Entailment.* **The journal of philosophy,** vol. 68, pp. 808–818.

1971b abstract. *R-mingle and relevant disjunction.* **The journal of symbolic logic,** vol. 36, p. 366. Accepted 1967.

1971c abstract. *Logics contained in R.* **The journal of symbolic logic,** vol. 36, pp. 365–366. Accepted 1967.

1972. *On relevantly derivable disjunctions.* **Notre Dame journal of formal logic,** vol. 13, pp. 476–480.

1972a abstract. *Negation disarmed.* **The journal of symbolic logic,** vol. 37, p. 445.

1972b abstract. *Canonical metavaluations.* **The journal of symbolic logic,** vol. 37, pp. 445–446.

1973. *Intuitionism, entailment, negation.* **Truth, Syntax, Modality,** ed. H. Leblanc, Amsterdam (North-Holland Publishing Company), pp. 168–198.

1973a. *Conservative extension in relevant implication.* **Studia logica,** vol. 31, pp. 39–46.

1973b. *On conserving positive logics.* **Notre Dame journal of formal logic,** vol. 14, pp. 224–236.

1973c abstract. *Simple Belnap lattices and their implicative extensions.* **The journal of symbolic logic,** vol. 38, pp. 350–351

1974. *Entailment is not strict implication.* **Australasian journal of philosophy,** vol. 52, pp. 212–231.

197+ *Coherence revisited.* Typescript, November, 1972.

197+a. *The pure calculus of entailment is the pure calculus of entailment.*

197+b. *Metacompleteness.* **Notre Dame journal of formal logic.**

197+c. *S5: the poor man's connexive implication.* Typescript, July, 1972.

197+d. *On conserving positive theories.* **Notre Dame journal of formal logic.**

MEYER, ROBERT K. and DUNN, J. MICHAEL

1968. *Entailment logics and material implication.* **Notices of the American Mathematical Society,** vol. 15, pp. 1021–1022.

1969. *E, R and γ.* **The journal of symbolic logic,** vol. 34, pp. 460–474.

See Dunn and Meyer.

MEYER, ROBERT K., DUNN, J. MICHAEL and LEBLANC, HUGUES

1974. *Completeness of relevant quantification theories.* **Notre Dame journal of formal logic,** vol. 15, pp. 97–121.

See Meyer and Leblanc.

MEYER, ROBERT K. and LEBLANC, HUGUES

1970 abstract. *A semantical completeness result for relevant quantification theories.* **The journal of symbolic logic,** vol. 35, p. 181.

See Meyer and Dunn and Leblanc.

MEYER, ROBERT K. and PARKS, ZANE

1972. *Independent axioms for the implicational fragment of Sobociński's three-valued logic.* **Zeitschrift für mathematische Logik und Grundlagen der Mathematik,** vol. 18, pp. 291–295.

Meyer, Robert K. and Routley, Richard

1972. *Algebraic analysis of entailment* (*I*). **Logique et analyse,** vol. 15, pp. 407–428.

1973. *Classical relevant logics* (*I*). **Studia logica,** vol. 32, pp. 51–66.

1973a. *An undecidable relevant logic.* **Zeitschrift für mathematische Logik und Grundlagen der Mathematik,** vol. 19, pp. 389–397.

197+. *Algebraic analysis of entailment* (*II*).

See Routley and Meyer.

Mihăilescu, E. G.

1938. *Recherches sur l'équivalence et la réciprocité dans le calcul des propositions.* **Annales scientifiques de l'Université de Jassy,** vol. 24, pp. 116–153.

Moh Shaw-Kwei

1950. *The deduction theorems and two new logical systems.* **Methodos,** vol. 2, pp. 56–75.

Montague, Richard and Henkin, Leon

1956. *On the definition of 'formal deduction'.* **The journal of symbolic logic,** vol. 21, pp. 129–136.

Monteiro, A.

1960. *Matrices De Morgan caracteristiques pour le calcul propositionnel classique.* **Anais de academia Brasileira de ciencias,** vol. 32, pp. 1–7.

Moore, G. E.

1920. *External and internal relations.* **Proceedings of the Aristotelian Society,** n.s., vol. 20, pp. 40–62.

Nelson, E. J.

1930. *Intensional relations.* **Mind,** n.s., vol. 39, pp. 440–453.

1933. *Deductive systems and the absoluteness of logic.* **Mind,** n.s., vol. 42, pp. 31–34.

1933a. *On three logical principles in intension.* **The monist,** vol. 43, pp. 268–284.

Ohnishi, M. and Matsumoto, K.

1962. *A system for strict implication.* **Proceedings of the symposium on the foundations of mathematics,** Katadu, Japan, pp. 98–108.

Pahi, B.

1966 abstract. *On the non-existence of finite characteristic matrices for some implicational calculi.* **The journal of symbolic logic,** vol. 31, pp. 682–683.

Parks, R. Zane

1972. *A note on R-mingle and Sobociński's three-valued logic.* **Notre Dame journal of formal logic,** vol. 13, pp. 227–228.

See Meyer and Parks, Parks and Chidgey.

PARKS, R. ZANE and CHIDGEY, JOHN R.

1972. *Necessity and ticket entailment.* **Notre Dame journal of formal logic,** vol. 13, pp. 224–226.

PARRY, WILLIAM T.

1933. *Ein Axiomensystem für eine neue Art von Implikation (Analytische Implikation).* **Ergebnisse eines mathematischen Kolloquiums,** vol. 4, pp. 5–6.

1939. *Modalities in the "Survey" system of strict implication.* **The journal of symbolic logic,** vol. 4, pp. 137–154.

PEANO, GIUSEPPE

1889. **Arithmetices principia, novo methodo exposita.** Turin.

POPLE, HAROLD E., JR.

1972. *A goal-orientated language for the computer.* Reprinted in **Representation and meaning experimenting with information processing systems,** eds. Simon and Siklossy (Prentice-Hall).

POPPER, KARL R.

1940. *What is dialectic?* **Mind,** n.s., vol. 49, pp. 403–426.

1943. *Are contradictions embracing?* **Mind,** n.s., vol. 52, pp. 47–50.

1947. *New foundations for logic.* **Mind,** n.s., vol. 56, pp. 193–235.

POST, EMIL L.

1921. *Introduction to a general theory of elementary propositions.* **American journal of mathematics,** vol. 43, pp. 163–185. Reprinted in van Heijenoort 1967, pp. 264–283.

1943. *Formal reductions of the general combinatorial decision problem.* **American journal of mathematics,** vol. 65, pp. 197–215.

POTTINGER, GARREL

1972. **A theory of implications.** University of Pittsburgh doctoral dissertation.

PRAWITZ, DAG

1964 abstract. *Normal deductions.* **The journal of symbolic logic,** vol. 29, p. 152.

1965. **Natural deduction. A proof-theoretical study.** Stockholm-Göteborg-Uppsala (Almqvist & Wiksell).

PRIOR, ARTHUR N.

1948. *Facts, propositions, and entailment.* **Mind,** n.s., vol. 57, pp. 62–68.

1955. **Formal logic.** New York (Oxford University Press). Second edition, 1962.

1956. *Logicians at play; or syll, simp and Hilbert.* **Australasian journal of philosophy,** vol. 34, pp. 182–192.

1960–61. *The runabout inference-ticket.* **Analysis,** vol. 21, pp. 38–39.

See Lemmon and Meredith and Meredith and Prior and Thomas, Meredith and Prior.

QUINE, WILLARD V. O.

1940. **Mathematical Logic.** New York (Norton). Revised edition, Cambridge, Massachusetts (Harvard University Press) 1951.

RASIOWA, H. and SIKORSKI, R.

1963. **The mathematics of metamathematics.** Warsaw (Panstwowe Wydawnictwo Naukowe). Second edition, revised, 1968.

RESCHER, NICHOLAS

1974. *Bertrand Russell and modal logic.* **Studies in modality.** Monograph no. 8, **American philosophical quarterly** monograph series, pp. 85–96.

ROSSER, J. BARKLEY

1953. **Logic for mathematicians.** New York (McGraw Hill).

ROUTLEY, RICHARD

1972. *A semantical analysis of implicational system I and of the first degree of entailment.* **Mathematische Annalen,** vol. 196, pp. 58–84.

1974. *A rival account of logical consequence.* **Reports on mathematical logic,** no. 3, pp. 41–51.

197+. *Analysis of the implication and implication-conjunction parts of relevant logics.* Cyclostyled.

See Meyer and Routley, Routley and Meyer, Routley and Montgomery, Routley and Routley.

ROUTLEY, RICHARD and MEYER, ROBERT K.

1972 abstract. *A Kripke semantics for entailment.* **The journal of symbolic logic,** vol. 37, p. 442.

1972a. *The semantics of entailment (II).* **Journal of philosophical logic,** vol. 1, pp. 53–73.

1972b. *The semantics of entailment (III).* **Journal of philosophical logic,** vol. 1, pp. 192–208.

1973. *The semantics of entailment (I).* **Truth, syntax, modality,** ed. H. Leblanc, Amsterdam (North-Holland Publishing Company), pp. 199–243.

197+. *The semantics of entailment (IV) E,* Π' and Π. **The journal of symbolic logic.**

197+a. **Relevant logics and their rivals.**

See Meyer and Routley.

ROUTLEY, RICHARD and MONTGOMERY, HUGH

1968. *On systems containing Aristotle's thesis.* **The journal of symbolic logic,** vol. 33, pp. 82–96.

ROUTLEY, RICHARD and ROUTLEY, VALERIE

1969. *A fallacy of modality.* **Noûs,** vol. 3, pp. 129–153.

1972. *Semantics of first degree entailment.* **Noûs,** vol. 6, pp. 335–359.

RUSSELL, BERTRAND

1906. Review of MacColl 1906. **Mind,** n.s. vol. 15, pp. 255–260.

1956. **Logic and knowledge; essays 1901–1950.** New York (Macmillan Company).

See Whitehead and Russell.

RYLE, GILBERT

1949. **The concept of mind.** New York (Barnes and Noble, Inc.).

1954. **Dilemmas.** Cambridge, England (University Press).

SCHILLER, FERDINAND C. S.

1930. **Logic for use.** New York (Harcourt, Brace).

1931. *The nature and validity of formal logic.* **Aristotelian Society supplementary volume,** vol. 10, pp. 11–26.

SCHMIDT, STEVEN

1970. *The revised zodiac.* **Time,** November 29. p. 44.

SCHOCK, R.

1962. *A definition of event and some of its applications.* **Theoria,** vol. 27, pp. 250–268.

SCHRÖDER, F. W. K. ERNST

1890–1895. **Vorlesungen über die Algebra der Logik (exacte Logik).** Vol. 1 (1890), vol. 2 (1891), vol. 3 (1895), Leipzig. Reprint, Bronx (Chelsea Publishing Company) 1965.

SCHÜTTE, KURT

1950. *Schlussweisen-kalküle der Prädikatenlogik.* **Mathematische Annalen,** vol. 122, pp. 47–65.

SCROGGS, SCHILLER JOE

1951. *Extensions of the Lewis system S5.* **The journal of symbolic logic,** vol. 16, pp. 112–120.

SEXTUS EMPIRICUS

200 (?) **Outlines of Pyrrhonism** and **Against the logicians.** English translation by the Rev. R. G. Bury, Cambridge, Massachusetts (Harvard University Press) 1929 and 1935.

SMILEY, T. J.

1959. *Entailment and deducibility.* **Proceedings of the Aristotelian Society,** n.s., vol. 59, pp. 233–254.

SOBOCIŃSKI, BOLESŁAW

1952. *Axiomatization of a partial system of three-valued calculus of propositions.* **The journal of computing systems,** vol. 1, pp. 23–55.

1956. *On well-constructed axiom systems.* **Polskie Towarzystwo naukowe na obzyźnie,** pp. 1–12.

STALNAKER, ROBERT C.

1968. *A theory of conditionals.* **Studies in logical theory,** ed. N. Rescher, Supplementary monograph to the **American philosophical quarterly,** Oxford (Basil Blackwell) pp. 98–112.

1970. *Probability and conditionals.* **Philosophy of science,** vol. 37, pp. 64–80.

See Stalnaker and Thomason.

STALNAKER, ROBERT C. and THOMASON, RICHMOND H.

1970. *A semantic analysis of conditional logic.* **Theoria,** vol. 36, pp. 23–42.

STEVENSON, CHARLES L.

1970. *If-iculties.* **Philosophy of Science,** vol. 37, pp. 27–49. Also published in **Logic and art. Essays in honor of Nelson Goodman,** ed. Richard Rudner and Israel Scheffler, Indianapolis (The Bobbs-Merrill Company) 1972, pp. 279–309.

STONE, M. H.

1936. *The theory of representations for Boolean algebras.* **Transactions of the American mathematical society,** vol. 40, pp. 37–111.

1937. *Topological representations of distributive lattices and Brouwerian logics.* **Časopis pro pěstování matematiky a fysiky,** vol. 67, pp. 1–25.

STRAWSON, P. F.

1952. **Introduction to logical theory.** London (Methuen and Co.).

1958. Review of von Wright 1957. **The philosophical quarterly,** vol. 8, pp. 372–376.

SUGIHARA, TAKEO

1955. *Strict implication free from implicational paradoxes.* **Memoirs of the Faculty of Liberal Arts, Fukui University, Series I,** pp. 55–59.

SUPPES, PATRICK

1957. **Introduction to logic.** Princeton (D. Van Nostrand Company, Inc.).

TARSKI, ALFRED

1930. *Über einige fundamentalen Begriffe der Metamathematik.* **Comptes rendus des séances de la Société des Sciences et des Lettres de Varsovie,** Classe III, vol. 23, pp. 22–29. English translation (*On some fundamental concepts of metamathematics*) by J. H. Woodger, in Tarski 1956, pp. 30–37.

1933. *Der Wahrheitsbegriff in den formalisierten Sprachen.* **Studia philosophica,** vol. 1 (1936), pp. 261–405. English translation (*The concept of truth in formalized languages*) by J. H. Woodger in Tarski 1956, pp. 152–278.

1956. **Logic, semantics, metamathematics.** English translations by J. H. Woodger. Oxford (Clarendon Press).

ULRICH, D. E.

1967. **Matrices for sentential calculi.** Wayne State University doctoral dissertation.

URQUHART, ALASDAIR

1972. *Semantics for relevant logics.* **The journal of symbolic logic,** vol. 357, pp. 159–69.

1972a abstract. *A general theory of implication.* **The journal of symbolic logic,** vol. 37, p. 443.

1972b. *The completeness of weak implication.* **Theoria,** vol. 37, pp. 274–282.

1972c. **The semantics of entailment.** University of Pittsburgh doctoral dissertation.

VAN FRAASSEN, BAS

1967. *Meaning relations among predicates,* **Noûs,** vol. 1, pp. 161–179.

1969. *Facts and tautological entailments.* **The journal of philosophy,** vol. 66, pp. 477–487.

1969a. *Logical structure in Plato's Sophist.* **The review of metaphysics,** vol. 22, pp. 482–498.

1973. *Extension, intension, and comprehension.* **Logic and ontology,** ed. M. Munitz, New York (New York University Press), pp. 101–131.

See van Fraassen and Lambert.

VAN FRAASSEN, BAS and LAMBERT, K.

1967. *On free description theory.* **Zeitschrift für mathematische Logik und Grundlagen der Mathematik,** vol. 13, pp. 225–240.

VAN HEIJENOORT, JEAN

1967. **From Frege to Gödel. A source book in mathematical logic, 1879–1931.** Cambridge, Massachusetts (Harvard University Press).

VON WRIGHT, GEORG HENRIK

1951. **An essay in modal logic.** Amsterdam (North-Holland Publishing Company).

1957. **Logical studies.** London (Routledge and Kegan Paul).

1959. *A note on entailment.* **The philosophical quarterly,** vol. 9, pp. 363–365.

WARD, M.

1942. *The closure operators of a lattice.* **Annals of mathematics,** vol. 43, pp. 191–196.

See Ward and Dilworth.

WARD, M. and DILWORTH, P. P.

1939. *Residuated lattices.* **Transactions of the American Mathematical Society,** vol. 45, pp. 335–354.

WHATELY, RICHARD

1851. **Logic.** London and Glasgow.

WHITE, T. H.
1954. **The book of beasts.** New York (G. P. Putnam's Sons).

WHITEHEAD, ALFRED NORTH and RUSSELL, BERTRAND
1910–13. **Principia mathematica,** vols. 1–3. Cambridge, England (University Press). Second edition, 1925.

WINDELBAND, W.
1893. **A history of philosophy: the formation and development of its problems and conceptions.** Translated by James H. Tufts. New York (Macmillan Company).

WISDOM, JOHN
1936. *Philosophical perplexity.* **Proceedings of the Aristotelian Society,** n.s., vol. 37, pp. 71–88. Reprinted in **Philosophy and psycho-analysis,** New York (Philosophical Library) 1953, pp. 36–50.

WOODRUFF, PETER
1969. **Foundations of three-valued logic.** University of Pittsburgh doctoral dissertation.

WOODS, JOHN
1964. *Relevance.* **Logique et analyse,** n.s., vol. 7, pp. 130–137.

ZINOV'EV, A. A.
1967. **Osnovy logiceskoj teorii naucynx znanij.** Moscow (Izdatel'stvo "Nauka"). Revised and enlarged English edition **(Foundations of the logical theory of scientific knowledge (complex logic))** with an appendix by G. A. Smirnov, E. A. Sidorenko, A. M. Fedina, and L. A. Bobrova, translated by T. J. Blakeley, Dordrecht (D. Reidel Publishing Company) 1973.

BARCAN MARCUS, RUTH C.

1946. *A functional calculus of first order based on strict implication.* **The journal of symbolic logic,** vol. 11, pp. 1–16.

COPI, IRVING M.

1954. **Symbolic.** New York (The Macmillan Company). Fourth edition, 1973.

FITCH, FREDERIC B.

1952. **Symbolic logic.** New York (The Ronald Press Company).

FREGE, GOTTLOB

1893. **Grundgesetze der Arithmetik, begriffsschriftlich abgeleitet.** Band I. Jena (Verlag Hermann Pohle). Partial English translation by Montgomery Furth in Gottlob Frege, **The basic laws of arithmetic, exposition of the system,** Berkeley and Los Angeles (University of California Press) 1964.

MEYER, ROBERT K.

1974. *Entailment is not strict implicatiom.* **Australasian journal of philosophy,** vol. 52, pp. 212–231.

QUINE, WILLARD V. O.

1950. **Methods of logic.** New York (Holt, Rinehart and Winston). Revised edition, 1959.

ROSSER, J. BARKLEY

1936. *Extensions of some theorems of Gödel and Church.* **The journal of symbolic logic,** vol. 1, pp. 87–91.

TURKEN, HY and THOMPSON, S. C.

1970. **The official encyclopedia of baseball** (fifth revised edition). Cranbury, New Jersey (A. S. Barnes).

THOMASON, S. K.

197+. *Toward a formalization of dialectical logic.*

INDEX OF NAMES

A SUBENTRY such as "1956" under "Ackermann" refers to the corresponding entry in the Bibliography. Since full names are given there, we include them here only for individuals not listed in the Bibliography, or to disambiguate. Italicized page numbers refer to material by (rather than about) the co-author named; e.g., *180–206* under "Dunn."

These indices were prepared by computer; entries were typed directly into the computer, and the programs analyzed and sorted them, and prepared a printer-ready index. Efficiency, accuracy, and completeness were heightened by the fact that this procedure permitted several successive "editions" of the indices without any labor except for that involved in entering the corrections and additions themselves. Programs and procedures are available from NDB. Thanks are due the University of Pittsburgh Computer Center for their fine facilities, to Collie Henderson for accurately typing in the name entries, and to Wolf and Robin Dwyer for catching a number of substantive mistakes.

INDEX OF SUBJECTS

SPECIAL symbols are listed at the end; see also under the heading "notation for." Greek letters are alphabetized after all other symbols.

System names are listed in the usual alphabetical order, so that "$\mathbf{E}_{\rightarrow}$" and "$\mathbf{LE}_{\rightarrow}$", for example, do not occur together. Special symbols occurring as suffixes on system names are alphabetized however the computer thinks best; it generally puts them (more accurately: their transliterations) ahead of the alphabet, so that $\mathbf{E}_{\rightarrow}$ precedes $\mathbf{E}_{\mathrm{fde}}$. We assume an auotmatic cross-reference among entries having the same boldface portion. In this connection, we remind the reader that according to our conventions on system nomenclature as explained in the Preface, there are in this volume four possible prefixes for any system: the null prefix, F, L, and Lμ.

SPECIAL SYMBOLS

ONLY symbols with more than local employment are listed. Also consult the Index of subjects under "notation for."

GRAMMATICAL SYMBOLS

		Page
p, q, r, s	propositional variables	6
A, B, C, D	formulas	6
$\rightarrow$	if-then; implication; entailment	6
$\rightleftarrows$	co-entailment (also used proof-theoretically)	9
$\overline{}$, $\sim$	negation	107, 193
&, AB	conjunction	151
$\vee$	disjunction (also used algebraically)	151
$\circ$	co-tenability (also used algebraically)	344
$+$	intensional disjunction	344
$\Box$	necessity	27
$\Diamond$	possibility	111
⥽	strict "implication"	101, 116
$\supset$	material or intuitionist "implication"	101, 151
$\equiv$	material "equivalence"	151
.	dots (for elimination of parentheses)	6
$\vdash$	consecution (also used proof-theoretically)	51

PROOF-THEORETICAL SYMBOLS

$\vdash$	provability; deducibility (also used grammatically)	49, 258
$\vdash_{T}$	provability in T	300
$\dashv$	unprovability	49
$\rightleftarrows$	mutual provable entailment (also used grammatically)	111

ALGEBRAIC SYMBOLS

$\leq$	lattice ordering; less-than-or-equals	190
$\vee$,ab	lattice greatest lower bound	191

Library of Congress Cataloging in Publication Data

Anderson, Alan Ross.
Entailment: the logic of relevance and necessity.

Bibliography: p. 493.
Includes indexes.
1. Logic, Symbolic and mathematical. I. Belnap, Nuel D., 1930. joint author. II. Title.
QA9.A634 1975 511'.3 72.14016
ISBN 0-691-07192-6